# Conference Computational Civil Engineering (CCE2023)



IOP Conference Series: Materials Science and Engineering
Volume 1304

Iasi, Romania
24-25 May 2023

ISBN: 978-1-7138-9568-8
ISSN: 1757-8981

**Printed from e-media with permission by:**

Curran Associates, Inc.
57 Morehouse Lane
Red Hook, NY 12571

**Some format issues inherent in the e-media version may also appear in this print version.**



Institute of Physics
Dirac House, Temple Back
Bristol BS1 6BE UK

Phone: 44 1 17 929 7481
Fax: 44 1 17 920 0979

techtracking@iop.org

**Additional copies of this publication are available from:**

Curran Associates, Inc.
57 Morehouse Lane
Red Hook, NY 12571 USA
Phone: 845-758-0400
Fax: 845-758-2633
Email: curran@proceedings.com
Web: www.proceedings.com

# TABLE OF CONTENTS

IOP Conf. Series: Materials Science and Engineering 1304 (2024) 011001 doi:10.1088/1757-899X/1304/1/011001

ORGANIZERS:

CIMNE CITY constructii UTCB UPT Universitatea Transilvania din Brașov

PARTNERS:

ORDINUL ARHITECȚILOR DIN ROMÂNIA IAȘI

IAȘI Unitec

SPONSORS:

Kaufland UNIC ROTAREX MAFFEIS engineering AGRICONSTRUCT Xella SYM RSC

IOP Conf. Series: Materials Science and Engineering 1304 (2024) 011001 doi:10.1088/1757-899X/1304/1/011001

**Proceedings of the**
**"Computational Civil Engineering" CCE2023**
**Iasi, Romania - May 24th – 25th, 2023**

**Editors:**
Petru Mihai
Vasile-Mircea Venghiac
Ioana Olteanu
Gabriel Barbat
Daniel Covatariu
Gabriela Covatariu

**Organizers**
Faculty of Civil Engineering and Building Services Iasi, Romania
Universitary Foundation "Gheorghe Asachi" of Iasi
CIMNE - International Centre for Numerical Methods in Engineering (CIMNE), UPC Barcelona, Spain

**Co-Organizers**
City University of London
Technical University of Civil Engineering Bucharest
Faculty of Civil Engineering Cluj-Napoca
Faculty of Civil Engineering Timisoara
Faculty of Civil Engineering Brasov
Faculty of Civil Engineering Constanta
Faculty of Civil Engineering, Cadastre and Architecture Oradea

**Partners**
Unitec Institute of Technology Auckland
Iasi City Hall
The Order of Romanian Architects – Iasi Branch

IOP Conf. Series: Materials Science and Engineering 1304 (2024) 011001 doi:10.1088/1757-899X/1304/1/011001

# The Committees of *CCE 2023 Conference*

| | |
|---|---|
| **Conference President**: | |
| Prof. Irina Lungu | International Affairs Prorector for "Gheorghe Asachi" Technical University of Iasi |
| **Chairperson**: | |
| Prof. Konstantinos Daniel Tsavdaridis | City University of London |

**Advisory Board**

| | |
|---|---|
| Prof. Dorina-Nicolina Isopescu | "Gheorghe Asachi" Technical University of Iasi |
| Assoc. Prof. Tania Mariana Hapurne | "Gheorghe Asachi" Technical University of Iasi |
| Prof. Nicu Marcoie | "Gheorghe Asachi" Technical University of Iasi |
| Prof. Radu Văcăreanu | Technical University of Civil Engineering Bucharest |
| Prof. Daniela Manea | Technical University of Cluj-Napoca |
| Prof. Raul Zaharia | "Politehnica" University of Timisoara |
| Prof. Ioan Tuns | Transilvania University of Brasov |
| Prof. Marcela Prada | University of Oradea |
| Lecturer Cosmin Filip, PhD | "Ovidius" University of Constanta |
| Prof. Emer. Mihai Budescu | Senior Expert in Civil Engineering |
| Prof. Constantin Ionescu | Senior Expert in Civil Engineering |
| Prof. Emer. Horia Alex Barbat | Senior Expert |
| Prof. Emer. Nicolae Taranu | "Gheorghe Asachi" Technical University of Iasi |
| Assoc. Prof. Vasilica Ciocan | "Gheorghe Asachi" Technical University of Iasi |

**Scientific committee**

| | |
|---|---|
| Prof. Thanasis C. Triantafillou | University of Patras |
| Prof. João Carlos Gonçalves Lanzinha | Univ. Beira Interior |
| Prof. Cristina Cosma | Wentworth Institute of Technology |
| Prof. Chérifa Abid | Aix-Marseille Université France |
| Prof. Cristina Mihaela Campian | Technical University Cuj Napoca |
| Prof. Fideliu Păuleț-Crăiniceanu | "Gheorghe Asachi" Technical University of Iasi |
| Prof. Mebarek-Oudina Fateh | University of Skikda, Algeria |
| Prof. Wynand Steyn | University of Pretoria |
| Prof. Chris Roth | University of Pretoria |
| Prof. dr. Mariusz Maslak | Cracow University of Technology |
| Prof. dr. Melito A. Baccay | Technological University of the Philippines |
| Prof. Andrei Burlacu | "Gheorghe Asachi" Technical University of Iasi |
| Assoc. Prof. Mihail Iancovici | Technical University of Civil Engineering of Bucharest |
| Assoc.Prof. Valentin-Vasile Ungureanu | Transilvania University, Brașov |
| Assoc.Prof. Daniel Aelenei | NOVA School of Science and Technology |
| Assoc. Prof. Pierre-Olivier Logerais | Université Paris-Est Créteil |
| Assoc. Prof. Gobesz Ferdinand-Zsongor | Technical University of Cluj Napoca |
| Assoc. Prof. Gabriel Oprișan | "Gheorghe Asachi" Technical University of Iasi |
| Assoc. Prof. Mebarek-Oudina Fateh | University of Skikda, Algeria |
| Assoc. Prof. Cristian Chereches | "Gheorghe Asachi" Technical University of Iasi |
| Assoc. Prof. Petru Mihai | "Gheorghe Asachi" Technical University of Iasi |
| Sen.Lect. Patricia Alonso Alonso | University of A Coruña |
| Sen.Lect. Lușa Tuleasca | Unitec Institute of Technology |

IOP Conf. Series: Materials Science and Engineering 1304 (2024) 011001 doi:10.1088/1757-899X/1304/1/011001

IOP Conf. Series: Materials Science and Engineering 1304 (2024) 011001 doi:10.1088/1757-899X/1304/1/011001

# Preface

Continuous evolving labour market influenced by the present world-wide challenges and crisis is strongly demanding for adequately qualified work force and innovative solutions competing for market success.

Research and innovation in constructions' related fields requires enhancing four dimensions of the transversal skills and competences: digital, green, resilience, and entrepreneurial skills.

Computational Civil Engineering (CCE) 2023 is dedicated to the research outputs feeding high quality education and training, fostering collaboration and cooperation among research groups with wide scientific interests, on the relevant topics recently endorsed by European Project calls within the 2021-2027 budgeted actions

Traditionally, CCE has been mainly organized, by the Faculty of Civil Engineering and Building Services of Iasi and "Matei Teiu-Botez" Academic Society in the previous editions. Recent joint research work started with the contribution of various Faculties of "Gheorghe Asachi" Technical University of Iași (TUIASI) and thus beginning with CCE 2023 will be jointly organized by TUIASI and "Gheorghe Asachi" University Foundation. The conference is organized once every 2 years since 2015, with physical attendance in the beautiful city of Iasi with the venue in the historical building designed by Gustave Eiffel in the early XXth century Grand Hotel Traian.

The 17th edition CCE2023 was organized by "Gheorghe Asachi" Technical University of Iasi (Faculty of Civil Engineering and Building Services, Faculty of Architecture and Faculty of Hydrotechnics, Geodesy and Environmental Engineering) in partnership with the Universitary Foundation "Gheorghe Asachi" of Iasi and the International Centre for Numerical Methods in Engineering (CIMNE) Barcelona Spain. We were very honoured to collaborate for conference organizing with the City University of London (UK), the Technical University of Civil Engineering Bucharest, the Faculty of Civil Engineering Cluj-Napoca, the Faculty of Civil Engineering Timisoara, the Faculty of Civil Engineering Brasov, the Faculty of Civil Engineering Constanta and the Faculty of Civil Engineering, Cadastre and Architecture Oradea.

Also, we were grateful for the support received from our partners Unitec Institute of Technology Auckland, Iasi City Hall and the Order of Romanian Architects – Iasi branch.

The scientific event was held on May 24$^{th}$-25$^{st}$ at Grand Hotel Traian from Iasi, Romania.

CCE2023 aimed to bring together leading academic scientists, researchers, practitioners, numerical analysis programs developers to exchange / share their experiences to the Computational Civil Engineering domain to share their research findings to the theoretical and practical levels.

The proposed topics in construction related fields for CCE2023 edition of the conference was the following:

- *Computer-Aided Design and Engineering*
- *Computational Methods in Civil Engineering*
- *Numerical Analysis of Civil Engineering Structures*
- *Computational Fluid Dynamics*
- *Artificial Intelligence in Civil Engineering*
- *Digital Technologies*
- *Building Information Modelling*
- *Structural Topology Optimisation and 3D Printing*
- *Greening Future in Civil Engineering*
- *Power and Energy Systems for Civil Engineering*

IOP Conf. Series: Materials Science and Engineering 1304 (2024) 011001 doi:10.1088/1757-899X/1304/1/011001

In plenary sessions were presented very interesting and up-to-date exposures of the keynote speakers, well-known researchers or practicians in Civil Engineering domain:

- prof. **Konstantinos Tsavdaridis** – City University of London (UK) - *The art of modular connections – a new way of thinking*
- prof. **Christina J. Hopfe** – Technical University Graz (Austria) - *Critical reflections on the use of simulation and decision-making*
- arh. **Andrei Padure** – Ai Build (UK) - *Computational design applications*
- prof. **Damir Hatic** – VIA University College (Denmark) - *Circular value creation in the existing building mass*
- dr. **Erika A. Pärn** – University of Cambridge (UK) - *The digital twins for the built environment*
- dr. **Gabriel Barbat**, International Centre for Numerical Methods in Engineering, Spain - *Objective numerical evaluation of structural failure via adaptive mesh and formulation refinement*

This year, the online organization offered an opportunity for all the participants make it easy for the researchers and experts from 12 countries to attend CCE2023.

The organizers want to express gratitude to all members of the Scientific Committee and the reviewers, and welcomes the speakers, papers presenters and participants of this conference and wish all of you a fruitful event.

Finally, but not at the end, thanks to all sponsors and individuals who have contributed to the organization and success of this special event CCE2021. Hope to meet them again to the next edition CCE2023 at the end of May.

More information about this scientific event can be found on the conference website www.cce.ci.tuiasi.ro.

*The editors*

# Peer Review Statement

All papers published in this volume have been reviewed through processes administered by the Editors. Reviews were conducted by expert referees to the professional and scientific standards expected of a proceedings journal published by IOP Publishing.

- **Type of peer review:** Double Anonymous
- **Conference submission management system:** Morressier
- **Number of submissions received:** 40
- **Number of submissions sent for review:** 40
- **Number of submissions accepted:** 35
- **Acceptance Rate (Submissions Accepted / Submissions Received × 100):** 87.5
- **Average number of reviews per paper:** 1.98
- **Total number of reviewers involved:** 44
- **Contact person for queries:**
  **Name:** Ioana Olteanu
  **Email:** ioana.olteanu@academic.tuiasi.ro
  **Affiliation:** "Gheorghe Asachi" Technical University of Iasi

IOP Conf. Series: Materials Science and Engineering 1304 (2024) 012001 doi:10.1088/1757-899X/1304/1/012001

# Time-History Analyses on the Effect of Partial Infill Walls on the Seismic Response of a Reinforced Concrete Frame Structure

**V M Venghiac[1], O M Banu[1], G Taranu[1] and S A Baetu[1]**
[1] The "Gheorghe Asachi" Technical University of Iasi, Faculty of Civil Engineering and Building Services, no. 1, Prof.dr.doc. D. Mangeron Street, Iasi, *Romania*

E-mail: sergiu-andrei.baetu@academic.tuiasi.ro

**Abstract.** Time-history analysis (THA) is based on direct step-by-step integration of the equation of motion in time domain considering as input either the recorded accelerograms of real or artificially generated earthquakes. Numerical simulations are sensitive in terms of material properties, geometry, boundary conditions and applied loading scenarios correct definition. Therefore, additional care is required in the pre-processing stage of the analysis when the numerical model is defined. Infill walls play an important role on the seismic behaviour of structures during earthquakes. The response of the structure to seismic action changes due to the presence of the infill walls. The paper presents the results obtained by means of THA on the effect of partial infill walls on the seismic response of a scaled-down RC frame structure subjected to four consecutive uniaxial seismic motions. The parameters of the research were the presence of the partial infill wall and the thickness of the wall. The obtained results are discussed from the point of view of maximum lateral displacements and response spectra expressed in terms of accelerations. The results contribute towards the setting up of the subsequent shake table tests of scaled-down RC frame models with different configurations that would validate and help calibrate the numerical model.

## 1. Introduction

The significant advances made in the microarchitecture of computer processors increasing their computational power and efficiency reflected in every aspect of modern life. Research in the field of civil engineering, in terms of numerical simulations, has seen a huge development due to the afore mentioned advances. Improved mathematical models of physical phenomena, coupled with more and more complex numerical models of civil engineering structures, resulted in accurate assessments of their behaviour under various loading conditions.

Numerical simulations are now at the heart of every major research project, from reliability analyses [1,2], to assessing the behaviour of different structures in difficult site conditions [3–5], to optimizing the design process and increase the safety of newly built or strengthened buildings [6–10], to creating intervention strategies and relief plans in case of occurring natural disasters [11].

Time-history analysis (THA) is based on direct step-by-step integration of the equation of motion in time domain considering as input either the recorded accelerograms of real earthquakes or synthetically / artificially generated ones to fit design elastic response-spectra of different design codes. Although this type of analysis is used mainly by researchers, as it requires a vast amount of knowledge on how to properly define de input motion, it successfully overcomes the drawbacks of other often employed methods such as: modal analysis used together with the behaviour factor q to account for the effect of energy dissipation, non-linear static (pushover) analysis which, although quite widespread, fails to account for the effect of load reversals on the capacity of a structure [12].

Numerical simulations, on the other hand, albeit being very useful, are sensitive in terms of material properties, geometry, boundary / support conditions and applied loading scenarios correct definition. Therefore, additional care is required in the pre-processing stage of the analysis when the numerical model is defined, especially in case of complex structures with geometries that require attention, various materials being used and loading scenarios that have to be applied in a certain order [13]. Although numerical simulations offer the advantage of varying a large number of parameters [14], they do, however, need experimental validation [15–17]. Still, numerical simulations could be used as a tool to predict the behaviour of the investigated structure during the laboratory experiments, in order to avoid damages to equipment.

Infill walls play an important role on the seismic behaviour of structures during earthquakes [18]. The response of the structure to seismic action changes due to the presence of the infill walls, although their effect is still disregarded in many countries [19]. They significantly reduce the deformation capacity of the structure, increasing its stiffness and, consequently, the intensity of the seismic force taken over by the structure. This may result in damages either at the level of the load bearing structure, e.g. the reinforced concrete (RC) frame, or at the level of the infill wall. Either way, the overall behaviour changes significantly from what was initially considered during the design stage [19].

The paper presents the results obtained by means of THA on the effect of partial infill walls on the seismic response of a scaled-down RC frame structure. The parameters of the research were the presence of the partial infill wall and the thickness of the wall. The model was subjected to four consecutive uniaxial seismic motions. The obtained results are discussed from the point of view of maximum lateral displacements and response spectra expressed in terms of accelerations. The results contribute towards the setting up of the subsequent shake table tests of scaled-down RC frame models with different configurations that would validate and help calibrate the numerical model.

## 2. Numerical Model

### *2.1. RC frame geometry without infill walls*

The geometry of the considered scaled-down RC frame ground floor structure is presented in Figure 1. While the material properties of the reinforcement are similar to the ones presented in [20], the selected concrete strength class was C20/25.

The considered partial infill-wall had a height of 80 cm and was assumed to be located along axes A and B (Figure 1, left). While previous shake table tests conducted on similar configurations [16] considered an infinite, in plane, rigidity of the partial infill wall, the present study accounts for the influence of the partial infill wall on the lateral stiffness of the model depending on the wall thickness.

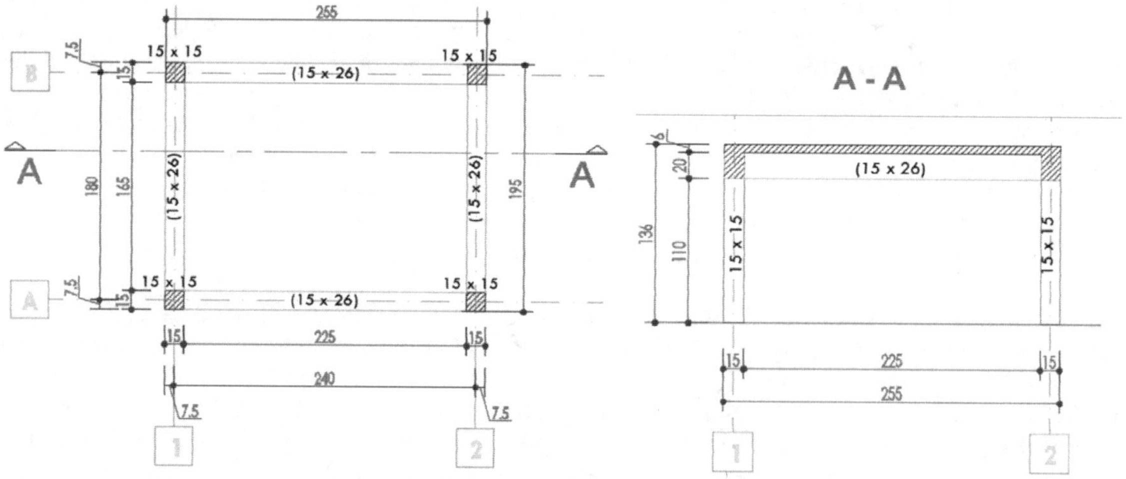


**Figure 1.** Geometry of the model (dimensions in cm): plan view (left), longitudinal section (right).

IOP Conf. Series: Materials Science and Engineering 1304 (2024) 012001 doi:10.1088/1757-899X/1304/1/012001

### 2.2. Finite element model

The 3D numerical model was generate using SAP2000. Standard material properties were considered for the C20/25 concrete that was applied to all elements: slab, beams, columns and infill walls. The beams and the columns were defined as linear frame elements and the slab and the infill walls as thin-shell 2D-surface elements. For this stage of the research, a full bond was assumed between the infill walls and the columns. Fixed supports were considered at the base of the columns and linear simple supports at the bottom edge of the infill walls.

The slab was 6 cm thick and was reinforced with 4 mm bars spaced at 100 mm in both directions. The columns were reinforced with four 14 mm bars as longitudinal reinforcement and 4 mm stirrups spaced at 100 mm. The longitudinal reinforcement was defined as 1D Reinforcement type element. The beams in both in-plane directions (Figure 1 left) were reinforced with four 12 mm and four 10 mm bars respectively as longitudinal reinforcement and 6 mm stirrups spaced at 100 mm in the shear spans and at 200 mm in the middle for both beams. The reinforcement in the columns was assumed to be fully bonded to concrete, whereas the reinforcement in the beams was defined only as equivalent cross-sectional area of steel.

The material properties for the three typologies are presented in Table 1. The models were named in accordance with the presence or absence of the partial infill walls. Hence, RCF stands for reinforced concrete frame, whereas W followed by a number indicates that a partial infill wall was considered with a thickness (in cm) corresponding to the number following letter W. The material properties of C20/25 concrete were considered in accordance with Eurocode 2 [21], where $f_{cd}$ and $f_{ctd}$ are the design compressive and tensile strength, respectively, and E is the longitudinal modulus of elasticity.

**Table 1.** Model typology and concrete characteristics

| Case nr. | Case name | Infill wall thickness [cm] | C20/25 concrete material properties | | |
|---|---|---|---|---|---|
| | | | $f_{cd}$ [MPa] | $f_{ctd}$ [MPa] | E [MPa] |
| 1 | RCF | - | | | |
| 2 | RCF-W6 | 6 | 13.33 | 1.03 | 29962 |
| 3 | RCF-W12 | 12 | | | |

### 2.3. Time history loading scenarios

Besides the self-weight, the model was subjected to ground motion accelerograms recorded for the Vrancea 1977 and Vrancea 1990 earthquakes. All seismic scenarios were applied uniaxially, along the longitudinal direction of the model in order to assess the influence of the partial infill walls on the seismic response of the structure. The EW and NS components of both earthquakes were applied in sequence for a total of 4 time-history analyses for each considered model.

The accelerograms of Vrancea 1977 and 1990 earthquakes are presented in Figure 2 and Figure 3, respectively. The two earthquakes were considered representative for Romania based on their magnitude and peak ground acceleration (PGA) values.

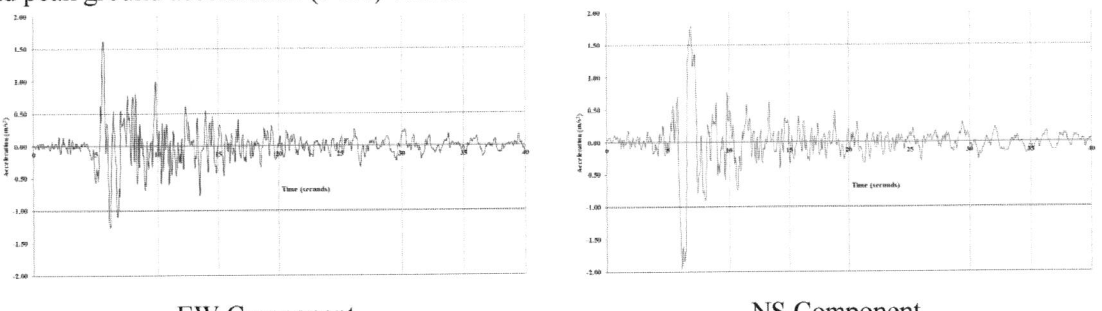

EW Component NS Component

**Figure 2.** Acceleration time histories and response spectrum for Vrancea 1977 earthquake.

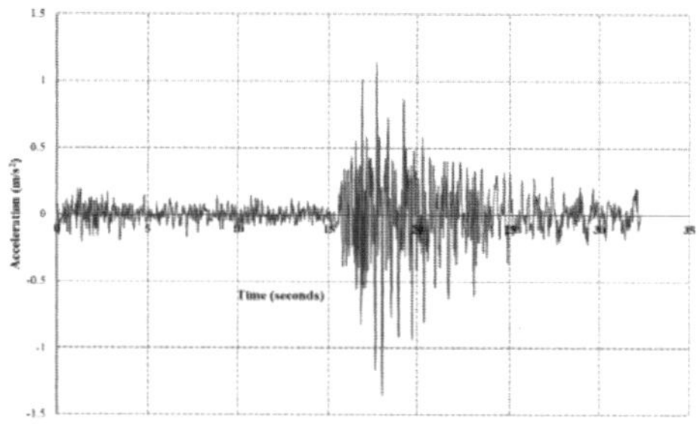

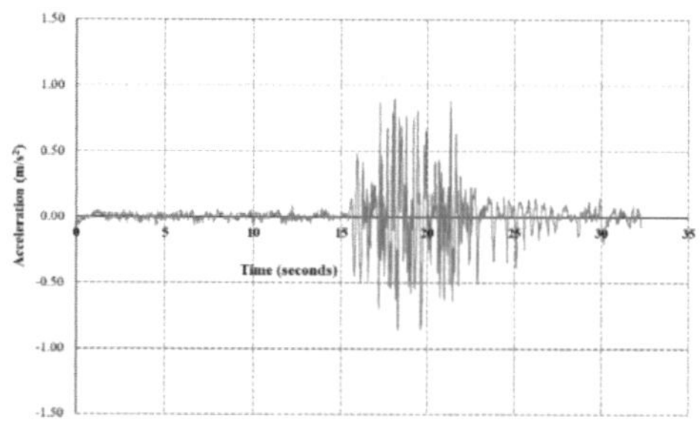

EW Component NS Component

**Figure 3.** Acceleration time histories and response spectrum for Vrancea 1990 earthquake.

The characteristics of the selected earthquakes are presented in Table 2. The analysis codes tried to include the earthquake year and the recording station code. Hence 901IAS1 stands for the Vrancea earthquake that took place in 1990 (the one on May 30$^{th}$), 901, and IAS1 represents the station code where the strong motion was recorded.

**Table 2.** Characteristics of selected earthquakes

| Earthquake | Analysis code | Depth [km] | Magnitude [Mw] | PGA [m/s$^2$] | |
|---|---|---|---|---|---|
| | | | | EW Component | NS Component |
| Vrancea 1977 | VN77 | 109 | 7.5 | 1.62 | 1.95 |
| Vrancea 1990 | 901IAS1 | 91 | 7.0 | 1.35 | 0.90 |

## 3. Results

### *3.1. Modal analysis*

The dynamic characteristics of the model were determined prior to running the time-history analyses. The obtained data is summarized in Table 3. The results for RCF analysis case are in agreement with previously reported data but for a different concrete strength class [20]. It can be observed the scaled down model is very rigid. The addition of the partial infill wall leads to a further increase in the stiffness of the model, lowering its fundamental period of vibration. Increasing the partial infill wall thickness results in a further decrease in the value of the fundamental period of vibration.

**Table 3.** Modal analysis results

| Case nr. | Case name | Fundamental period of vibration [seconds] | Fundamental frequency of vibration [Hz] |
|---|---|---|---|
| 1 | RCF | 0.05242 | 19.077 |
| 2 | RCF-W6 | 0.05161 | 19.377 |
| 3 | RCF-W12 | 0.04649 | 21.508 |

Figure 4 presents the deformed shape of the considered models obtained from the modal analysis. It can be observed that adding the infill wall results in a change of direction for the fundamental mode of vibrations. While for RCF case the first mode was a translational one along the longitudinal direction (Figure 1), for both RCF-W6 and RCF-W12 case the first mode of vibration took place along the transversal direction (Y in-plane axis of the model).

Moreover, in case of RCF-W6 case the second and thirds modes were local ones, involving the deformation of the partial infill wall. Mode 4, on the other hand, was a global torsional one. For the RCF-W12, however, case the second mode of vibration was a torsional one. It can, therefore, be

IOP Conf. Series: Materials Science and Engineering 1304 (2024) 012001 doi:10.1088/1757-899X/1304/1/012001

concluded that not only the presence of the partial infill wall changes the dynamic characteristics of the model, but the vibrational modes themselves and their order are changed, too. The thickness of the partial infill wall has also a significant influence on the presented aspects.

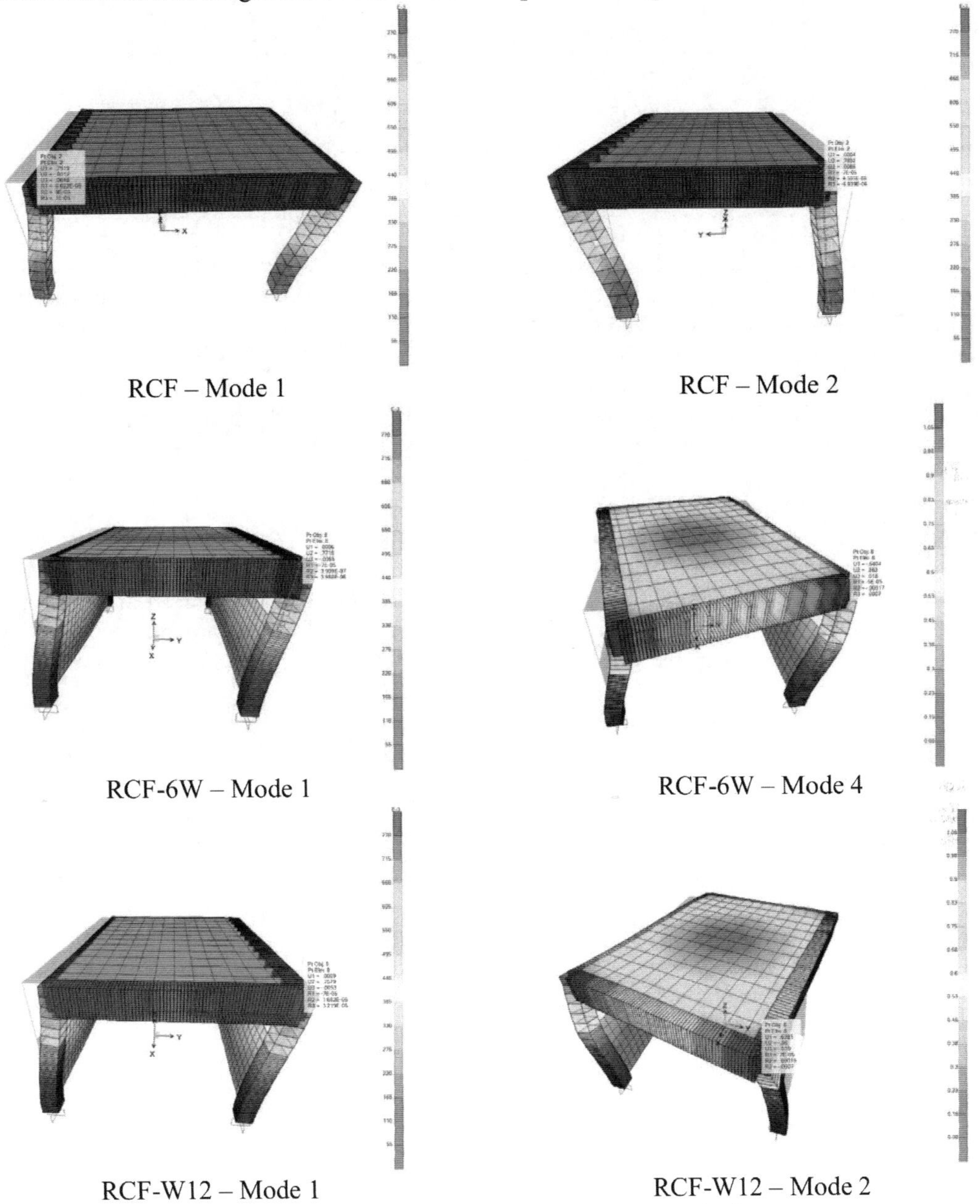

**Figure 4.** Modes of vibration for the considered analysis cases.

### *3.2. Maximum lateral displacements*

The maximum lateral displacements recorded in a building subjected to seismic motions is an important parameter because it helps assessing the inter-storey drift. The obtained results, for each model, are summarized in Table 4.

**Table 4.** Maximum lateral displacement and storey drift

| Earthquake scenario | Lateral displacement [mm] | | | Storey drift [%] | | |
|---|---|---|---|---|---|---|
| | RCF | RCF-W6 | RCF-W12 | RCF | RCF-W6 | RCF-W12 |
| VN77EW | 0.13 | 0.04 | 0.04 | 0.012 | 0.004 | 0.004 |
| VN77NS | 0.18 | 0.05 | 0.04 | 0.016 | 0.005 | 0.004 |
| 901IAS1EW | 0.20 | 0.04 | 0.04 | 0.018 | 0.004 | 0.004 |
| 901IAS1NS | 0.24 | 0.03 | 0.03 | 0.022 | 0.003 | 0.003 |

It can be observed that in case of RCF the maximum lateral displacement increases with each applied seismic load case. It could be concluded that there might be an accumulation of damage inside the structure. This assumption could be validated by the analysis of the input and response spectra. For both cases when the partial infill wall was considered the obtained values of the maximum lateral displacements were 3-8 times smaller, depending on the considered scenario, compared to the more flexible RCF case. The obtained results indicate that the thickness of the partial infill wall did not play any role on the values of the maximum lateral displacements.

*3.3. Response spectra of seismic motions*

The response spectra on the considered seismic motions, expressed in terms of spectral acceleration, are presented in Figure 5. Taking a look at the SA for the Vrancea 1977 earthquake it can be observed the peak values would correspond to longer period structures, longer than 1 second, for the NS component. On the other hand, the EW component seems to affect structures with a fundamental period of vibration lower than 1 second. Moreover, the amplitude of the SA for the EW component was lower than for the NS component.

The SA spectra of the Vrancea 1990 earthquake indicate that both components would affect lower period structures. The amplitude of SA is almost similar for both EW and NS components, at around 0.25 seconds. Hence, taking into account the dynamic characteristics of the RCF model, Table 3, the results presented in Table 4 in terms of maximum lateral displacements could be explained by a closer match between the large amplitude of SA of the Vrancea 1990 earthquake and the fundamental period of vibration of the model, coupled with a higher flexibility of the RCF model compared to the RCF-W models.

The amplitude of the spectral acceleration corresponding to the fundamental period of vibration for each analysis case are summarized in Table 5.

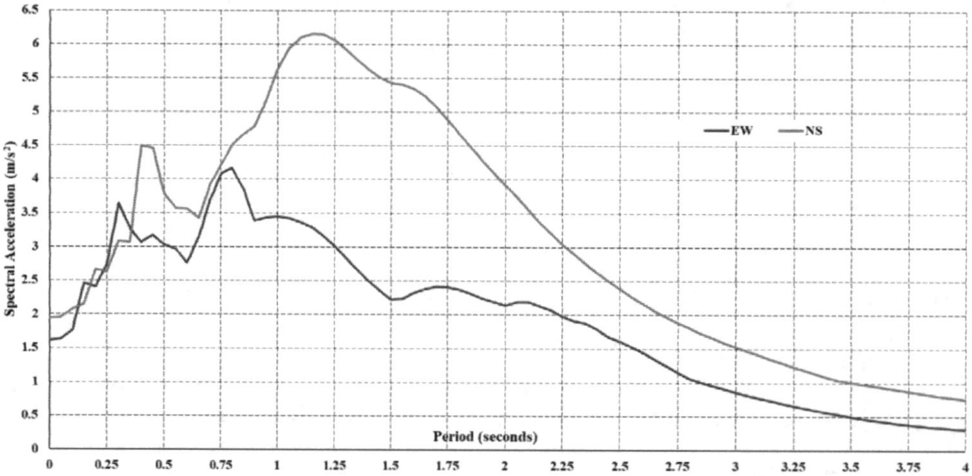


Vrancea 1977 earthquake

IOP Conf. Series: Materials Science and Engineering 1304 (2024) 012001 doi:10.1088/1757-899X/1304/1/012001

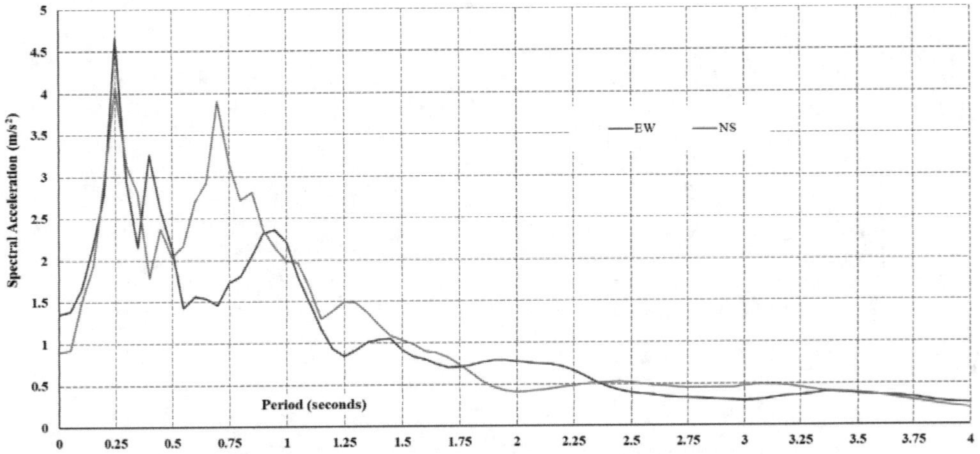


Vrancea 1990 earthquake

**Figure 5.** Spectral accelerations of the Vrancea 1977 and 1990 earthquakes.

**Table 5.** Amplitude of spectral acceleration [m/s²]

| Earthquake scenario | RCF | RCF-W6 | RCF-W12 |
|---|---|---|---|
| VN77EW | 1.640 | 1.638 | 1.625 |
| VN77NS | 1.959 | 1.957 | 1.945 |
| 901IAS1EW | 1.402 | 1.398 | 1.371 |
| 901IAS1NS | 0.953 | 0.944 | 0.885 |

*3.4. Response spectra of the considered analysis models*

The response spectra of the considered analysis models, for a 5% structural damping, are presented in Figure 6.

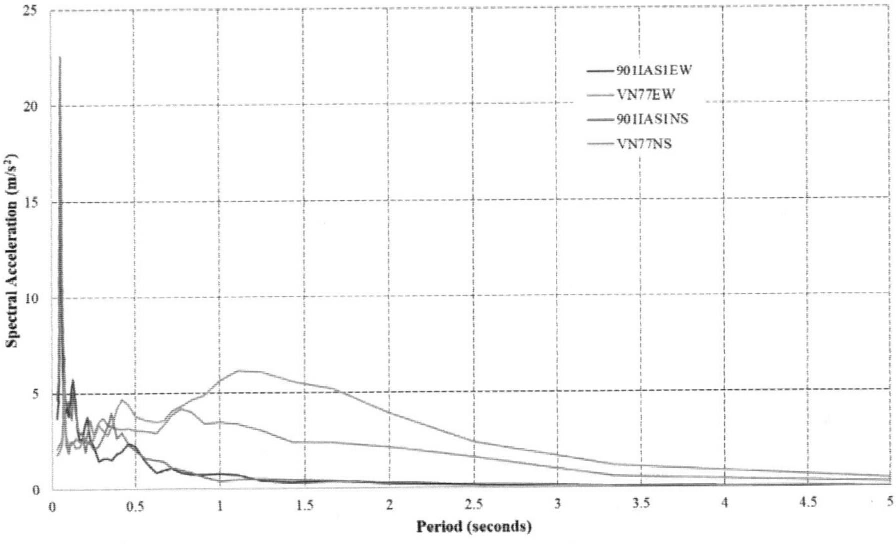


RCF model

IOP Conf. Series: Materials Science and Engineering 1304 (2024) 012001 doi:10.1088/1757-899X/1304/1/012001

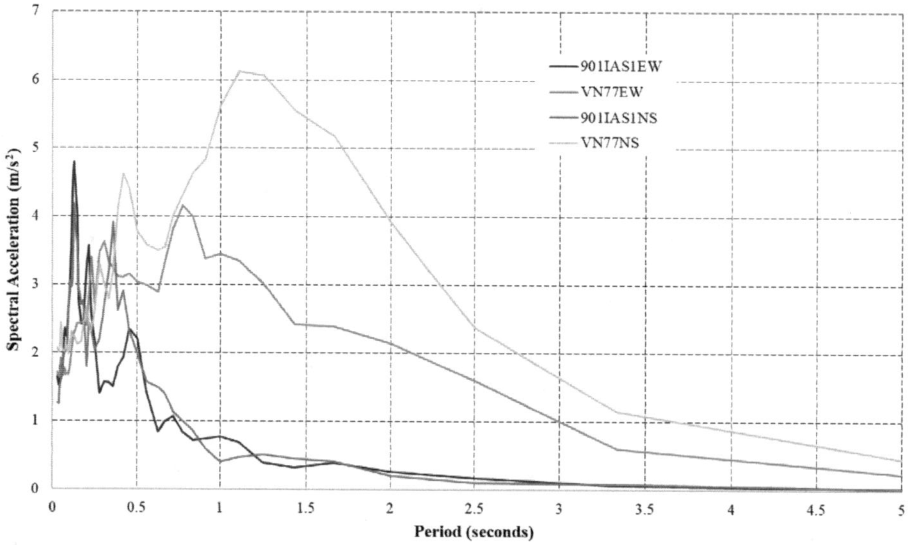


RCF-W6 model

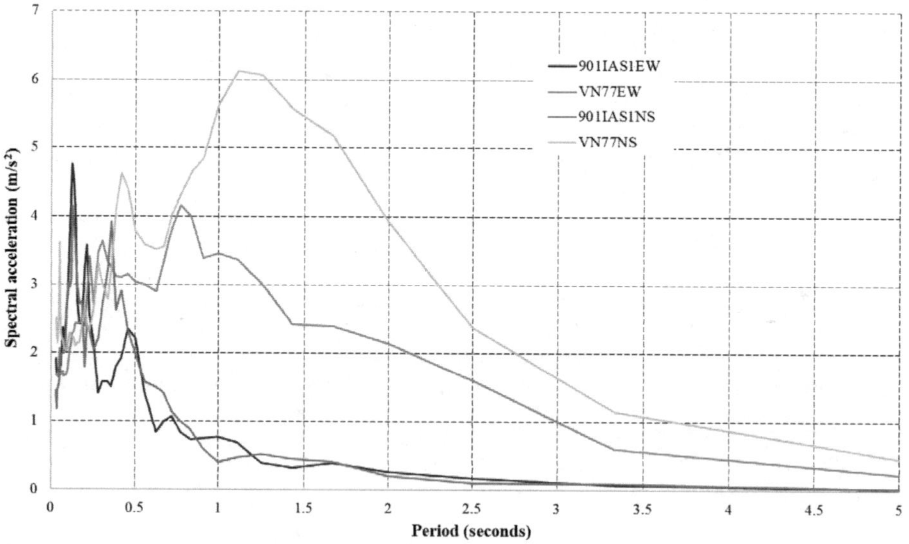


RCF-W12 model

**Figure 6.** Spectral accelerations of the considered models.

The amplitudes of the SA determined from the response spectra are summarized in Table 6.

**Table 6.** Amplitude of spectral acceleration from the response spectra [m/s$^2$]

| Earthquake scenario | RCF | RCF-W6 | RCF-W12 |
|---|---|---|---|
| VN77EW | 2.177 | 1.910 | 1.827 |
| VN77NS | 2.533 | 2.350 | 2.838 |
| 901IAS1EW | 1.867 | 1.803 | 1.654 |
| 901IAS1NS | 1.598 | 1.624 | 1.569 |

By comparing the values presented in Table 6 with the data summarized in Table 5, it can be concluded that there are amplifications of the input SA amplitude for each considered model. Considering the large in plane stiffness of the slab and its inertia coupled with the high stiffness of each model, these amplifications of the amplitude of input SA were expected.

## 4. Conclusions

The paper presents the results obtained by means of time-history analyses on the seismic behaviour of a reinforced concrete frame structure subjected to different earthquake scenarios considering the presence and the thickness of the partial infill walls. The infill walls were positioned along the longitudinal direction of the model and the selected strong motion components were applied sequentially along the same direction.

The presence of partial infill walls change the fundamental mode of vibration as well as the order of subsequent modes of vibration. In both cases where infill walls were considered, the second mode of vibration is a torsional one and not a translational one. Thinner walls lead to the occurrence of local modes of vibration, meaning additional support conditions are required in the numerical model.

None of the selected seismic events produce large lateral displacements of the model in both configurations – with and without partial infill walls. This may be due to the high stiffness of the model, taking into account the dynamic properties of each investigated geometry.

All models exhibited amplification in the magnitude of the spectral acceleration. Taking into account the high in-plane stiffness of the slab, present in all models, and the overall stiffness of the structure, derived from the dynamic characteristics, the higher amplitudes of the response spectral accelerations are expected.

The obtained results will be used to prepare subsequent shake table tests of scaled down models of reinforced concrete structures having similar geometry.

## 5. References

[1] Grubišić M, Ivošević J and Grubišić A 2019 Reliability Analysis of Reinforced Concrete Frame by Finite Element Method with Implicit Limit State Functions *Buildings* 9 119

[2] Kohns J, Stempniewski L and Stark A 2022 Fragility Functions for Reinforced Concrete Structures Based on Multiscale Approach for Earthquake Damage Criteria *Buildings* 12 1253

[3] Alzabeebee S and Forcellini D 2021 Numerical Simulations of the Seismic Response of a RC Structure Resting on Liquefiable Soil *Buildings* 11 379

[4] Biglari M, Hadzima-Nyarko M and Formisano A 2022 Seismic Damage Index Spectra Considering Site Acceleration Records: The Case Study of a Historical School in Kermanshah *Buildings* 12 1736

[5] Teng F, Zhang Y, Yan W, Wang X and Li Y 2022 Numerical Simulation of Reinforced Concrete Piers after Seawater Freeze–Thaw Cycles *Coatings* 12 1825

[6] Tang Z, Peng Z and Liu X 2022 Numerical Study on Optimal Design and Seismic Capacity of Double-Span RC Frame Structures with Exterior Verandahs *Buildings* 12 1901

[7] Pastia C and Luca S-G 2012 Seismic Protection of a Realistic Bridge Structure Using Passive Friction Devices *Bull. Polytech. Inst. Iași. Constr. Archit. Sect.* LVIII (LXI 77–84

[8] Mathe A, Cătărig A and Moldovan I 2015 Statics And Kinematics Of Semirigid Steel Frames Under Seismic Action *J. Appl. Eng. Sci.* 5 59–64

[9] Alexa P and Mathe A 2016 Energy state of multi-story structures equipped with passive protection *Proceedings of the International Colloquium on Stability and Ductility of Steel Structures, SDSS 2016* pp 851–7

[10] Pastia C and Luca S-G 2017 Passive Viscous Fluid Devices for Seismic Energy Dissipation *Bull. Polytech. Inst. Iași. Constr. Archit. Sect.* 63(67) 115–24

[11] Valluzzi M, Sbrogiò L and Saretta Y 2021 Intervention Strategies for the Seismic Improvement of Masonry Buildings Based on FME Validation: The Case of a Terraced Building Struck by the 2016 Central Italy Earthquake *Buildings* 11 404

[12] Betti M, Galano L and Vignoli A 2015 Time-History Seismic Analysis of Masonry Buildings: A Comparison between Two Non-Linear Modelling Approaches *Buildings* 5 597–621
[13] Ereiz S, Duvnjak I, Damjanović D and Bartolac M 2021 Analysis of Seismic Action on the Tie Rod System in Historic Buildings Using Finite Element Model Updating *Buildings* 11 453
[14] Pardalopoulos S I, Pantazopoulou S J and Manolis G D 2022 On the Modeling and Analysis of Brittle Failure in Existing R/C Structures Due to Seismic Loads *Appl. Sci.* 12 1602
[15] Rosca O V, Canarache R, Budescu M and Oanea D 2014 The Experimental Testing and Numerical Model Calibration of a Steel Structure *Bull. Polytech. Inst. Iași. Constr. Archit. Sect.* LX (LXIV) 9–18
[16] El Khouri I, Garcia R, Mihai P, Budescu M, Ţăranu N, Toma I-O, Guadagnini M, Escolano-Margarit D, Entuc I S, Oprisan G, Hajirasouliha I and Pilakoutas K 2018 Shake Table Tests on Frames Made with Normal and FRP-Confined Rubberised Concrete *16th European Conference on Earthquake Engineering (16ECEE)* (Thessaloniki, Greece)
[17] Lee K-S, Lee B-G and Jung J-S 2022 Seismic Strengthening of R/C Buildings Retrofitted by New Window-Type System Using Non-Buckling Slit Dampers Examined via Pseudo-Dynamic Testing and Nonlinear Dynamic Analysis *Appl. Sci.* 12 1220
[18] Kim M and Yu E 2021 Experimental Study on Lateral-Load-Resisting Capacity of Masonry-Infilled Reinforced Concrete Frames *Appl. Sci.* 11 9950
[19] Messaoudi A, Chebili R, Mohamed H and Rodrigues H 2022 Influence of Masonry Infill Wall Position and Openings in the Seismic Response of Reinforced Concrete Frames *Appl. Sci.* 12 9477
[20] Toma I O, Mihai P, Venghiac V-M, Taranu G, Baetu S-A, Toma A M and Petrescu T C 2020 Numerical Simulations of the Seismic Behavior of a Damaged RC Frame Retrofitted with Composite Fabric *17th World Conference on Earthquake Engineering, 17WCEE* (Sendai) pp 3f – 0019
[21] EN1992-1-1:2004 2004 Eurocode 2: Design of concrete structures - Part 1-1: General rules and rules for buildings

**Acknowledgement**
The research presented in this paper was conducted as part of the PN-III-P2-2.1-PED-2021-0677 research grant "Sustainable Concrete for Energy Efficient Buildings" funded by Executive Agency for Higher Education, Research, Development and Innovation Funding – UEFISCDI.

IOP Conf. Series: Materials Science and Engineering 1304 (2024) 012002 doi:10.1088/1757-899X/1304/1/012002

# Numerical Investigations on the Seismic Behavior of a Reinforced Concrete Frame Structure

**G Taranu[1], V M Venghiac[1], O M Banu[1] and V Florea[1]**
[1] The "Gheorghe Asachi" Technical University of Iasi, Faculty of Civil Engineering and Building Services, no. 1, Prof.dr.doc. D. Mangeron Street, Iasi, *Romania*

E-mail: oana-mihaela.banu@academic.tuiasi.ro

**Abstract.** Recent seismic events that took place during the past two decades and struck densely populated areas caused significant material losses and, sometimes, casualties. Taking into account the ageing of the building stock around the world coupled with unavoidable degradations, urgent measures need to be taken to either strengthen or retrofit such structures based on data obtained from analytical, experimental or numerical investigations. The present paper presents the results of numerical simulations, using time-history analyses, on a scaled-down reinforced concrete (RC) frame structure subjected to two multidirectional seismic scenarios. The obtained results are discussed from the point of view of maximum displacements, accelerations and equivalent stresses. Considering the dynamic characteristics of the model, neither considered seismic event were able to produce large lateral displacements or accelerations. Damages were localized at the beam-column joints in case of Turkey 2023 earthquake scenario.

## 1. Introduction

Recent moderate to severe seismic events that took place during the past two decades and struck densely populated areas caused significant material losses and, in some unfortunate cases, casualties. Investigations launched following the seismic event highlighted the fact that some of the totally or partially collapsed structures were either substandard or were not built according to the design plans. Substandard structures refer to the type of structures built in accordance to older code regulations (focusing on providing the elements with enough strength to resist lateral loads) that do not fulfil the requirements of modern design codes (to ensure a certain level of ductility in addition to strength requirements) [1]. Taking into account the ageing of the building stock around the world coupled with unavoidable degradations [2], urgent measures need to be taken to either strengthen or retrofit such structures in order to prevent additional losses due to subsequent seismic events [3].

Reinforced concrete structures are present in nearly all populated areas with a variety of structural typology and purpose, from individual houses to condominiums, schools, hospitals, a.s.o. Such diverse typology means different behaviour during a seismic event, a behaviour that should be correctly assessed and understood [1,4]. Reinforced concrete frame structures provide both the required ductility and the strength necessary to resist lateral loads such as those induced by earthquakes. However, proper design of beams, column and joints should be applied in order to ensure the safety of occupants and to limit the damage of the structure [5].

Numerical simulations have the advantage of being relatively simple to be applied from individual elements to complex, large-scale, structural systems. At the same time, they offer the possibility of changing a large number of parameters from geometrical dimensions, material behaviour, boundary conditions and/or input actions [6,7]. In order for such simulations to provide trustworthy results, they should be based on accurate analytical models and should be validated by laboratory investigations on structural elements or scaled down models [8–10].

While time-history analyses (THA) may not be significant from a practitioner's perspective, it may offer insightful information to researchers from the point of view of seismic behaviour of investigated component [11] or structure [12,13] subjected to different earthquake induced vibrations. This is because each component of the seismic motion may have a different frequency content that could produce damages to the structure to a lesser or greater extent depending on the dynamic properties of the structure itself coupled with local soil conditions. In view of the very recent disastrous earthquake that took place in Turkey, the present paper presents the results of numerical simulations on a scaled-down reinforced concrete (RC) frame structure subjected to two multidirectional seismic scenarios. The obtained results are discussed from the point of view of maximum displacements, accelerations and equivalent stresses.

## 2. Numerical Model

*2.1. RC frame geometry and materials*

The geometry of the considered scaled-down RC frame ground floor structure is presented in Figure 1. While the material properties of the reinforcement are similar to the ones presented in [14]. The selected concrete strength class was C16/20 with the characteristics shown in Table 1.

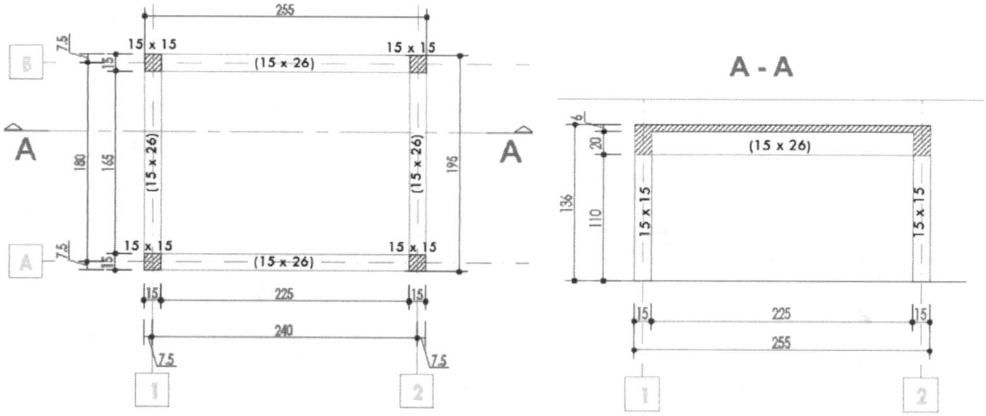


**Figure 1.** Geometry of the model (dimensions in cm): plan view (left), longitudinal section (right).

**Table 1.** Material properties of concrete

| Element | Material | Young's modulus (MPa) | Poisson's ratio | Tensile strength (MPa) | Compressive strength (MPa) |
|---|---|---|---|---|---|
| column<br>beam<br>floor | Concrete C16/20 | 28600 | 0.2 | 1.9 | 16 |

*2.2. Finite element model*

The 3D numerical model was generate using ANSYS 2021R1. The model consisted of 6786 solid elements using SOLID186 option. The contact between the reinforcement and the concrete was modelled using CONTA174 and TARGET170 elements in ANSYS and Coulomb friction model was assumed. Taking into account the geometry of the model, Figure 1, and the fact that the maximum aggregate size was 16 mm, the SOLID186 brick element dimension was considered as 50 mm. This would result in a fine mesh to capture any local stress concentrations but coarse enough to include the concrete aggregate.

The generated numerical model and mesh discretization are presented in Figure 2. The structure was considered to have fixed supports at the base of the columns.

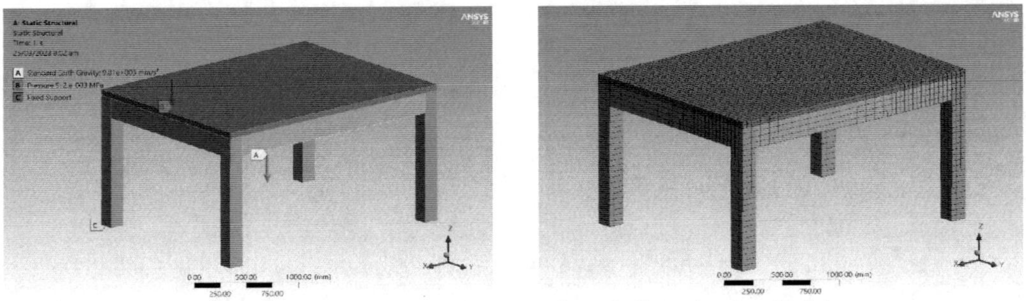

**Figure 2.** Numerical model (left) and mesh discretization (right).

### *2.3. Loading scenarios*

Besides the self-weight, the model was loaded with a 0.03 MPa surface load, uniformly distributed over the entire surface of the slab, to simulate the live load. While Vrancea 1977 is a reference seismic event for Romania, the disastrous effects of Turkey 2023 [15] earthquake prompted for its use in the numerical simulation in order to assess its effects on the considered scaled-down model. The time histories of both Vrancea and Turkey earthquakes, with their components along East-West (EW), North-South (NS) and vertical (Z) directions, are presented in Figure 3 and Figure 4, respectively.

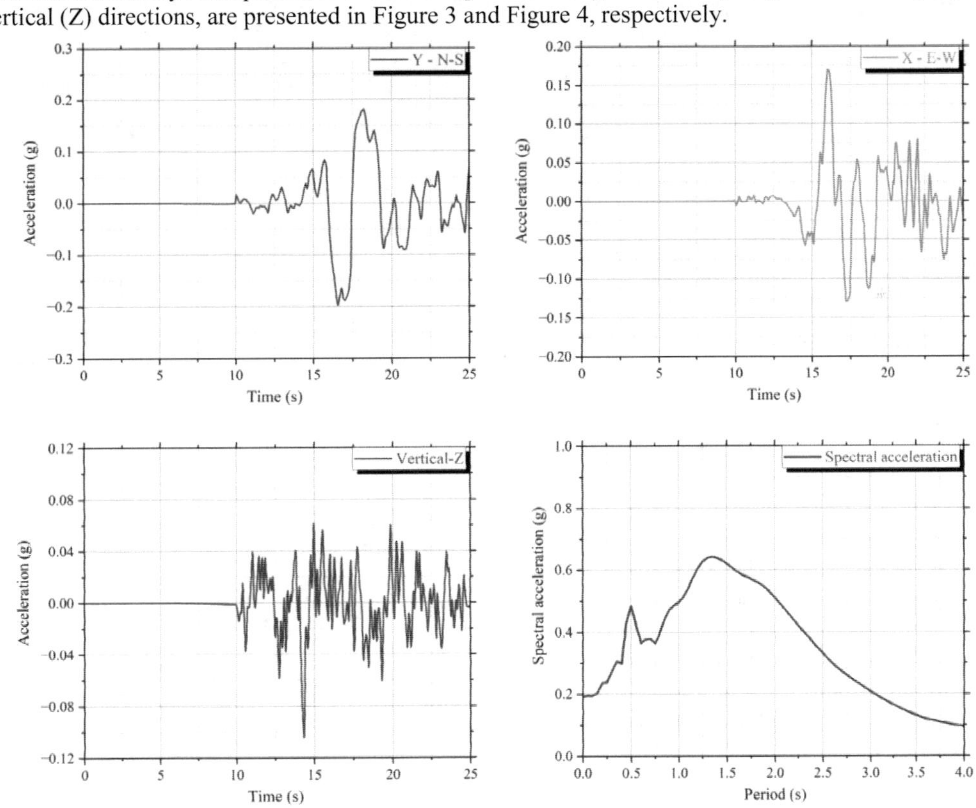


**Figure 3.** Acceleration time histories and response spectrum for Vrancea 1977 earthquake [15].

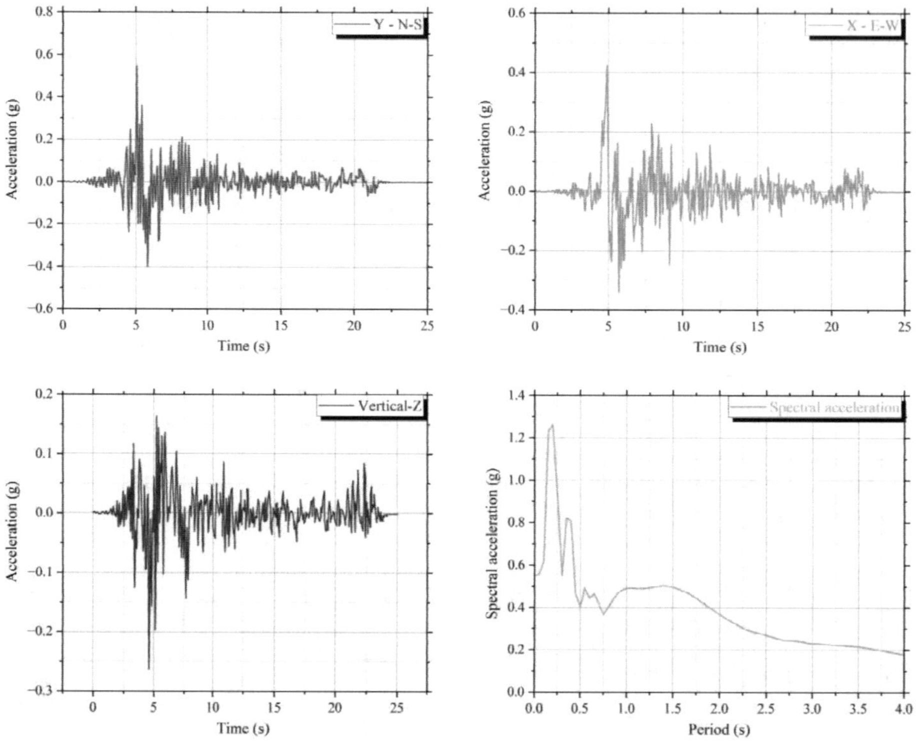


**Figure 4.** Acceleration time histories and response spectrum for Turkey 2023 earthquake [15].

**Table 2.** Time history analyses description

| Case nr. | Case name | Description |
|---|---|---|
| 1 | T-H Vrancea_1977 | Y (N-S accelerogram) + X (E-W accelerogram) + vertical Z |
| 2 | T-H Turkey_2023 | Y (N-S accelerogram) + X (E-W accelerogram) + vertical Z |

The input data presented in Figures 3 and 4 was obtained from the Engineering Strong Motion Database, an European collection of accelerograms [15]. In order to reduce the computational time, the acceleration data was sampled at 0.05 seconds time step from the original processed input file. The THA cases were run independently on the scaled down structural model. The considered scenarios are presented in Table 2. For both considered earthquake scenarios the E-W component was applied along the longitudinal direction of the model, X axis in Figure 2, whereas the N-S component was applied along the transversal direction of the model, Y axis in Figure 2. The numerical model will be improved during the subsequent stages of the research to account for different angles of incident seismic waves with respect to X and Y axes of the model.

The load case combinations were generated in accordance to Romanian design code CR 0-2012 [16].

## 3. Results

### *3.1. Modal analysis*

The dynamic characteristics of the model were determined prior to running the time-history analyses. The obtained data is summarized in Table 3. The results are in agreement with previously reported data

IOP Conf. Series: Materials Science and Engineering 1304 (2024) 012002 doi:10.1088/1757-899X/1304/1/012002

but for a different concrete strength class [14]. It can be observed the scaled down model is very rigid with a fundamental frequency of vibration nearing the value corresponding to a rigid body. The first two modes of vibration were purely translational. The fundamental mode of vibration resulted in translation along X axis, the longitudinal axis of the model, Figure 5, whereas the second mode of vibration resulted in translation along the Y axis of the model, the transversal axis, Figure 5.
Taking into account the response spectra presented in Figure 3 and Figure 4 as well as the fundamental period of vibration of the model, it can be anticipated that the Turkey 2023 earthquake scenario will have a stronger influence on the model in terms of accelerations, displacements and even induced stresses in the material.

**Table 3.** Modal analysis results

| Mode | Frequency (Hz) | Period (sec) | X direction mass participation (%) | Y direction mass participation (%) | Total mass (kg) |
|---|---|---|---|---|---|
| 1 | 23 | 0.043 | 93.7 | 1.39e-09 | 1520 |
| 2 | 23.3 | 0.0429 | 1.39e-09 | 93.8 | 1520 |

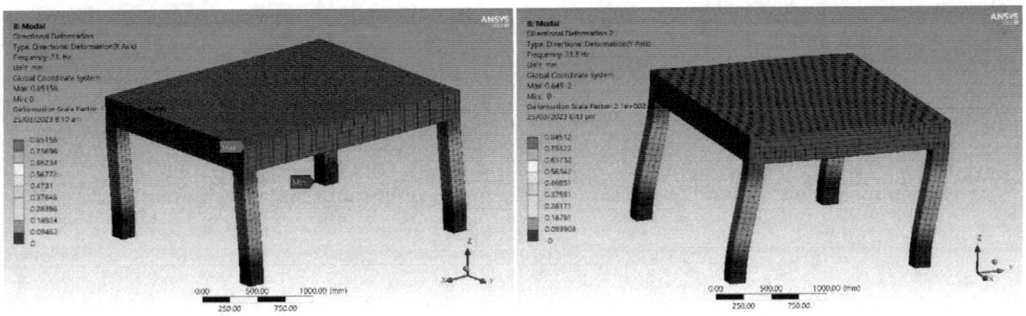

**Figure 5.** Modes of vibration: fundamental (left); second mode of vibration (right).

*3.2. Maximum displacements*
The maximum lateral displacements recorded in a building subjected to seismic motions is an important parameter because it helps assessing the inter-storey drift. This would give valuable information in terms of the seismic performance of a structure. The combined effect of all three components of the seismic motion for both considered earthquake scenarios is presented in Figure 6.

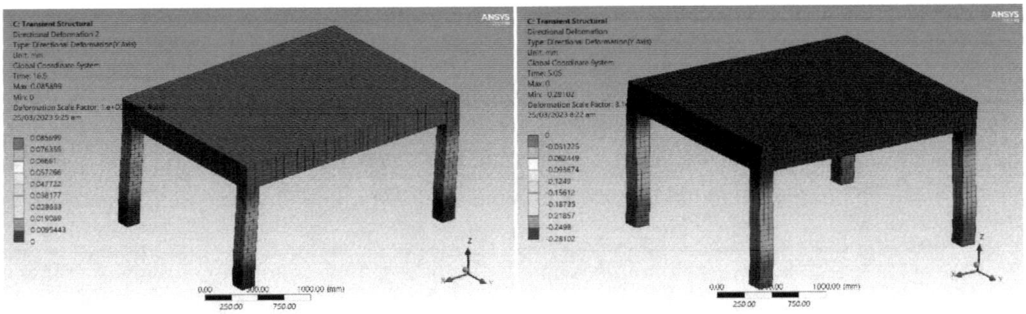

a) Vrancea 1977 earthquake b) Turkey 2013 earthquake
**Figure 6.** Maximum displacements for the selected seismic scenarios.

It can be observed that the maximum lateral displacement was almost four times larger for Turkey 2023 earthquake compared to Vrancea 1977. At the same time, the maximum displacement for Turkey 2023 earthquake took place along Y axis of the model corresponding to the N-S component, Figure 4, with larger amplitudes of the seismic motion in terms of input acceleration. Similarly, the maximum displacement for Vrancea 1977 earthquake was also obtained along the Y axis of the model. While both E-W and N-S components had similar amplitudes for the input acceleration the decreased stiffness of the model along Y axis resulted in the maximum lateral displacement being obtained along this axis.

### *3.3. Maximum accelerations*

Considering the inertia forces that develop in a structure during a seismic event, the acceleration levels may offer insightful information on the overall seismic behaviour of the structure. The obtained results, shown in Figure 7, indicate an almost 10-fold increase in the values of acceleration recorded at the level of slab for the Turkey 2023 earthquake compared to Vrancea 1977 event. Considering the spectral accelerations of the two earthquakes (Figures 4 and 3, respectively) the obtained results follow the pattern of the input motions taking into account the fact that the model was very rigid.

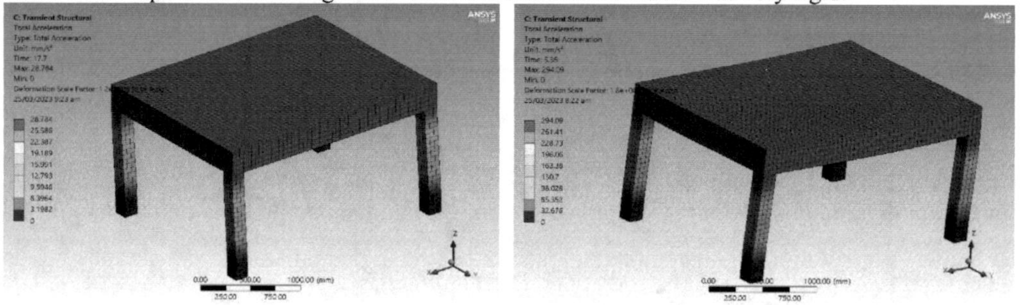

a) Vrancea 1977 earthquake b) Turkey 2013 earthquake

**Figure 7.** Maximum acceleration for the selected seismic scenarios.

### *3.4. Equivalent von Misses stress*

Since the input seismic motion was a complex one, with components along all three axes of the model, the strength check could be performed only in terms of equivalent stress. Equivalent von Misses stress was chosen to express the magnitude of the obtained results, Figure 8. The stress levels for the two earthquake scenarios indicate that Turkey 2023 even would produce equivalent stresses that would exceed the tensile strength of concrete and, therefore, cracks would be expected in the structure. The localization of these cracks were obtained at the base of the columns and at the top part of the columns, near the connection point with the beam elements. A similar localization was obtained for the Vrancea 1977 scenario.

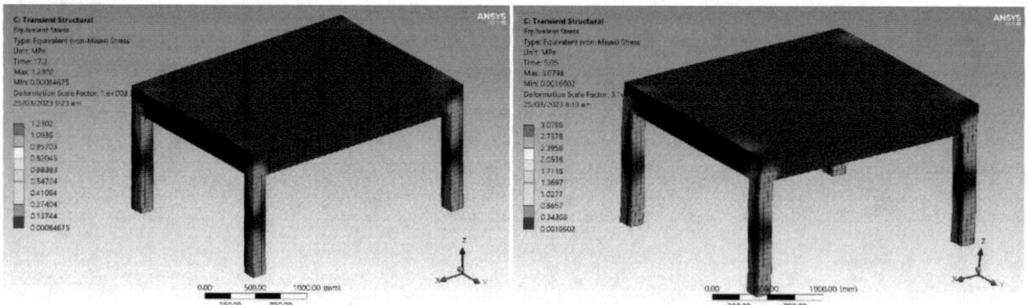

a) Vrancea 1977 earthquake b) Turkey 2013 earthquake

**Figure 8.** Maximum equivalent stresses for the selected seismic scenarios.

It would be interesting to point out the fact that the maximum displacement, maximum acceleration and maximum equivalent stress values were all obtained at the time stamp 5.05 seconds for Turkey 2023 earthquake. Taking into account the acceleration records presented in Figure 4, the obtained results are in accordance to the input data.
For Vrancea 1977 seismic event, both maximum displacements and accelerations were obtained at the time stamp 16.5 seconds whereas the maximum value of the equivalent stress showed a little delay corresponding to a time stamp of 17.2 seconds.

*3.5. Compliance with Romanian seismic design code specifications*
Figure 9 presents a comparative analysis in terms of maximum relative displacements and the allowable displacement specified in the Romanian seismic design code, P100-2013 [17]. It can be observed that neither seismic events would generate relative lateral displacements that would exceed the allowable value for serviceability limit state (SLS). The structure being rigid, it would try to dissipate the input energy by means of internal friction mechanisms instead of lateral vibrations. This is why, in case of Turkey 2023 seismic even, the tensile strength of C16/20 concrete was exceeded and damages would be expected at the base of the columns and at the beam-column joints.

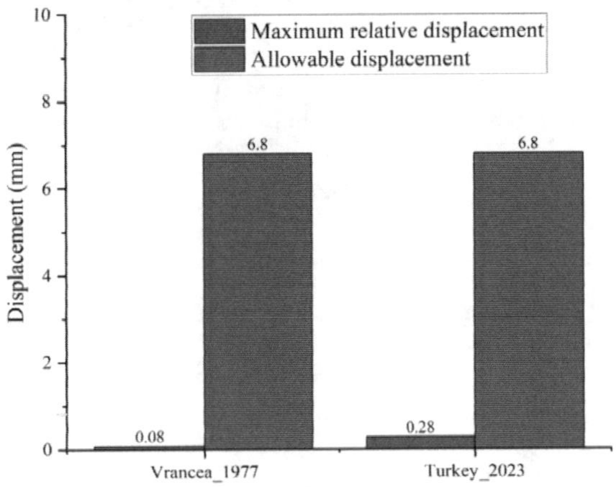


**Figure 9.** Maximum relative displacement: calculated vs. allowable.

## 4. Conclusions

The paper presents some preliminary results on the numerical investigations of the seismic behaviour of a reinforced concrete frame structure subjected to different earthquake scenarios by means of time history analysis. The earthquakes are chosen based on their relevance for Romania, Vrancea 1977, and in view of the recent disastrous seismic even in Turkey 2023. The numerical model of a scaled-down ground-floor RC frame building is generated in ANSYS 2021R1 environment. Based on the obtained results, the following conclusions can be drawn:
Neither seismic events produce large lateral displacements of the model that would exceed the allowable values corresponding to serviceability limit state specified in the Romanian seismic design code. The total maximum displacement is obtained for along the direction on the strong motion component with the highest intensity. In case of Vrancea 1977 earthquake, taking into account that both horizontal components have similar intensities, the maximum displacement is obtained along the weaker axis of the model. Similar results are obtained for the maximum acceleration.
The results obtained from the TH analysis of Turkey 2023 earthquake indicate that the tensile strength of concrete is exceeded during the early stage of seismic motion. The localization of the damage is obtained at the base of the columns and near the beam-column connection joints.

## 5. References

[1] Pardalopoulos S I, Pantazopoulou S J and Manolis G D 2022 On the Modeling and Analysis of Brittle Failure in Existing R/C Structures Due to Seismic Loads *Appl. Sci.* 12 1602

[2] Chen S and Jiang H 2023 Experimental Investigation on Seismic Performance of Non-Uniformly Corroded RC Moment-Resisting Frames *Materials (Basel).* 16 2649

[3] Lee K-S, Lee B-G and Jung J-S 2022 Seismic Strengthening of R/C Buildings Retrofitted by New Window-Type System Using Non-Buckling Slit Dampers Examined via Pseudo-Dynamic Testing and Nonlinear Dynamic Analysis *Appl. Sci.* 12 1220

[4] Bugueño I, Carvallo J and Vielma J C 2022 Influence of Directionality on the Seismic Response of Typical RC Buildings *Appl. Sci.* 12 1534

[5] Yu J, Zhang E, Xu Z and Guo Z 2022 Seismic Performance of Precast Concrete Frame Beam-Column Connections with High-Strength Bars *Materials (Basel).* 15 7127

[6] Hasgul U 2020 Investigation of Parameters Affecting the Equivalent Yield Curvature of Reinforced Concrete Columns *Materials (Basel).* 13 1594

[7] Sococol I, Mihai P, Toma I-O, Olteanu-Donţov I and Venghiac V-M 2021 Stress-Strain Relation Laws for Concrete and Steel Reinforcement Used in Non-Linear Static Analytical Studies of the Moment Resisting Reinforced Concrete (RC) Frame Models *Bull. Polytech. Inst. Iaşi. Constr. Archit. Sect.* 67 17–29

[8] Rosca O V, Canarache R, Budescu M and Oanea D 2014 The Experimental Testing and Numerical Model Calibration of a Steel Structure *Bull. Polytech. Inst. Iaşi. Constr. Archit. Sect.* LX (LXIV) 9–18

[9] Rosca O V 2016 Some Considerations Regarding the Assessment of Dynamic Characteristics *Bull. Polytech. Inst. Iaşi. Constr. Archit. Sect.* 62(66) 91–100

[10] El Khouri I, Garcia R, Mihai P, Budescu M, Ţăranu N, Toma I-O, Guadagnini M, Escolano-Margarit D, Entuc I S, Oprisan G, Hajirasouliha I and Pilakoutas K 2018 Shake Table Tests on Frames Made with Normal and FRP-Confined Rubberised Concrete *16th European Conference on Earthquake Engineering (16ECEE)* (Thessaloniki, Greece)

[11] Cui Z, Alipour A and Shafei B 2019 Structural performance of deteriorating reinforced concrete columns under multiple earthquake events *Eng. Struct.* 191 460–8

[12] Domizio M, Ambrosini D and Curadelli O 2017 Nonlinear dynamic numerical analysis of a RC frame subjected to seismic loading *Eng. Struct.* 138 410–24

[13] Toma A-M, Taranu G, Nishida T, Máthé A É, Câmpian C-M, Baccay M and Toma I-O 2020 Assessing the Seismic Performance of a R.C. Frame Structure by Numerical Simulations - An Efficient Tool for a Sustainable Future *Proceedings Of International Conference Building Services And Energy Efficiency* (Sciendo) pp 231–40

[14] Toma I O, Mihai P, Venghiac V-M, Taranu G, Baetu S-A, Toma A M and Petrescu T C 2020 Numerical Simulations of the Seismic Behavior of a Damaged RC Frame Retrofitted with Composite Fabric *17th World Conference on Earthquake Engineering, 17WCEE* (Sendai) pp 3f – 0019

[15] Luzi L, Lanzano G, Felicetta C, D'Amico M C, Russo E, Sgobba S, Pacor F and 5 O W G 2020 Engineering Strong Motion Database (ESM) (Version 2.0). Istituto Nazionale di Geofisica e Vulcanologia (INGV). *Ist. Naz. di Geofis. e Vulcanol.*

[16] M.D.L.P.A. 2012 CR 0-2012: Cod de proiectare. Bazele proiectarii constructiilor (in Romanian)

[17] M.D.L.P.A. 2013 P100-1-2013: Cod de proiectare seismica, partea I. Prevederi de proiectare pentru cladiri (in Romanian)

**Acknowledgement**

The research presented in this paper was conducted as part of the PN-III-P2-2.1-PED-2021-0677 research grant "Sustainable Concrete for Energy Efficient Buildings" funded by Executive Agency for Higher Education, Research, Development and Innovation Funding – UEFISCDI.

# High-temperature behavior of zeolite-containing mortar. Experiment and modelling

**M S Alexa-Stratulat[1], G Bunea[1], O M Banu[1], S G Maxineasa[1] and I O Toma[1]**
[1]The "Gheorghe Asachi" Technical University of Iasi, Faculty of Civil Engineering and Building Services, no. 1 Prof.dr.doc. D. Mangeron Str., 700050, Iasi, *Romania*

E-mail: ionut.ovidiu.toma@tuiasi.ro

**Abstract.** High temperature experiments were carried out on cylindrical-shaped specimens of mortar containing different concentrations of either cement-replacing zeolite or aggregate-replacing zeolite. The behavior was compared with that of plain mortar. Furthermore, finite-element models were designed, based on material characteristics obtained on the real specimens, to test the temperature exposure effect when similar conditions were applied.

## 1. Introduction

The construction industry has a considerable impact on both available resources and the environment. There is a growing interest in finding valid alternatives for the usual construction materials. Zeolites have received a lot of attention due to their pozzolanic activity[1], which stems from high concentrations of $Al_2O_3$ and $SiO_2$ [2]. They are naturally occurring, volcanic, rocks that can be also synthetized, and have been used as either aggregate [3] or cement replacement [4,5].

Using zeolites in concrete and mortar has been shown to have an impact on many properties of the resulting material, with the benefits or disadvantages depending on the concentration [1]. Compressive strength [6], freeze-thaw resistance [7], increased resistance to chloride and sulphate attacks [8] are some of the problems tackled by using zeolites.

This study focused on thermal behavior, which was tested to be, generally, better for mortars containing zeolites [9]. On the other hand, Uzal and Turanli [10] experimentally showed that cements containing clinoptilolite tuff have less pores larger than 50nm when compared to ordinary Portland cement specimens, which means a higher probability for the appearance of exfoliation. Nevertheless, similar to other properties, there are many parameters that can influence, directly or indirectly, the performance at elevated temperatures, such as: purity, aluminum content, the type and structure of the zeolites.

Testing a new material to fire resistance can be an expensive endeavor. Getting as much information as possible from the material, rather than the element, characterizations might prove important from an economic point of view. An in-house developed experimental stand was used to measure and record the heat transfer through mortar samples subjected to elevated temperature. Based on the recorded temperature data, the numerical models were calibrated to obtain a good correlation between simulations and experiments, in order to validate the models as instruments for high-temperature exposure studies. Based on the validated numerical models, a comparison of previously tested mortar specimens, only by using simulations, in a controlled and error free environment, was conducted. This is believed to be a valid approach for the preliminary stage of larger studies on the thermal behavior of new materials subjected to elevated temperatures.

## 2. Materials and methods

Reference mortar specimens were cast using CEM II/B-M (S-LL) 42.5R rapid hardening Portland cement, (0-4) mm sand as fine aggregate and tap water. The cement had a 65-79% clinker amount, with 21-35% blast furnace slag and limestone. The reference specimen was referred to as "*Ref.*". The fine aggregate and the cement were replaced in various percentages, and designated as follows:

- *AZx* = recipes with aggregates replaced by "x" percentage of zeolite aggregates (x = 10, 20, 30).
- *PZx* = recipes with cement replaced by "x" percentage of zeolite powder (x = 10, 20, 30).

In all cases, the percentage denotes a replacement by volume. This was translated into mass quantities to be used in the actual mixing procedure. Nevertheless, in the case of zeolite aggregates, the density of particles is similar to those of sand, thus the replacement was performed by mass from the beginning. For the PZ specimens, the densities used in computing the mass values were: $\rho_{cement} = 3150 kg/m^3$ and $\rho_{zeolite\_powder} = 2377 kg/m^3$.

Reference water-to-cement ratio (w/c) was 0.6. Based on the existing literature, it is expected that using zeolite, in either form (aggregate or powder), will change the water needed for the mix to have good workability. Thus, when zeolite aggregate was used, the w/c was increased alongside the percentage replacement in order to balance its water absorption. When replacing the cement with lower density zeolite powder, to achieve the same water-to-binder ratio (w/b) as the reference would imply lowering the water quantity. However, for the same reason described above, the same amount of water was employed, resulting in an increased w/b. Quantities used for a reference amount of 500g of cement are given in table 1.

**Table 1.** Mix proportions for 500g of cement.

| Recipe | Cement[g] | Zeolite powder[g] | Zeolite aggregate[g] | Sand[g] | Water[ml] | w/b |
|---|---|---|---|---|---|---|
| Ref. | 500 | - | - | 2050 | 300 | 0.60 |
| AZ10 | 500 | - | 205 | 1845 | 325 | 0.65 |
| AZ20 | 500 | - | 410 | 1640 | 350 | 0.70 |
| AZ30 | 500 | - | 615 | 1435 | 375 | 0.75 |
| PZ10 | 450 | 37.3 | - | 2050 | 300 | 0.62 |
| PZ20 | 400 | 74.6 | - | 2050 | 300 | 0.63 |
| PZ30 | 350 | 111.9 | - | 2050 | 300 | 0.65 |

The zeolite used in this study was Clinoptilolite (87-90%) and has a composition (both powder and aggregate) provided by seller and presented in table 2. It is important to note that Clinoptilolite is stable up to 450°C and has a pH of 8.7. The maximum particles size for the zeolite aggregates was 4mm, and the powder consisted of particles with a mean diameter of 29μm.

**Table 2.** Chemical composition of the clinoptilolite.

| Compound | Percentage (%) |
|---|---|
| $SiO_2$ | 68.75-71.3 |
| $Al_2O_3$ | 11.35-13.10 |
| CaO | 2.86-5.20 |
| $K_2O$ | 3.17-3.40 |
| $Fe_2O_3$ | 1.90-2.10 |
| MgO | 1.18-1.20 |
| $Na_2O$ | 0.82-1.30 |
| L.I. | 8.73-8.86 |

The specimens were cast in cylindrical steel molds with a diameter of 70mm and a height of 30mm. Every recipe was used to create 12 cylinders, that are used in the current research. Provided that existing results indicate a good amount of consistency across the specimens of the same mix, and given the focus of this study, only one specimen of each recipe was used. Density and thermal transfer properties were assessed close before the high-temperature testing.

The study consisted of two parts. First part was the experimental evaluation of the high-temperature effect on the specimens. The second part was the finite element simulation aimed at both calibration and comparison.

### *2.1. Experimental assessment*
The experiments were carried on a stand created for this purpose. A sketch the arrangement is presented in figure 1. The heat source was an air-heating gun capable of reaching temperatures up to 550°C, as per the producer's technical data. The specimens were positioned 1cm away from the gun. Before starting the high-temperature exposure, a K-type thermocouple was employed to measure the temperature of the hot air, in order to start the experiment from a known state. This would also help comparing results with data obtained by simulations. Temperature was monitored on the side opposite to the heated one, in the center point, by means of a FLIR BCam thermal imaging camera.

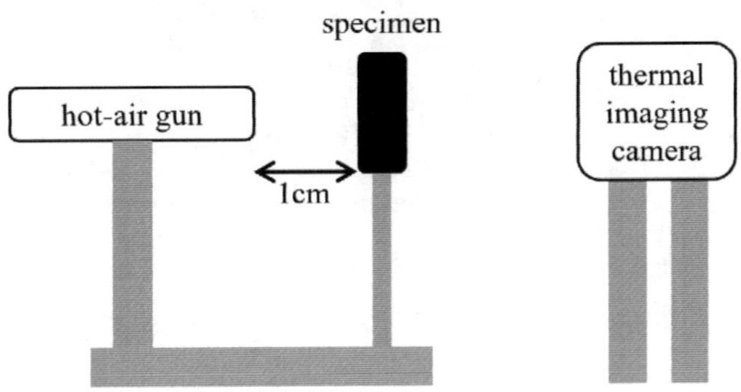


**Figure 1.** Experimental stand sketch.

### *2.2. Simulations assessment*
Simulations were performed using the LISA 8.0 FEM analysis software. It provides ample analysis options even in the free version, that allows 1300 nodes. For the present purpose, the number of nodes and, subsequently, the number of elements associated to it was enough for a thorough assessment.

Each specimen was modeled starting from an initial circular surface mesh consisting of 96 elements and 107 nodes (figure 2a), and of appropriate radius. The nodes were arranged in a concentrical manner. Furthermore, the extrusion process to the needed thickness was employed with a number of 10 subdivisions (figure 2b). This resulted in a total of 960 elements and 1177 nodes for each specimen.

a) b)

**Figure 2.** Finite element model: a) starting cross-section; b) extruding result.

Temperature was applied on the central surface, on a central zone of d=17.5mm diameter. This was chosen to be in agreement with the experimental setup. The heat source was modeled as a part of the body, extruded from the center of one of the faces (named "hot surface"). This part was kept constant at the desired high-temperature (figure 3). For the opposite circular surface (named "cold surface") and the lateral surface we applied the property of convection with the outside air, using the following parameters: $T_{air}$ = 22°C, transfer coefficient = 5 W/(m$^2$·K).

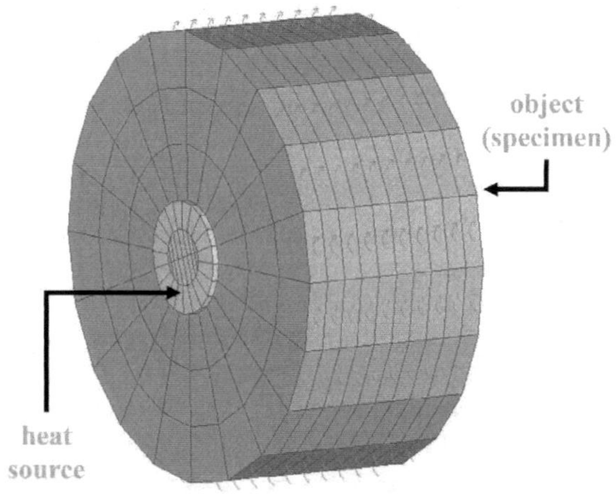


**Figure 3.** Model of the specimen with heat source.

## 3. Results and discussions

In the first instance, specimens' dimensions, mass and thermal properties (thermal conductivity and specific heat) were measured. Table 3 holds the values obtained for the particular specimen used from each recipe. These values were used as input parameters for the simulations.

**Table 3.** Measurement values of parameters used for simulations.

| | Recipe | | | | | | |
|---|---|---|---|---|---|---|---|
| Parameter | Ref. | AZ10 | AZ20 | AZ30 | PZ10 | PZ20 | PZ30 |
| Density [kg/m$^3$] | 2069 | 1950 | 1951 | 1898 | 2047 | 2022 | 1993 |
| Thermal conductivity [W/(m·K)] | 2.0037 | 1.5989 | 1.3962 | 1.2405 | 1.8417 | 1.8236 | 1.7802 |
| Specific heat [J/(kg·K)] | 830 | 793 | 828 | 852 | 816 | 853 | 821 |

The density of the zeolite containing specimens is lower, than the reference mix, as expected. The influence is greater for zeolite aggregates than for zeolite powder, since the porosity should be higher. In a similar fashion, the thermal conductivity decreases linearly with increasing the zeolite concentration. Replacing aggregates results in a higher decrease in thermal conductivity than replacing cement (figure 4a)). It can be noted that this could also be an indicator of the lack of water inside the pores of zeolites.

The AZ specimens show an increased need of energy for increasing temperature when the replacement percentage is higher. Nevertheless, the AZ10 specimens is the only mix that has a lower specific heat than the reference one. The AZ20 presents a similar value to the reference and AZ30 a higher one. In the case of PZ mixes, the behavior is not linear. While PZ10 and PZ30 show similar, though lower than the reference, values of the specific heat, the PZ20 spikes above that (figure 4b)).

The modifications described above need further research, as they may be the result of hydration dynamics in the presence of the zeolite aggregate/powder.

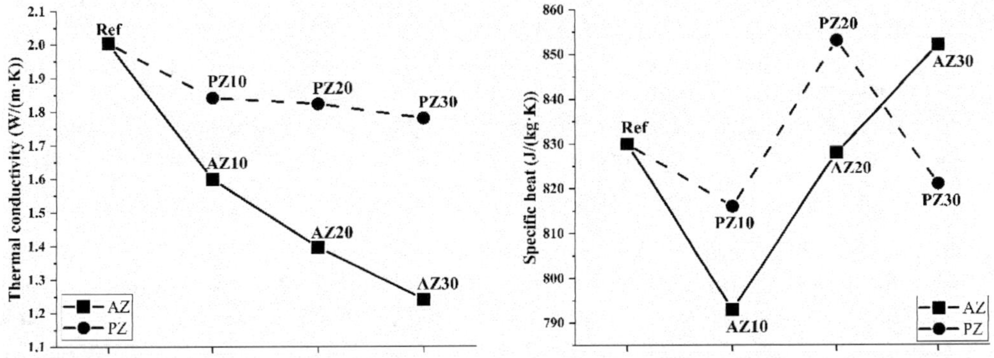


a) b)

**Figure 4.** a) Thermal conductivity and b) specific heat variations of mortar specimens.

The porous structure of zeolite aggregates could play an important role in the thermal behavior of the mortars using them. A small amount of zeolite aggregate might result in lower porosity than the reference mix, following that the water in the zeolite pores will lead to a higher amount of hydration products. A bigger replacement percentage could result in a higher porosity, since the cement quantity is fixed in a certain specimen. With a specific heat value of 1000 J/(kg·K), the air is harder to heat up than mortar, thus more air results in a higher specific heat.

On the other hand, zeolite powder might produce less porous mortars than its aggregate counterpart, based on the density comparison. In this case, the above attempted explanation, on how the specific heat is influenced by entrapped air, holds for PZ10 and PZ30. While the exception showed by PZ20 requires further study in the microstructure of the specimens in order to be explained, this is not the aim of this paper.

### *3.1. Experimental results of high-temperature exposure*

For every specimen in the study, the model was calibrated around the experimental results. The heating of the specimens was done at a constant distance, but the device used to obtain the high temperature was prone to errors. The aim of the study was to perform a comparison of the involved materials even in the presence of such an inconvenience.

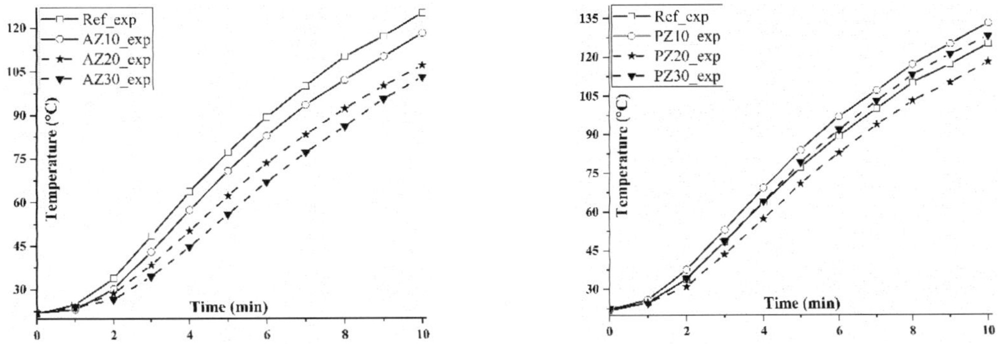


a) b)

**Figure 5.** Experimental results for a) AZ and b) PZ high-temperature exposure.

IOP Conf. Series: Materials Science and Engineering 1304 (2024) 012003 doi:10.1088/1757-899X/1304/1/012003

In figure 5, the results of high-temperature exposure are presented. The AZ specimens show an increasingly insulating behavior with increased replacement percentage (figure 5a)). This is to be expected, taken into account the values for thermal conductivity and specific heat of the mixes. For the PZ recipes, the maximum temperatures reached after 10 minutes of heating are closely related to the specific heat properties shown above (figure 5b)).

This data gave us an insight into the behavior at high-temperature for the specimens, and it also provides an incentive to rectify, if needed, the comparison charts with the reference. We wanted to show that, even if a somewhat unprecise experimental setup is used, valuable and reliable data can be obtained with additional fine-tuning. Each curve presented in figure 5 was used as a basis for comparison with simulation results.

*3.2. Simulation results of high-temperature exposure*

The simulations have been performed for every specimen that was previously experimented on. Using the precise characteristic data (density, thermal conductivity and specific heat), models were created in order to demonstrate that the experiments were correct, albeit with an exposure temperature that was not exactly known. This step in the study was important for validating the process rather than the data. In order to achieve this, several exposure temperatures in the 400°C - 500°C interval were used for each specimen, so that, in the end, to find a value that will result in a temperature curve as close as possible to the experimentally determined one.

Simulations results that are best fitting the measured behavior are presented in figure 6. It can be observed that there is great correlation between results in every case. This was achieved for exposure temperatures listed in table 4.

**Table 4.** Temperature values of calibrated simulations.

| Specimen | Ref. | AZ10 | AZ20 | AZ30 | PZ10 | PZ20 | PZ30 |
|---|---|---|---|---|---|---|---|
| Temperature [°C] | 430 | 430 | 480 | 500 | 480 | 450 | 450 |

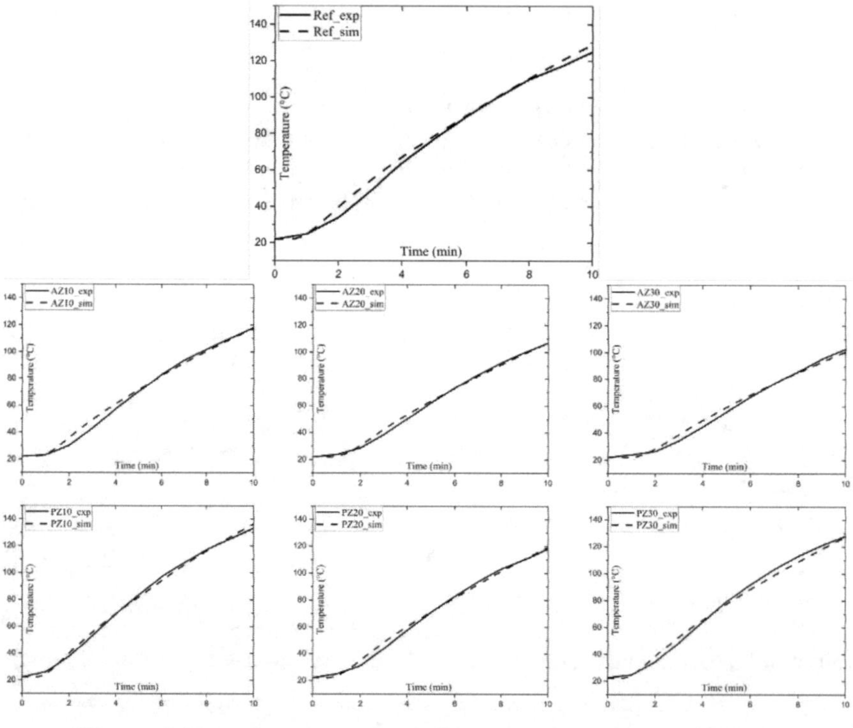


**Figure 6.** Experiment vs. simulation at calibrated temperatures.

IOP Conf. Series: Materials Science and Engineering 1304 (2024) 012003 doi:10.1088/1757-899X/1304/1/012003

Variation of temperature values presented in table 4 can be explained either by errors in the air-heating gun, but they can also represent the result of inhomogeneities in the mortar materials. The temperature was applied on a surface, while it was measured in a point. The material that this point was representing could have induced a lower or a higher speed of temperature modification.

The final process was modeling all the specimens with the same applied temperature to them (denoted as *fcs* - for comparison simulations). The models were the identical to those used for the previous simulations, with only the applied temperature being changed to 500°C in all cases. Thus, we sought an accurate comparison between the specimens. Time of exposure was increased to 15 minutes in order to highlight differences in curves of similar slopes.

As it can be observed from figure 7a), the AZ specimens show a behavior similar to the one captured by experiment, in both the relation between them, as well as the one with the reference. Even if the calibrating temperatures were quite different, the relationship was not fundamentally changed. This might be due to the big differences among the recipes as far as the thermal properties are concerned.

The PZ recipes (figure 7b)) show a higher deviation from experiment. The order of the biggest temperature that is reached after 10 minutes exposure, between the replacement percentage, is the same. Nevertheless, the actual difference between them is altered. The PZ10 and PZ30 demonstrate similar performances and PZ20 is the least affected, in agreement with the experiment. The simulations demonstrated that the reference is the most impacted by the exposure, when compared to PZ, contrasting the experiment. The thermal parameters of PZ10 and PZ30 were found to be more closely related to each other and also to the reference, thus a change in calibrating temperature is shown to be important.

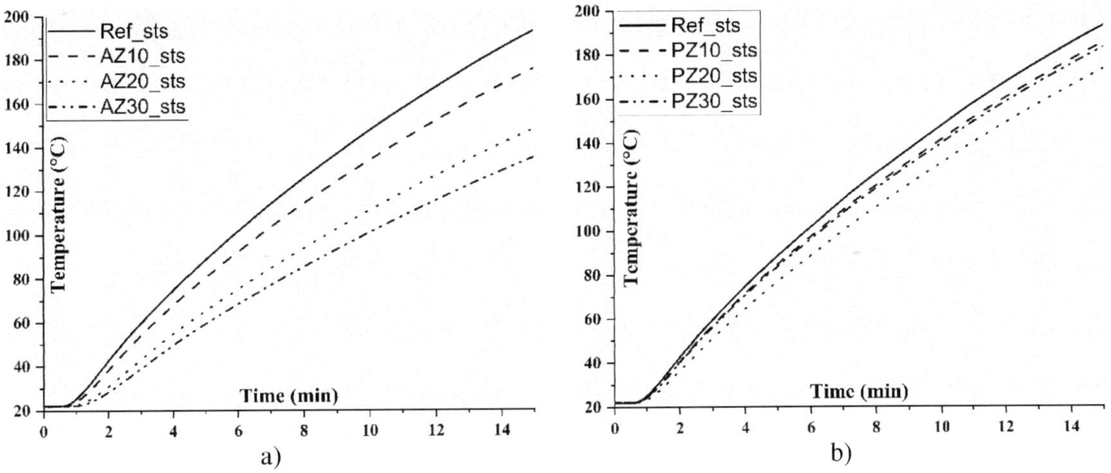


**Figure 7.** Same temperature exposure (500°C) simulations of a) AZ and b) PZ for 15 minutes.

## 4. Conclusions

The study focused on demonstrating the viability of using a free simulation software for assessing the performance of small-scale specimens when exposed to high temperature. By coupling the modeling results with experimental data, we showed that a less precise setup gives enough information about the process, so that simulation can be further used for an accurate characterization of involved materials. This type of analysis should prove useful as a first step in the researching of new materials and their thermal load performance. Bridging the gap between testing small-scale and large-scale specimens would be beneficial with respect to many types of involved resources. It is noteworthy that such simulations, based on only one specimen from a particular recipe, can be of great importance given that inhomogeneities are not influencing the result.

With respect to the materials used in the study, it can be concluded that zeolites, particularly in aggregate form, represent a valuable replacement in increasing high-temperature performance. The best result indicated a decrease of almost 30% in the temperature achieved after 15 minutes of exposure.

## 5. References

[1] Tran Y T, Lee J, Kumar P, Kim K-H, Lee S S 2019 *Natural zeolite and its application in concrete composite production Composites Part B*: Engineering 165 354–64
[2] Mumpton F A 1977 *Mineralogy and Geology of Natural Zeolites* (De Gruyter)
[3] Karakurt C, Kurama H, Topçu İ B 2010 *Utilization of natural zeolite in aerated concrete production* Cement and Concrete Composites 32 1–8
[4] Poon C S, Lam L, Kou S C, Lin Z S 1999 *A study on the hydration rate of natural zeolite blended cement pastes* Construction and Building Materials 13 427–32
[5] Iswarya G, Beulah M 2021 *Use of zeolite and industrial waste materials in high strength concrete – A review* Materials Today: Proceedings 46 116–23
[6] Ahmadi B, Shekarchi M 2010 *Use of natural zeolite as a supplementary cementitious material* Cement and Concrete Composites 32 134–41
[7] Nagrockiene D, Girskas G, Skripkiunas G 2014 *Cement freezing–thawing resistance of hardened cement paste with synthetic zeolite* Construction and Building Materials 66 45–52
[8] Madhuri P V, Kameswara Rao B, Chaitanya A 2021 *Improved performance of concrete incorporated with natural zeolite powder as supplementary cementitious material* Materials Today: Proceedings 47 5369–78
[9] Domínguez I A C, Abdellaoui Y, Abatal M, Patiño-Carachure C 2018 *Characterizations and Industrial Applications for Cement and Concrete Incorporated Natural Zeolite Zeolites and Their Applications* ed M N Rashed and P N Palanisamy (InTech)
[10] Uzal B, Turanlı L 2012 *Blended cements containing high volume of natural zeolites: Properties, hydration and paste microstructure* Cement and Concrete Composites 34 101–9

**Acknowledgements**

A part of the research presented in this paper was conducted as part of the PN-III-P2-2.1-PED-2021-0677 research grant "Sustainable Concrete for Energy Efficient Buildings" funded by Executive Agency for Higher Education, Research, Development and Innovation Funding – UEFISCDI.

This research was supported, in part, by InoHubDoc POCU/993/6/13/153437.

IOP Conf. Series: Materials Science and Engineering 1304 (2024) 012004 doi:10.1088/1757-899X/1304/1/012004

# Seismic Response of a Tall Reinforced Concrete Frame Structure Subjected to both Horizontal and Vertical Strong Motions

**A M Toma[1], M Movila[1], O V Rosca[1] and I O Toma[1]**
[1] The "Gheorghe Asachi" Technical University of Iasi, Faculty of Civil Engineering and Building Services, no. 1, Prof.dr.doc. D. Mangeron Street, Iasi, *Romania*

E-mail: ionut.ovidiu.toma@tuiasi.ro

**Abstract.** The continuous development of urban areas coupled with the ageing of built-stock and the variety of structures built following different design code recommendations may result in different behaviours during seismic events. Hence, reliable information is needed in order to assess the risk of severe damage or collapse during earthquakes. Accurate numerical simulations coupled with analytical models provide essential information on the seismic performance of civil engineering structures. The results obtained from realistic numerical simulations may serve as entries in data-bases designed to train and calibrate early warning systems (EWS). Their continuous development is of paramount importance taking into account their potential to save lives and prevent economic loses in case of a severe seismic event. The paper presents the preliminary results obtained from numerical simulations on the seismic response of a tall reinforced concrete (RC) frame structure subjected to both horizontal and vertical string motions produced by an earthquake. The obtained results, in terms of response spectra in terms of accelerations, may be used for training EWS sensors to identify a seismic event and transmit warning messages at local, regional and even national level.

## 1. Introduction

The continuous development of urban areas coupled with the ageing of built-stock and the variety of structures built following different design code recommendations caught the attention of decision makers and researchers around the world. While the primary role of the former is to draft resilience plans in case of natural disasters [1], with particular focus on strong earthquakes, the responsibility of the latter may reside into offering accurate information on the behaviour of different types of structural systems in case of a severe earthquake [2]. The first step towards reducing the associated risk, seismic risk in this case, of a natural disaster consists in correctly evaluating the vulnerability of at-risk structures and the possible consequences that may rise after a severe seismic event. The accurate assessment of seismic vulnerability of buildings may result in reducing the associated risk by taking a-priori preparatory measures, e.g. retrofitting, strengthening, and by drafting relief and rescue plans and policies in the aftermath of such an event [3]. The responsibility of designers and contractors is to follow seismic design code regulations which are continuously updated based on field observations, recorded data and extensive research work [4].

Accurate numerical simulations [5–7] coupled with analytical models are efficient tools in assessing the seismic behaviour of civil engineering structures [8]. Their main advantage is the possibility of varying several parameters to account of all possible scenarios without the need of costly and time-consuming experimental tests. Another advantage consists in the continuous updating of the seismic scenarios based on the recorded data from recent severe earthquakes from around the world: Japan (Tohoku, 2011), Nepal (2015), Puerto Rico (2020), Croatia (Zagreb, 2020), Haiti (2021), Turkey (Gaziantep, 2023). The results obtained from realistic numerical simulations may serve as entries in data-bases designed to train and calibrate early warning systems (EWS) [9,10]. Their continuous development are of paramount importance taking into account their potential to save lives and prevent economic loses in case of a severe seismic event. While such systems are already well developed and

functional in countries such Japan, South Korea and U.S.A., their use in Romania is rather limited to institutions deemed of national interest (e.g. nuclear power plants, oil refineries). Early warning of population is very limited, the main limiting factor being the warning infrastructure itself [11].
In view of the recent seismic events and taking into account the ageing of the built stock, the present paper presents the preliminary results obtained from numerical simulations on the seismic response of a tall reinforced concrete (RC) frame structure subjected to both horizontal and vertical strong motions produced by an earthquake. The obtained results, in terms of response spectra of accelerations, may be used for training EWS sensors to identify a seismic event and transmit warning messages at local, regional and even national level.

## 2. Numerical Model

The chosen building is based on the RC frame structure type. It consists of a tall ground floor, 4.35 m in height, 7 storeys each measuring 2.75 m in height, and a technical room housing the elevator equipment. It is commonly met in urban areas in Romania, following a modular design pattern adopted during the mid-1980s.

### *2.1. RC frame structure*

The 3D view of the numerical model is shown in Figure 1. Columns have different cross-section depending on their position along the height of the structure, changing every two storeys. The numerical model considered also the presence of the elevator shaft, acting as a central core, and the staircase.
The floor plan is shown in Figure 2. Although the rim of the floor may be symmetric with respect to Y axes, the presence of the staircase and the elevator shaft resulted in the structure being non-symmetric with respect to both in-plane axes. Therefore, a coupled displacement and rotation fundamental mode of vibration is expected, with the dominant component being the lateral displacement along Y axis.
A C18/22.5 concrete strength class was used in all structural and non-structural elements. The design compressive strength considered in the initial project was 14 MPa and the modulus of elasticity was 28.5 GPa.

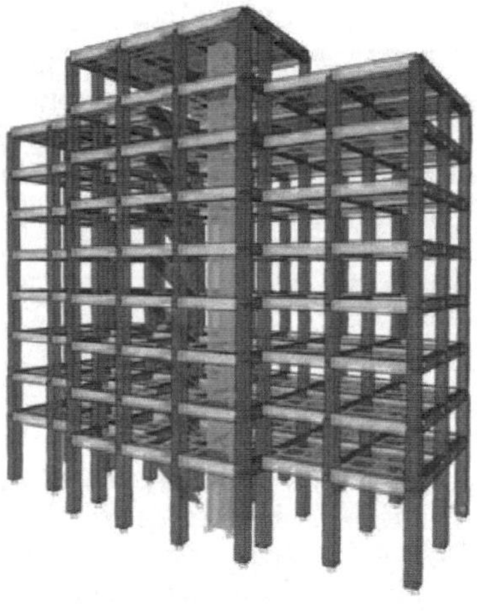

**Figure 1.** 3D view of the numerical model.

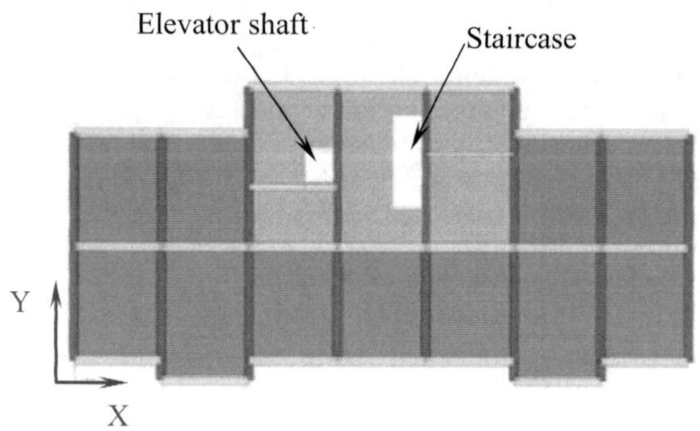


**Figure 2.** Plan configuration.

### *2.2. Loading scenarios*

The structure was subjected to both horizontal and vertical ground motions recorded for Vrancea 1977 [12] and Turkey 2023 earthquakes [13]. The former is one of the reference earthquakes taken into account in terms of seismic design in Romania while the latter became significant in view of the destructions and high number of casualties in its wake. The time histories of both Vrancea and Turkey earthquakes, with their components along East-West (EW), North-South (NS) and vertical (Z) directions, are presented in Figure 3 and Figure 4, respectively. The characteristics of the two seismic events are summarized in Table 1.

**Table 1.** Characteristics of Vrancea 1977 and Turkey 2023 seismic events.

| Event | Magnitude(Mw) | Component | PGA ($m/s^2$) | PGV (m/s) | PGD (m) |
|---|---|---|---|---|---|
| Vrancea 1977 [12] | 7.5 | EW | 1.66 | 0.31 | 0.10 |
| | | NS | 1.96 | 0.74 | 0.20 |
| | | Z | 0.98 | 0.11 | 0.03 |
| Turkey 2023 [13] | 7.8 | EW | 5.82 | 5.65 | 1.31 |
| | | NS | 5.84 | 5.84 | 1.52 |
| | | Z | 6.63 | 6.63 | 0.78 |

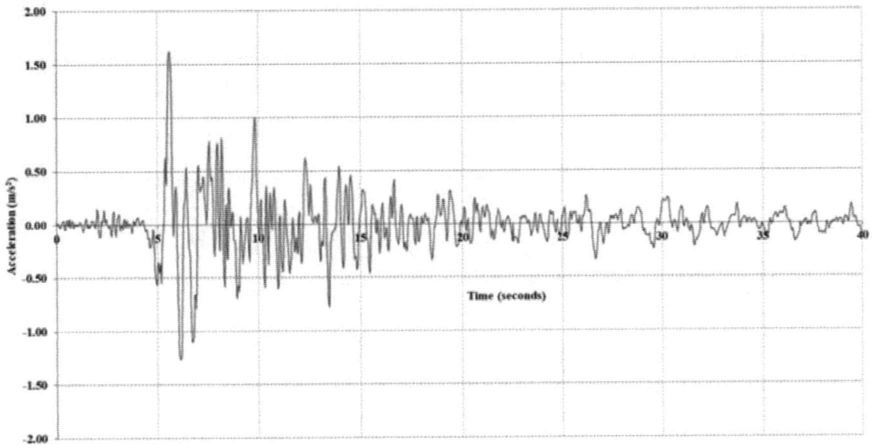


a) EW Component

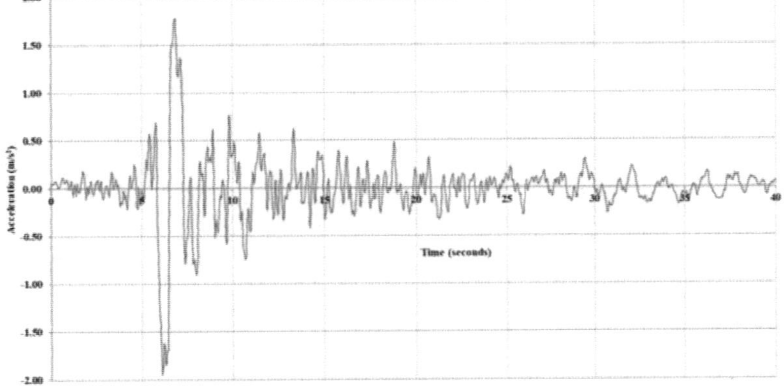


b) NS Component

IOP Conf. Series: Materials Science and Engineering 1304 (2024) 012004 doi:10.1088/1757-899X/1304/1/012004

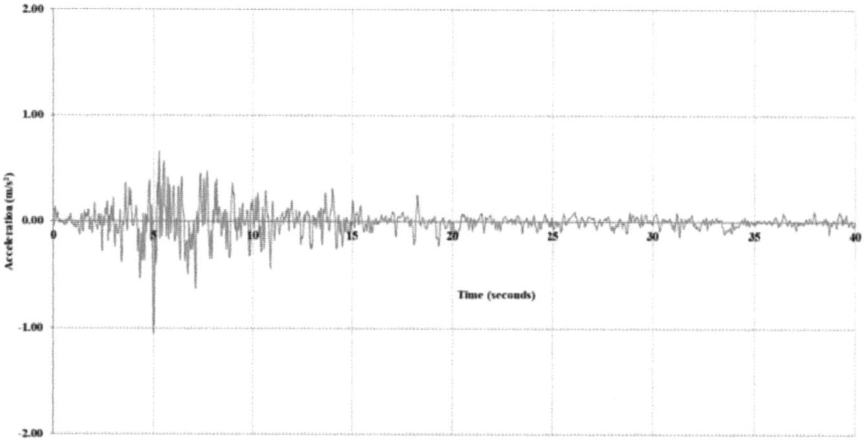


c) Z Component

**Figure 3.** Acceleration time histories for Vrancea 1977 earthquake [12].

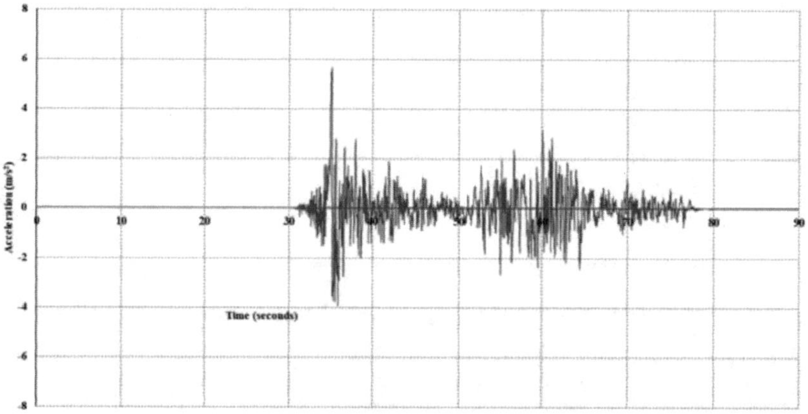


a) EW Component

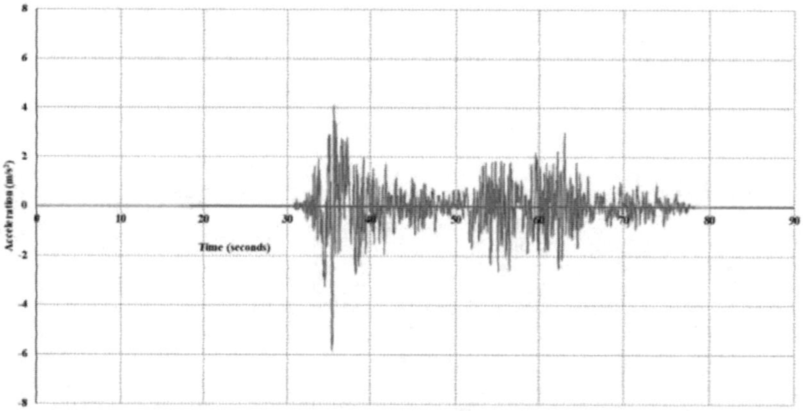


b) NS Component

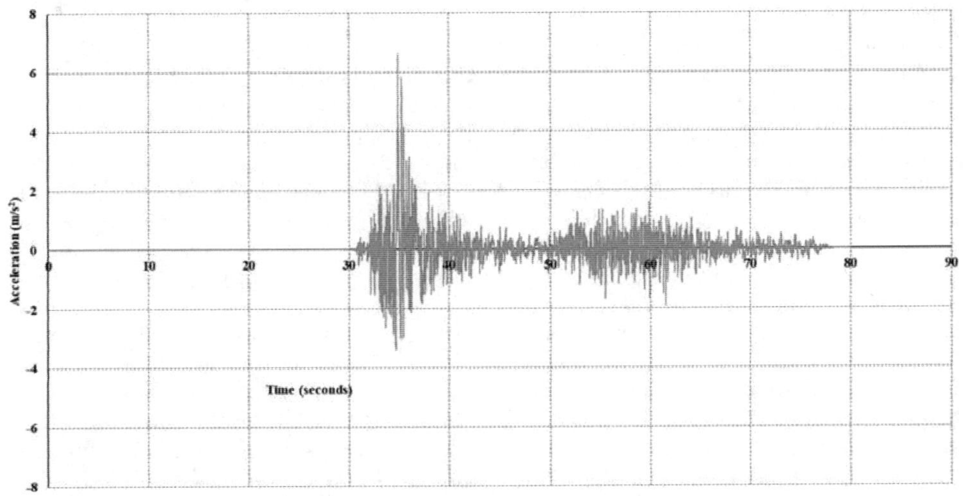


c) Z Component

**Figure 4.** Acceleration time histories for Turkey 2023 earthquake [13].

The structure was assumed to be located near the epicentre of both earthquakes such that the vertical ground motion was also significant. Consequently, the following loading cases were considered: both EW and NS components acting simultaneously, all three components acting simultaneously, EW and NS components having a 0.5 seconds delay compared to the Z component. The last scenario was intended to simulate the fact that different seismic waves travel at different velocities and therefore, they reach the site at different time stamps.

The EW component of both earthquakes was applied along X axis of the model (Figure 2) while the NS component was applied along Y axis (Figure 2). The considered loading cases are summarized in Table 2. However, the numerical model will be improved during the subsequent stages of the research to account for the real location of the building with respect to the traveling direction of seismic waves. Consequently, different angles will be applied to the seismic motions with respect to the X an Y axes of the model.

**Table 2.** Considered seismic loading scenarios.

| Event | Loading case designation | Component along the axes of the model | | |
|---|---|---|---|---|
| | | X | Y | Z |
| Vrancea 1977 [12] | VN-E-N | EW | NS | - |
| | VN-E-N-Z | EW | NS | Z |
| | VN-E-N-Z2 | EW* | NS* | Z |
| Turkey 2023 [13] | TK-E-N | EW | NS | - |
| | TK-E-N-Z | EW | NS | Z |
| | TK-E-N-Z2 | EW* | NS* | Z |

* Components were applied with a 0.5 seconds delay

## 3. Results

The obtained results are presented and discussed in terms response spectra determined for the spectral acceleration (SA) of the investigated structure, corresponding to each storey of the building. The level of SA on the response spectrum corresponding to the fundamental period of vibration can offer insightful information on the expected damage of the structure due to the seismic motion. At the same time, amplification phenomena can be rendered evident.

Response spectra are very useful tools for the analysis of performance of structures and/or equipment during seismic events. Therefore, based on the fundamental period of vibration or fundamental frequency of vibration, researchers can estimate the response of a structure by reading the value of the SA from the response spectra of the ground motion. This is a quick manner of assessing whether or not the resonance phenomenon is to be expected.

The fundamental period of vibration of the considered structure was determined as 0.7863 seconds.

### *3.1. Ground motion spectra*

The response spectra of Vrancea 1977 earthquake expressed in terms of spectral acceleration (SA) are shown in Figure 5. The SA spectra are plot with respect to the period of vibration. The fundamental period of vibration of the structure is shown as vertical line. From the intersection of the vertical line with each of the response spectra, the expected acceleration values of the ground can be determined. The obtained values are summarized in Table 3. The vertical line corresponds to the fundamental period of vibration of the selected structure and it indicates the intensity of the ground acceleration an oscillator having the same dynamic characteristics of the considered structure would be subjected to. For small values of damping, such as the ones usually considered in civil engineering, the values of spectral acceleration (SA) are almost identical with the values of the pseudo-spectral acceleration (PSA).

The response spectra of Turkey 2023 earthquake expressed in terms of spectral acceleration (SA) are shown in Figure 6.

Comparing the results presented in the two figures it can be concluded that the NS component of Vrancea 1977 earthquake would affect longer period structures with a fundamental period of vibration of 1.15 seconds, the Tukey 2023 seismic event would significantly affect structures with a fundamental period of vibration of 0.3 seconds for both its EW and NS components. Moreover, the seismic event of 2023 being a shallow depth earthquake, it would result in higher values of the spectral acceleration due to significantly lower damping of seismic waves amplitudes.

**Table 3.** Spectral accelerations for Vrancea 1977 and Turkey 2023 seismic events.

| Event | Fundamental period of vibration (seconds) | Spectral acceleration (m/s$^2$) | | |
|---|---|---|---|---|
| | | EW component | NS component | Z component |
| Vrancea 1977 [12] | 0.7863 | 4.139 | 4.423 | 1.378 |
| Turkey 2023 [13] | | 8.174 | 10.641 | 1.592 |

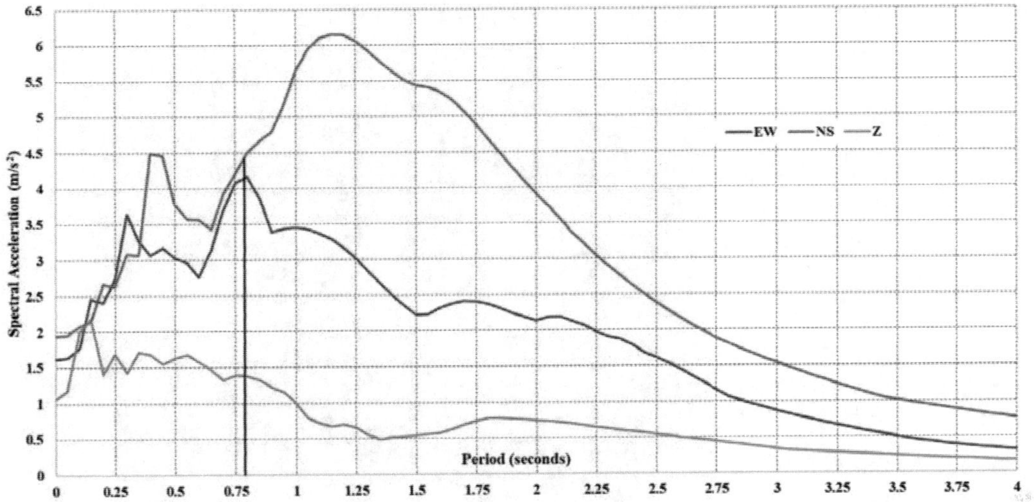


**Figure 5.** Response spectra of Vrancea 1977 earthquake.

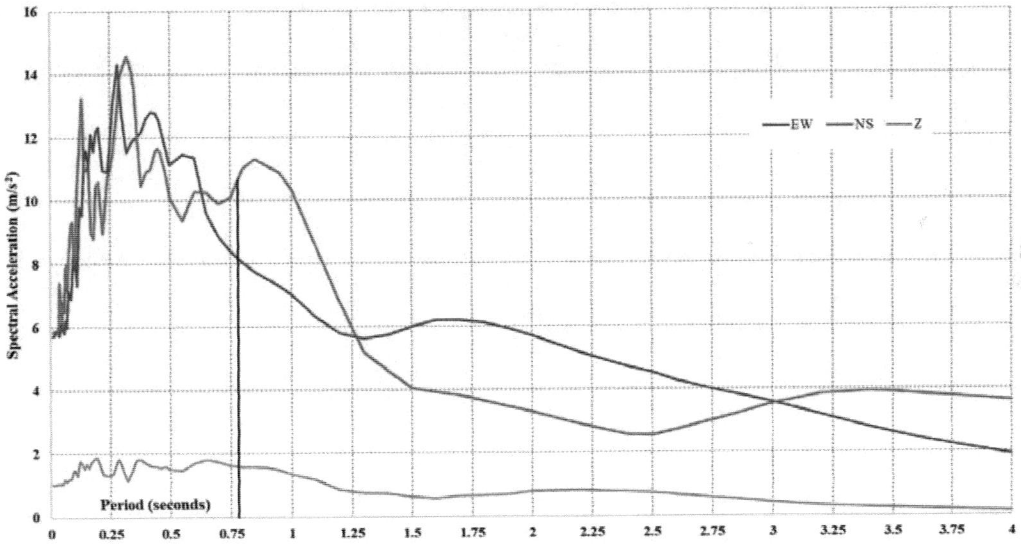


**Figure 6.** Response spectra of Turkey 2023 earthquake.

*3.2. Response spectra of the structure*

The response spectra, in terms of accelerations, at the level of each storey was obtained for each of the loading scenario presented in Table 2. Moreover, the obtained response spectra were obtained with respect to each directional component of the seismic motions EW, NS and Z components, assuming a 5% damping.

From the analysis of the response spectra, the values of the spectral acceleration corresponding to the fundamental period of vibration of the structure were identified. The location of the selected nodes for the assessment of the response spectra of the structure is shown in Figure 7.

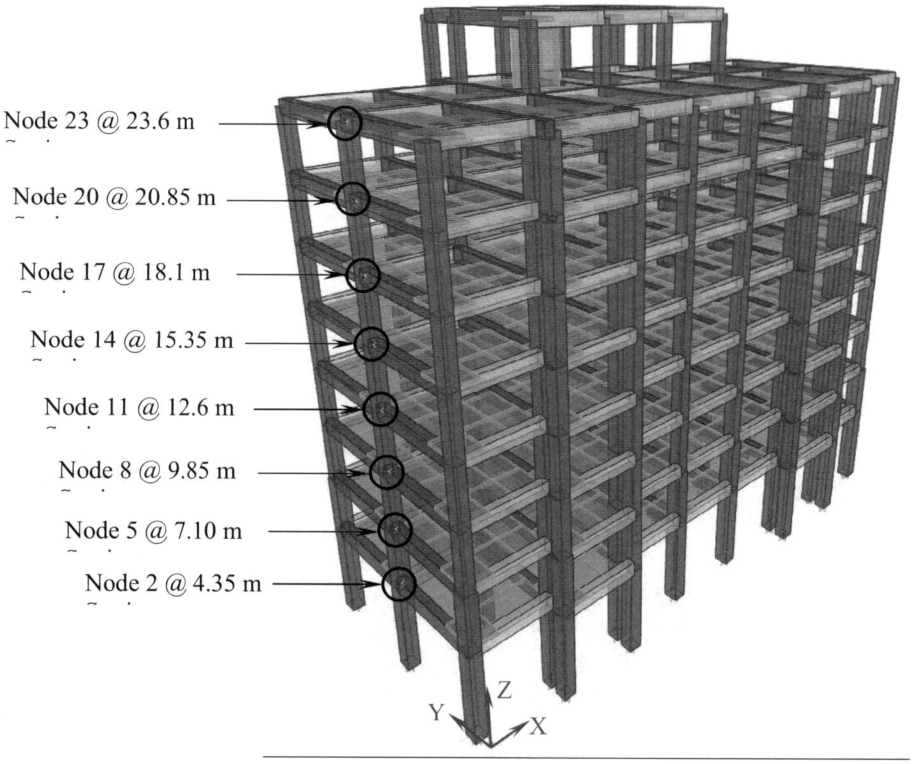


**Figure 7.** Selected nodes for assessing the response of the structure.

The response spectra of the considered structure obtained at the 8 selected nodes along X, Y and Z directions, for the Turkey 2023 earthquake, are presented in Figure 8.

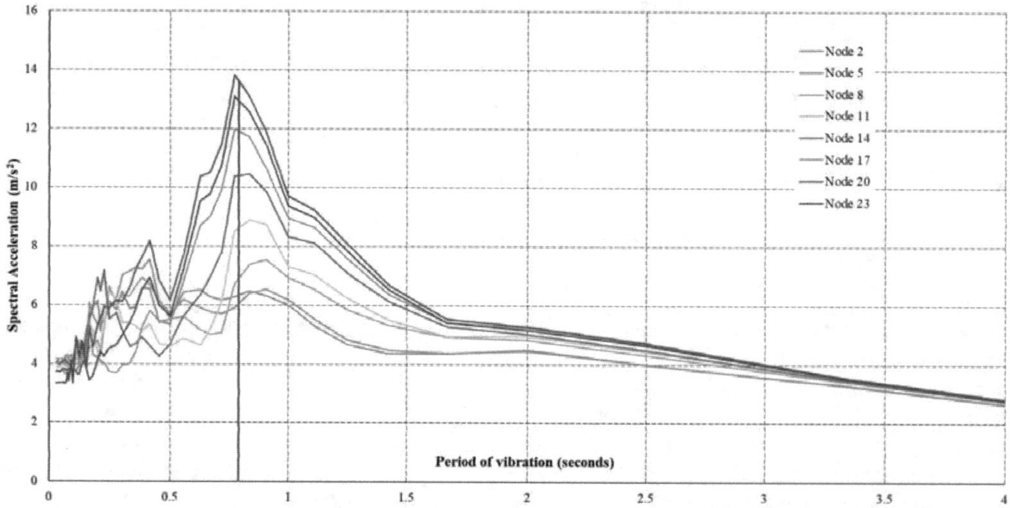


**a.** Response spectral acceleration at the selected nodes in X direction (EW component).

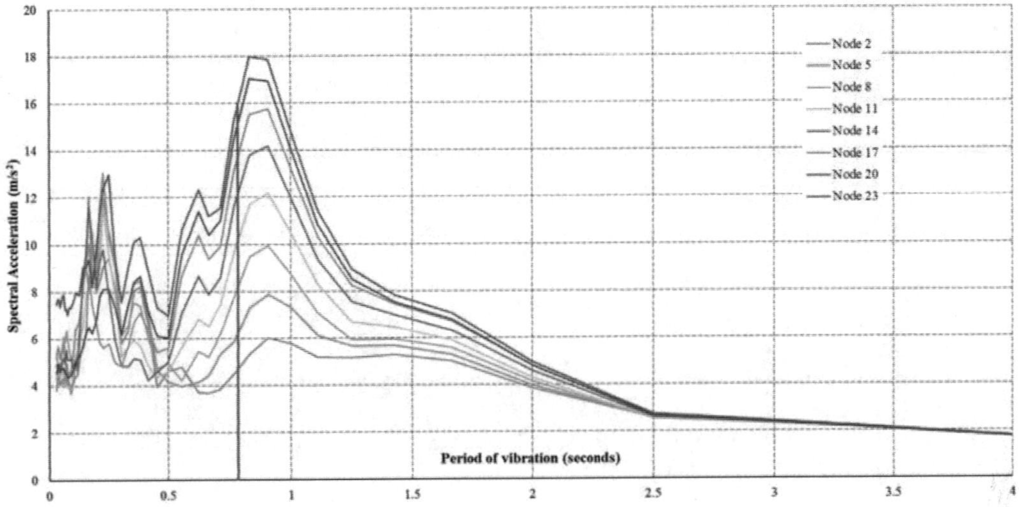


**b.** Response spectral acceleration at the selected nodes in Y direction (NS component).

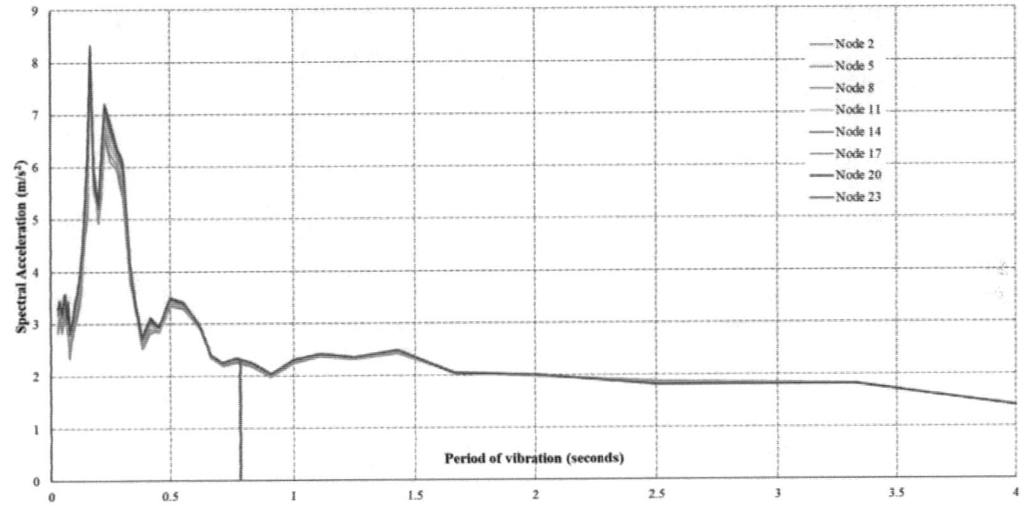


**c.** Response spectral acceleration at the selected nodes in Z direction (Z component).

**Figure 8.** Response spectral acceleration at the selected nodes for Turkey 2023 seismic action.

By looking at the results obtained for the X direction (EW component) it can be observed that they correspond to the peak values of the SA for the fundamental period of vibration of the structure. The values are almost equal to the ones along Y direction of the model (NS component) even though the theoretical applied magnitude of the ground acceleration, considering the response spectra shown in Figure 6, for the EW component is lower than the NS component. This could be attributed to the fact that the structure is stiffer in X direction and therefore less damping of the input acceleration was achieved.

*3.3. Directional spectral accelerations*

The obtained values of spectral acceleration corresponding to the fundamental period of vibration of the structure are summarized in Table 4. The table contains the data obtained for all considered nodes, Figure 7, along the three axes, for each earthquake scenario (Table 2) in case of Vrancea 1977 earthquake.

**Table 4.** Response spectral accelerations for all considered nodes (unit: m/s$^2$)

| Node | VN-E-N | | | VN-E-N-Z | | | VN-E-N-Z2 | | |
|---|---|---|---|---|---|---|---|---|---|
| | X | Y | Z | X | Y | Z | X | Y | Z |
| **2** | 0.24 | 0.45 | 0.01 | 0.24 | 0.45 | 0.12 | 0.24 | 0.44 | 0.11 |
| **5** | 0.34 | 0.68 | 0.02 | 0.34 | 0.68 | 0.12 | 0.34 | 0.68 | 0.11 |
| **8** | 0.48 | 0.97 | 0.03 | 0.48 | 0.97 | 0.12 | 0.48 | 0.97 | 0.12 |
| **11** | 0.62 | 1.26 | 0.03 | 0.62 | 1.26 | 0.12 | 0.62 | 1.25 | 0.12 |
| **14** | 0.75 | 1.53 | 0.03 | 0.75 | 1.53 | 0.12 | 0.75 | 1.53 | 0.13 |
| **17** | 0.86 | 1.77 | 0.04 | 0.86 | 1.77 | 0.13 | 0.87 | 1.77 | 0.13 |
| **20** | 0.95 | 1.97 | 0.04 | 0.95 | 1.97 | 0.13 | 0.96 | 1.96 | 0.13 |
| **23** | 1.01 | 2.11 | 0.04 | 1.01 | 2.11 | 0.13 | 1.01 | 2.10 | 0.13 |

Table 5 contains the data obtained for all considered nodes, Figure 7, along the three axes, for each earthquake scenario (Table 2) in case of Vrancea 1977 earthquake.

**Table 5.** Response spectral accelerations for all considered nodes (unit: m/s$^2$)

| Node | TK-E-N | | | TK-E-N-Z | | | TK-E-N-Z2 | | |
|---|---|---|---|---|---|---|---|---|---|
| | X | Y | Z | X | Y | Z | X | Y | Z |
| **2** | 6.30 | 4.96 | 0.15 | 6.32 | 4.73 | 2.24 | 6.26 | 4.82 | 2.21 |
| **5** | 6.12 | 5.84 | 0.22 | 6.03 | 6.25 | 2.26 | 5.99 | 5.97 | 2.21 |
| **8** | 7.00 | 7.97 | 0.30 | 6.88 | 8.24 | 2.28 | 6.80 | 7.65 | 2.22 |
| **11** | 8.67 | 10.10 | 0.38 | 8.62 | 10.20 | 2.30 | 8.62 | 9.76 | 2.23 |
| **14** | 10.54 | 11.85 | 0.43 | 10.40 | 12.20 | 2.31 | 10.32 | 11.28 | 2.23 |
| **17** | 12.10 | 13.56 | 0.46 | 11.91 | 13.76 | 2.32 | 11.73 | 13.16 | 2.24 |
| **20** | 13.13 | 14.88 | 0.47 | 12.96 | 15.16 | 2.32 | 12.88 | 14.40 | 2.24 |
| **23** | 13.83 | 15.78 | 0.47 | 13.63 | 15.98 | 2.32 | 13.50 | 15.36 | 2.24 |

By comparing the data presented in Table 4 with the data corresponding to Vrancea 1977 earthquake from Table 3, it can be noticed that the values of the response SA did not exceed the values of the SA of the seismic motion. This could be explained by the fact that the structure was built after the 1977 seismic event when the design codes were updated based on the information collected in the aftermath of the earthquake. Therefore, the structure would be able to withstand a similar earthquake with minor to moderate damage and, possibly, no casualties.

On the other hand, the data presented in Table 5 suggests that the accelerations at level 5 (node 14) and above would exceed the values of the SA of the input seismic motion, suggesting an amplification of the input signal. This could result in moderate to severe damage and higher probability of casualties. Moreover, introducing the vertical component in the numerical simulation, it could be observed that there is a significant amplification of the SA in the vertical direction from the point of view of the response of the structure.

Taking into account the presented information and the fact that the early warning systems (EWS) are getting more and more accurate and fast in alerting the decision makers and general public in case of a severe seismic event, it is now clearer than ever the need for further developing such EWS. The continuous expansion of sensor arrays to monitor seismic motions and record the data could be beneficial to the development of EWS from the point of view of large datasets that can be used to program the EWS in identifying seismic motions and produce a local or national warning.

## 4. Conclusions

The paper presents some preliminary results of a larger research project aimed at developing and testing new early warning system sensors to detect seismic motion. The data presented in this paper highlights the need of such developments by considering a common structural typology present in the urban areas of Romania. The structure was designed and built after the Vrancea 1977 earthquake. The numerical model considers the real dimensions of the structural elements and the material properties specified in the project.

The structure is subjected to a combination of seismic waves acting in the longitudinal and transversal direction of the structure coupled with the vertical component of the seismic motions. The results are presented and analysed from the point of view of spectral accelerations of the input motions and the response of the structure.

Based on the obtained results it can be concluded that the structure behaves well in case of a similar seismic event to Vrancea 1977. Response spectral acceleration values are well below the corresponding accelerations of the input motion. However, should a seismic even similar to Turkey 2023 occur, significant amplifications of the input motion is observed in terms of spectral acceleration values. This is due, in part, to the fact that the 2023 earthquake occurred at a shallow depth whereas the 1977 seismic event took place at 96 km below the surface of the earth. Last but not least, the spectra of the two seismic events are different from one another, hence influencing in a different manner the response of the structure.

## 5. References

[1] Tian Y, Ren J, Xu Z and Qi M 2023 A Cost–Benefit Analysis Framework for City-Scale Seismic Retrofitting Scheme of Buildings *Buildings* **13** 477

[2] Suwansaya P and Warnitchai P 2023 Simplified Procedure for Rapidly Estimating Inelastic Responses of Numerous High-Rise Buildings with Reinforced Concrete Shear Walls *Buildings* **13** 670

[3] Biglari M, Hadzima-Nyarko M and Formisano A 2022 Seismic Damage Index Spectra Considering Site Acceleration Records: The Case Study of a Historical School in Kermanshah *Buildings* **12** 1736

[4] Pavel F and Vacareanu R 2023 Review of Methodologies for Displacement Checks in Modern Seismic Design Codes *Buildings* **13** 940

[5] Sococol I, Mihai P, Toma I-O, Olteanudonţov I and Venghiac V-M 2021 Stress-Strain Relation Laws for Concrete and Steel Reinforcement Used in Non-Linear Static Analytical Studies of the Moment Resisting Reinforced Concrete (RC) Frame Models *Bull. Polytech. Inst. Iaşi. Constr. Archit. Sect.* **67** 17–29

[6] Ţăranu G 2022 Case Study – A Comparative Analysis of Two Scenarios of a RC Frame Structural System for a Duplex House *Bull. Polytech. Inst. Iaşi. Constr. Archit. Sect.* **67** 9–22

[7] Gruia L-C 2022 On the Seismic Resilience of Interwar Buildings *Bull. Polytech. Inst. Iaşi. Constr. Archit. Sect.* **67** 119–32

[8] Abdel Raheem S E, Fooly M Y M, Abdel Shafy A G A, Taha A M, Abbas Y A and Abdel Latif M M S 2019 Numerical simulation of potential seismic pounding among adjacent buildings in series *Bull. Earthq. Eng.*

[9] Allen R M and Stogaitis M 2022 Global growth of earthquake early warning *Science (80-. ).* **375** 717–8

[10] Chandrakumar C, Prasanna R, Stephens M and Tan M L 2022 Earthquake early warning systems based on low-cost ground motion sensors: A systematic literature review *Front. Sensors* **3** 1–16

[11] Serea E, Temneanu M C, Donciu C and Berdan M 2022 Influence of Household Tools in

Regional Configuration of Earthquake Early Warning Systems *EPE 2022 - Proc. 2022 12th Int. Conf. Expo. Electr. Power Eng.* 376–81

[12] Marmureanu G, Cioflan C O and Marmureanu A 2011 Intensity Seismic Hazard Map of Romania by Probabilistic and (Neo)Deterministic Approaches, Linear and Nonlinear Analyses *Rom. Reports Phys.* **63** 226–39

[13] Luzi L, Lanzano G, Felicetta C, D'Amico M C, Russo E, Sgobba S, Pacor F and 5 O W G 2020 Engineering Strong Motion Database (ESM) (Version 2.0). Istituto Nazionale di Geofisica e Vulcanologia (INGV). *Ist. Naz. di Geofis. e Vulcanol.*

**Acknowledgement**
This research was funded by European Regional Development Fund, grant number 7227/19.11.2021 (SMIS code 137414)

# The impact of blast vibration on the site response

**Gulnisa OZKAN[1], Kaveh DEHGHANIAN[2*]**

[1,2]Civil Engineering Department, Istanbul Aydin University, Istanbul, Turkey

[1]gulnisaozkan@stu.aydin.edu.tr, [2]kavehdehghanian@aydin.edu.tr

[*]Corresponding author:kavehdehghanian@aydin.edu.tr

**Abstract**. In civil engineering, one of the main problems is the stability of structures. Blast vibrations can be one of several elements influencing the stability of these structures. In order to prevent any effects on the structures, the maximum suggested values of the peak particle velocity and frequency from the blast are generally determined using the existing appropriate government rules and industry standards as reference. In this paper, the effect of these vibration on the site response has been investigated. Parameters such as peak ground acceleration, Pseudo spectral acceleration and maximum stress and strain has been compared for two different soil models. The results indicate that the soft soil has a significant impact and amplifies the input parameters considerably.

**Keywords: Blast, site response analysis, maximum strain, spectral acceleration**

## 1. Introduction

Explosive materials, which have been used for different purposes since ancient times, have started to be used in the construction sector over time, apart from military and defense purposes. Explosive materials, which cause great damage when used for military and defense purposes, have been found worth using in rock excavations when used for the right purposes. With the developing technology, drilling equipment has developed. explosive materials started to be mass produced and became affordable. Thus, excavation has made by drilling and blasting method started to be used on a large scale and even became widespread. In large-scale construction projects (Tunnels, airports, dams, high standard highways, etc.) where mechanized excavation (earthmoving works) is not feasible, it has been preferred because it is advantageous in terms of time and cost [1]. The ability to disable primary crushers and target mass displacement is also an ultimate engineering ideal. The planning and execution phases of this type of project with drill-blasting are of a different dimension compared to other excavation methods. It usually consists of drilling of holes, charging and blasting, loading, transportation, crushing-grinding [2]. Before starting these operations, which we call blasting excavation, the geological characteristics of the site, the energy required for excavation, the amount of production and the required time are calculated. Then, planning is made and a list of equipment supply list is made in accordance with the required number and purpose. During the working period, blasting records are kept and working methods are continuously examined and improved if necessary. These drilling-blasting activities, which are methods of soil loosening or direct excavation, have become widely used because they are more advantageous in terms of cost and time compared to conventional methods such as mechanized excavation (earthmoving works), which are limited by the material-mass properties of soils and the technology of use. Today, drilling and blasting is used in areas such as foundation excavations and tunneling activities. It is applied together with the control of environmental factors.

### *1.1. Effect of Environmental Factors*

There are four different disadvantages of bastings which are rock particles throwing around, dust emission, ground shaking and noise. These negativities of blasting activities, which have started to be carried out in urban areas in parallel with population growth and urbanization, cause environmental problems. Explosive excavation is inevitable in activities such as tunnels, subways, foundation excavations, dams, etc. The increase in the size and capacity of these activities also affects the amount of explosive material. For this reason, since they are often close to settlements, ground shaking and noise cause environmental problems. Considering the safety of life and property and psychological problems, it should be taken into account that such problems should be prevented or kept below the danger limits during blasting. The most important feature expected from a good blasting is to be safe in terms of environmental impacts. Standards have been established in most countries for environmental problems arising from blasting excavations. However, complaints about ground shaking and the noise have not been prevented [3].

When the explosive placed in a hole is detonated, it releases a huge amount of energy through pressure (up to 50 GPa) and temperature (up to 5000 K). 20-30% of this energy is used for fragmentation and displacement of the rock mass, while the remaining energy is used for impacts such as ground shaking, noise, excessive fracturing and back cracks [3]. In other words, a very small fraction of the energy released by blasting is absorbed for fragmentation and translation. The damped energy spreads as seismic waves from the blasting source and causes negative environmental impacts. Blasting-induced ground shaking generates transient (short-term) and irregular ground motion. The velocity of motion in soil particles starts from zero, reaches its highest value and gradually dampens. The most important issue in ground shaking investigations is the maximum particle velocity. Because as the highest velocity value increases, the shaking intensity of the structure will also increase. In places close to the blasting zone, the characteristics and nature of ground shaking are affected by blast design, hole layout, explosive quantity, firing interval and firing direction. In short, particle velocity is an important damage indicator. Away from the blasting zone, the characteristics and nature of ground shaking are influenced by the characteristics of the rock or soil environment. The frequency and terrain coefficients of ground shaking are also important factors in determining whether or not damage occurs [4].

### *1.2. Damage Classification*

Blasting damage criteria developed by different researchers working on blasting in the world are still applied today. These researches have been analysed under two main headings: Identification, measurement and parameter analysis of blasting-induced vibration and noise; appropriate blasting design by determining damage criteria for various structures and matching these criteria with post-blasting parameters. Due to the variation of the criteria in different countries, a unified table has been prepared as a source for reference and evaluation in applications. Some of these parameters are the damage criteria of the US Bureau of Mines (USBM) [5] and the German DIN 4150 Standarts [6]. The damage classification scheme established by the USBM is given in Table 1.

**Table 1.** Damage Classification [5]

| Damage Classification | Definition of Damage |
|---|---|
| Threshold Damage (Start of Damage) | Paint cracking and swelling, small plaster cracks in the junction of the building elements, prolonged old cracks |
| Light Damage | Plaster swelling and falling plaster, string on stone walls and capillaries in the windows, cracks from hair to 3 mm thickness, relaxed mortar pouring. |
| Fundamental Damage | Wide cracks on the walls, cracks in the belts, weakening of the carrier elements of the structure, stone walls on the stone walls, brick falls in chimneys, decrease in loading capability |

*1.2.1. Environmental Noise Control Regulation of Republic of Turkey, Ministry of Environment, Urbanization and Climate Change* Turkey has a regulation with reference values such as USBM damage criteria and German DIN 4150 standard under the title of "Environmental Vibration Assessment", the principles regarding the control of environmental vibration caused by various vibration sources are given: In order to prevent blasts in mines, quarries and similar activity areas from damaging the surrounding structures, the vibration level to be measured on the ground, except for the nearest structure, cannot exceed the values given in Table 2. Measurements are made in three directions and the highest of these is taken. Vibrations are measured as peak values in 1/3 octave bands [7].

**Table 2.** The highest permissible values of the ground vibrations that will be created outside the closest structure of vibrations to be caused by blast in mines and quarries and similar areas in Turkey [7]

| Structural Type | The highest rate of vibration on the base of building (mm/s) | | | In the flooring of the upper floor for all frequencies (mm/s) |
|---|---|---|---|---|
| | f=1-10 Hz | f=10-50 Hz | f=50-100 Hz | |
| Industrial structures | 20 mm/s | 40 mm/s | 50 mm/s | 40 mm/s |
| Durable structures such as houses, bricks and concrete | 5 mm/s | 15 mm/s | 20 mm/s | 15 mm/s |
| Other | 2 mm/s | 8 mm/s | 10 mm/s | 8 mm/s |

## 2. Equivalent Linear Method (EQLM)

The Equivalent linear analysis, based on the SHAKE program, is a practical type of analysis developed to accurately model soil behavior under dynamic loads. Shear modulus and damping ratio are used as iterative parameters. Thus, an equivalent linear site - response analysis is performed which shows the dynamic properties of the soil [8]. Equivalent linear analysis, which reflects the nonlinear behavior of the soil, was developed in 1972 by Schnabel et al. to capture the inelastic response of the soil [9]. Huang et al in 2001 showed that equivalent linear analysis gives higher maximum acceleration values than linear analysis [10]. The reason for this is that the calculations can be performed at higher frequencies [11]. Equivalent linear analysis method is among the most widely used methods for analyzing site response under dynamic loads [12]. The major disadvantage of equivalent linear analysis is that the results become too biased in the face of increasing strains. In soft or liquefaction-prone soils, there are significant differences between the results of transient nonlinear analysis and equivalent linear analysis [13]. Therefore, equivalent linear site-response analysis is not used at high strain levels. In case of high strain levels, it is recommended to perform nonlinear site-response analysis to obtain more accurate results [14].

Unit weight (γ), shear wave velocity (Vs), damping ratio (D) and seismic loading are the input parameters used in the equivalent linear analysis. For each soil layer, the shear modulus (G) and damping ratio (D) are calculated based on the shear strain [15]. There are also nonlinear soil parameters that are usually described by the shear modulus reduction ($G/G_{max}$ - γ) and damping ratio (D -γ) curves. These soil parameters can be calculated by laboratory tests with undisturbed specimens. The limit of γ max is recommended to be 1% for the equivalent linear method to be useful in a limited shear stress range [16,17]. The equivalent linear analysis method models the nonlinear and hysteretic stress-strain behavior of soil layers under cyclic loading. Where, the shear modulus is the secant shear modulus ($G_{sec}$).

$$Gsec = \frac{\tau c}{\gamma c} \tag{1}$$

In the equation, $\tau_c$ is the shear stress and $\gamma_c$ is the shear strain amplitude. In other words, the secant shear modulus varies depending on the strain amplitude. The equivalent linear damping ratio (D) is the damping ratio that shows the hysteretic behavior of the soil layer in a single stress-strain cycle without repeated loading [17]. Also, in this type of analysis, the representative values for the damping ratio (D) and the secant shear modulus ($G_{sec}$) are varied until they are consistent with the strain level in the soil [9].

## 3. Material and Method

### *3.1. Identification of blasting-induced ground motion in time domain*

Ground motions and noises measured in three dimensions during the blasting operations were recorded by the device. Vibration meter devices can record at 5 millisecond (0.0005 second) intervals for the desired duration. In these studies, which are generally recorded in the range of 2-4 seconds, the device measures particle velocity. The output of the device consists of four columns. The first three columns show the ground motion while the fourth column shows the noise (dB). The value with the highest particle velocity among the values measured in three dimensions is the maximum particle velocity (PPV). The data obtained as a result of the measurements are taken from the devices and transferred to the computer. Acceleration records were started by creating time history from the programs specific to each brand of device. The maximum particle velocity values selected were chosen to be suitable for "structures such as houses, bricks and concrete" and "industrial structures", which are determined by the limit values of the regulation prepared by the Ministry of Environment, Urbanization and Climate Change of the Republic of Turkey [7]. In this standard, concrete buildings and industrial buildings are categorized based on their functions: Concrete structures are located in cities and industrial buildings are supposed to have only industrial functions.

### *3.2. Properties of soils used in the study*

In this study, two soil profiles were defined, one using the drilling data obtained from the field and one using the equation. The soil profiles were subjected to blasting data records. These blasting data were selected according to Environmental Noise Control Regulation of Republic of Turkey, Ministry of Environment, Urbanization and Climate Change. Under standard conditions for "durable structures such as houses, bricks and concrete" which has around 1-10 Hz limit value blasting data is rc1, 10-50 Hz limit value blasting data is rc2 and 50-100 Hz limit value blasting data is rc3. Again, according to these standards for industrial buildings, which has around 1-10 Hz limit value blasting data is i1, 10-50 Hz limit value blasting data is i2 and 50-100 Hz limit value blasting data is i3. The behavior analysis of soils under these dynamic loadings was modeled with DeepSoil v7.0. The DeepSoil program used for this study is based on the principle of one-dimensional soil behavior analysis. DeepSoil program, first developed in 1998, is capable of both frequency and time domain analysis and equivalent linear and nonlinear analysis [8]. A total of 12 equivalent linear analyses were performed using 6 different blasting data.

A total of 2 soil models were used in the study. In the soil models used, the unit volume weight $\gamma$=24.0 kN/m$^3$ [19] was selected for the bedrock, the shear wave velocity was defined as 900 m/s and the damping ratio was defined as 5 in the soil models formed with rock. Soil 1 comprises a two-layer soil and soil 2 has 13 layers and has been defined as $Vs= 120\times(1+d)^{0.32}$. d shows the depth in this formula. Soil 1 is adopted from a Multichannel Analysis of Surface Waves (MASW) [20]. The soil 2 has been adopted from Dehghanian and Yılmaz [21]. The parameters determined for the soils are given in table 3.

**Table 3.** Soil properties [19,20,21]

| Soil | Layer | Depth (m) | Vs (m/s) | $\gamma$ (kN/m$^3$) |
|---|---|---|---|---|
| Zemin 1 | 1 | 6 | 484.00 | 16.2 |
| | 2 | 36 | 858.00 | 19.8 |
| Zemin 2 | 1 | 3 | 187.00 | 20 |
| | 2 | 6 | 223.67 | 20 |
| | 3 | 9 | 250.72 | 20 |
| | 4 | 12 | 272.67 | 20 |
| | 5 | 15 | 200.00 | 20 |
| | 6 | 18 | 307.88 | 20 |
| | 7 | 21 | 322.67 | 20 |
| | 8 | 24 | 336.14 | 20 |
| | 9 | 27 | 348.56 | 20 |
| | 10 | 30 | 360.10 | 20 |
| | 11 | 33 | 370.90 | 20 |
| | 12 | 36 | 381.07 | 20 |

One-dimensional dynamic field behavior analyses were defined by dividing the soils into sub-layers rather than single layers due to the detailed representation of the shear wave velocity variation while creating the soil models defined in Deepsoil v7.0, a C++ based one-dimensional seismic hazard analysis program.

## 4. Analyses

Deepsoil program soil profiles exhibit different behavior under static and dynamic loading. Soil profiles subjected to dynamic loading show stress-strain behavior. Dynamic loadings that generate shear stresses and shear deformations affect the soil behavior. The resulting deformations cause nonlinear behavior and the soil behavior supporting the foundations of the structure can affect the structural damage that may occur. When performing field response analyses, stiffness and damping ratio reduction curves as well as shear wave velocity values are used as dynamic soil parameters in equivalent linear analyses. For rocky and sandy soils, Seed and Idriss [22] developed attenuation curves. These attenuation curves were prepared for 3 different cases according to relative densities. These are; "Lower Limit" curves for relative frequency values lower than 30%, "Mean Limit" curves for values between 30% - 60% and

"Upper Limit" curves for values higher than 60%. For clayey soils, Vucetic and Dobry [23] have prepared the attenuation curves. These attenuation curves depend on the plasticity index.

### *4.1. PGA*

Figure 1 shows the variation of Peak Ground Acceleration (PGA) with respect to the depth for soil model 1 and 2 respectively for concrete and industrial buildings. It can be seen that PGA is maximum at the surface. As the depth and the stiffness of the soil increases, the Peak acceleration decreases. In soil model 2, a soft layer has been embedded between two stiff layers. It can be seen that the PGA shows a sudden increase in 15 m depth. The trend is similar for both building types.

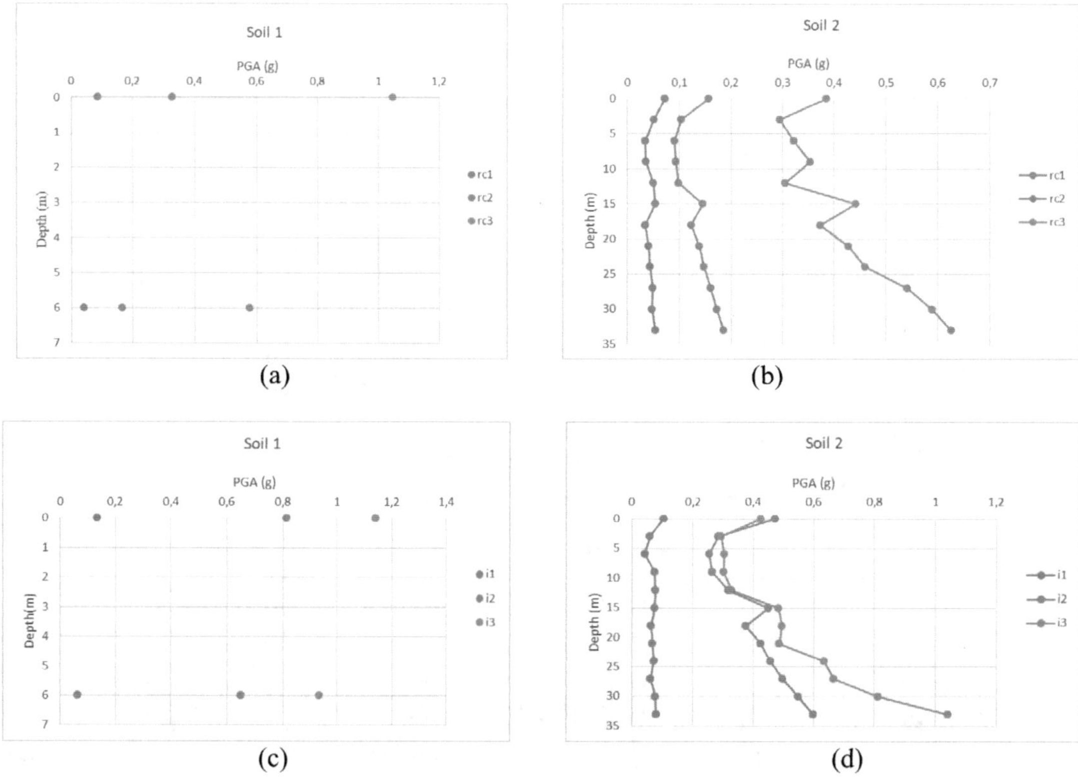


**Figure 1.** (a) PGA at limit values for durable structures such as houses, brick and concrete on Soil 1, (b) PGA at limit values for durable structures such as houses, brick and concrete on Soil 2, (c) PGA at limit values for industrial structures on Soil 1, (d) PGA at limit values for industrial structures on Soil 2

IOP Conf. Series: Materials Science and Engineering 1304 (2024) 012005 doi:10.1088/1757-899X/1304/1/012005

### *4.2. Amplification Factor*

Figure 2 shows the amplification ratio of two soil models. It can be seen that the PSA ratio in the soil surface is greater than the bedrock in all cases or in other words the soft soil amplifies PSA.

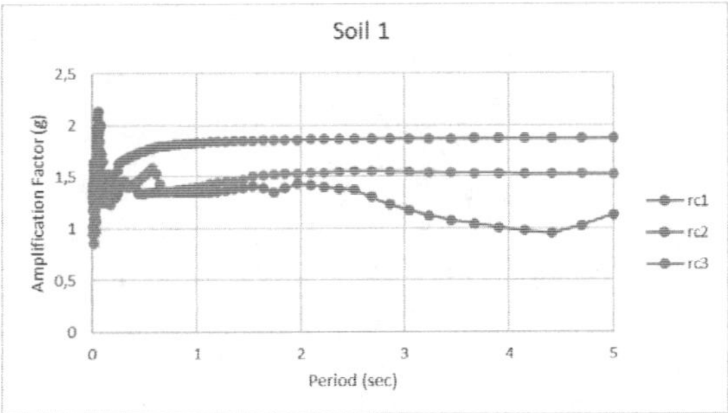


(a)

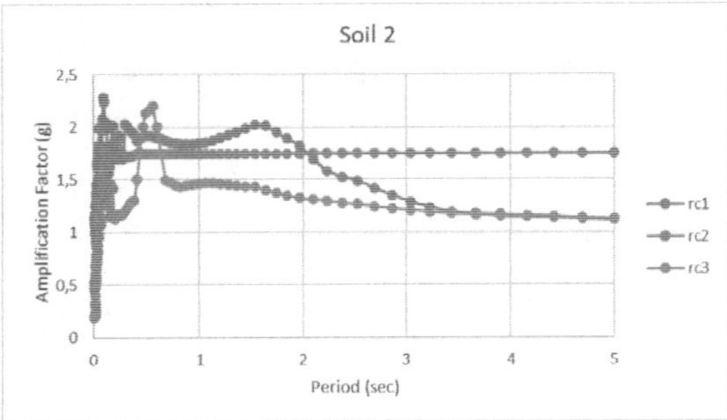


(b)

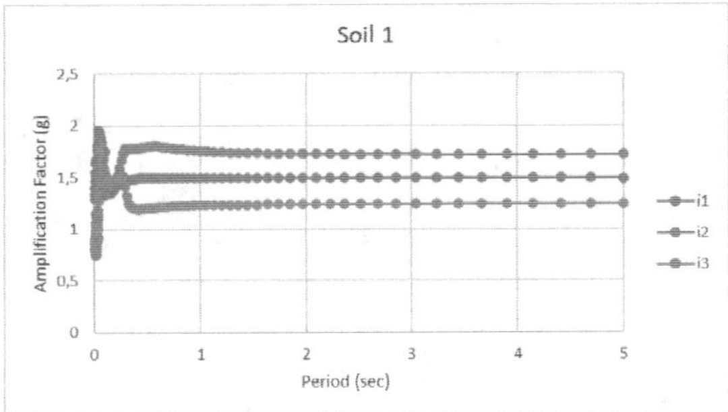


(c)

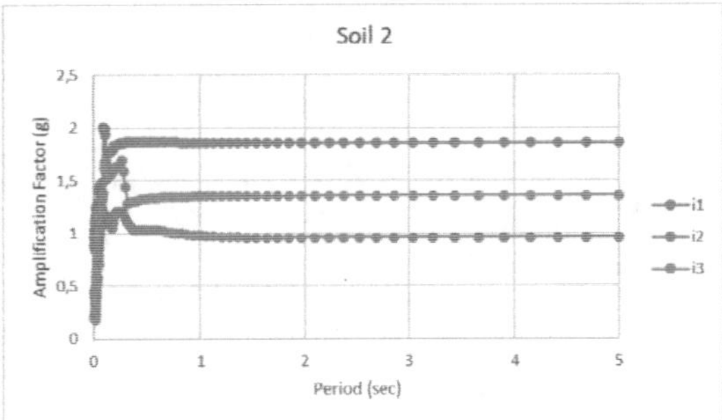


(d)

**Figure 2.** (a) Amplification Factor at limit values for durable structures such as houses, brick and concrete on Soil 1, (b) Amplification Factor at limit values for durable structures such as houses, brick and concrete on Soil 2, (c) Amplification Factor at limit values for industrial structures on Soil 1, (d) Amplification Factor at limit values for industrial structures on Soil 2

### *4.3. Max Stress and strain*

Figure 3 shows the distribution of maximum stress with respect to depth. Figure 4 demonstrate the variation of maximum strain with respect to depth as well. It can be concluded that in both figures, the maximum of the parameters occurs at the surface and the values decrease with the increase in depth. For soil 2 there is a sudden increase at 15 m depth in which a soft layer is located between two stiffer layers.

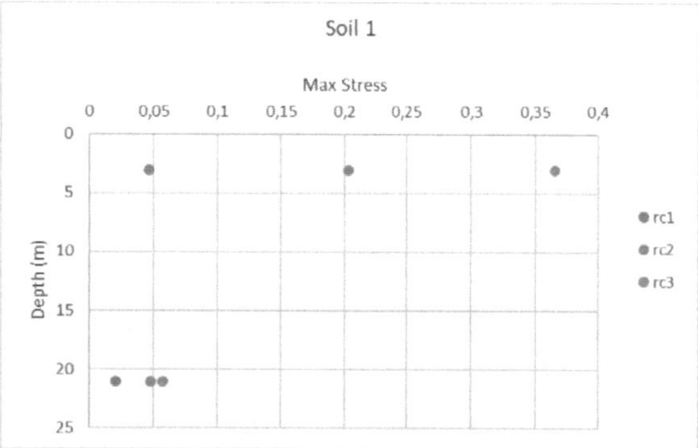


(a)

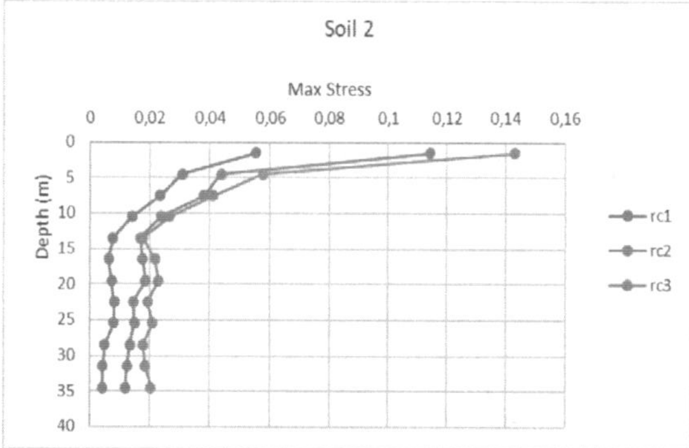


(b)

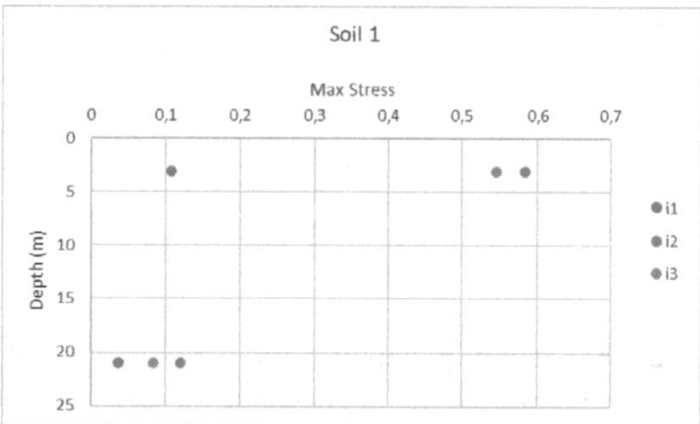


(c)

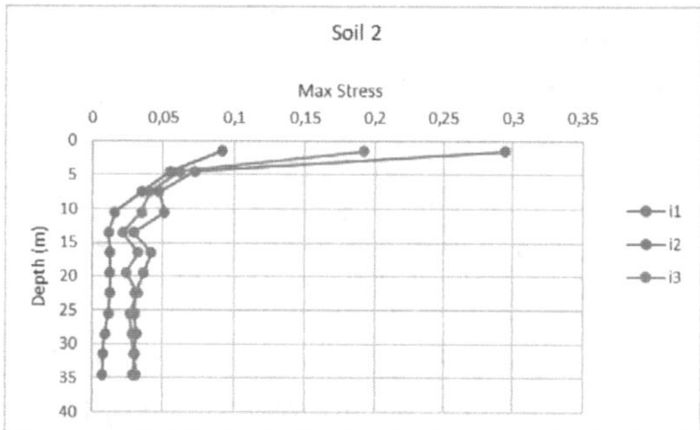


(d)

**Figure 3.** (a) Maximum Stress at limit values for durable structures such as houses, brick and concrete on Soil 1, (b) Maximum Stress at limit values for durable structures such as houses, brick and concrete on Soil 2, (c) Maximum Stress at limit values for industrial structures on Soil 1, (d) Maximum Stress at limit values for industrial structures on Soil 2

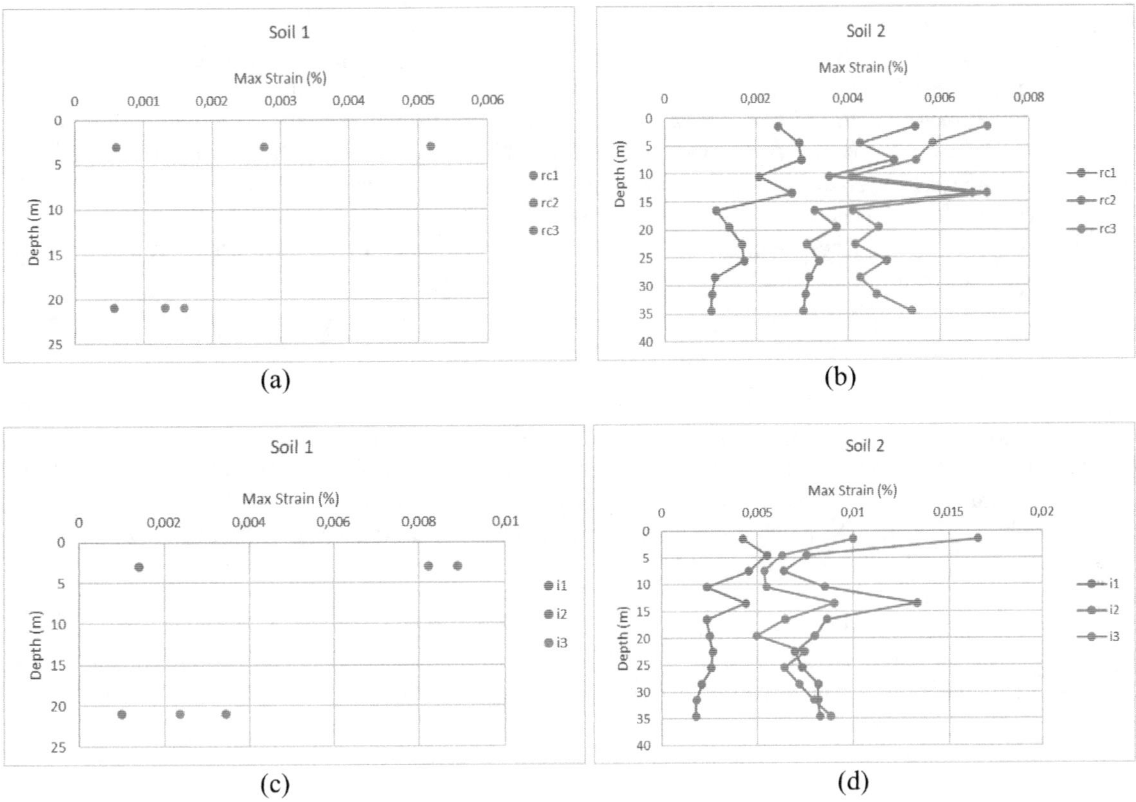


**Figure 4.** (a) Maximum Strain at limit values for durable structures such as houses, brick and concrete on Soil 1, (b) Maximum Strain at limit values for durable structures such as houses, brick and concrete on Soil 2, (c) Maximum Strain at limit values for industrial structures on Soil 1, (d) Maximum Strain at limit values for industrial structures on Soil 2.

## 5. Conclusions

Based on the results of the site response analysis, it is evident that the soft layer, affects the natural frequency of the blast waves. It was shown that, while maintaining all other characteristics, the natural frequency of the blast wave progressively switches to substantially lower frequencies as the Vs decreases. Natural frequencies tended to be lower for more vertical slopes as opposed to greater for flatter slopes. Another significant finding was that the maximum strain and stress are affected considerably by the soil layer specifications. The amplification results show a high amplification ratio specially at low frequencies. The behaviour differs for industrial and concrete structures due to the variation of the highest rate of vibration on the base of building as it is indicated in table 2.

**Acknowledgment**

The data used in this paper is obtained by help Prof. Dr. Ali Kahriman. The authors wold like to thank him for his support.

**References**

[1] Kahriman A 1995 Sivas Investigation of Optimum Blasting Conditions for Celestite Ore and Siderocks of Ulas Region and Their Association with Rock Properties (*Ph.D. - Doctoral Program*) Cumhuriyet University, Sivas, Turkey (in Turkish)

[2] Ceylanoglu A, Kahriman A and Demirci A 1993 The Importance of Drilling-Blasting, Its Uses and Its Relevance to Mining Engineering *1. Drilling and Blasting Symp.* (Ankara, Turkey) pp 127-138 (in Turkish)

[3] Hagan T N 1973 Rock Breakage by Explosives, *Int. Proc. of the National Symp. on Rock Fragmentation* (Adelaide) pp 1-17

[4] Kahriman A, Gorgun S, Tuncer A and Karadoğan A K 1998 Effects of Vibrations Caused by Blast Excavations on Urban Settlements and Measures to be Taken *Geology and Urbanization Symp* (Istanbul,Tukey)

[5] Iskind D E, Stagg M S, Kopp J W and Dowding C H 1980 Structure response and damage produced by ground vibration from surface mine blasting United States.

[6] DIN 4150-3 1999 Structural vibration-Effects of vibration on structures Deutsches Institut für Normung e. V. Document Number:Din 4150-3

[7] Republic of Turkey Ministry of Environment, Urbanization and Climate Change 2022 Gazette No 32029 *Environmental Noise Control Regulation*

[8] Hashash Y M A, Musgrove M I, Harmon J A, Ilhan O, Xing G, Numanoglu O, Groholski D R, Phillips C A and Park D 2020 DEEPSOIL 7.0, User Manual

[9] Schnabel P B, Lysmer J and Seed H B 1972 in Shake-91 In SHAKE: a computer program for earthquake response analysis of horizontally layered sites (Issue December) *University of California, Berkeley*

[10] Huang H C, Shieh C S and Chiu H C 2001 Linear and nonlinear behaviors of soft soil layers using Lotung downhole array in Taiwan Terrestrial *Atm. and Oceanic Sci.*

[11] Arslan H and Siyahi B 2006 A comparative study on linear and nonlinear site response analysis *Environmental Geo.* pp 1193–1200

[12] Yoshida N, Kobayashi S, Suetomi I and Miura K 2002 Equivalent Linear Method Considering Frequency Dependent Characteristic of Stiffness and Damping *Soil Dyn. and Earthq. Eng.* **Vol 22 No 3** pp 205-222

[13] Electronic Power Research Institute (EPRI) 1988 Site Response to Earthquake Ground Motion NP-5747 project 2556-7 (California) pp 1-306

[14] Rathje E M, Kottke A R and Trent W L 2010 Influence of input motion and site property variabilities on seismic site response analysis *J. of Geotech. and Geoenv. Eng.* 136(4) pp 607–619

[15] Kaklamanos J, Baise L G, Thompson E M and Dorfmann L 2015 Comparison of 1D linear, equivalent-linear, and nonlinear site response models at six KiK-net validation sites *Soil Dyn. and Earthq. Eng.* pp 207–219

[16] Assimaki D, Gazetas G and Kausel E 2005 Effects of local soil conditions on the topographic aggravation of seismic motion: Parametric investigation and recorded field evidence from the 1999 Athens Earthquake *Bulletin of the Seismological Society of America* 95 pp 1059–1089

[17] Stewart J P and Kwok A O L 2008 Nonlinear seismic ground response analysis: Code usage protocols and verification against vertical array data *Geotechnical Earthquake Engineering and Soil Dynamics IV* pp 1-24

[18] Bardet J, Ichii K and Lin C H 2000 EERA A Computer Program for Equivalent-linear Earthquake site Response Analyses of Layered Soil Deposits

[19] Geotechnical Software Suite GEO5- User's Guide Manual Version 12 @ Fine Ltd. 2011

[20] Toker Drilling and Civil Eng. Cons. j-s comp. 2020 Soil Investigation Report Based on Drilling for Urban Transformation in Uskudar, Turkey (in Turkish)

[21] Dehghanian K 2016 A Semi-empirical method for estimation of maximum seismic shear strain in imperforated ground excited by vertically propagating shear waves (*Ph.D. - Doctoral*

*Program*) Middle East Technical University
[22] Seed H B, and Idriss I M 1970 Soil moduli and damping factors for dynamic response analyses [Report No. EERC 70-10]
[23] Vucetic M and Dobry R 1991 Effect of soil plasticity on cyclic response *J. of Geotech. Eng.* 117(1) pp 89–107

IOP Conf. Series: Materials Science and Engineering 1304 (2024) 012006 doi:10.1088/1757-899X/1304/1/012006

# Evaluation of bridge load capacity using proof-load tests

**M Brinissat [1], R P Ray [2], R Kuti [3]**
[1,2] Department of Structural and Geotechnical Engineering, Széchenyi István University, 9026 Győr, Hungary.
[3] Department of Machine Design, Faculty of Mechanical Engineering, Informatics and Electrical Engineering, Széchenyi István University, Egyetem tér 1,9026 Győr, Hungary.

E-mail: brinissat.marame@hallgato.sze.hu

**Abstract.** A proof-load test is often necessary to accurately measure the field behavior of a bridge structure. A new segmental steel bridge has been constructed over the Tisza River in Hungary. The bridge comprises of two independent steel box girders assembled by a welding system. Engineers performed proof load tests before opening the structure to evaluate its load-carrying capacity and deck deflection. This paper presents several aspects of the load testing and evaluation of the bridge structure under static and dynamic loads using a rating factor (RF) approach compared to a baseline finite element model. The load test results helped to calibrate the model and verify the method. The study reveals that the tested bridge has an elastic behavior, a high capacity, and adequate dynamic performance.

## 1. Introduction

Steel box girder bridges provide adaptable and space-saving support systems in urban areas. Prefabricated segments delivered to the site provide the initial elements of the structure. Fabricators then erect and assemble them in the field without requiring complex falsework and only a minimum of temporary support. The Hungarian national standard code requires load-carrying capacity investigations using static and dynamic load tests for spans greater than 100 m before opening for service [1]. The primary objective of a proof load test is to provide direct evidence that a bridge can support its prescribed factored live loads without suffering any possible deficiencies after construction. To achieve this, dedicated sensors are typically installed temporarily to record measurement data, while in-situ dynamic load testing and short-term monitoring are carried out under service load [2]. Some deficiencies, if left undetected, can cause damage or even a sudden collapse. Such collapses occurred while testing Serbia's steel truss bridge over the Morawa River, Switzerland's road bridge near Salez, and France's suspension bridge in Maurin [3]. The deficiencies in a bridge during construction result from the premature yield of components whose specified properties do not meet design specifications immediately after the end of construction. Primarily, they are undetected defects caused by inappropriate design or assembly, so they appear without notice during the design and erection of the structure. They may reveal themselves during a comprehensive inspection or proof load testing. If they are detected, the contractor must initiate repair operations. If they do not affect the load-carrying capacity [4], reduced service may start while the contractor completes repair work.

As previously stated, a proof-load test produces load conditions that allow engineers to rate the bridge's performance and aid in detecting defects before going into service [2]. The load tests also establish a more accurate load-carrying capacity based on the rating factor (RF) which indicates the actual capacity of the bridge structure under live loads. It can also help to estimate load distribution and impact factors [5]. Essentially, it is an experimental evaluation of the actual behavior of the bridge to identify both observable and hidden defects that may restrict or reduce the bridge's serviceability.

This study evaluates the load-carrying capacity for the new box girder bridge located over the Tisza River using load rating factors (RFs). A Finite Element (FE) model reproduces the load-displacement response of the entire superstructure. The FE model uses several static and dynamic load test combinations to evaluate the structure more accurately. The final section compares RFs, the FE model, and measured data.

## 2. Methodologies for load-carrying capacity evaluation of the bridge

Before opening to traffic, a visual inspection was conducted on the new bridge over the Tisza River. The inspection results indicated that the structure is in good condition, with no visible signs of deterioration on any of the assessed bridge elements. However, visual inspection makes it challenging to quantitatively evaluate the actual capacity of a bridge subjected to traffic loads. Due to the increase in traffic levels, the bridge damage, and the rating design code requirements. Accurate evaluation serves both present load-carrying requirements and provides a baseline for future condition-based maintenance. Therefore, field load tests conducted directly after the visual inspection, in November 2021, provided data to validate the FE-Model and evaluate the bridge's capacity.

The load rating procedure evaluates how much a bridge can safely carry when subjected to the weight of a specific vehicle, typically a truck. This is expressed as a rating factor (RF) [6], with RFs greater than one indicating that the bridge can safely support the loads being considered.

## 3. Proof load testing of the bridge

The proof load tests of the bridge are conducted near the end of construction according to Hungarian standards [7]. At that time, engineers check the structure's load capacity before opening the bridge for service. During the proof load test of the bridge, instruments recorded deck deflections and fiber stress under static and dynamic loadings. The static load test had a negligible dynamic impact on the bridge and correctly represented actual loading. Test instruments measured mid-span deflections in all spans and stress conditions along the primary span (148m). Additional optical measurements verified the positions of foundations and support pillars. Trucks loaded the right-side deck for the static tests with 400 kN loads, shown in Figure 1(a). For dynamic load tests, two trucks side-by-side drove the length of the bridge at uniform speeds from 10 to 90 kph. The positions of the trucks appear in Figure 1(b).

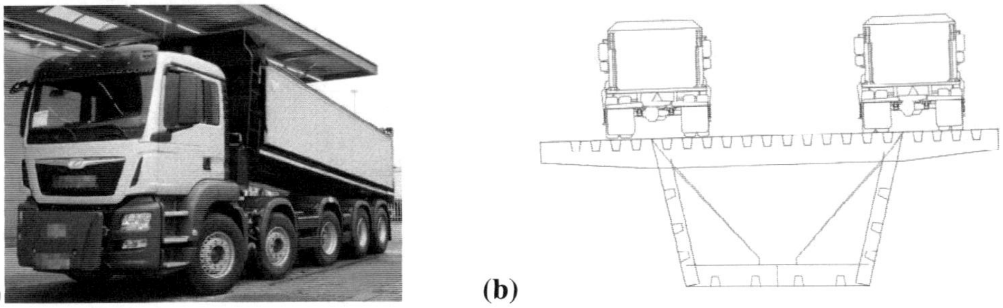

(a) (b)

**Figure 1.** Load testing; (a). truck used during tests, (b). side-by-side dynamic load.

## 4. Analysis using the FEM and load testing

### *4.1. FE Structural Modeling*

The bridge comprises two independent directional decks and seven spans made from steel orthotopic box plate girders. Its total length is 758m, with a central span of 148 m, and each deck is 14m wide. The spacing between interior diaphragms is 4m, and the primary material is welded S355 steel. Elastomeric bearings support the superstructure. The bridge report and project drawings provided input data for the FEM. The structural analysis software Midas Civil modeled the right side of the bridge, according to the details presented in [8]. The numerical modeling phase assumed no cracking or defects existed since visual inspections verified the quality of construction.

IOP Conf. Series: Materials Science and Engineering 1304 (2024) 012006 doi:10.1088/1757-899X/1304/1/012006

The numerical model simulated static and dynamic loads by applying nodal loads. The static load consisted of nodal loads uniformly distributed according to the arrangement of the trucks, as shown in Figure 2. The dynamic load used time-dependent nodal loads that followed the progress of the moving trucks. Dynamic load application followed the speed of the trucks.

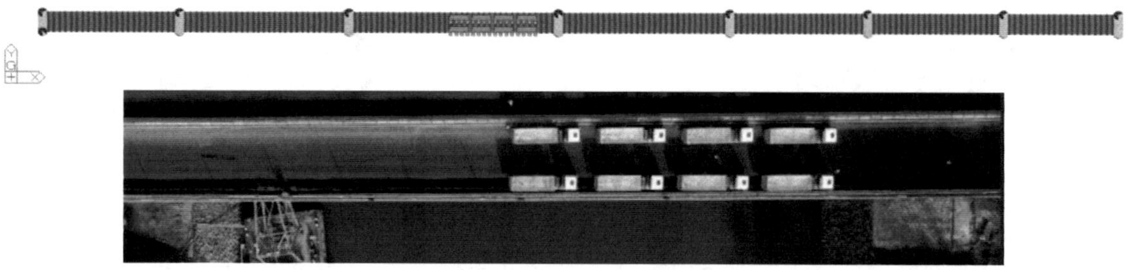

**Figure 2.** FE model with static loads (pink lines) on the bridge deck (top view of the 3rd span).

*4.1.1. Static loading results*

During the static test, the vertical displacements along the box plate girders were measured at mid-span to evaluate the behavior of the continuous superstructure. The trucks reproduced loading in two traffic lanes near mid-span, above the support pillars (reverse moment), and along alternating lanes on several spans (torsion).

The vertical displacement results from the static tests and the calibrated model appear in Figure 3. The graphs show close agreement between the modeled and measured displacements, and the bridge's structural behavior showed no anomalies.

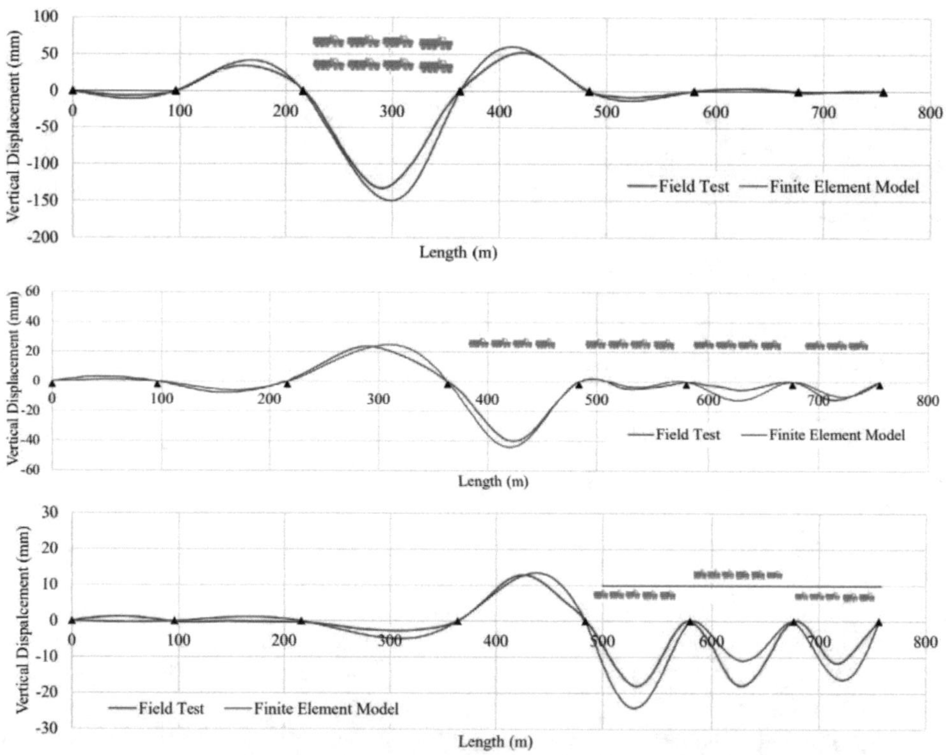


**Figure 3.** Bridge vertical displacements under static loadings.

IOP Conf. Series: Materials Science and Engineering 1304 (2024) 012006 doi:10.1088/1757-899X/1304/1/012006

The load distribution factor (DF) of the bridge is an essential parameter for structure behavior and load rating, which can be calculated using displacement responses all over the bridge [9]. Equation (1) defines the distribution factor:

$$DF_i = \frac{\delta_i}{\sum_{j=1}^{n} \delta_j} \tag{1}$$

Where $i$ and $j$ are the girder number, and $\delta$ is the displacement of the girders.

The distributions of displacements across the box girders appear in Table 1 for three load cases. According to the previous formula, we chose the highest distribution factors:

**Table 1.** Load distribution factors.

| **Load cases** | Two lanes loaded | One lane loaded | Side-by-side loads |
|---|---|---|---|
| **DF right girder plate-web** | 0.5 | 0.39 | 0.42 |
| **DF left girder plate-web** | 0.5 | 0.61 | 0.58 |

Table 1 shows that the distribution factors had the highest values when only one traffic lane was loaded. Indeed, it is better to adopt the two-lane and the side-by-side loading scenarios because the loads will be distributed here into all supporting plate-web girders since the deck slab is a wide plate.

*4.1.2. Dynamic loading results*

In the dynamic test, two trucks with 400 kN and 5 axles passed above the bridge at 10, 30, 50, 70, and 90 km/h. Figure 4 shows the variation of vertical displacement for a speed of 30 km/h, recorded at the mid-spans of Span 1 and 2. Once the bridge is loaded the graph begins with sagging displacements and progresses to a large peak under the front wheel of the trucks. When the truck gets to the following span, hogging displacements begin. As the truck continues, deflections rapidly decrease.

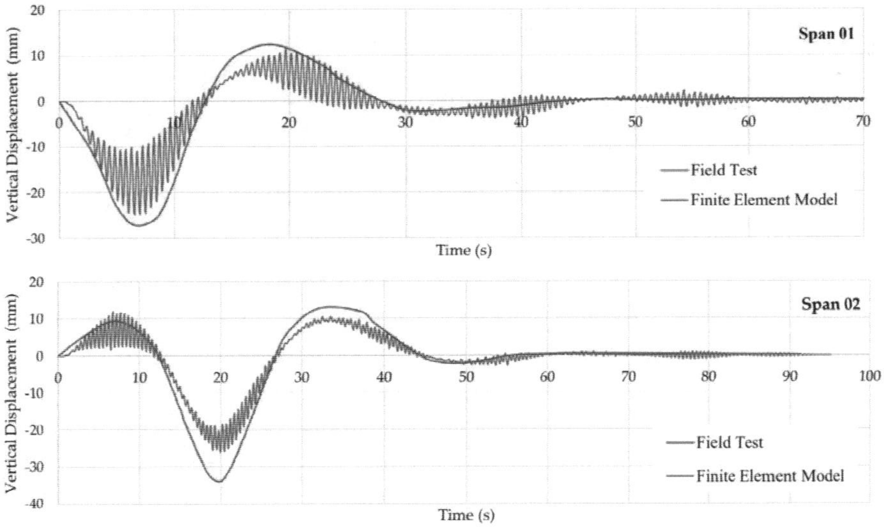


**Figure 4.** Dynamic response of the bridge for the first two spans at a speed of 30 km/h.

Figure 5 from the FE model shows the variation of vertical acceleration at mid-span 1 under a moving load speed of 90 km/h. The red, green, and blue lines indicate the location of the moving load as it crosses Span 1, 2, and 3. The Figure shows significant vibration as the truck traverses Span 1. The vibration attenuates as the truck moves on to Spans 2 and 3. All dynamic load analyses revealed such characteristics, illustrating a practical result from moving loads.

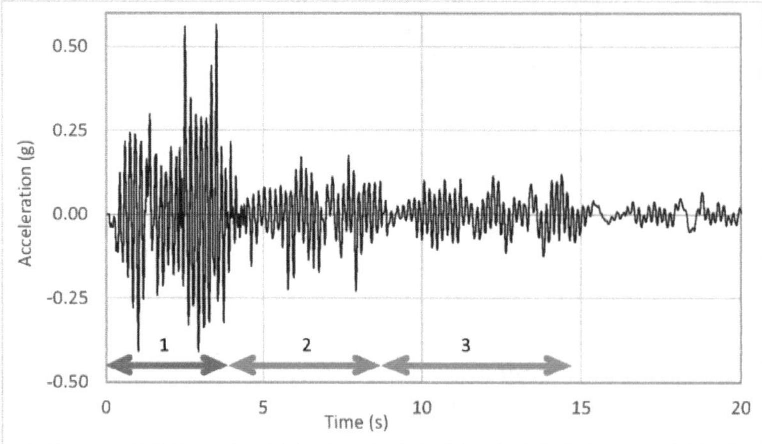


**Figure 5.** Vertical acceleration for the first span at a speed of 90 km/h.

The impact factors were also determined during the dynamic test. These factors are affected by the span's free length, the type of bridge elements, the material properties, boundary conditions, the nature of loading, and the type of response [8]. Vertical displacements at the mid-spans provided the basis for computing the impact factor. The ratio of maximum dynamic displacements ($\delta_{dyn,max}$) at different speeds to the maximum pseudo-static displacement ($\delta_{stat,max}$) of 10 km/h speed at the same points appears in Equation 2.

$$\delta = \frac{\delta_{dyn,max}}{\delta_{stat,max}} \tag{2}$$

Figure 6 summarizes the results showing that the impact factors generally rise with speed and achieve a value of 1.21 in the field test and 0.98 in the FE model, for speeds up to 90 km/h. The maximum values compare well to the recommended maximum permitted value of 1.7 by EUROCODE 1 [10]. The reference values agree with the loading test results.

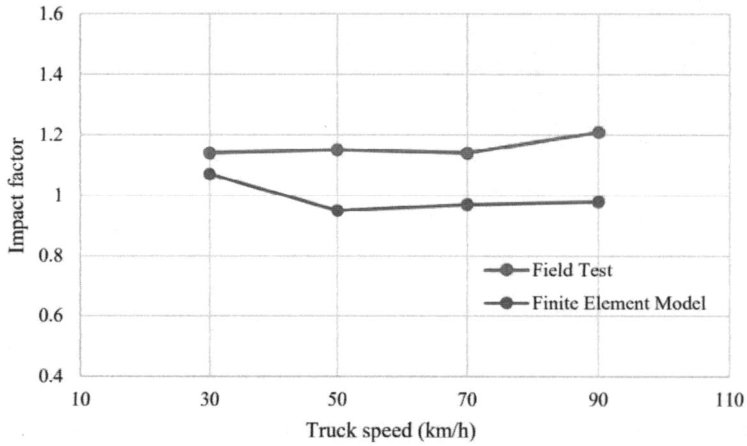


**Figure 6.** Impact factors as a function of truck speed.

## 5. Evaluation of the load-carrying capacity

The loading tests achieved their primary objective by obtaining a reliable estimation of the bridge capacity. Load rating factors helped to evaluate the load capacity of the bridge.

The factors usually represent the proportion of live load capacity and live load demand shown in Equation 3, based on the stress design formula [5]:

$$RF = \frac{\sigma_{all} - \sigma_{DL}}{\delta \sigma_{LL}} \quad (3)$$

Here: RF, equals load rating factor; $\sigma_{all}$, admissible stress of the structural material based on design code and test; $\sigma_{DL}$, stress under dead load; $\sigma_{LL}$, stress under live load, and finally δ, impact factor presented in Figure 6.

The stresses induced by truck loads in both field tests and simulations were utilized to calculate the corresponding RF, according to Equation (3). Table 2 shows that the bridge has enough capacity and the RF produced from the FE model tends to be more conservative. The standard load-carrying capacities determined using the FE model exceed the load-carrying capacities based on the measurement data, especially for higher speeds. Higher capacity means the FE model exhibits more stiffness than the measured structure. The difference may originate from many factors, such as different material stiffness in the field or differences in component cross-sections not considered during the analysis of new bridge.

**Table 2.** Bridge rating factors as a function of truck speed.

| **Speed (km/h)** | **30** | **50** | **70** | **90** |
|---|---|---|---|---|
| **RF Field Test** | 13.49 | 12.50 | 9.98 | 9.07 |
| **RF FEM** | 13.92 | 21.27 | 20.06 | 20.10 |

## 6. Conclusion

The calibrated FE model results capture the bridge's actual state after comparing it to the proof test data. Vertical displacement analysis yields impact factors within the design limits of a structure behaving elastically. Based on these findings, the bridge structure is functioning as designed and can be open to traffic action.

FE model rating factors indicate that the bridge structure may have greater strength than that predicted by proof testing. Some of the FE model assessments do not precisely agree with measured behavior. The model overestimates the rating factors for higher speeds compared to the proof load assessment, so it may sometimes not provide exact information relating to a structure's 'as built' behavior.

The proof-load tests verified the actual behavior of the bridge structure for safe serviceability. Thus, the collected data during the load testing will provide information for future monitoring and help to detect any loss of capacity or service capability.

## 7. References

[1] e-UT 09.02.34 2020 *Dynamic load capacity measurement: Analysis of measurement results* Hungarian roads association, (in Hungarian)

[2] Lantsoght E, Van Der Veen C, Hordijk D A, De Boer A 2017 *Development of recommendations for proof load testing of reinforced concrete slab bridges* Engineering Structures, vol. 152, pp. 202–210, doi: 10.1016/j.engstruct.2017.09.018.

[3] Schacht G, Krontal M, Marx S, Bolle G, Wismar H 2011 *Loading Tests of Existing Concrete Structures-Historical Development and Present Practise* in fib Symposium, Prague.

[4] Vavruš M, Bujňák J, Koteš P 2019 *Experimental verification of real behavior of bridge structures using proof-load tests* Pollack Period., vol. 14, no. 1, pp. 75–84, doi: 10.1556/606.2019.14.1.8.

[5] Marefat M S, Ghahremani-Gargary E, Ataei S 2004 *Load test of a plain concrete arch railway*

*bridge of 20-m span* Constr. Build. Mater., vol. 18, no. 9, pp. 661–667, doi: 10.1016/j.conbuildmat.2004.04.025.

[6] AASHTO 2021 *The Manual for Bridge Evaluation* American Association of State Highway and Transportation Officials, 3rd Edition, November.

[7] e-UT 08.01.61 2002 *Bridge testing I. Load capacity testing of built road bridges* Hungarian roads association, (in Hungarian)

[8] Brinissat M, Ray R P, Kuti R 2023 *Evaluation of the Szapáry Long-Span Box Girder Bridge Using Static and Dynamic Load Tests* Infrastructures, 8(5), 91, doi: 10.3390/ infrastructures8050091.

[9] Kim H J, Kim H K, Park J Y 2013 *Reliability-based evaluation of load carrying capacity for a composite box girder bridge* KSCE J. Civ. Eng., vol. 17, no. 3, pp. 575–583, doi: 10.1007/s12205-013-0603-7, pp. 833–854.

[10] EN 1991-3 2003 *Eurocode 1: Basis of design and actions on structures : Part 2 Traffic loads on bridges*.

IOP Conf. Series: Materials Science and Engineering 1304 (2024) 012007 doi:10.1088/1757-899X/1304/1/012007

# Comparative analysis of two strengthening solutions for the structural rehabilitation of a historical masonry building

**G Taranu[1*], I Olteanu[2], I-S Entuc[3], O-E Colt[4], G Bunea[5], I-O Toma[6]**
[1*, 2, 3, 4, 5, 6] The "Gheorghe Asachi" Technical University of Iasi

Corresponding author: george.taranu@academic.tuiasi.ro

**Abstract.** The paper presents a comparative analysis based on the Finite Element Method (FEM) for a masonry building located in Iasi, the North-Eastern side of Romania. This specific region is known to be found in a seismic area having a design peak ground acceleration of 0.25g. The building was assessed as being in the 1st seismic risk class, according to the Romanian norms, with a high probability of collapse in case of future earthquakes. Being a historical monument, the structural strengthening solutions are very limited. The study consists of modal and seismic analyses for the masonry structure. A comparison of the advantages and disadvantages of two approaches of structural rehabilitation solutions is done. The structural strengthening solution "A" consists of the addition of reinforced concrete thin wall elements, while the other - called solution "B" - consists of an interior steel profile skeleton mechanically and chemically connected to the adjacent masonry walls.

## 1. Introduction

For centuries, masonry buildings represent an important problem to be investigated by civil engineers. The number of masonry buildings is high due to the abundance and accessible manufacturing process of the main used materials, like bricks and mortars. Since Roman times, we have had bricks and different types of mortars, which made possible the construction of the greatest buildings in history, e.g. the Pantheon, Roman aqueducts, Hagia Sophia a.o. In Europe, including Romania, masonry was the main building material for an important period of time. A lot of public buildings and/or individual homes were made using brick masonry. Considering the limited regulation for these types of buildings in seismic areas, the intermittent occurrence of medium and large earthquakes, at a recurrence interval of 20 or 50 years has produced a lot of damage to structural systems.

Some of the buildings with a poor structural system have already collapsed due to the inadequate design of the masonry walls and floors to horizontal actions. The problem derived from poor structural engineering knowledge at that time. In Romania for example, in the last 50 years, seismic action was better approached, so today civil engineers have a higher degree of knowledge about masonry structural behaviour to seismic action. In the last decades, European Norms were adopted and local annexes were elaborated for masonry structures [1,2], seismic actions [3,4] and for the investigation and strengthening processes [5]. The preservation/conservation of some historical buildings is important because the city's value is given by the originality and integrity of these old constructions.

To identify the optimal and most appropriate solution for strengthening masonry, appropriate analyses are required. At the same time, numerical analyses of masonry structures are a complex problem, due to the special mechanical characteristics of the masonry. Being a laborious process, as it is presented in literature [6–13] and not accessible to all engineers, the regulations provide analytical instruments utilized to assess the seismic vulnerability of the buildings. Moreover, they give recommendations to improve the structural behaviour of the masonry building including numerous types of strengthening solutions, e.g. reinforced concrete elements, injected mortars, composite materials or adding steel profiles. The engineers have to evaluate and propose a balanced solution based on their experience [14–18], the requirements of the regulations and the client's possibilities. Starting from previous experiences with experimental research [19–22] and numerical analyses, the authors investigated a historical

IOP Conf. Series: Materials Science and Engineering 1304 (2024) 012007 doi:10.1088/1757-899X/1304/1/012007

masonry building located in Iasi, in North – Eastern part of Romania, almost 200 km from the Vrancea's epicentre seismic area – the origin of the most important Romanian earthquakes (figure 1).

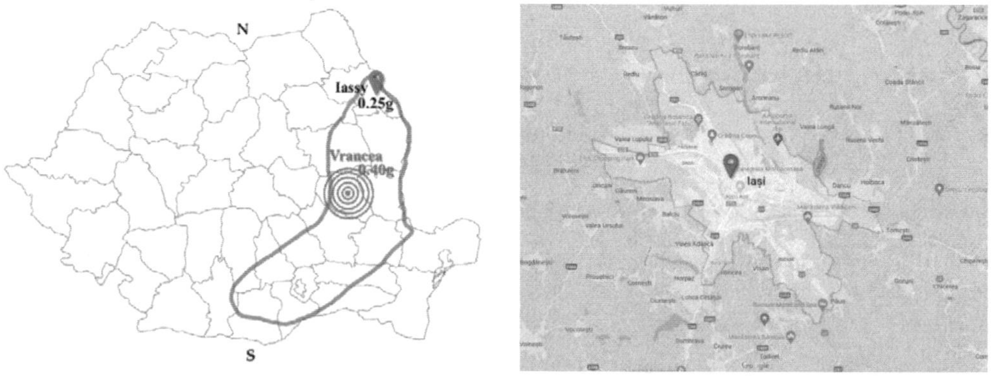


a b

**Figure 1.** Location of the building: a) Vrancea seismic area and Iasi location in N-E of Romania; b) Building location in the city centre

**Figure 2**. Aerial view of the building

The building has over 100 years since its construction. In the last decade, it was not used. In the South and North parts, there are other similar buildings, but their period of construction is different. The building from the South part was strengthened 15 years ago with reinforced concrete elements. The studied building named C22 – C24 has multiple visible damages observed on the façades and interior. Therefore, immediate seismic risk assessment is required.

## 2. Materials and structural elements integrity

Geotechnical drillings and diggings had to be made with the purpose of gathering all the necessary information regarding site's soil characteristics and the existent foundation system. Following the laboratory determination, the soil stratification: contains two layers: from 0 to -3.20 m - a filling soil layer, and from -3.20m to -10.00m – a yellow dusty clay/silty clay, plastic to consistent, moisture sensitive. The maximum value of the plastic pressure is $p_{pl}$=245kPa.

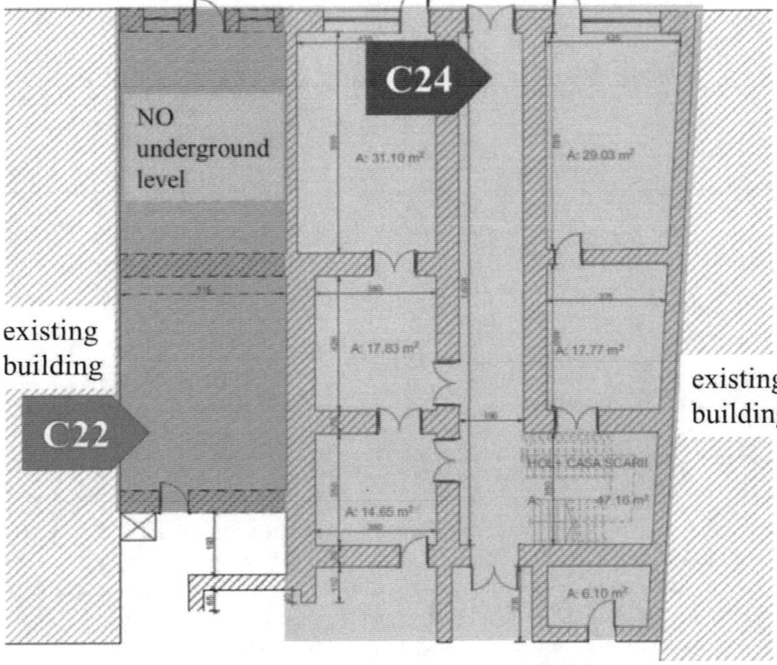


**Figure 3**. Building architectural plan – underground level

**Figure 4**. Front view of the building

The C24 building presented in Figures 3 and 4, has an underground level, with a foundation system made of stone masonry and a maximum foundation depth of -3.90 m. The C22 building has no underground level, the existing foundation, in this case, is constructed under the minimum permissible value of approximately 60 cm and is composed of stone masonry and a wood beam. In Table 1 the main structural elements are presented with their characteristics.

**Table 1.** Structural elements and their materials [22–24]

| Structural element | Material | Modulus of elasticity E (MPa) | Compressive strength $f_c$ (MPa) |
|---|---|---|---|
| Foundation | Stone masonry with sand-lime mortar | 3000 | 3 |
| Wall | Ceramic brick masonry with sand-lime mortar | 1000 | 1 |
| 1st Floor | Ceramic brick masonry vaults | 2000 | 2 |
| 2nd Floor | Wooden beams | 9000 | 15 |
| Roof | Wooden beams | 9000 | 15 |

In the evaluation process of the seismic risk factor for the building, a series of visual and quantitative assessments have been made. The structural conformation of the building has 90 cm thick walls at the underground level and a floor made of masonry vaults, which rest on I-shaped steel profile beams, as presented in figure 5. At the first level, the thickness of the walls is 70 cm and 50 cm, and the above slab is made of wood beams. On the second level, the masonry walls have a thickness of 50 cm with the slab made of wooden beams. The roof of the building is made up entirely of timber beams.

a b

**Figure 5**. a) Underground level of the building; b) upper building masonry wall

*2.1 Degradations of the building*

In the visual inspection, numerous types of damages were observed, as follows: water infiltration, vertical and inclined cracks, steel deep corrosion of the beams on top of the underground level, areas with missing plaster, wood elements deformations. Figure 6 presents different types of damages observed at the first and second floor of the building.

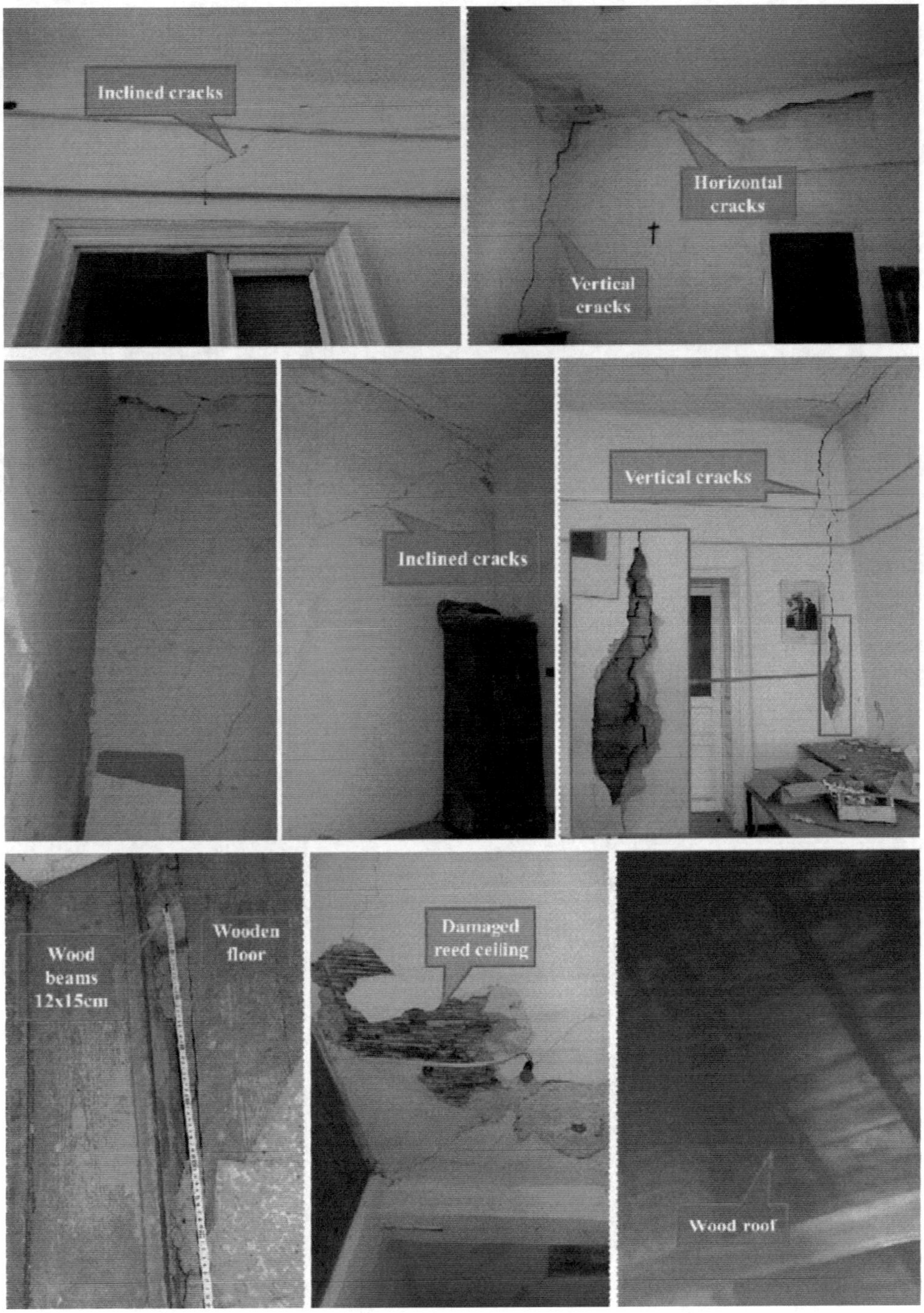


**Figure 6**. Different types of structural damages and their visual aspect

*2.2 Seismic risk evaluation*

As per the Romanian seismic assessment code of existing buildings, [5], the score for each seismic factor was computed considering a "limited knowledge" of the structure. The mechanical properties of the materials correspond to a trust factor of 1.35. The evaluation sought to identify the highly vulnerable areas of the structure and to verify the criteria regarding the requirements for stability, strength, rigidity and ductility.

The value of the degree of fulfilment of the seismic composition conditions, $R_1$ from 1 to 100, is established based on the score assigned to each category of composition conditions. In the case of the analysed structure, the score was 28, corresponding to a high seismic risk noted $R_{sI}$. In the case of $R_2$, the degree of structural integrity allocation is established based on the score assigned to each category of conditions, regarding the assessment of the degradation state for the structural elements. The factor range is between 1 and 100. The obtained score for the studied building was 30 which corresponds also to a high seismic risk $R_{sI}$.

The degree of structural seismic response $R_3$ characterises the strength and ductility of the structure, as a whole, and the strength and stability of the non-structural components, in relation to the seismic requirements for the present and future. For the evaluation of the $R_3$ index, a linear numerical model of the structure was analysed, which led to a low value of the factor of 18% from 100%. That means the analysed structure does not have the sufficient structural capacity to withstand a future earthquake and a strengthening intervention is absolutely necessary. In this case, two solutions were proposed and they were compared based on linear numerical analyses.

## 3. Numerical modelling of the building

Generally, the linear modelling of masonry structures with 2D shell elements does not provide accurate results along the thickness of the walls but in the case of historical monuments, this information sometimes is necessary. However, when a global analysis is required to be conducted for obtaining a general tendency and stress distribution, with some hypotheses and assumptions, a linear numerical model with shell finite elements for masonry walls can represent a good instrument. In this paper, the structural system of the building, presented in figure 7, was modelled using 2D finite elements in Robot Structural Analysis software [25]. First, the actual state of the structure was considered, with wood elements added as linear finite elements, and masonry walls - as 2D shell elements, Figure 7.

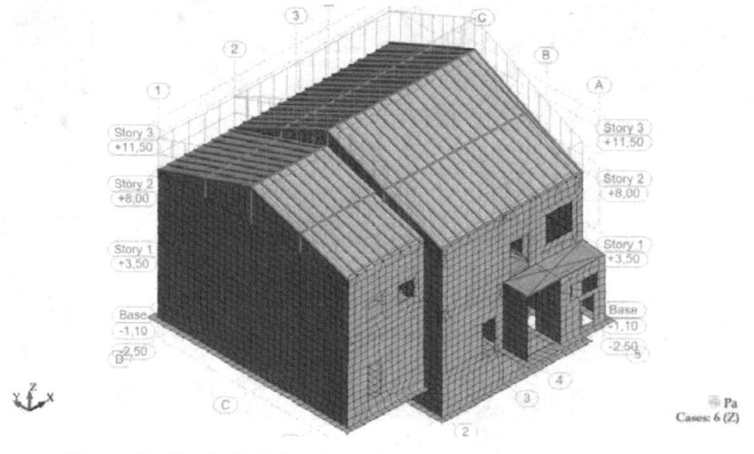


**Figure 7**. 2D shell finite element model of the existent building

The load cases and load values selected according to Romanian design codes are presented in Table 2. These were combined also according to CR 0-2012 Romanian code [26]. For the seismic analyses, a peak ground acceleration of 0.25g was considered, with a corner period of 0.7s. The structural behaviour factor $q$ was selected to be equal to 1, for a low dissipative masonry structure.

**Table 2.** Load cases

| Case | Name | Uniformly surface-distributed load q (kPa) | According to code |
|---|---|---|---|
| 1 | Dead weight | Automated | EC 1 [27] |
| 2 | Dead load | 2 | EC 1 [27] |
| 3 | Live Load | 2 | EC 1 [27] |
| 4 | Wind | 0.6 | CR 1-1-4/2012 [28] |
| 5 | Snow | 2.4 | CR 1-1-3/2012 [29] |
| 6 | Modal analysis | - | |
| 7 | Seismic X direction | - | P100-1/2013 [4] |
| 8 | Seismic Y direction | - | |
| 9 | Seismic Z direction | - | |

Figure 8 and 9 present the schematic arrangement of the 2D finite element models of the strengthened structure. In the first solution called "A" (figure 8), additional reinforced concrete elements were added in the vertical and horizontal direction, with the purpose of achieving a spatial skeleton of RC elements bonded by masonry walls. In case "B" (figure 9), the solution consists in adding an interior steel skeleton made of "U" and "IPE" profiles. The vertical steel profiles are linked to each other on both sides of the walls with threaded metal rods, which assure the tightening and best contact with the existing structure. In order to achieve a horizontal diaphragm, a composite slab was proposed, made of steel profiles and reinforced concrete.

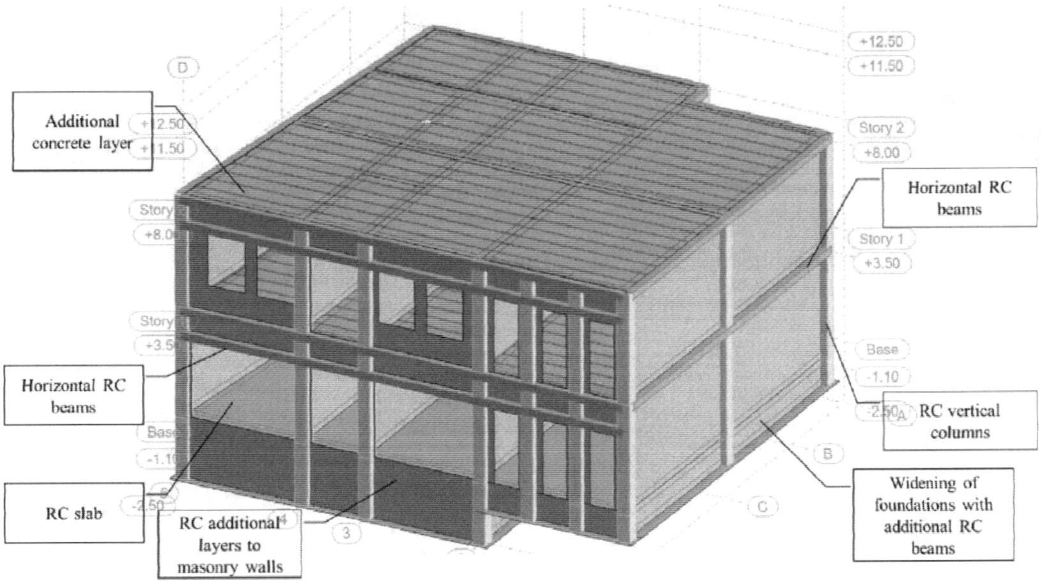


**Figure 8.** 2D finite element models for the strengthened structure using solution A with RC elements

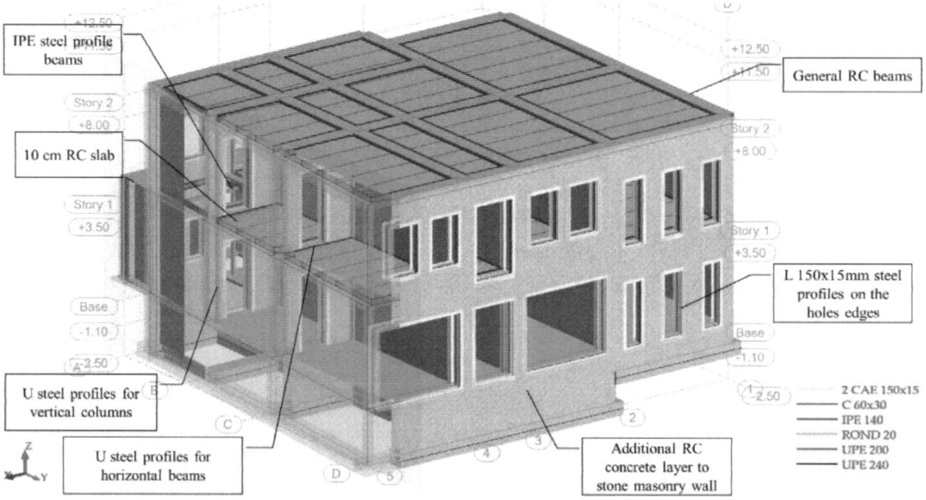


**Figure 9.** 2D finite element models for the strengthened structure using solution B with steel profiles elements

## 4. Results and discussion

From the numerical analyses, the following results were observed and compared: modal analysis results, the pressure on the foundation soil, stress map distribution and deformations. From the modal analyses, the dynamic characteristics are presented in table 3.

**Table 3.** Modal analysis results

| Model/Case | Frequency (Hz) | Period (sec) | X direction mass participation (%) | Y direction mass participation (%) | Total mass (t) |
|---|---|---|---|---|---|
| Initial-1 | **2.59** | **0.38** | **64.65** | **0.14** | **2186.73** |
| Initial-2 | **4.19** | **0.24** | **0.12** | **79.69** | |
| A-1 | **3.015** | **0.33** | **73.01** | **0.37** | **2573.15** |
| A-2 | **3.96** | **0.25** | **0.31** | **82.05** | |
| B-1 | **2.85** | **0.35** | **73.73** | **0.003** | **2342.43** |
| B-2 | **3.70** | **0.27** | **0.05** | **79.89** | |

Figures 10 present the first vibration mode for each analysed model. A small difference of the fundamental frequency it was observed. Figures 11-16 present results with map stress distribution for each scenario.

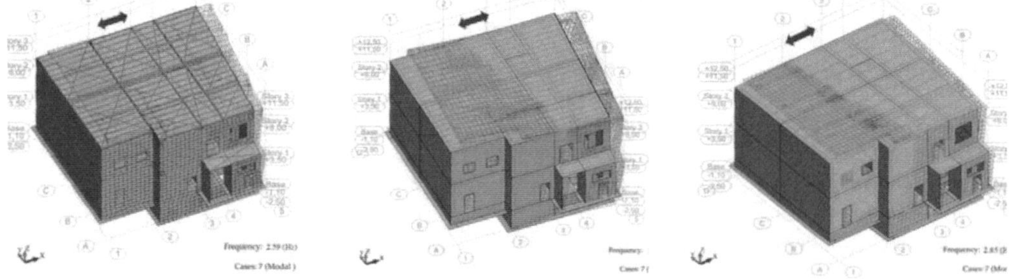

a b c

**Figure 10**. First vibration mode: a) un-strengthened initial model; b) solution A model; c) solution B model

IOP Conf. Series: Materials Science and Engineering 1304 (2024) 012007 doi:10.1088/1757-899X/1304/1/012007

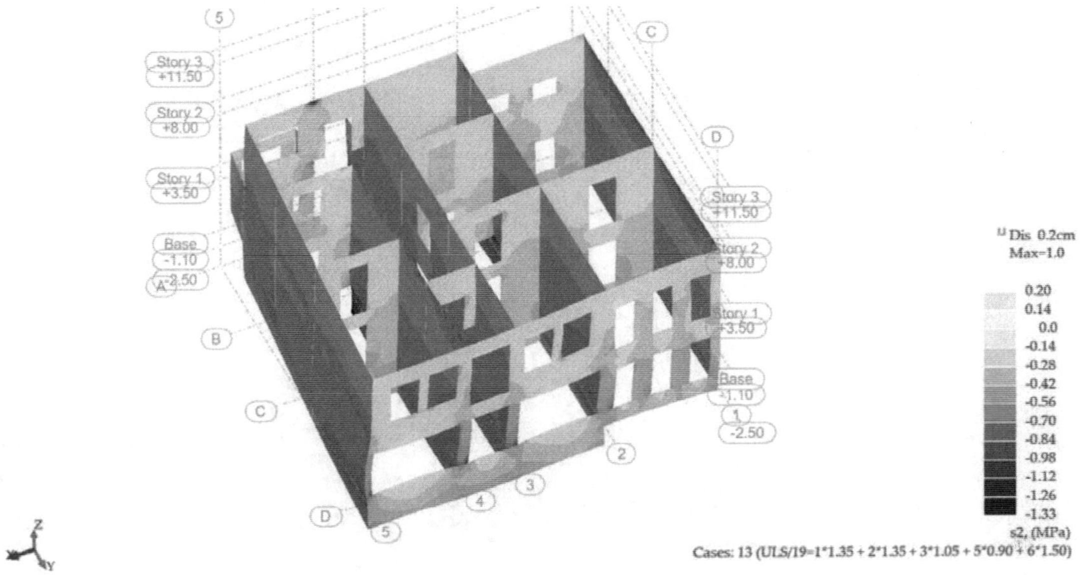


**Figure 11**. The initial structural model: - normal stress distribution with maximum value from gravitational combination;

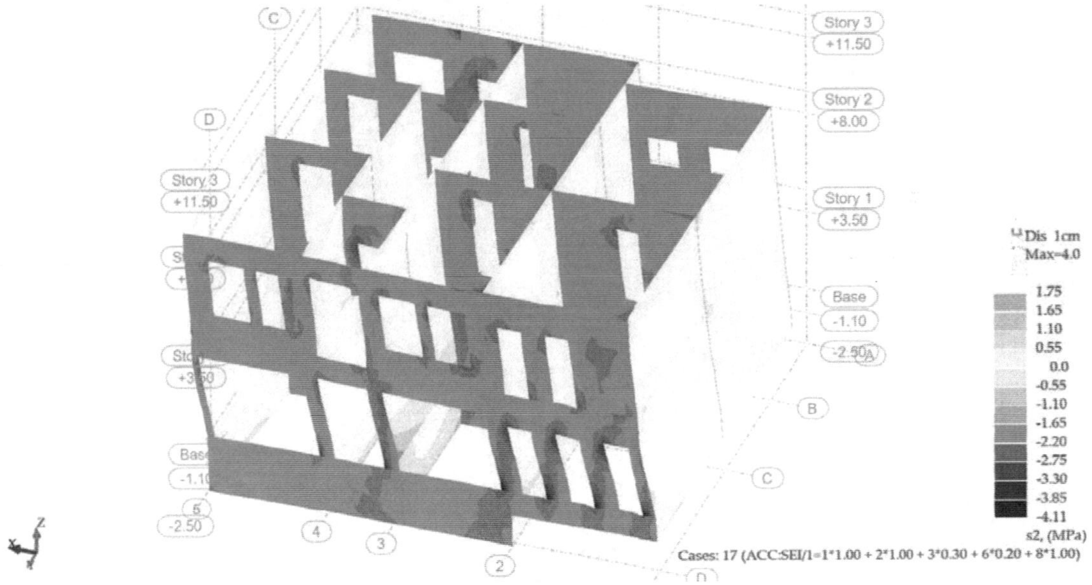


**Figure 12**. The initial structural model: - normal stress distribution with maximum value from seismic combination;

IOP Conf. Series: Materials Science and Engineering 1304 (2024) 012007 doi:10.1088/1757-899X/1304/1/012007

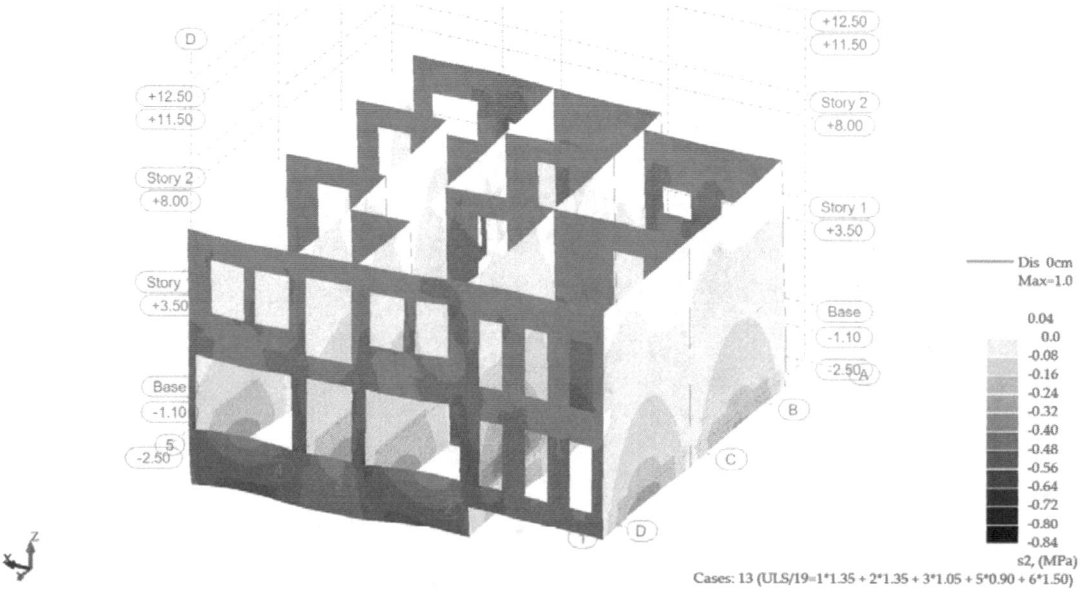


**Figure 13**. The structural numerical model strengthened with solution "A": - normal stress distribution with maximum value from gravitational combination

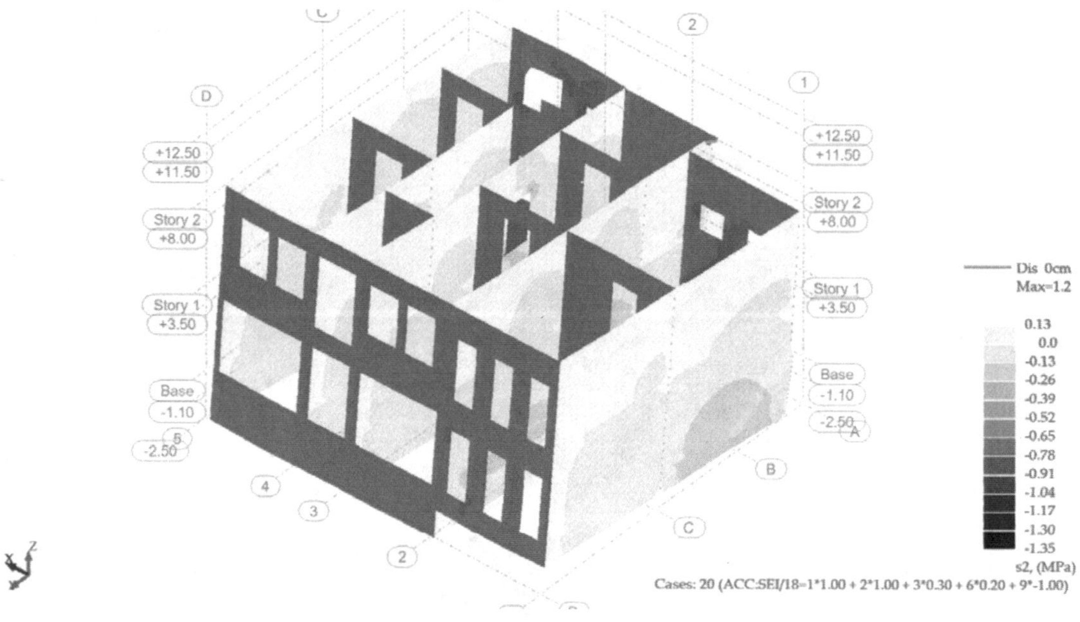


**Figure 14**. The structural numerical model strengthened with solution "A": - normal stress distribution with maximum value from seismic combination;

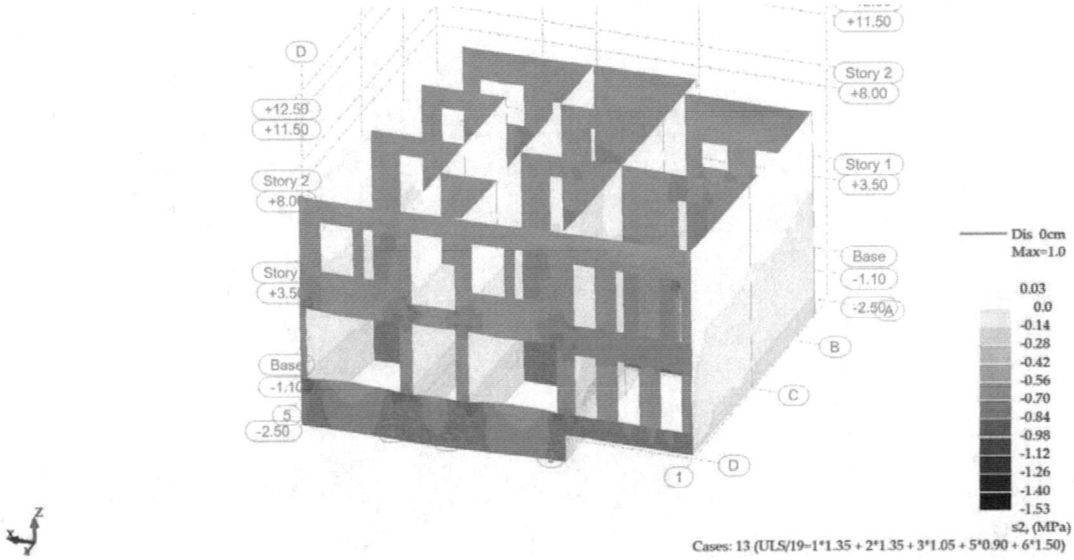


**Figure 15**. The numerical structural model strengthened with solution "B": - normal stress distribution with maximum value from gravitational combination

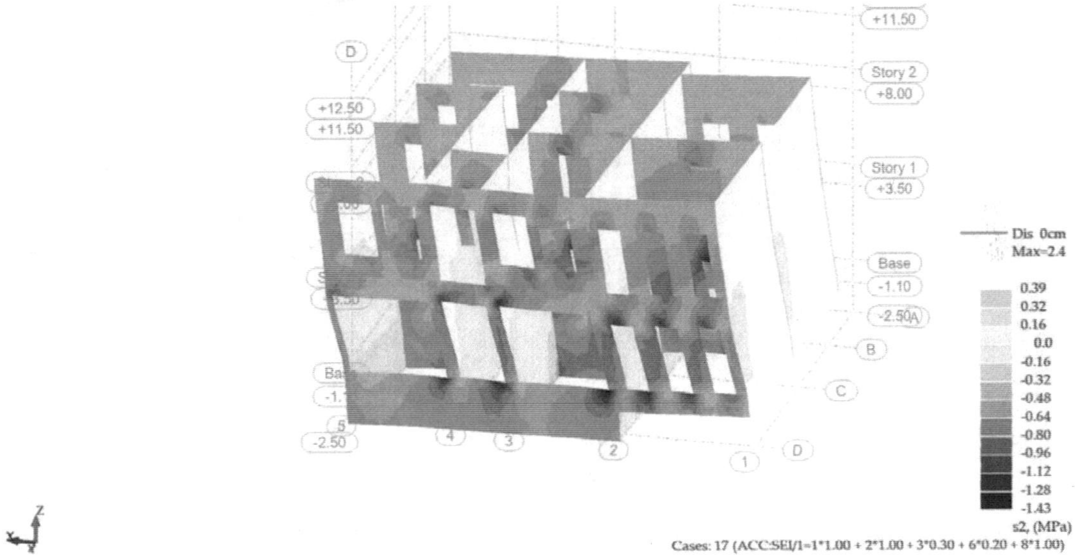


**Figure 16**. The numerical structural model strengthened with solution "B": - normal stress distribution with maximum value from seismic combination

In figure 17 it is shown a comparative graph between maximum values of normal stress for each analysed scenario. Figure 18 presents a comparative graph between maximum absolute displacements in seismic combination.

IOP Conf. Series: Materials Science and Engineering 1304 (2024) 012007 doi:10.1088/1757-899X/1304/1/012007

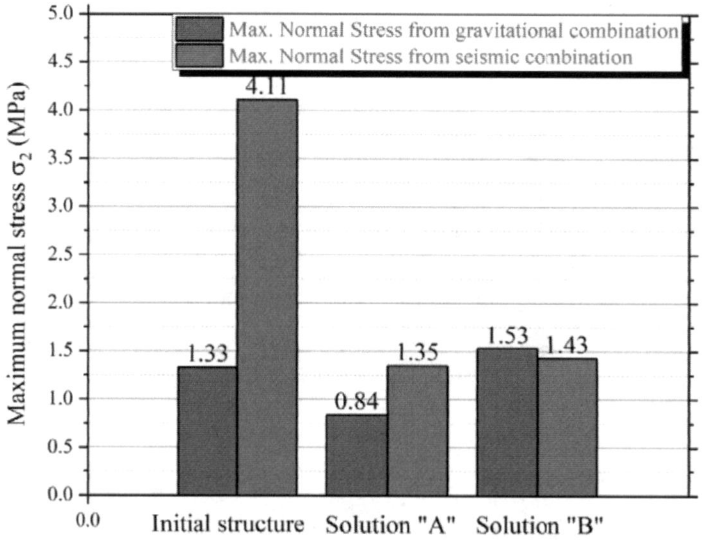


**Figure 17**. Comparative graph: - maximum normal stress;

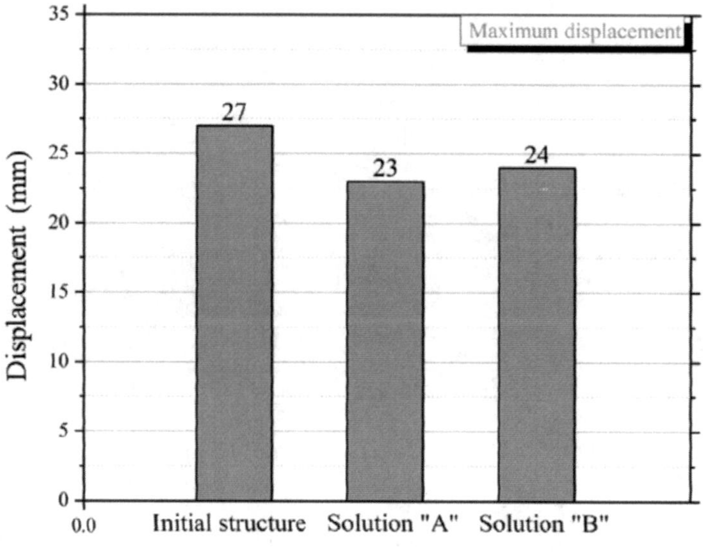


**Figure 18**. Comparative graph: - maximum displacement in seismic combination SLS

## 5. Conclusions

The paper presents a comparative numerical analysis of a historical masonry structure class B, located in Iasi - a city from N-E Romania, characterised by a design peak ground acceleration of 0.25g according to the current seismic design norm. The initial existing structure has a high seismic risk to collapse under a seismic action having a peak ground acceleration of around 0.25g being in seismic risk class I. Therefore, a strengthening solution is required. With the purpose of analysing the general behaviour of the two strengthening solutions of the structure under different loadings and the corresponding stress distribution, a linear numerical model with 2D shell elements was proposed. It was used for selecting the suitable structural rehabilitation solution after comparison.

Comparing the numerical analysis results, it was observed that in both proposed solutions, the maximum normal stress obtained using the seismic combination decreased by 287% compared to the initial state of the existing building, so the strengthened building will be in seismic risk class III. From the economic and technical point of view, solution "A" is more accessible to local entrepreneurs, but it takes more time and more labour. However, it has the advantage of higher fire and corrosion resistance. On the other hand, the second solution with steel profiles needs a smaller amount of time to be completed and it is demountable. Thus, the original structure will be better preserved and will not be affected by additional work. The main disadvantages of the steel profile solution are corrosion risk and relatively small fire resistance. Besides these, workers must have special certified technical skills in order to mount the structure. From the structural point of view, both solutions are appropriate, but future economic analyses are necessary in order to choose the most suitable one for the client. This aspect is part of future work and was not analysed in the current paper.

## 6. References

[1] European Comittee for Standardization 2005 *EN 1996-1-1: Eurocode 6: Design of masonry structures - Part 1-1: General rules for reinforced and unreinforced masonry structures*

[2] M.D.R.A.P 2013 Masonry structures design code - CR 6 - 2013

[3] European Comittee for Standardization 2004 *EN 1998-1: Eurocode 8: Design of structures for earthquake resistance – Part 1: General rules, seismic actions and rules for buildings*

[4] M.D.L.P.A Romania 2013 *P100-1/2013: Seismic design code - Part I: Design rules for buildings (in Romanian)*

[5] M.D.R.A.P 2019 Code for earthquake design P100-3/2019 Provisions for the seismic assessment of existing buildings

[6] Lourenço P B 2002 Computations on historic masonry structures *Progress in Structural Engineering and Materials* **4** 301–19

[7] Pleşu R, Taranu G and Budescu M 2013 FEM analyses of a 1:2 scale historical unreinforced masonry building *Computational Civil Engineering – CCE2013, International Symposium, Iasi, Romania*

[8] Dumitraş M, Cobîrzan N and Dumitraş D 2013 Considerations Regarding the Analysis of Masonry Walls to Horizontal Loads Using Equivalent Frame Method Building *Bulletin of the Polytechnic Institute of Jassy, Construction. Architecture Section* **LIX**

[9] Soveja L, Budescu M and Gosav I 2013 Modelling Methods for Unreinforced Masonry Structures *Bulletin of the Polytechnic Institute of Jassy, Construction. Architecture Section* **LIX**

[10] Brando G, Vacca G, di Michele F, Capasso I and Spacone E 2022 Experimental and Numerical Mechanical Characterization of Unreinforced and Reinforced Masonry Elements with Weak Air Lime Mortar Joints *Sustainability* **14**

[11] Asteris P G, Chronopoulos M P, Chrysostomou C Z, Varum H, Plevris V, Kyriakides N and Silva V 2014 Seismic vulnerability assessment of historical masonry structural systems *Eng Struct* **62–63** 118–34

[12] Venghiac V M, Olariu C P and Budescu M 2017 Structural Rehabilitation Analyses for a Romanian Cultural Heritage Building Located in Seismic Area *Advanced Engineering Forum* **21** 196–203

[13] Giordano A, Mele E and de Luca A 2002 Modelling of historical masonry structures: comparison of different approaches through a case study *Eng Struct* **24** 1057–69

[14] Budescu M, Taranu G, Plesu G and Taranu N 2010 An innovative hybrid solution to improve the seismic behaviour of a shear wall building module *Proceedings of the 3rd International Conference on Advanced Materials and Systems, ICAMS 2010*

[15] Ţăranu G, Lungu I, Ţăranu N and Budescu M 2013 Mechanical characteristics of glass fibre reinforced composites with cement and recycled anhydrite matrix *Revista Romana de Materiale/ Romanian Journal of Materials* **43**

[16] Taranu N, Oprisan G, Budescu M, Taranu G and Bejan L 2010 Improving structural response of masonry vaults strengthened with polymeric textile composite strips *International Conference on Engineering Mechanics, Structures, Engineering Geology, International Conference on Geography and Geology - Proceedings*
[17] Taranu N, Oprisan G, Budescu M, Taranu G, Munteanu V and Bejan L 2010 Strengthening solutions based on textile composite for masonry structures *Proceedings of the 3rd International Conference on Advanced Materials and Systems, ICAMS 2010*
[18] Venghiac V-M, Neagu C-P, Taranu G and Rotaru A 2023 Time History Analyses of a Masonry Structure for a Sustainable Technical Assessment According to Romanian Design Codes *Sustainability 2023, Vol. 15*
[19] Pleşu R, Ţăranu G, Covatariu D and Grădinariu I-D 2011 Strengthening and Rehabilitation Conventional Methods for Masonry Structures *Bulletin of the Polytechnic Institute of Jassy, Construction. Architecture Section* **LIV**
[20] Taranu G, Toma I-O and Taranu N 2022 Shake table test of a 1:2 scale historical unreinforced masonry building *1st Croatian Conference on Earthquake Engineering, 1 CroCEE, Zagreb, Croatia*
[21] Raluca Pleşu 2013 *Ph.D Thesis: Contributii privind metodele de reabilitare a structurilor din zidarie/ Contributions regarding the rehabilitation methods of masonry structures* (Iasi: The "Gheorghe Asachi" Technical University of Iasi, Romania)
[22] George Taranu 2021 *Strengthening of Masonry Arches and Vaults at Historical Monuments/Consolidarea Arcelor si Boltilor din Zidarie la Monumentele Istorice* ed Carmen Loghin (Iasi: Politehnium)
[23] Covatariu D, Toma I-O and Budescu M 2011 Experimental investigation on bonding characteristics of low-strength mortars used to repoint the joints of the damaged historical masonry structures *Bulletin of the Polytechnic Institute of Jassy, Construction. Architecture Section* **LVII**
[24] Covatariu D, Ţaran R-G and Toma I-O 2013 The strengthening of the damaged historic masonries by using specials mortars *Bulletin of the Polytechnic Institute of Jassy, Construction. Architecture Section* **LIX**
[25] Autodesk 2022 Robot Structural Analysis Professional
[26] M.D.L.P.A 2012 *CR 0-2012: Cod de proiectare. Bazele proiectarii constructiilor/ Romanian design code. The basics of construction design*
[27] European Committee for Standardization 2002 Eurocode 1: Actions on structures - Part 1-1: General actions - Densities, self-weight, imposed loads for buildings EN 1991-1-1
[28] M.D.L.P.A 2012 CR 1-1-4/2012: Cod de proiectare. Evaluarea actiunii vantului asupra constructiilor/ Romanian Design code. The evaluation of wind loads on structures
[29] M.D.L.P.A 2012 CR 1-1-3/2012: Cod de proiectare. Evaluarea actiunii zapezii asupra constructiilor/ Romanian design code. The evaluation of snow loadings on structures

# Modelling and experimental analysis of high-temperature effect on rubberized concrete

**D Covatariu[1], M S Alexa-Stratulat[1], G Covatariu[1]**
[1]The "Gheorghe Asachi" Technical University of Iasi, Faculty of Civil Engineering and Building Services, no. 1 Prof.dr.doc. D. Mangeron Str., 700050, Iasi, *Romania*

E-mail: daniel.covatariu@academic.tuiasi.ro

**Abstract.** The study presents results concerning the comparison of modelling and experimental work on rubberized concrete exposed to high temperatures. Crumb rubber was used in different concentrations to replace coarse aggregate. Experiments were made on flat-cylindrical specimens extracted from larger samples in order to be analysed. Simulations have been employed to calibrate measurements and to further analyse the specimens in an effort to prove their worth as a viable mechanism in studying larger-scale phenomena.

## 1. Introduction

The road transportations development has generated an increasing number of tyres; the statistics (in countries with heavy traffic) indicate that a number of tyres equal to the number of the population is used/replaced annually. In Europe, annually, are produced about three million tyres and 70% of them are discarded when no longer used. Thus, the number of used tires increases over time generating a serious negative influence on the environment.

One of the environmental sustainability's principles frequently claimed in Civil Engineering domain is the reutilisations of the construction waste in the same domain. It could prevent the environmental pollution and to contribute to the building's rising with lower costs. The use of crumbed rubber grains as an aggregates replacement in concrete has an important subject of research about 20 years ago [1]. Many countries avoid storing or disposing of used tyres waste in deposits areas, providing an important stimulant to seek new manners to reuse [2]. The impact of using recycled rubber as a partial replacement of the aggregate in concrete's structure has long been documented by compression, tension and bending tests [3, 4, 5, 6].

The fire test involves the fire resistance of the structural components and the reaction to fire of flammable components [7]. This could be important in the case of different ignition of stone aggregates and replacement of rubber aggregates (this paper shows).

Testing a new fire resistance material can be expensive (as a configuration for large-sized specimens). That's why validating a fire test model becomes important to be considered for future research projects.

## 2. Materials and methods

In order to perform the mechanical tests, reference concrete specimens (C30/37 class) were cast by using CEM I/A-LL 42.5 R cement, sand fine aggregates (0-4 mm - 582 kg/m$^3$ and 4-8 mm - 388 kg/m$^3$), coarse aggregates (8-16 mm - 647 kg/m$^3$) and plain water. The specimens were cast in cylindrical shapes, having 100 mm diameter and a height of 200 mm, according to SR EN 12390-1/2013 (Figure 1a).

Subsequently, have been casted new specimens, in which concrete's initial composition was modified by replacing 40%, 60%, respectively 80% of the total volume of fine aggregate 0-4 mm with rubber granules. The density of the rubber granules was similar to those of the replaced fine aggregate (sort 0-4 mm).

For high temperature testing, flat-cylindrically specimens were taken from initial samples (by mechanical cutting), having 30 mm heigh and 100 mm diameter (Figure 1b).

**Figure 1.** Cylinder-shape concrete specimens a) initial and b) cut samples

The study regarding the high-temperature behaviour of rubberized concrete specimens consisted of two main parts. In the first instance, a high-temperature exposure experiment was performed on all specimens. The next part consisted of simulations that were firstly focused on calibration (based on experimental data) and then used for comparing the behaviour of the specimens.

### *2.1. Experimental assessment*

The experiment was performed using a metal stand, housing the heat source and the specimen, as sketched in figure 2. Heat was produced by an electrical air-heating gun, deemed capable of reaching temperatures of maximum 550ºC. The distance between the specimen and the gun was set to 1 cm. Temperature was measured by means of a FLIR BCam thermal imaging camera, located far enough as for it not to be affected by heat. The experiment was started once the temperature provided by the heat source was measured to be constant. This assessment was done using a K-type thermocouple positioned in the proximity of the air-gun's exit.

The temperature was measured on the side opposite to the heated one, in its center-point. Data was recorded every minute, for a total of 10 minutes.

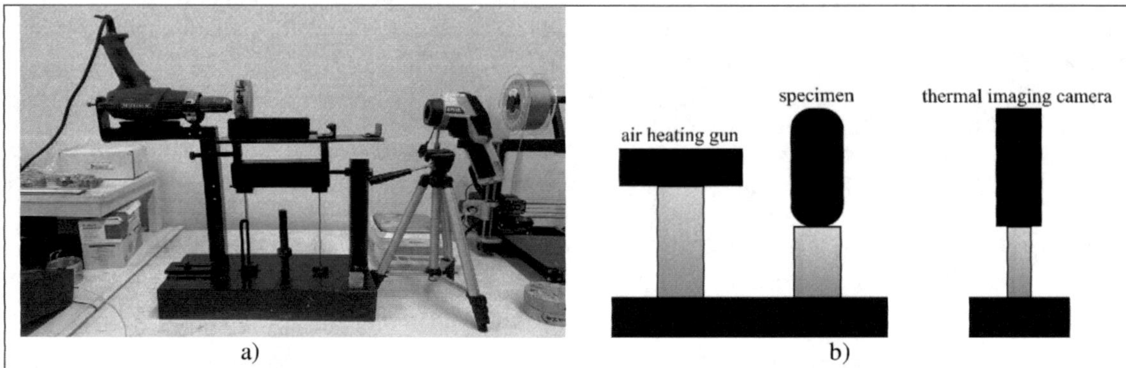


**Figure 2.** Set-up for experimental stand a) photo and b) sketch

### *2.2. Simulations assessment*

For the purpose of finite element method (FEM) modelling, it was used the free version of LISA 8.0 software. This allows maximum 1300 nodes, which is enough for the types of specimens employed in the study. Each concrete cylinder was modelled beginning with a circular mesh of 107 nodes / 96 faces. On that, the extrusion was applied to achieve the specific thickness of each individual specimen. For this, a total of 10 subdivisions was used, thus obtaining 1177 nodes and 960 elements (figure 3).

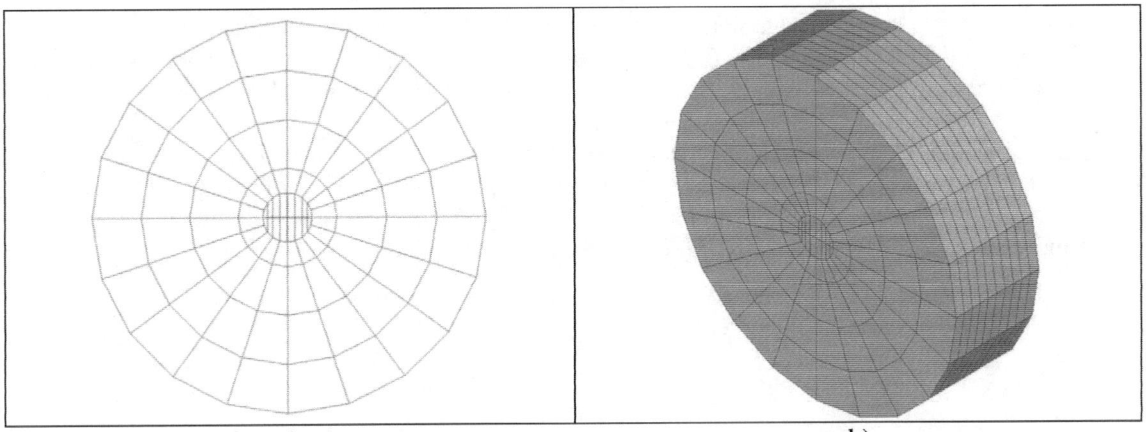

a) b)

**Figure 3.** Finite element model: a) circular cross-section and b) 3D

The heat source was modelled starting from a central surface on the face corresponding to the cylinder's exposed area, with a diameter of d = 2.5 cm. Through the extrusion process, an object was created out of this surface, and it was provided with a constant, desired, temperature (figure 4). The model also assumed the presence of convection on the lateral surface of the cylinder and on the face opposite to the heated one. For this, the following parameters were used: $T_{air}$ = 22°C and transfer coefficient (h) = 5 W/(m²·K).

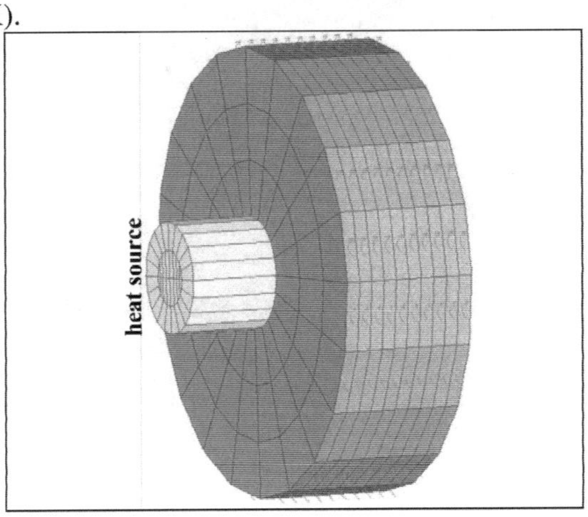


**Figure 4.** Specimen with heat source model

## 3. Results and discussions

The cylinders used in the study were initially measured for mass, dimensions and thermal properties. These values were important for simulation purposes. Results of these measurements are presented in table 1. The notations in the heading are as follows: ρ = density [kg/m³], λ = thermal conductivity [W/(m·K)] and $c_p$ = specific heat [J/(kg·K)].

**Table 1.** Parameters of specimens used for simulation.

| Recipe | ρ | λ | $c_p$ |
|---|---|---|---|
| PC | 2242 | 2.034 | 781 |
| RC40 | 2188 | 1.543 | 713 |
| RC60 | 2037 | 1.772 | 821 |
| RC80 | 1981 | 1.435 | 849 |

IOP Conf. Series: Materials Science and Engineering 1304 (2024) 012008 doi:10.1088/1757-899X/1304/1/012008

Values for density are quite predictable, since rubber is lighter than river coarse aggregates. The thermal properties measurements reveal that, although we would expect a similar linear variation with rubber concentration of both λ and $c_p$, the heterogenous nature of the specimens results in a different behaviour.

Figure 5 hold the plots for thermal properties previously measured. The results do not, necessarily, highlight the properties of the recipe as a whole, but rather the local characteristics of the larger specimen. However, this is important from a practical point of view, since local properties are of great importance, for example, when an element experiences a thermal load.

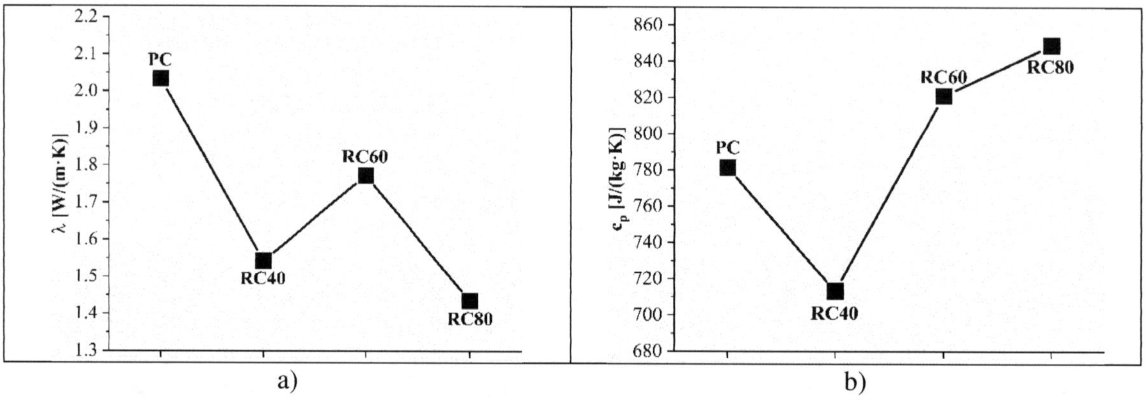


**Figure 5.** a) Thermal conductivity and b) specific heat variations of mortar specimens

*3.1. Experimental data*

Experimental data was plotted against the exposure time and designated as *specimen_exp* (figure 6). Using rubber as aggregate replacement results in higher thermal inertia. This can clearly be seen for RC40 and particularly for the 80% specimen. The RC60 cylinder does not follow the same trend, which is consistent with the thermal properties determined beforehand. In this case, the temperature increase with exposure time is close to the PC variation up to 115°C / 6min, when the curves start to diverge. After this point, the RC60 specimen's temperature tends to resemble the measurements for RC40.

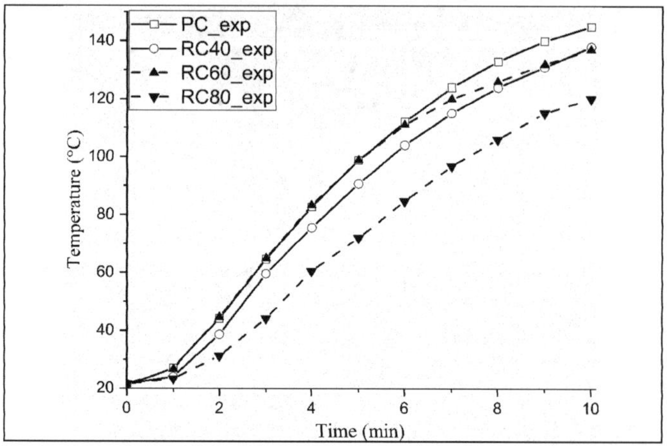


**Figure 6.** Temperature increase obtained experimentally

Having similar experimental data in the presence of the setup described above, as is the case of PC vs. RC60 (and, to a smaller extern, RC40), implies the need for corrections. Nevertheless, these results

IOP Conf. Series: Materials Science and Engineering 1304 (2024) 012008 doi:10.1088/1757-899X/1304/1/012008

are important in order to test the viability of simulations, which will be further employed for comparing the specimens used in the study.

### *3.2. Simulations results*

Models were designed for each specimen exposed to high-temperature. The input data needed for computing the transient thermal analysis consisted of: ρ, λ and $c_p$. These are known quantities, precisely measured at the start of the investigation. The first aim of the simulations was validating the experimental work, which is regarded as accurate with the exception of the exposure temperature. Therefore, simulations were calibrated to fit the experimental data by changing the input temperature, to the point of obtaining satisfactory convergence of experiment and simulation. The following values in table 2 were found to provide suitable results.

**Table 2.** Calibrated temperatures.

| Specimen | PC | RC40 | RC60 | RC80 |
|---|---|---|---|---|
| Temperature [°C] | 390 | 365 | 340 | 390 |

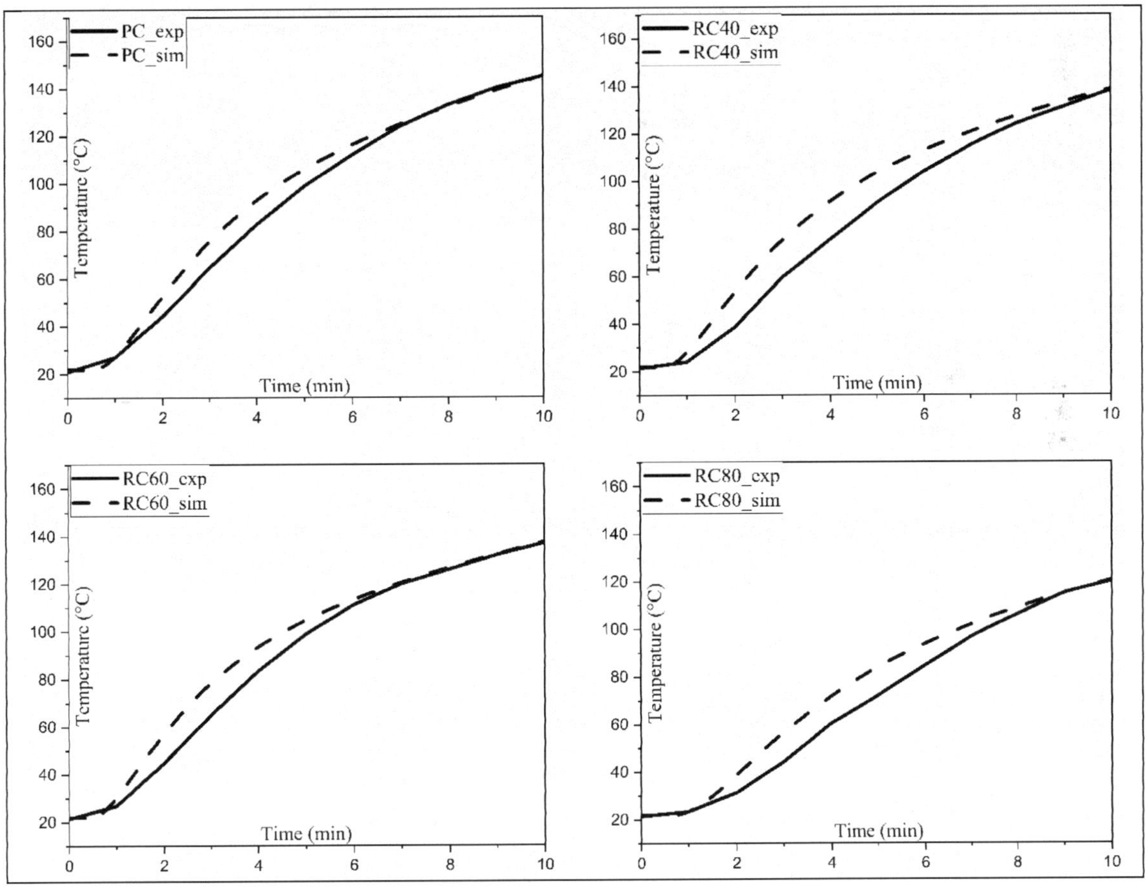


**Figure 7.** Experiment vs. simulation results when using calibrated temperatures

Variation of temperatures with exposure time are presented in figure 7. Each graph holds the plots for the experimental data and the simulated results to match them. All cases demonstrate that the correlation between the two curves is acceptable, with the lower temperature regime somewhat more distorted. The simulation data converges towards the same curve as the experimentally determined one, but the initial ramp-up of the heating in the experiment might be prone to more errors due to more complicated convection phenomena than the ones introduced in the simulation. Nevertheless, results are proving the simulation to be effective, thus the final process was employed.

Using the same models as before, with only changing the input temperature to 500°C for all specimens, the accurate comparison between PC, RC40, RC60 and RC80 was pursued. It can be observed that results are consistent to the differences in the calibration temperatures. The experimental errors caused RC60 high-temperature exposure to mimic firstly the PC's behaviour and tend to the RC40's one afterwards is not present here. Thus, by simulation, the proper curve was plotted. The temperature-time curves for PC, RC40 and RC60 are quite similar in values, in both experiment and simulation. From a measurement perspective, such results need very small-error experimental setups to be obtained.

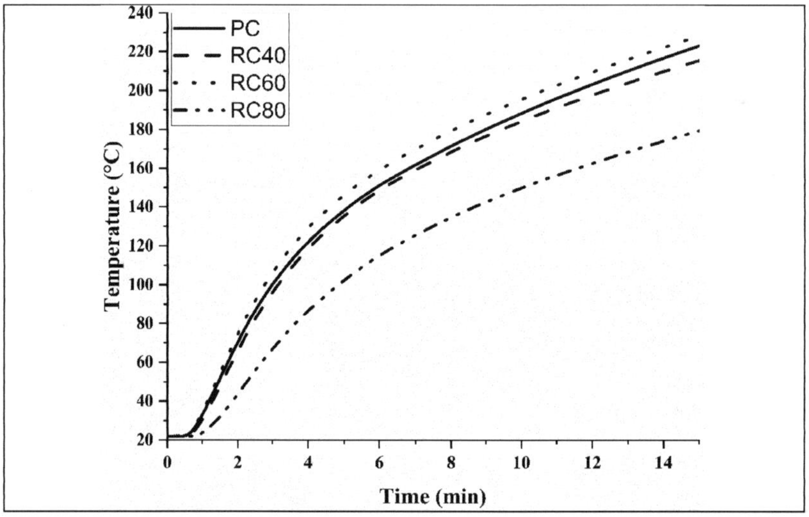


**Figure 8.** Simulation results when using the same input temperature (500°C)

## 4. Conclusions

The paper investigates the interplay between experiment and simulation in the case of high-temperature exposure of concrete and rubberized concrete. Since fire testing is extensive and expensive, alternative routes by studying smaller scale specimens with viable results is an important matter. If successfully implemented, such alternatives could result in great cost-cutting and a wider experimental reach in terms of mix recipes and new materials.

## 5. References

[1] Ganjian E, Khorami M, Maghsoudi AA 2009 Scrap-tyre-rubber replacement for aggregate and filler in concrete, *Constr Build Mater*, **23**(5):1828–36;

[2] Najim KB 2005 Modulus of elasticity and impact resistance of chopped worn-out tires concrete, *Iraq J. Civil Eng*, **1**(6):83–96;

[3] Sukontasukkul P 2009 Use of crumb rubber to improve thermal and sound properties of pre-cast concrete panel, *Constr Build Mater*, **23**(2):1084–92;

[4] Najim K B, Hall M R 2010 A review of the fresh/hardened properties and applications for plain (PRC) and self-compacting rubberised concrete (SCRC), *Construction and Building Materials* **24,** 2043–2051;

[5] Turgut P, Yesilata B 2008 Physico-mechanical and thermal performances of newly developed rubber-added bricks, *Energy Build,* **40**(5):679–88;

IOP Conf. Series: Materials Science and Engineering 1304 (2024) 012008 doi:10.1088/1757-899X/1304/1/012008

[6] Sukontasukkul P 2009 Use of crumb rubber to improve thermal and sound properties of pre-cast concrete panel, *Constr Build Mater*, **23**(2):1084–92;
[7] Hull T R 2008 Challenges in fire testing: reaction to fire tests and assessment of fire toxicity, *Advances in Fire Retardant Materials*, Woodhead Publishing Series in Textiles, https://doi.org/10.1533/9781845694701.2.255, pp. 255-290

## 6. Acknowledgements:

This paper was realized with the support of project EFECON – ECO-INNOVATIVE PRODUCTS AND TECHNOLOGIES FOR ENERGY EFFICIENCY IN CONSTRUCTION, POC/71/1/4 - Knowledge Transfer Partnership, Cod MySMIS: 105524, ID: P_40_295, Project co-financed by the European Regional Development Fund and having the support of the "Network of excellence in applied research and innovation for doctoral and postdoctoral programs / InoHubDoc", project co-funded by the European Social Fund financing agreement no. POCU / 993 / 6 / 13 / 153437.

IOP Conf. Series: Materials Science and Engineering 1304 (2024) 012009 doi:10.1088/1757-899X/1304/1/012009

# Different aspects regarding slope stability improvement using piles

**A Praporgescu [1] and A Popa [1]**
[1] Technical University of Cluj-Napoca, 25 Baritiu St., Cluj-Napoca, *Romania*,

E-mail: alexandru.praporgescu@dst.utcluj.ro

**Abstract.** The problem of slope stability is a common one in the field of geotechnical engineering. There is more than one method that can be adopted in order to improve the stability of a slope. During the last decade, using piles for improving the mechanical behaviour of slopes has become the main approach of dealing with unstable slopes. In the current paper, the stability of a slope was analysed by using the tridimensional finite elements method using different scenarios of consolidation with timber piles. A parametric analysis was carried out for two different slopes: sand slope and soft soil slope. There were also analysed two different constitutive models: Mohr-Coulomb and Hardening Soil Small with different location of the consolidation system. The stability analyses employed both timber piles and concrete piles which were modelled in Plaxis 3D software by using two different approaches: embedded piles and volume piles. In the end, the most favourable position for the pile consolidation system is highlighted in order to attain the maximum stability of the slope. Also, different aspects of the constitutive models, modelling the piles and pile material are discussed from the point of view of slope stability improvement.

## 1. Introduction

Landslides are natural phenomena that can occur in both natural or anthropic slopes. In some cases, they can result in material and property damage, or even in loss of human lives [1]. The problem of soil movement can be partially or totally resolved by slope stabilizing techniques which aim to increase slope stability and safety of the dangerous sites. In the recent decades, one of the most used method for achieving such results is using piles in order to consolidate the slopes that are suspected to become unstable [2], [3]. The benefit of using such a system can be quantified by using different methods of slope stability analysis, such as limit equilibrium methods (LEM), bidimensional finite elements methods (2D FEM) or tridimensional finite elements methods (3D FEM). All types of approaches have advantages and disadvantages, but it was showed that using a 3D method gives more insight into the complex task of modelling the real behaviour of a slope consolidated with piles [4]. Introducing piles inside the soil domain of analysis considerably increases the difficulty of the problem, as it opens the door for other important aspects such as: interaction between the piles and the soil, slope geometry, piles position, the modelling approach of the piles, constitutive model adopted and so on.

Another important aspect would be the impact of such a system on the environment and its ability to be sustainable. As nowadays most of the geotechnical structures are made out of reinforced concrete (which has a great equivalent $CO_2$ footprint), piles made out of wood could be an alternative to alleviate these disadvantages.

In these regards, in the current paper timber piles are analysed in comparison with concrete piles in order to evaluate the differences in behaviour and the possibility of using timber piles for slope stability improvement. The modelling of piles and soil was also approached through different ways. The soil was modelled by both Mohr-Coulomb (MC) model and Hardening Soil Small (HSS) model. The piles were modelled by using the embedded pile approach and the volumetric approach (volume piles), both options being available in the Plaxis 3D software.

IOP Conf. Series: Materials Science and Engineering 1304 (2024) 012009 doi:10.1088/1757-899X/1304/1/012009

## 2. Geometric and geotechnical characteristics of the analysed slopes

In order to apply all the aspects described above, a slope geometry was adopted. The geometry is shown in figure 1 and consisted in two different homogeneous configurations: sand and soft soil.

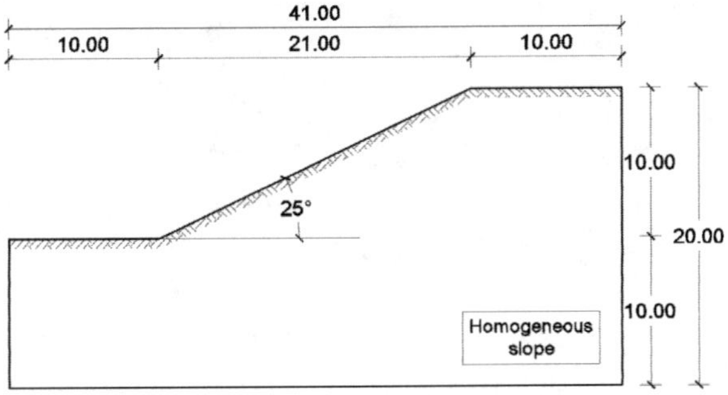


**Figure 1.** Profile of the analysed slope.

**Table 1.** Geotechnical parameters for sand in MC model.

| $\gamma_{nat}$ (kN/m$^3$) | $\gamma_{sat}$ (kN/m$^3$) | φ' (°) | c' (kPa) | Ψ (°) | ν (-) | E (kPa) | $\sigma_{tension}$ (kPa) |
|---|---|---|---|---|---|---|---|
| 18 | 20 | 41 | 0 | 15 | 0.25 | 36000 | 0 |

**Table 2.** Geotechnical parameters for soft soil in MC model.

| $\gamma_{nat}$ (kN/m$^3$) | $\gamma_{sat}$ (kN/m$^3$) | φ' (°) | c' (kPa) | Ψ (°) | ν (-) | E (kPa) | $\sigma_{tension}$ (kPa) |
|---|---|---|---|---|---|---|---|
| 15 | 16 | 27 | 15 | 0 | 0.30 | 4650 | 0 |

**Table 3.** Geotechnical parameters for sand in HSS model.

| | |
|---|---|
| $\gamma_{nat}$ (kN/m$^3$) | 18 |
| $\gamma_{sat}$ (kN/m$^3$) | 20 |
| φ' (°) | 41 |
| c' (kPa) | 0 |
| Ψ (°) | 15 |
| $\nu_{ur}$ (-) | 0.20 |
| $E_{50}^{ref}$ (kPa) | 30000 |
| $E_{oed}^{ref}$ (kPa) | 30000 |
| $E_{ur}^{ref}$ (kPa) | 90000 |
| m (-) | 0.55 |
| $p_{ref}$ (kPa) | 100 |
| $K_0^{nc}$ (-) | 1-sin(φ')=0.343 |
| $R_f$ (-) | 0.90 |
| $\sigma_{tension}$ (kPa) | 0 |
| $G_0$ (kPa) | 112500 |
| $\gamma_{0,7}$ (-) | 0.0002 |

**Table 4.** Geotechnical parameters for soft soil in HSS model.

| | |
|---|---|
| $\gamma_{nat}$ (kN/m$^3$) | 15 |
| $\gamma_{sat}$ (kN/m$^3$) | 16 |
| φ' (°) | 27 |
| c' (kPa) | 15 |
| Ψ (°) | 0 |
| $\nu_{ur}$ (-) | 0.20 |
| $E_{50}^{ref}$ (kPa) | 4300 |
| $E_{oed}^{ref}$ (kPa) | 1800 |
| $E_{ur}^{ref}$ (kPa) | 14400 |
| m (-) | 0.90 |
| $p_{ref}$ (kPa) | 100 |
| $K_0^{nc}$ (-) | 1-sin(φ')=0.546 |
| $R_f$ (-) | 0.90 |
| $\sigma_{tension}$ (kPa) | 0 |
| $G_0$ (kPa) | 25000 |
| $\gamma_{0,7}$ (-) | 0.0003 |

The geotechnical characteristics for each type of soil were chosen according to H. F. Schweiger [5] for both Mohr-Coulomb and Hardening Soil Small models and are presented in table 1, table 2, table 3 and table 4.

Taking into account that the optimum location of the piles was analysed, 12 possible locations of the row of piles were involved, ranging throughout the entire length of the slope, as shown in figure 2. Also, the length of the slope profile was considered being equal with $L_{profile}$=20.00 m during the analyses using embedded beams and $L_{profile}$=3.40 m during the analyses using volume piles.

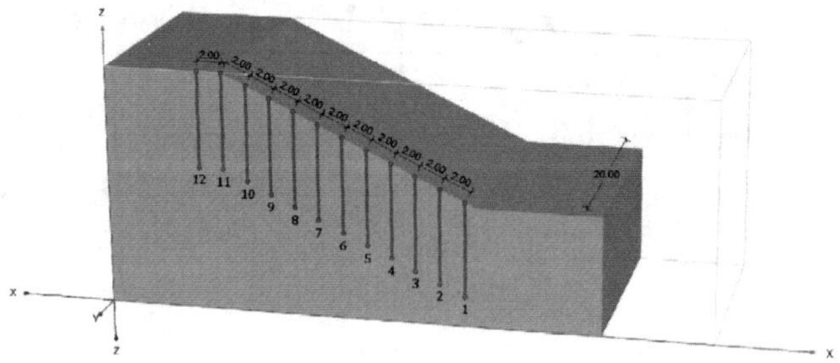


**Figure 2.** Possible positions of the piles analysed.

All the piles that were modelled in the stability analyses (timber or concrete) were given the same geometrical characteristics, according to table 5.

**Table 5.** Geometrical characteristics of the modelled piles.

| d (m) | $A_{pile}$ (m$^2$) | I (m$^4$) | $L_{inter-axe}$ (m) | $L_{pile}$ (m) |
|---|---|---|---|---|
| 0.40 | 0.1257 | $1.257 \cdot 10^{-3}$ | 1.00 | 8.00 |

Regarding the modelling of the soil slope using the Plaxis 3D software, the dimension of the finite elements was chosen as “fine” according to the software settings in order to reach equilibrium between the accuracy of the results and the analysis duration.

## 3. Different aspects regarding the constitutive models and modelling of piles

In numerical analyses, the behaviour of a soil mass can be greatly influenced by the adopted constitutive model for the soil. In order to compare the results, all the stability analyses that were carried out in this paper were done by employing both Mohr-Coulomb and Hardening Soil Small model.

The Mohr-Coulomb model is an elastic perfectly-plastic model that requires only 5 geotechnical parameters (see table 1 and table 2). Determining these parameters can be done either on site or in the laboratory. The basic idea behind the elasto-plastic modelling of the soil is that the strains are divided into the elastic and plastic ones. In order to corelate the state of stress with the elastic strains, Hooke's law is used. According to plasticity theory, the deformation state is proportional with the first order derivative of the plasticity function divided by the derivative of the stresses. Also, a plastic potential function is introduced, in order not to overestimate the dilatancy phenomenon. Being a first order model, it is usually used for general modelling of soil, dam stability analysis, slope stability and is not recommended for modelling clays or soft soils due to overestimation of the rigidity of normally and under consolidated soils [6].

On the other hand, the more advanced models which take into account the increase in rigidity at small strains may prove to be advantageous in closely modelling the soil-structure interaction. One of these models is Hardening Soil Small model, which represents an extension to the base model Hardening Soil integrated in Plaxis software.

The HSS model tries to model basic phenomena that take place in the soil mass, such as: increase in density, dependency of the rigidity on the stress levels, creep, dilatancy and rigidity variation depending on the tangential strains. The Hardening Soil Small model is one of the simplest small strain models which uses almost the same geotechnical parameters as Hardening Soil model (see table 4 and table 5). HSS model is usually adopted in deformation problems during the analysis of retaining walls, retaining geotechnical structures, etc. with almost no evidence of being used in slope stability problems.

Regarding the modelling of the piles, Plaxis 3D offers two alternatives of doing it: embedded piles or volume piles.

Embedded piles are elements that after discretization become linear elements having 3 nodes, each with 6 degrees of freedom. During a slope stability analysis, the piles from the slope body will be loaded by lateral forces. From this point of view, embedded pile elements have some limitations: being linear elements they cannot overlap and the soil-structure interaction is ensured by introducing slide elements in the axial direction of the elements. Unfortunately, slide elements cannot be introduced on the horizontal direction and the piles do not take into consideration the friction between them and the soil when they are laterally loaded. Even though this aspect is important for consolidation piles, acceptable behaviour of embedded piles is demonstrated in the literature [7]. The advantages of embedded piles are visible when there is a great number of piles that needs to be modelled, as the computational time is much more decreased than for using same amount of volume piles.

On the other hand, volume piles are elements that allow a 1:1 representation of a pile in the tridimensional space. The most important aspect is that the interface between the soil and the elements can be modelled through specific surface elements. The contact between different types of surfaces is simulated by affecting the strength at the interface level through the reducing factor $R_{inter}$. For the current paper, different types of contacts were analysed: sand-wood, soft soil-wood, sand-concrete, soft soil-concrete, for which the values of $R_{inter}$ were chosen according to [8] (table 6).

After the discretization of the domain, the volume piles will be a part of the mesh, the dimensions of the finite elements will be restricted to smaller values. Even though this aspect may increase the accuracy, it will also significantly increase the computation time which is a major disadvantage of this type of element.

Regarding the degree of freedom of the piles heads, all the piles modelled in the current paper were considered having free head. This is the most detrimental situation for the pile head fixity which will generate the smallest value for the factor of stability.

**Table 6.** Values of $R_{inter}$ factor depending on different types of materials.

| Pile material | Soil type | $R_{inter}$ |
|---|---|---|
| Wood | Sand | 0.88 |
| Wood | Soft soil | 0.70 |
| Concrete | Sand | 0.98 |
| Concrete | Soft soil | 0.95 |

## 4. Parametric analysis of the optimal piles position with embedded pile modelling

The first parametric analysis comprised finding the optimal position of the pile consolidation system. All the slope stability analyses were conducted by adopting both constitutive models discussed above (MC and HSS) and both types of piles: timber and concrete piles. While the geometrical characteristics of the piles were the same, the material properties of the piles were chosen according to table 7 and table 8 depending on the pile material.

**Table 7.** Material properties of wooden piles for 12% moisture content.

| $\gamma$ (kN/m$^3$) | E (kPa) | $R_{f,\perp}$ (kN/m$^2$) |
|---|---|---|
| 10.79 | $1.33 \cdot 10^6$ | 12000 |

**Table 8.** Material properties of concrete piles for C25/30 class concrete

| $\gamma$ (kN/m3) | E (kPa) | $\nu$ (-) |
|---|---|---|
| 25 | $31 \cdot 10^6$ | 0.15 |

All the piles in this analysis were modelled only by using embedded beam elements. The variation of the safety factor ($F_s$) in relation to the position of piles for different constitutive models and pile material is presented in figure 3 for the sand slope and in figure 4 for the soft soil slope.

As it can be seen in figure 3 for the sand slope, the values of $F_s$ are varying depending on the position of the piles and on the adopted constitutive model.

For the timber piles where the soil was modelled using Mohr-Coulomb model, the highest values of the stability factor were registered on the positions 6 and 8, approximately representing the middle of the slope. Considering the value of $F_s$ obtained for the unreinforced slope, in the case of timber piles modelled with MC model, only three locations of the piles generated greater values: 6, 7 and 8. The analyses showed that when the timber piles are placed on other locations, the stability of the slope is decreased and therefore the chosen solution is not suitable. The results obtained by using the HSS model show a similar variation for the factor of stability, but with smaller differences between the smallest and the greatest values. In this combination of factors, the greatest value for $F_s$ was generated on position no. 7, being the only location where the factor of safety has a greater value than the factor of safety of the unreinforced slope.

Taking into consideration the behaviour of the concrete piles for the same sand slope, the variation of $F_s$ is similar to the one obtained for the timber piles. In this case, only two positions of the piles generated $F_s$ values greater than the ones for the unreinforced slope and those are positions no. 6 and 7 (approximately the middle of the slope). This result is valid for both constitutive models employed. On all the other locations, placing the concrete piles will result in a decreasing stability factor and therefore other consolidation solutions may prove efficient.

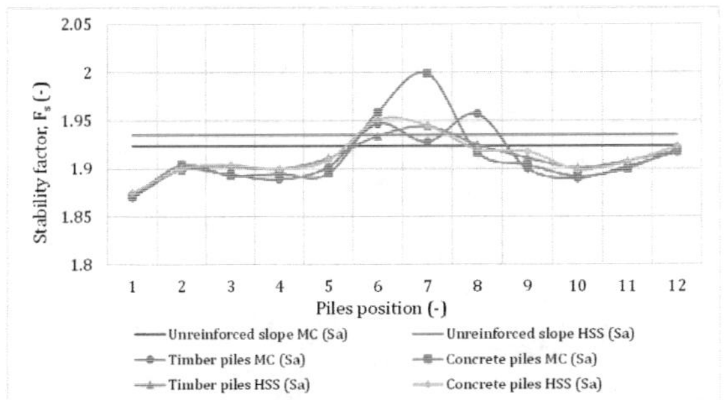


**Figure 3.** Variation of $F_s$ depending on embedded piles position for a slope made of sand

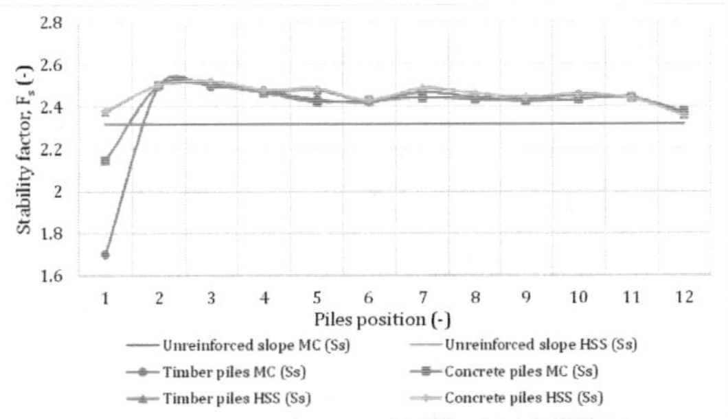


**Figure 4.** Variation of $F_s$ depending on embedded piles position for a slope made of soft soil

Comparing the results generated by the two constitutive models regarding the sandy slope, it can be observed that the results follow more or less a similar variation. Looking at the smallest values for $F_s$ both models generated very close values with the greater difference being in comparing the largest values of the stability factor. In this case it can be considered that Mohr-Coulomb model sensibly overestimates the values of $F_s$, but the general behaviour is very similar.

For the soft soil slope, figure 4 shows that the variation of the factor of stability with the position of the piles can almost be approximated by a constant variation.

The values of $F_s$ for the unreinforced slope are almost equal for MC and HSS models, which is showed by the overlapping horizontal lines in figure 4. It can also be observed that almost all values of the factor of stability are superior to those generated by the unreinforced slope (except position no. 1 for timber and concrete piles modelled with MC model). This means that the consolidation with piles is beneficial for the soft soil slope regardless of the position. The maximum values of the stability factor were obtained on the 2$^{nd}$ position for timber piles with MC model, on 3$^{rd}$ position for timber piles with HSS model and on 3$^{rd}$ position for concrete piles with both constitutive models which indicates the optimal position of the system in the lower third of the slope.

From the constitutive model point of view, the values generated by Mohr-Coulomb model and Hardening Soil Small model are very similar and the differences are neglectable, which means that both models are appropriate for the analysis of the stability level of a clay slope. Comparing the behaviour of timber and concrete piles, the values of $F_s$ are seemingly not affected by the material of the piles. In these regards, a consolidation with timber piles could be feasible from the stability point of view.

## 5. Parametric analysis of the optimal piles position with volume pile modelling

Volume pile elements represent a way to accurately model a pile in a tridimensional environment. Using this approach requires the user to model the piles by using the real dimensions of the elements and therefore volume piles can better simulate the interaction between the piles and the soil.

To investigate the influence of a pile consolidation system modelled with volume piles on the stability of a slope, a parametric analysis was carried out. The geometric characteristics of the slope are the same as in the previous analysis and it was considered a homogeneous slope made of sand and soft soil with the geotechnical characteristics according to table 1 and table 2 for Mohr-Coulomb model and according to table 3 and table 4 for Hardening Soil Small model. The geometrical characteristics of the piles were kept the same and the material was timber and concrete respectively.

Taking into account the disadvantages of modelling piles as volume piles, the length of the slope profile was set to a smaller value (4.00 m) in order for the analysis to comprise only 3 piles. This modification lead to decreased analysis time which allowed more analyses to be conducted in a shorter amount of time. Another aspect is that only position 1, 6, and 11 were analysed, representing the bottom, middle and the top of the slope. This alteration was made also due to the increased time of analysis required when using volume piles. The results are presented in figure 5 and figure 6.

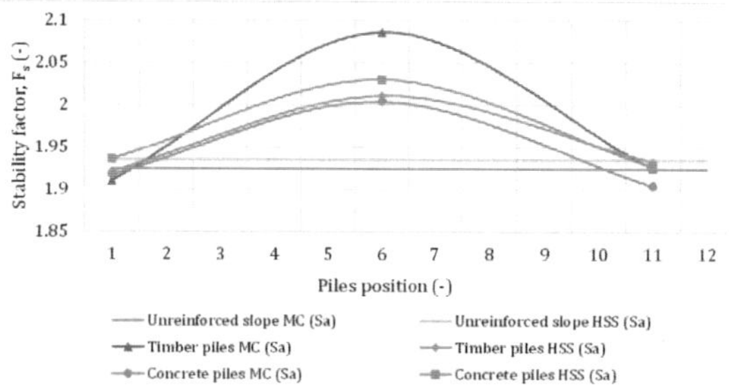


**Figure 5.** Variation of $F_s$ depending on volume piles position for a slope made of sand

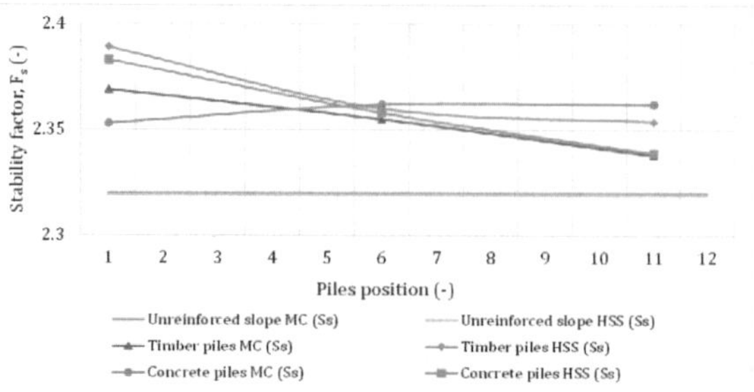


**Figure 6.** Variation of $F_s$ depending on volume piles position for a slope made of soft soil

The variation of $F_s$ which is showed in figure 5 for the three positions studied, expresses the fact that in the case of the sand slope, the optimal location of placing the piles is somewhere at the middle of the slope. This result is generated by both MC and HSS constitutive models. Also, even though the values of the factor of safety are not the same, they can be considered sensibly close for timber and concrete piles respectively.

It can be observed form figure 5 that for position no. 6, all the values of $F_s$ are greater than those obtained for the unreinforced slope, while on the positions no. 1 and 11, the values of the stability factor for the consolidated slope are smaller or equal with those of the unreinforced slope. This signifies the fact that placing a consolidation system at the bottom or the top of the sandy slope would not have a beneficial effect on the stability of the slope, therefore they are not economical.

Analysing figure 6 which shows the values of the stability factor for the soft soil slope, it can be seen that all of them are greater than the values of $F_s$ for the unreinforced slopes. This makes the consolidation system worthwhile for the stability of the slope, regardless the position of the piles. The optimal location for the consolidation system is, according to figure 6, at the bottom of the slope. This observation can be made for all the studied cases with the exception of the concrete piles analysed with Mohr-Coulomb model. For this case, the smallest value of $F_s$ was generated at the bottom of the slope and for the other two positions, the variation was constant (same value of $F_s$).

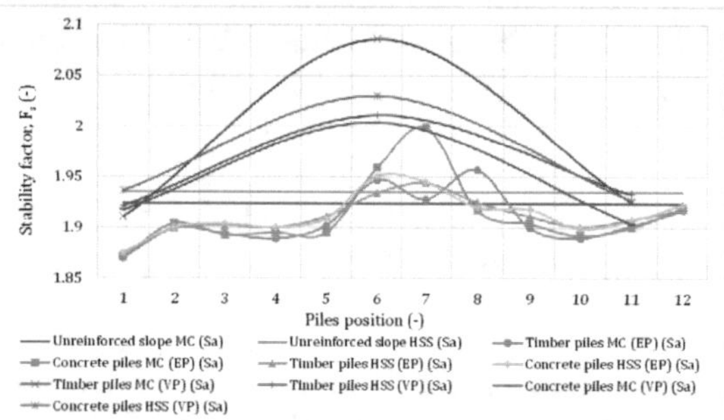


**Figure 7.** Variation of $F_s$ depending on piles position for a slope made of sand

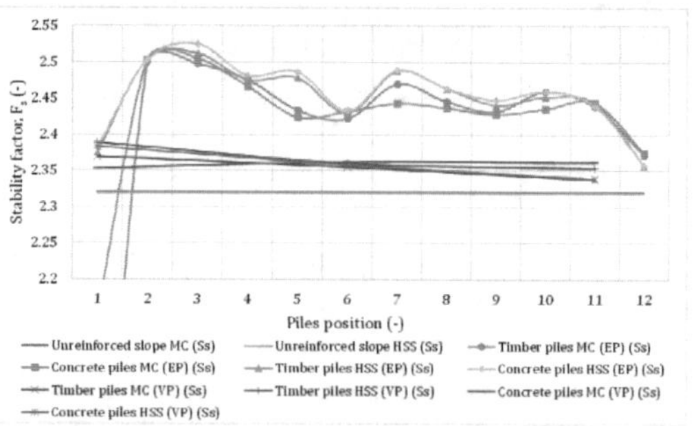


**Figure 8.** Variation of $F_s$ depending on piles position for a slope made of soft soil

All the results obtained in the numerical analyses by modelling the piles as volume piles for both types of slope (sand and soft soil) follow the same trend as the ones generated by the embedded piles analyses. Even though the magnitude of the values is different, the results show that for a sandy slope, the consolidation system is more efficient if it is placed somewhere at the middle of the slope while for a soft soil slope, the system should be placed somewhere at the bottom of the slope (lower third of the slope) for maximum stabilisation effect. In figure 7 and figure 8, the results obtained by using volume piles (VP) were plotted against those obtained by using embedded piles (EP). It can be observed that in the case of sandy slope, embedded piles were more conservative and should be used in these types of situations in order to be on the safe side. On the other hand, the soft soil slope volume piles solution gave the most pessimistic result and represents the safer way of approaching the pile modelling.

## 6. Conclusions

The current paper presents the results obtained from several parametric analyses which involved the stability computation for two homogeneous slopes (sand and soft soil), by employing different constitutive models (Mohr-Coulomb and Hardening Soil Small), for different types of piles (timber and concrete) which were modelled by using two different methods available in Plaxis 3D software (embedded piles and volume piles). The conclusions that were reached are the following:

- The optimal position of the pile consolidation system depends on the type of soil in the slope body. For slopes made of sand, the best placement of the system would be at the middle of the slope and for slopes made of soft soil, in the lower third according to the numerical analyses presented. It was also observed that sandy slopes are more sensible to a consolidation system with some generated values for $F_s$ being smaller than those of the unreinforced slope.
- From the stability point of view, the difference between Mohr-Coulomb model and Hardening Soil Small model is not very dramatic. The most important difference lies in the number of parameters required in order to apply the two models. As MC model needs less parameters, which are more easily available than those for the HSS model, it is the recommended constitutive model to be used in routine stability analyses, which is in accordance with the literature.
- The behaviour of timber and concrete piles was also analysed. It was shown that, from the stability point of view, the behaviour is very similar and timber piles could be a more sustainable alternative which benefits from the low equivalent $CO_2$ footprint.
- The method of modelling the piles in the 3D domain can have an impact on the behaviour of the consolidation system. Even though the volume piles represent the more accurate way of modelling the piles, they are limited by the very increased time necessary for the numerical analysis and the dimension of the finite elements. It was observed that for models containing more than three volume piles, the necessary analysis time can increase with as much as 3 to 4 times than the same situation modelled using embedded piles. It was also shown that for sandy slopes embedded piles generated more conservative results and should therefore be used, while for soft soil slopes the volume piles generated the most pessimistic result and should be used in such situations.

## 7. References

[1] Tang H, Wasowski J, Juang C H 2019 *Geohazards in the three Gorges Reservoir Area China–Lessons learned from decades of research* Eng. Geo. 105267
[2] Wei W B, Cheng Y M 2009 *Strength reduction analysis for slope reinforced with one row of Piles* Comp. and Geotech. 36 1176-85
[3] Guo W D 2013 *$P_u$-Based solutions for slope stabilizing piles* Int. J. Geomech. 13 292-310
[4] Abdelaziz A, Hafez D, Hussein A 2015 *The effect of pile parameters on the factor of safety of piled-slopes using 3D numerical analysis* HBRC J. 13 277-85
[5] Schweiger H F 2014 *Influence of EC7 design approaches on the design of deep excavations with FEM* Geotech. 37 169-76
[6] Obrzud R F, Truty A 2020 *The hardening soil model-a practical guidebook,* ed. Z Soil (Switzerland)
[7] Dao T P T 2011 *Validation of Plaxis embedded piles for lateral loading,* (Delft)
[8] Potyondy J G 1961 *Skin friction between various soils and construction materials* Geotech. 11 339-53

IOP Conf. Series: Materials Science and Engineering 1304 (2024) 012010 doi:10.1088/1757-899X/1304/1/012010

# Earthquake behavior of flat slab structures, particularities and opportunities for use in the building stock of the Chisinau municipality

**A A Sîli**

Gheorghe Asachi Technical University of Iasi, Bd. Profesor Dimitrie Mangeron 67, 70050 Iasi, Romania

E-mail: anatolie.sili@student.tuiasi.ro

**Abstract**. The Republic of Moldova is characteristic for active seismic phenomena and there is always a direct danger of strong earthquakes. The building stock of the country has a complex character, this aspect is manifested in the municipality of Chisinau, where we have different buildings in terms of materials used, structural systems and height regime. New trends in construction are focused on buildings with a load-bearing frame structural system and a frame with a flat slab structure. The usual design practice is characteristic for the implementation of beam-column systems, where the plates are supported by the beams and the beams are supported by the columns. Nowadays, for the residential buildings design, a more elegant and ergonomic floor system is used - the flat slab system. Therefore, the slabs are subjected to increased loads, which requires special attention in the design and implementation of these types of structures.

## 1. Introduction

The construction sector is an important field in the economy of the Republic of Moldova. It represents an active sector of the economy, even if last year, according to the NBS data, it showed a decline. Of the total amount of 15.7 billion MDL, which constituted the total cost of the construction works completed in 2022, about 52.5% is accounted for by the newly executed constructions. [1] The housing fund has the largest share in this volume of completed works. In rural areas, the residential stock is mainly composed of individual dwellings with a low height regime. In cities, the presence of sectors with individual dwellings can be detected, but at the same time, there are urban regulated areas for collective development with an average height regime or development with a large number of levels. But these multi-storey buildings, in most district centers, give an insignificant share of the total building stock. The municipalities of Chisinau and Balti have a higher share of multi-storey buildings in the structure of the building stock. The Chisinau municipality is the largest city in the country and one of the largest urban agglomerations in the region. Performing multiple roles, it is an attractive city center for investments and, accordingly, for the construction of new buildings. Analyzing the housing fund of the municipality of Chisinau, we reveal its complex structure. Individual residential houses stand out for the variety of materials used, whether local materials such as small blocks of lacustrine limestone, natural stone, brick or newer materials such as ACC blocks, ceramic blocks, etc. By analyzing the height regime of these types of housing, a thing can be highlighted, they vary from ground floor structures to structures with multiple floors, most commonly ground floor and 3 levels. Due to the rise in prices for land intended for construction and of building materials, single attached houses, as well as terraced houses, have gained popularity. Thus, in the territories of the municipality, where, according to urban planning regulations, the construction of such structures is allowed, we received the percentage of land occupancy (LPP) and the maximum regulated land use coefficient (LUC), as a result, an increased number of inhabitants in these areas. For multi-level individual dwellings, these values vary, LPP=35-55 % and LUC=0.5-1, while for residential multi-storey buildings, LPP=65-8 % and LUC=2-8. [2]

Collective buildings or residential quarters in the municipality of Chisinau are differentiated depending on the period of construction according to several criteria:

- the dominant height regime;
- main material of the load-bearing structure;
- interior design solutions;
- urban planning solutions for the location of buildings on the land;
- aesthetic appearance.

Residential buildings built in 1950-1970 in the municipality of Chisinau are of medium height and with load-bearing walls structural system. Thus, the structural frame consisted of reinforced concrete frames with masonry of small limestone blocks. The main structural elements were made of small blocks of lacustrine limestone with dimensions of 190mm x 188mm x 390mm or large blocks with dimensions of 190mm x 240mm x 490mm. For the interior systematization of the residential buildings the flats with small rooms were common. The need to provide the population with housing and the difficult economic situation determined this trend in the construction industry.

Residential buildings built in 1970-1990 are distinguished by the use of new technologies of that period, namely the use of precast concrete elements. This method reduced construction lead time and thus managed to cover much of the high housing demand. The residential buildings were made using standard design projects, which were adjusted according to the construction conditions in certain areas (Figure 1). Therefore, the serial flat buildings became the dominant housing typology, and a group of serial projects were designed as flats serial number 143, 102, MS etc.

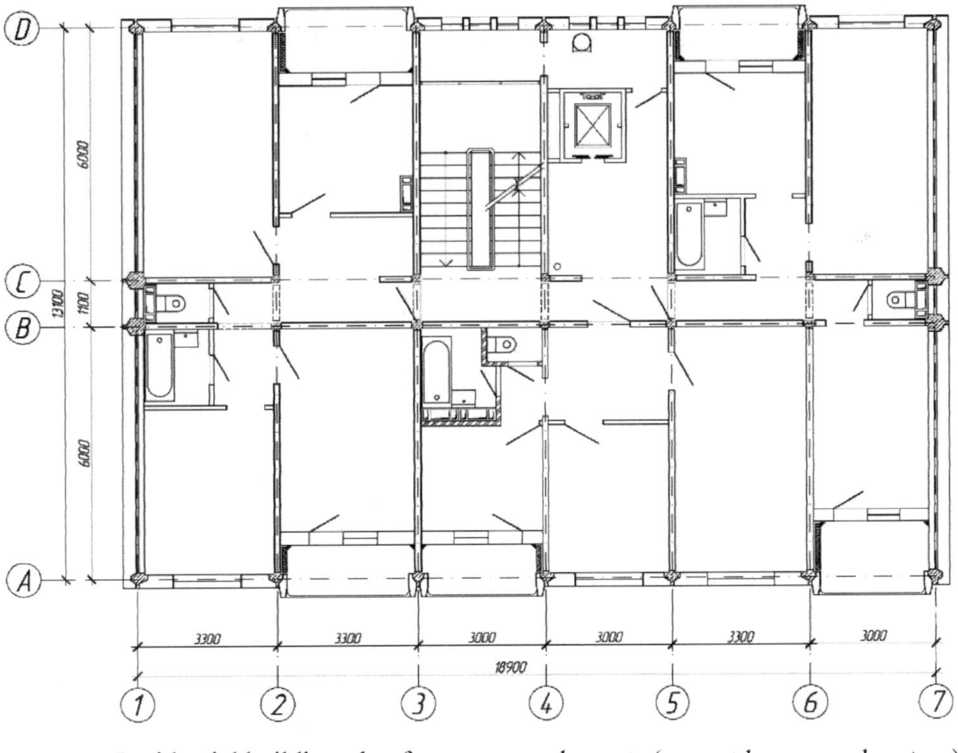


**Figure 1.** Residential building plan from precast elements (precast large-panel system).

Simultaneously with the described above building systems, structures with a load-bearing skeleton made of precast elements or tower-type buildings made of monolithic reinforced concrete were built. The construction of new residential areas during the period 1990-2020 was dominated by buildings with load-bearing frame structural system. The main load-bearing structural elements of the frame being columns and beams. This constructive system has proved its efficiency and resistance over time to seismic actions. At the same time, the load-bearing structures with a system of flat slabs have become popular since the 2000s. It is distinctive by the fact that the connection between the column and the slab is not carried out through a monolithic beam, but directly through a column-slab joint.

## 2. Calculation method for a classic structural frame and a flat slab system

The Republic of Moldova is located in an area where seismic activities occur with a certain frequency. In the period 1940-2023 there were a series of earthquakes that had disastrous economic consequences. The intensity of these seismic activities was 5.0 on the Richter scale in 2000, 7.2 on the Richter scale in 1977 and 7.4 on the Richter scale in 1940. [3] According to the seismic zoning map of the Republic of Moldova, the territory is divided into 3 seismic zones, therefore the southern part having a magnitude of 8, the center having a magnitude of 7, and the northern and eastern regions with a magnitude of 6. According to the seismic intensity the city of Chisinau is located in the area with a magnitude of 7 degree. The design of buildings and their construction is carried out considering this important factor. Thus, seismic design of concrete structures will provide sufficient energy dissipation capacity under cyclic loads without significantly reducing the resistance to horizontal and vertical forces. [4] The design and construction of buildings must be done in accordance with the current rules, and the execution process must be controlled and carried out in compliance with regulations, technological provisions that do not allow deviations in the construction process.

Earthquake-resistance strength is a characteristic of buildings and structures to be able to withstand the sudden ground shaking that characterizes the degree of their resistance to earthquakes. The mentioned above characteristic describes the resistance degree of the buildings to the earthquake. This index is very important in building seismic exploration, a branch of civil engineering that specializes in the behavior of buildings and structures under seismic impact.

In this sense, in order to obtain a clear picture of behavior of buildings, calculations were made of the resistance of the building structures, a building with a classic load-bearing frame, beam-column and a building on flat slab system (Figure 2). The classic load-bearing frame system demonstrated good stiffness, stability and good behavior during seismic actions, the biggest disadvantage of this system would be the flares in the lower part due to the beams (Figure 3). The structural calculation for these 2 buildings was carried out in the calculation program LIRASAPR2016.

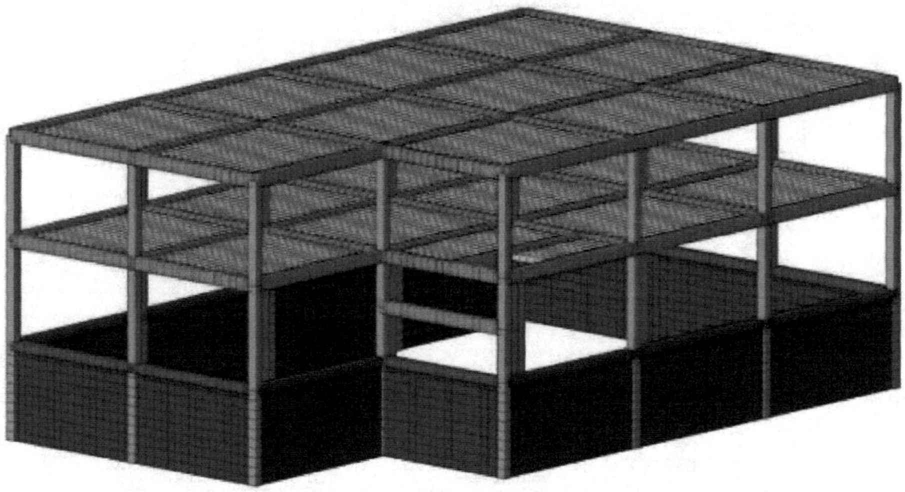

(a)

IOP Conf. Series: Materials Science and Engineering 1304 (2024) 012010 doi:10.1088/1757-899X/1304/1/012010

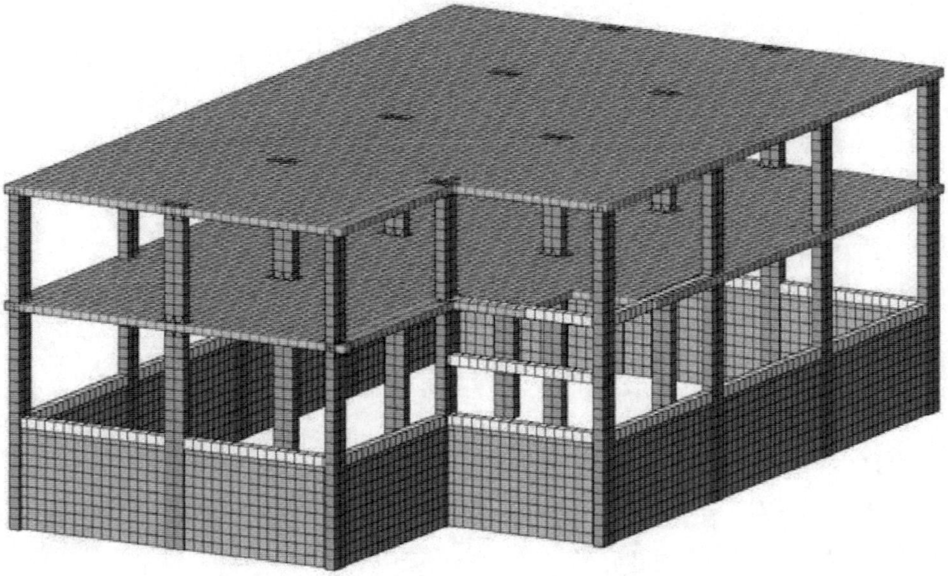

(b)

**Figure 2**. Analyzed structures. (a) Load-bearing frame system. (b) Flat slab system.

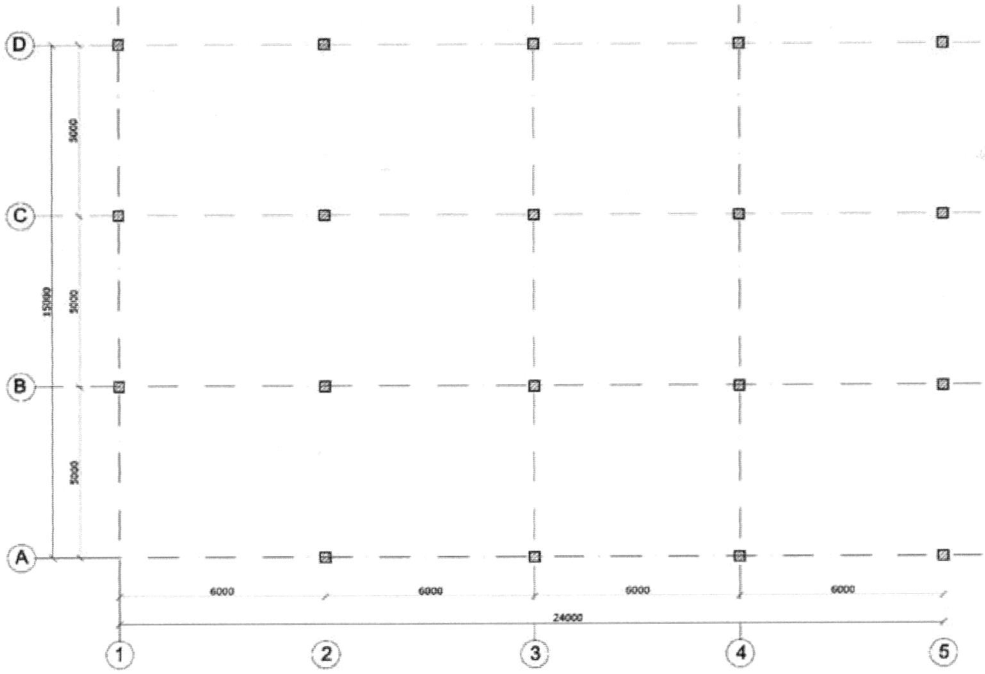


**Figure 3**. Load-bearing frame structural system scheme.

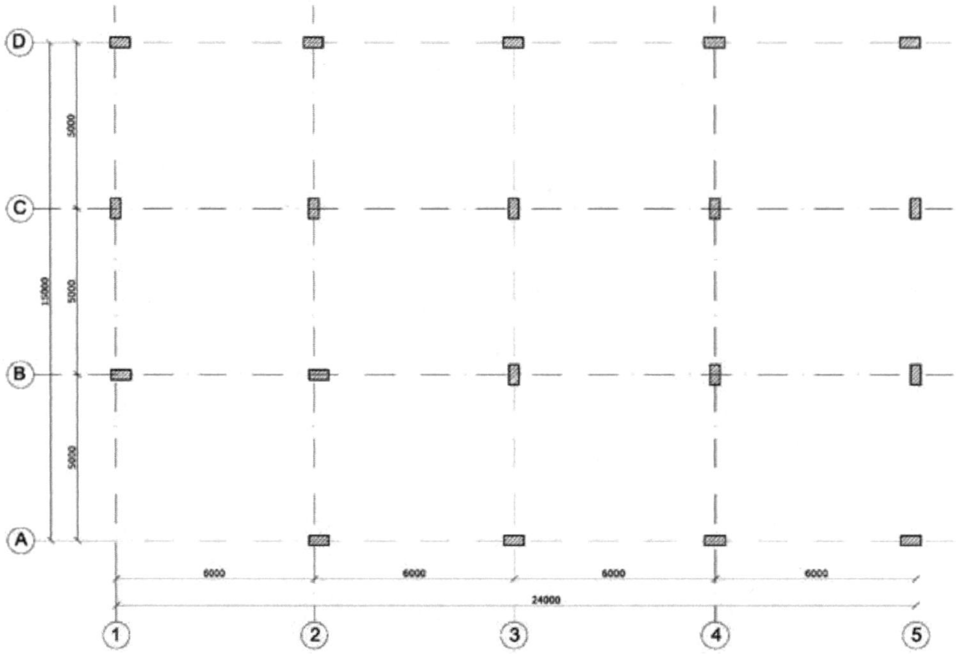


**Figure 4. Flat slab structure system scheme.**

The flat slab structure systems are significantly more flexible than traditional frame structures, thus becoming more vulnerable to seismic loading conditions [5] (Figure 4). In U.S.A, C.A.P. Turner constructed flat slabs in 1906 mainly using intuitive and conceptual ideas, which was start of this type of construction. Many slabs were load-tested between 1910-1920 in U.S.A. [6] This type of structure is widely used in European countries where seismic phenomena are frequent, being traditionally used in Italy, Spain, Portugal, both, for residential buildings and for socio-cultural purposes [7]. They are very popular among architects, designers, engineers and customers. Such an increased demand for this type of buildings is since this system offers freedom in the internal systematization of the houses, the possibility and flexibility in changing the floor plan, the easy drawing of engineering networks and advantages in aesthetic aspect (Figure 5). For floors of residential buildings, offices or car parks, a maximum opening of 7 m is recommended, and walls resting directly on floor slabs should be made of lightweight elements. [8]

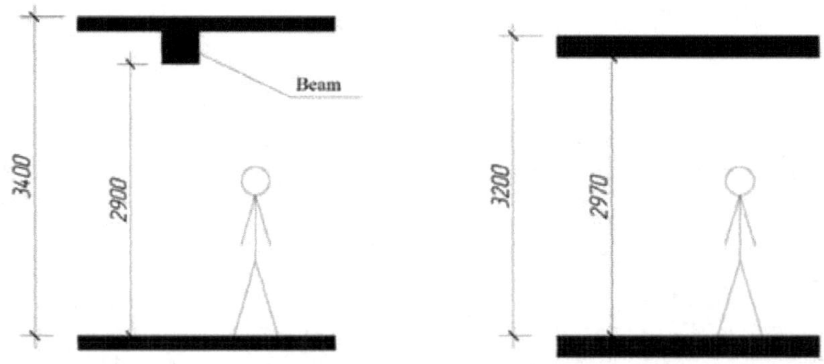


**Figure 5.** Sections depending on the structural system.

The general dimensions of the analyzed building with a classic load-bearing frame system are 15m x 24m. The bearing elements of the construction have the following dimensions: columns - 300mm x 300mm; beams – 300mm x 400mm; slabs – 150mm and the floor height of 3300mm (Figure 3).

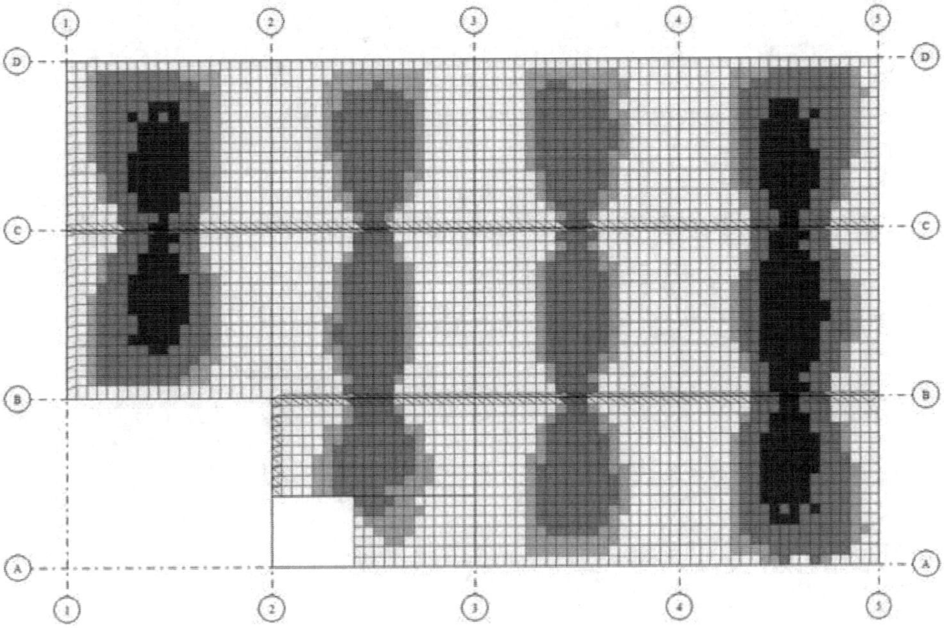


**Figure 6.** Inferior loads values in the Mx slab

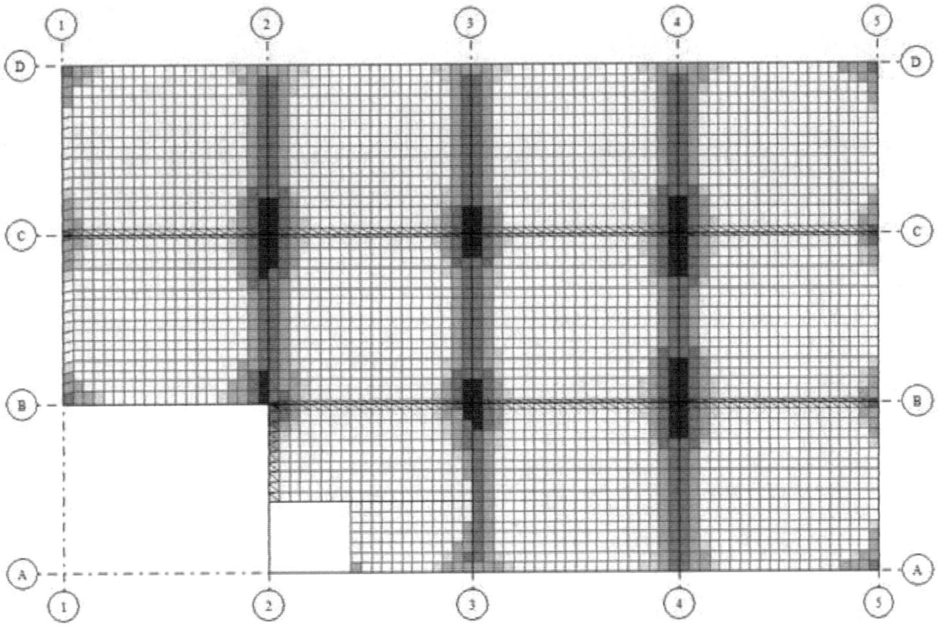


**Figure 7.** Superior loads values in the Mx slab

The general dimensions of the analyzed building with a flat slab structure system are 15m x 24m. The bearing elements of the construction have the following dimensions: columns - 300mm x 600mm; flat slab – 230mm; and floor height of 3300mm (Figure 4).

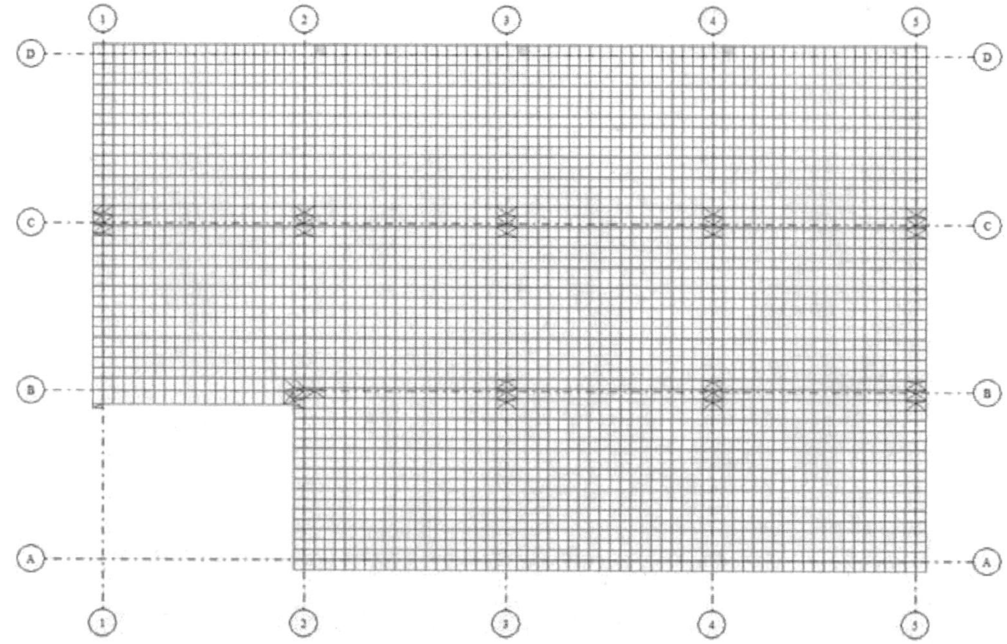

**Figure 8.** Inferior loads values in the Mx slab

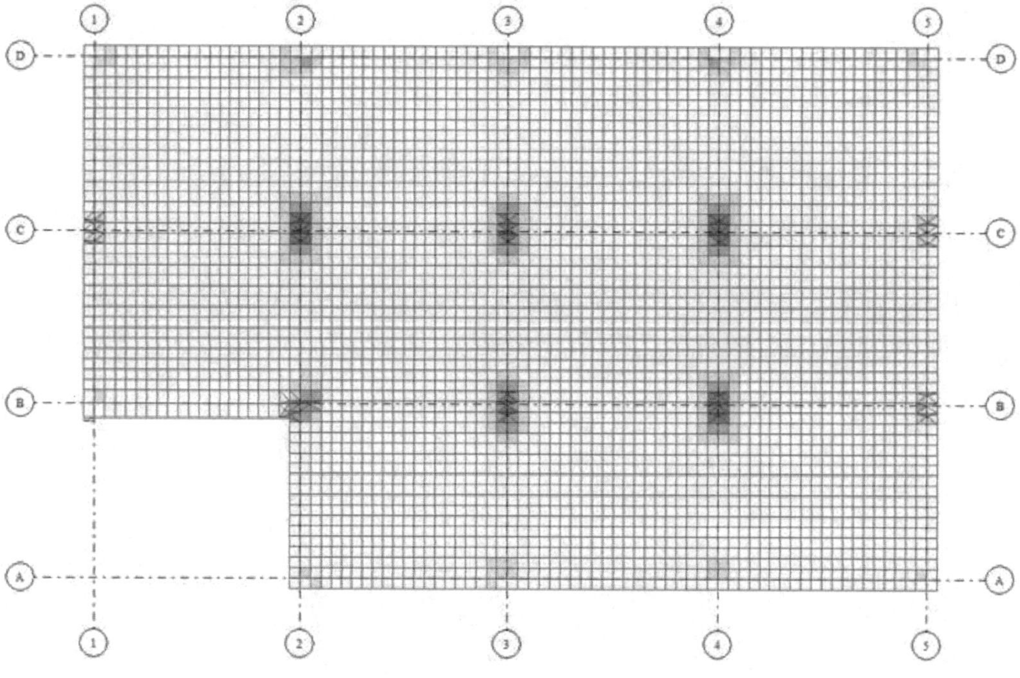

**Figure 9.** Superior loads values in the Mx slab.

Analyzing the diagrams and calculation schemes with load-bearing structural system, can be noticed the classical way of working of the structure under the action of loads and seismic load. The maximum values of the loads to which the beam is subjected appear in the lower part, in the middle of the span (Figure 6), and in the superior part at the beam-column joint (Figure 7). According to the same principle, the loads act on the monolithic slab, in the lower part the maximum load values are noticed in the middle of the span (Figure 8) and at the superior part in at the spans or at the colums (Figure 9). The values of axial forces in the columns decrease vertically, having the maximum value in the embedded zone.

Analyzing the diagrams of the internal forces to which the load-bearing elements are subjected, the more loaded areas at the joint between the column and the plate in the upper part of the flat slab are highlighted. Additionally, the maximum loads are noted in the middle of the flat slab from the bottom side. For the vertical elements, the most stressed area is the one between the part anchored in the ground and the ground floor, vertically, the value of the loads to which the elements are subjected shows a decrease.

In order to improve the performance of the building with flat slab system slabs to the seismic loading, it is necessary to evenly distribute the piers with the working areas in all directions, so that possible displacements are compensated without major stresses.

The displacements of the structure are analyzed with reference to the base of the construction and the highest elevation of the last floor. The horizontal movements of the structure in the longitudinal and transverse direction at the floors level shall not exceed 1/500 of the height. [9]

Analyzing the displacement of the building under the action of seismic loads, we ascertain the following:

- In case of the flat slab structure system:
  on the x direction:
- x direction 0.............3.62mm;
- y direction -0.123.........0.0202mm;
  on the y direction:
- x direction -0.179.........0.315mm;
- y direction -0.00769......4.47mm.

- In case of the load-bearing frame structural system:
  on the x direction:
- x direction -0.00181....6.63mm;
- y direction -0.0593......0.213mm;
  on the y direction:
- x direction -0.961.......0.844mm;
- y direction -0.0115......7.58mm.

## 3. Conclusion

In order to obtain structures that meet durability requirements, it is necessary to take into account the proper design of the structure, the choice of materials, the recommended constructive provisions, execution, quality control, inspections and classifications. [10]

Analyzing the results obtained out of the calculations, we reveal the good behavior of the structure with a flat slab system. The displacements of the flat slab system structure are smaller than in the case of the load-bearing frame structural system. The stress values in the load-bearing elements describe classical stress diagrams.

Flat slab structures have advantages over the classic load-bearing frame structure, such as useful floor height, shorter lead times, and lower steel consumption. As a result of calculation, a lower consumption of reinforcement for the structure with the flat slab system compared to the load-bearing frame system was obtained. Also, in the case of skeleton structures, it is necessary to reinforce additionally the beam.

In order to obtain an efficient flat slab structure, it is necessary to evenly place the vertical bearing elements in all directions. Furthermore, depending on the opening between the vertical bearing elements and the loads to which the slab is subjected, the thickness of the slab is determined. With openings of 4.0x - 6.0m, the thickness of the slab is recommended to be 160mm - 230mm. The recommended ratio between the opening and slab thickness is 1/30.

IOP Conf. Series: Materials Science and Engineering 1304 (2024) 012010 doi:10.1088/1757-899X/1304/1/012010

**References**

[1] https://statistica.gov.md/ro
[2] Local Urban Planning Regulation of the City of Chisinau, SC "LINIA NOVA" SRL, Chisinau 2007
[3] The Institute of Geology and Seismology of the Republic of Moldova, www.idsi.md
[4] Postelnicu T, Pascu R, Zamfirescu D, Popa V 'Seismic design code P100. part I-P100-1/2006 Design provisions for buildings, September 2006'
[5] Nipan Bhandar Kayastha, Rama Debbarma 'Seismic performance of reinforced concrete building with flat slab (*International Conference on Sustainable Materials and Structures for Civil Infrastructures* (SMSC12019))'
[6] Kush Kumar Chaudhary, Mavoori Hitesh Kumar 'Comparative seismic analysis of conventional slab and flat slab with and without shear wall (*International Research Journal of Engineering and Technology (IRJET)*, Volume 08 Issue 04 April 2021, e-ISSN: 2395-0056)'
[7] Coelho E, Candeias P, Anamateros G, Zaharia R, Taucer F, V. Pinto A 'Assessment of the seismic behaviour of RC flat slab building structures (*World Conference on Earthquake Engineering*, Vancouver, B.C., Canada, August 1-6 2004, Paper No.2630)'
[8] Guidelines for the design of slabs in seismic areas, Reference GP 118-2012, Bucharest, February 2012
[9] Fendrikov V.L., Kachkurkin Yu.A. 'Collection of norms and rules for the design of reinforced concrete frame buildings with flat slabs, Krasnodar 2006'
[10] Eurocode 2: Design of concrete structures. SR EN 1992-1-1

IOP Conf. Series: Materials Science and Engineering 1304 (2024) 012011 doi:10.1088/1757-899X/1304/1/012011

# Case study on reinforced concrete frame structure adaptability

**I Olteanu[1], C-I Anton[2], G Țăranu[1]**
[1] Gheorghe Asachi Technical University of Iasi, Faculty of Civil Engineering and Building Services
[2] master student at Gheorghe Asachi Technical University of Iasi, Faculty of Civil Engineering and Building Services

E-mail: ioana.olteanu@academic.tuiasi.ro

**Abstract.** The limited space to build from scratch significantly increased the need to rehabilitate for the re-functionalization of different buildings. The reinforced concrete frame structure is easily adaptable and has several advantages in this direction. When changing the function of a building an evaluation of the existing damages is done as well as a numerical simulation with the new loads. The article presents the case of changing a neighbourhood clinic into a kindergarten. The seismic risk coefficients are computed (R1, R2 and R3) in order to establish the seismic risk class for the existing building and also an evaluation of the damage state. A comparison between the seismic force computed when the building was designed and according to current codes is performed. In order to identify the failure mechanism pushover analysis was performed in both X and Y direction.

## 1. Introduction

The constant change in modern society and the limited space to build from scratch significantly increased the interest to reuse old buildings, no matter their original destination. Instead of tearing down a structure, some beneficiaries are opened to restore them to a purpose that is of interest today. This trend contributes beside the economic development, to urban revitalization by giving abandoned buildings a new life. Such an example is the transformation of La Fabrica, an out-of-use monument to Catalonia's industrial past, into Ricardo Bofill home and office [13]. The building was initially constructed during Catalonia's first golden period of industrialisation in the early 1920s, then new structures and extensions were added gradually when production demanded, finally serving as a repair shop during the World War I. Contributing to this strand of the architectural canon was of great interest to Bofill, with his refreshing take on urbanism [Figure 1]. In the underground there is a gallery, meanwhile at the ground floor a huge conference and exhibition room, with 10-metre-high ceilings which was initially the factory's processing area. The raw cement walls which retain the memories of the previous use of the building were kept unchanged [14].

In order to be able to do that professionals should reinvent themselves and acquire new skills for the reconfiguration of old buildings. Buildings are designed to withstand the time challenge. This is the main reason why, sometimes during their lifecycle functional change is needed in order to adapt to the community's needs or the beneficiary's view. In order to support the community, special funding are offered by the Government and by the European Committee for the rehabilitation of existing infrastructure. An example of such funding is the National Recovery and Resilience Plan of Romania, which is a strategic document regarding the investment priorities and reforms needed for recovery and sustainable growth, in alignment with the green and digital transitions of the European Commission.

**Figure 1a.** Old photo with the La Fabrica

**Figure 1b.** La Fabrica converted into the house of Ricardo Bofill

**Figure 1c.** Conference Hall view, interior of Ricardo Bofill house

In order to access these funds a particular documentation needs to be submitted which follows national and international regulation. In the Romanian legislation, among the essential documents is the expertise report which is done by an authorised expert. The report gives information regarding the qualitative and quantitative state of the construction and, if it is the case, some recommendations regarding the structural

IOP Conf. Series: Materials Science and Engineering 1304 (2024) 012011 doi:10.1088/1757-899X/1304/1/012011

rehabilitation measures that need to be taken into consideration to strengthen the structural system. This report is done following the P100 part III, from 2019 regulations [4].
The article presents the Romanian methodology for an expertise report which studies the damaged state of a neighbourhood clinic in order to convert it into a kindergarten hence in the area several private medical offices were opened in the last years and the building has not been used for some time. On the other hand, the need for a new kindergarten in the area is stringent since the city of Iasi has extended a lot in that direction, in the Copou hill direction and further to Breazu and Rediu villages. In the second part of the paper, a nonlinear static analysis is performed for one of the buildings in order to compare the results with those obtained through the simplified computation method.

## 2. Romanian methodology to evaluate existing buildings

According to P100-3/2019 there are three methods to assess the building state, differentiated by the detailing degree of the computation methods and the considered checking: level 1 (low complexity), level 2 (medium complexity) and level 3 (high complexity) [4]. The corresponding method is chosen based on the following criteria: technical requirement from the design and execution period of the project, building complexity from structural point of view (openings, number of floors, horizontal and vertical regularities), available data which gives the knowledge level, the function of the building, its importance and value, seismic hazard conditions according to P100-1/2013, structural system, fundamental requirements for the buildings, the purpose of the technical report and other relevant conditions.
Level 2 methodology is most frequently used for current buildings. It contains a qualitative assessment and a quantitative one. The qualitative evaluation implies to check a list of elements referring to the structural layout based on the used materials, meanwhile the quantitative one is based on an elastic structural computation which uses specific coefficients for each material. The earthquake effect is approximated by a set of conventional forces which are then applied on the structure. The lateral forces are chosen so that the obtained displacements are obtained through the linear computation to simulate the displacements imposed to the structure by the seismic load. The checking relation depends on the failure mode (ductile or brittle) of the structural elements to different internal forces. The structural computation can be done considering one of the methods presented in P100-1/2013 meaning the equivalent static seismic force or the modal computation with response spectrum [3].
The qualitative assessment measures the extent to which the rules for general compliance of structures and detailing of structural and non-structural elements are respected in the analysed constructions. The nature of the structural deficiencies and their extent represent essential criteria for the decision of structural intervention and rehabilitation solutions. The following aspects are considered: the loading path towards the foundation, building redundancy check, construction layout, the interaction with other constructions and elements. Each criteria is based on the scientist's experience and the visual inspection points, which are in the end summed up. Depending on the final score, the R1 (risk coefficient) is defined for the building which includes it in one of the four risk classes. R1 represents the degree of fulfilment of the conditions of structural compliance, of the composition of the structural elements and of the constructive rules for structures that withstand effect of the seismic action.
R2 risk coefficient represents the degree of structural damage that expresses the proportion of structural damages produced by seismic action and other causes. The damaged state is evaluated in a table, based on the visual inspection considering the following criteria: the earthquake action effect, the vertical actions, deformations effects, poor execution, and environment effect. In the end a total score between 0 and 100 is obtained which includes the building in one of the four seismic risk classes [3,4].
The degree of seismic structural insurance represents the ratio between the capacity and the seismic structural requirement, expressed in terms of strength for level 1 and 2 methodologies or in terms of displacements for level 3 methodology. R3 risk coefficient is determined for the ultimate limit states.

## 3. Building description, initial results and discussions

The analysed structure was designed initially to be a day nursery, which was finished in 1971. The building was designed according to P13-63 seismic regulation. In 1983 it was converted into a neighbourhood clinic, which was closed in 2013. Between 2013 and 2022 the building was closed and

abandoned. For the initial space conversion, no structural rehabilitation was performed on the building, only lightweight modification of the interior space. The benefits of a reinforced concrete frame structure were extensively used in this case. In 2020, the building was studied to establish the damage degree and the rehabilitation measures necessary in order to convert it into a kindergarten. The building is a complex one, made of 2 bodies, A and B, with a technical basement, ground floor and one floor. The structural system is represented by a reinforced concrete frame structure with precast reinforced concrete elements as exterior envelope. The dimensions of the main structural elements are: 25x45 cmxcm beams, 35x45 cmxcm columns and slabs of 13 cm thickness.
In case of the seismic force, the literature review showed that the seismic norms had improved along the building lifecycle [5, 7]. If initially the building was designed after a code which considered the seismic force a percentage from the building weight, the today evaluation is performed according to P100-2013, which is based on Eurocode 8, improved by the recent research results in the field [1, 3, 6].
Table 1 presents synthetically the coefficients considered in both codes for the seismic load assessment (P13/63 and P100/2013) [2, 3, 4]. It can be noticed that some coefficients had been introduced, some of them had been replaced and some values had been changed. If a comparison is made between these coefficients, it can be seen that with respect to the P13-63, the current code is experiencing a reduction of the seismic force with only 4.24%. These results are consistent with the findings on site regarding the limited damage suffered by the building along the years.

**Table 1.** Coefficients used for the seismic load assessment.

| | P13-63 | P100-2013 |
|---|---|---|
| $\gamma_I$ (importance factor) | - | 1.2 |
| $k_s$ (coefficient regarding the seismic protection degree) / $a_g$ (peak ground acceleration) | 0.05 | 0.25 |
| $\beta_k$ (the normalised spectrum of the elastic response for the horizontal components of the ground acceleration) | 2.75 | 2.5 |
| $\Psi$ (reduction coefficient) | 1.20 | - |
| q (behaviour coefficient) | - | 4.725 |
| $\varepsilon_k / \lambda$ (correction coefficient) | 1 | 1 |
| | 0.165 (100%) | 0.158 (95.76%) |

After the visual inspection only, minor damages were observed. Among these were cracks of the exterior and interior plaster, the woodwork of the openings and the terrace finishes which lead to water infiltration in the top slab, Figure 2. Significant problems were identified at the sidewalk which also led to infiltrations to the foundation.
Based on the visual inspection R1 and R2 coefficients were computed. When checking the structural integrity of building B, a R1 coefficient of 85 points was obtained. In assessing this factor the following parameters were considered: conditions referring to the structural configuration (the loads path, the presence of soft stories, irregularities in plan and elevation, the torsional effect, the foundation layout), conditions referring to the interactions with other structures (distances from other buildings, the presence of short columns, the non-structural elements are adequately connected to the structural system), conditions referring to the elements formation and conditions referring to the slabs. According to the received grading, the building is considered to be in IIIrd seismic risk class. In case of R2 coefficient, when the different possible structural and non-structural damages were assessed and based on the visual inspection a value of 87 was obtained. Considering that the maximum value is 100, it is noticed that, from this point of view the structure is also included in IIIrd seismic risk class.

**Figure 2.** Initial building state

The structure was introduced in SAP2000 computer software and the static analysis was performed. In the numerical simulation a characteristic snow load of 2.5kN/m$^2$ was considered. The self-weight of the structure was computed automatically by the software, meanwhile the considered dead load for the current floor was of 1.5 kN/m$^2$ and 8.5 kN/m$^2$ for the top floor, and the live load was of 3 kN/m$^2$ for the current floor and 1.5 kN/m$^2$ for the top floor. The results were represented by the internal forces magnitudes. These values were used to compute the R3. A value of 67 was obtained, which is at the lower level of IIIrd seismic class.

According to P100-2019, the IIIrd seismic risk class contains structures that withstand seismic load and can present structural damage, without affecting the structural safety, but the non-structural damages may be significant.

The modal analysis was performed in order to assess the deformed shapes and the modal participation factor. Figure 3 presents the structural deformation for the first two eigen modes. It can be noticed that for eigen mode 1 the deformation is represented by translation on the short direction, in eigen mode 1 torsion appears, meanwhile in eigen mode 3 translation on the long direction was obtained.

Considering this low value of R3 coefficient and the fact that the building will be rehabilitated in order to build a kindergarten, a pushover analysis was performed in SAP2000 software in order to assess the most vulnerable elements and the capacity curves on X and Y direction of the analysed building.

## 4. Nonlinear static analysis, results and discussions

The pushover analysis is used in the evaluation of the seismic performance of existing structures, but also in the design stage. It is considered an appropriate method for the Performance based design of structures. The method involves the application of a series of inelastic static analyses on a structure, using a loading pattern defined on the basis of the first eigen mode or other shapes of loading or in the terms of accelerations. The loading pattern remains unchanged in the analysis process, but the intensity increases until the structure reaches the maximum displacement imposed by the designer [8-10]. The results of the pushover analysis are represented by the failure mechanism, respectively the evaluation of the structure's capacity, plotting the variation of the displacement from the top of the structure with respect to the shear force at the base of the structure [11-12].

The capacity curve for the analysed building, on X and Y direction are presented in Figure 4. It can be observed that the maximum efforts are withstood in X direction due to both the geometric configuration of the structure and the positioning of the cross-sections. A maximum force of around 6000kN can act on the structure, leading to a displacement of 12mm in the longitudinal direction of the building, causing damage in the ground floor columns. As expected, in the short direction - Y, the building is considered to be more vulnerable, as the maximum displacement is only 0.5mm.

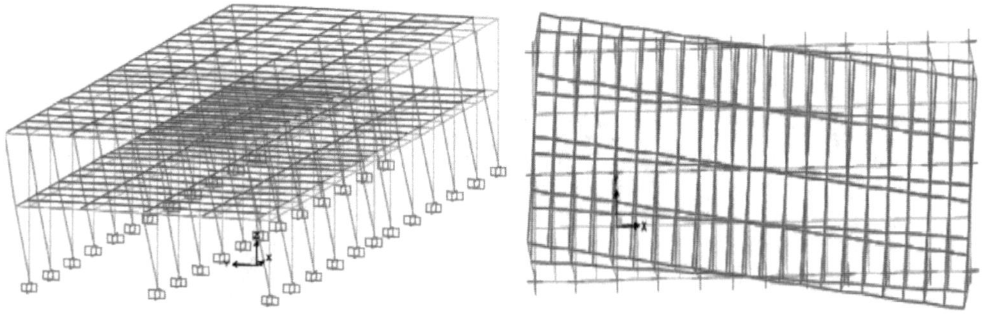

**Figure 3.** Deformed shape for eigenmode 1 and 2

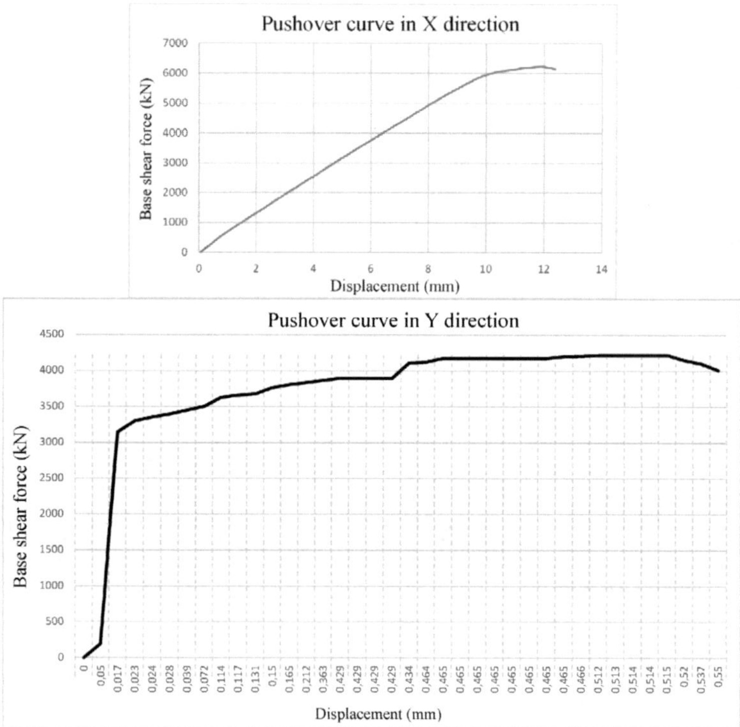


**Figure 4.** Capacity curves for the pushover analysis on X and Y direction respectively

In case of the failure mechanism, it was observed that for both load cases – pushover in X and Y direction, the first cracks are developing in the beams from the first floor, and then continue to the rest of the beams. As the force increases, as well as the displacement of the structure the cracks for the immediate occupancy level are obtained at the base of the ground floor columns, Figure 5. The failure mechanism that was observed is in accordance with the EC8 and P100/2013 concept which recommend to design soft beams-strong columns.

IOP Conf. Series: Materials Science and Engineering 1304 (2024) 012011 doi:10.1088/1757-899X/1304/1/012011

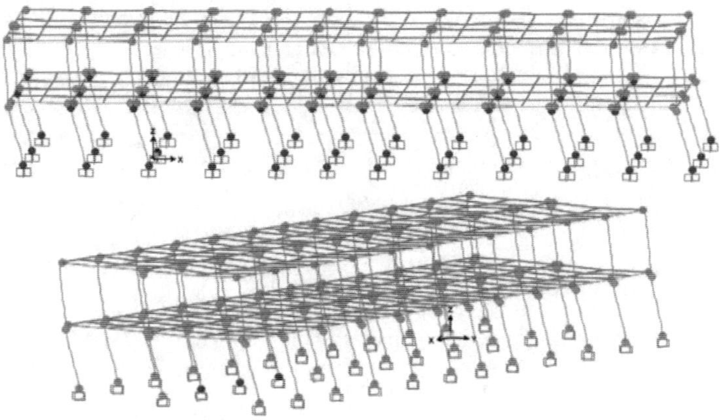

**Figure 5.** Failure mechanism from pushover analysis in X and Y direction respectively

## 5. Conclusions

Considering the changes, a society undergoes in order to adapt to current trends, the built environment should be designed in order to offer the possibility for the buildings to take different functionalities, depending on the necessities. Moreover, the limited space to build from scratch in major cities and the restrictions from the local authorities regarding new buildings, as well as the available funding possibilities for the rehabilitation of existing buildings encourages the revitalisation of existing ones which are currently left abandoned or are not used.

In the case study presented it can be seen that the reinforced concrete frame structures can be considered to be very easily adaptable no matter what functionality is considered. The large open space between the structural elements offers a wide space which can be arranged easily in any shape with lightweight materials, as is the autoclaved concrete.

**Figure 6.** Finalized works

In the present study it was also observed that, even though the structure was built according to a 1963 regulation, the restrictions applied then are similar to those from current codes which are in accordance to the European legislation, Eurocode 8. A decrease of the seismic force with 4.24% was obtained from the computation. Due to this similarity, and to the fact that the building was well maintained, no structural rehabilitation was required. The works involved for the re-functionalization of the space referred to re-plastering, changing the woodworks and some changes in the layout of the walls adequate for a kindergarten, Figure 6.

In the pushover analysis it was observed that the structure is more vulnerable in the short direction, Y and that the structure is well confirmed as the failure mechanism is in accordance with the weak beams-strong column principle from EC8. The results from the pushover analysis are in accordance with the simplified methodology applied to assess the seismic coefficients R1, R2 and R3 which place the structure in IIIrd seismic class, meaning that the structural elements of the building can withstand future earthquakes without significant damage, meanwhile the non-structural elements might be affected by a seismic action.

## 6. References

[1] EC8 (2002) Eurocode 8. Design of structures for earthquake resistance, Brussels
[2] P13-63 - Normativ conditionat pentru proiectarea constructiilor civile si industrial in regiuni seismice (Conditional regulation for the design of civil and industrial buildings in seismic regions)
[3] P100-2013 – Cod de proiectare seismica- Partea III – Prevederi de proiectare pentru cladiri (Seismic design code - Part III - Design provisions for buildings)
[4] P100-2019 – Cod de proiectare seismica- Partea I – Prevederi de proiectare pentru cladiri (Seismic design code - Part I - Design provisions for buildings)
[5] Olteanu I. 2011 *Evaluarea comportării structurilor în cadre din beton armat supuse la acţiuni seismice (Vulnerability assessment for reinforced concrete frame structures subjected to seismic loading)*, Ed. POLITEHNIUM Iasi, ISBN 978-973-621-363-2, 221 pages.
[6] Olteanu I., Dinu D.I., Soveja L., Budescu M. 2018, *The Necessity of Performing Current Maintenance Work on Buildings*, Bulletin of the Polytechnic Institute of Jassy. Construction. Architecture Section, I--SSN 1224-3884, 64 (68), Fasc. 2.
[7] Budescu M., Soveja L., Olteanu I. 2016, *Some remarks regarding seismic vulnerability for orthodox churches*, The 1940 Vrancea Earthquake. Issues, Insights and Lessons Learnt, DOI 10.1007/978-3-319-29844-3_1, Springer International Publishing, Switzerland.
[8] Venghiac VM, Neagu CP, Taranu G, Rotaru A 2023 *Time History Analyses of a Masonry Structure for a Sustainable Technical Assessment According to Romanian Design Codes*, Sustainability 15 (4)
[9] Taranu G 2021 *Comparative study on the structural behaviour of a 30m height building with a RC frames structure located in a high intensity seismic area,* IOP Conference Series: Materials Science and Engineering, Computational Civil Engineering (CCE 2021), Vol 1141
[10] Taranu G 2020 *Case Study - Comparative Analysis of an RC Structure, Part of a Complex of Adjacent Buildings,* Bulletin of the Polytechnic Institute of Jassy. Construction. Architecture Section, I--SSN 1224-3884, 66, Fasc. 3.
[11] Păuleţ-Crăiniceanu F., Luca S.G., Pastia C, Florea V -2019 *Improved Strategy For Optimal Control Of Civil Engineering Structures*, Computational Civil Engineering Conference (Cce2019), Iop Conference Series: Materials Science And Engineering (Vol. 586, No. 1, P. 012044). Iop Publishing
[12] Mohamed Nour El-Din Abd-Alla 2007 *Application Of Recent Techniques Of Pushover For Evaluating Seismic Performance Of Multistory Buildings* , Master Thesis, Faculty Of Engineering, Cairo University Giza, Egipt
[13] Ricardo Bofill's cement factory home La Fábrica | House & Garden (houseandgarden.co.uk)
[14] Explore La Fabrica House — The transformation of the cement factory into a magnificent mansion by the Spanish architect - Living + Nomads – Travel tips, Guides, News & Information!

**ACKNOWLEDGEMENTS**
This paper was realised with the support of project EFECON – ECO-INNOVATIVE PRODUCTS AND TECHNOLOGIES FOR ENERGY EFFICIENCY IN CONSTRUCTION, POC/71/1/4 - Knowledge Transfer Partnership, Cod MySMIS: 105524, ID: P_40_295, Project co-financed by the European Regional Development Fund.

IOP Conf. Series: Materials Science and Engineering 1304 (2024) 012012 doi:10.1088/1757-899X/1304/1/012012

# Finite element analysis on composite steel-concrete shear walls with centered and staggered openings

**V Todea[1], V Popescu[1], D Dan[1], S C Florut[1], V Stoian[1]**
[1] Politehnica University Timişoara, Faculty of Civil Engineering, Department of Civil Engineering and Building Engineering Services, 2nd T. Lalescu, Timişoara, *Romania*

E-mail: daniel.dan@upt.ro

**Abstract.** Present paper presents a comparative study made on the seismic performance of composite steel-concrete shear walls with rectangular openings, by performing numerical analyses using ATENA 3D software. Two specimens were designed at a reduced scale 1:3 having similar arrangements in the cross-section and the same geometry. First element was conceived with centered openings while the second one was designed with staggered openings. Steel connectors disposed as horizontal stiffeners were used to design and to assure the full connection between the structural steel profiles embedded in the edges and the concrete core. The openings that were considered on each story of the walls, were having the dimension of a common architectural door. A part of the results recorded from the experimental tests were used to calibrate the numerical models. This study aims to establish the seismic performance and to investigate the influence of the openings and the geometric position on the overall seismic behavior of the composite shear walls. Key parameters which describe the seismic performance of structural elements such: bearing capacity, deformation capacity, stiffness degradation, cracks and strain development are compared between the specimens and discussed. The results showed that different arrangements of the openings could significantly modify or increase the seismic performance of the composite structural walls.

## 1. Introduction

Composite steel-concrete walls are frequently met in case of low to high-rise buildings to brace the structures against seismic activity. In general, the composite walls are made from a rectangular concrete web panel reinforced by a series of hot-rolled or welded structural steel profiles. These profiles could be partially or totally embedded in the cross-section, the connection with the concrete web panel being assured by steel connectors [1]. Many times, due to architectural reasons, structural walls need to be conceived with different types of openings such as doors or windows, to assure the building functionality. In most of the cases, the openings are geometrically aligned but sometimes staggered openings are required on the height of the walls (figure 1). In other cases, due to functional modifications of the existing buildings, new openings are needed to be cut out into the structural walls, to assure the ventilation or heating services paths. The openings could significantly modify the seismic behavior of structural walls and have to be considered in the design process of the structures [2], [3]. In the literature the subject related to the seismic behavior of composite steel-concrete shear walls with centered or staggered openings is not thoroughly investigated, additional investigations and analytical or numerical studies are needed to extend the knowledge and the existing database of such structural elements. To investigate the influence of the geometrical position of openings on the seismic behavior of composite steel-concrete shear walls, a comparative study made on two specimens designed at a reduced scale 1:3 is presented in this paper, as part of a large theoretical and experimental study that is in progress now. The present study is based on the numerical results recorded using the non-linear structural analysis software ATENA 3D Engineering. The seismic performance of one composite steel-concrete shear wall with centered openings CSRCW-21-CO and one with staggered openings CSRCW-21-SO is investigated and the results are compared in terms of: load bearing capacity, deformation capacity, lateral stiffness degradation, cracks, and strain development. Also, the boundary conditions and material laws used to calibrate the numerical models with the experimental results are described in this article.

**Figure 1.** Example of low-rise building with staggered openings

## 2. Details of the designed specimens

The investigated specimens present partially I-shaped steel profiles embedded in the edges and measure three meters height, one meter in width and one hundred millimeters in thickness, being designed at a reduced scale 1:3 to simulate the first three storeys of an element part of a lateral resisting system. The core of the walls was encased into a rectangular reinforced concrete block of foundation which measure 1,50 m in length, 400 mm in height and 350 mm width. The CSRCW-21-CO specimen has centered aligned door openings (which measure at the reduced scale 300x700 mm equivalent to real dimensions of 900x2100 mm) while in the case of the CSRCW-21-SO specimen, the openings were moved eccentrically left-right on the height of the wall, from the vertical axis to the edges by 95 mm (figure 2).

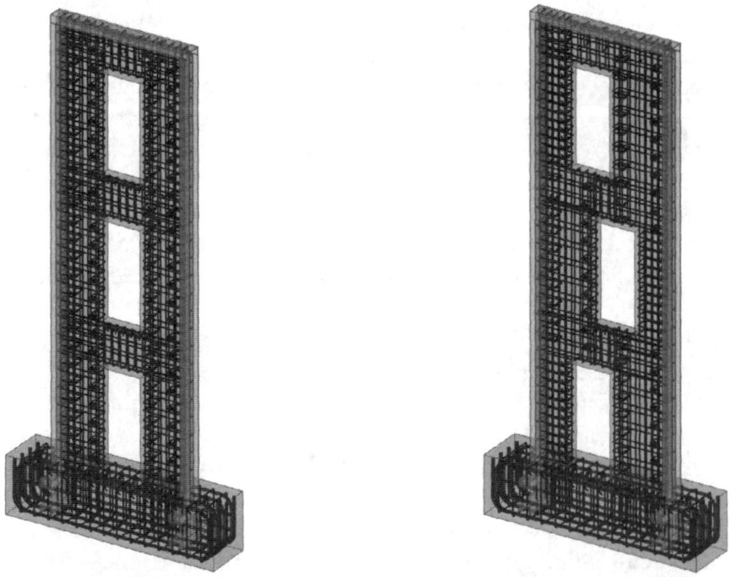

a. CSRCW-21-CO b. CSRCW-21-SO

**Figure 2.** The geometry and reinforcements of the investigated specimens

The structural I-shaped steel profiles embedded in the edges were manufactured using 5,65 mm welded steel plates. The composite connection between the reinforced concrete web panel of the walls and the structural steel profiles was designed as full connection. The connection was assured by a series of steel plates perforated by two circular holes, through which two longitudinal bars were passing as can be seen in figure 3. Hot-rolled steel bars with 8 mm diameter were used to reinforce the concrete web panel of both specimens. The shear capacity of the vertical piers was increased using enclosed stirrups which were welded to the steel profiles on the edges and were spaced at every 60 mm on the height of the walls. The specimens were constructed with a U-shaped steel profile on the top to assure a more uniform distribution of the local stresses induced by the vertical load. More details about walls detailing are described in [4].

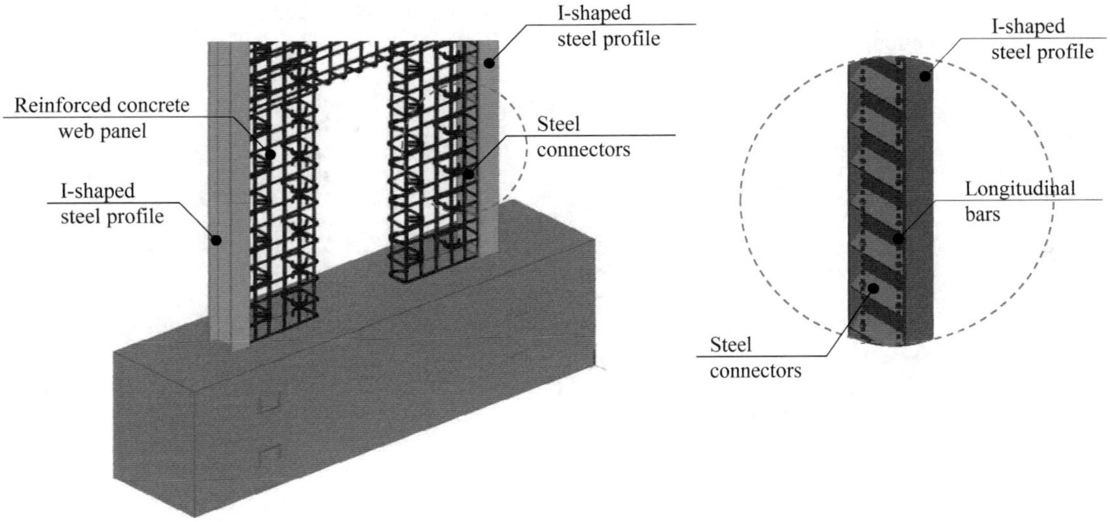


**Figure 3.** Details of composite connection between steel profiles and reinforced concrete core

## 3. Material properties

Hot-rolled S500 with ribbed periodical sections steel bars were used to reinforce the composite walls. The I-shaped steel profile welded plates partially embedded in the edges of the wall, were designed as S460 steel grade. The yield characteristic strength ($f_{yk}$), maximum strength ($f_u$) and the Young elastic modulus ($E_s$) of the proposed steel plates and reinforcements, were obtained by performing tensile tests. The mechanical properties obtained are listed in table 1. The concrete material was designed as C30/37 concrete strength class. The mechanical parameters of concrete were determined based on compressive tests performed on three cubic samples at 28 days. Based on the results, the mean strength $f_{cm}$ result around 49.23 MPa.

**Table 1.** Mechanical parameters of steel

| Steel type | Yield strength $f_{yk}$ (MPa) | Ultimate strength $f_u$ (MPa) | Elasticity modulus $E_s$ (MPa) |
|---|---|---|---|
| Reinforcements | 530.25 | 636.02 | 225290 |
| Plates | 461.34 | 554.58 | 206820 |

## 4. Numerical analysis

The geometry of the composite walls was defined and modeled using 3D volume elements called macroelements. The contact between each macroelement was assumed to be perfectly rigid. Eight macroelements were used to model the overall geometry of the walls. The 3D solid tetra elements with six nodes of integration and 80 mm size were used to assure the FE mesh of the models. To prevent stress concentrations at the points where horizontal and vertical external forces were applied, two 30 mm thick metal plates were defined. A surface spring was attached on the bottom side of the walls beam foundation to simulate the surface support generated by the lab foundation (figure 4). The spring was set as infinite rigid in compression and inactive in taking the negative translations (simulating a rocking foundation) effect produced by the overturning bending moment at the base of the walls. The negative translations were prevented by the secondary surface spring attached on the top part of the beam foundation edges, simulating the axial stiffness of the local steel anchoring rods.

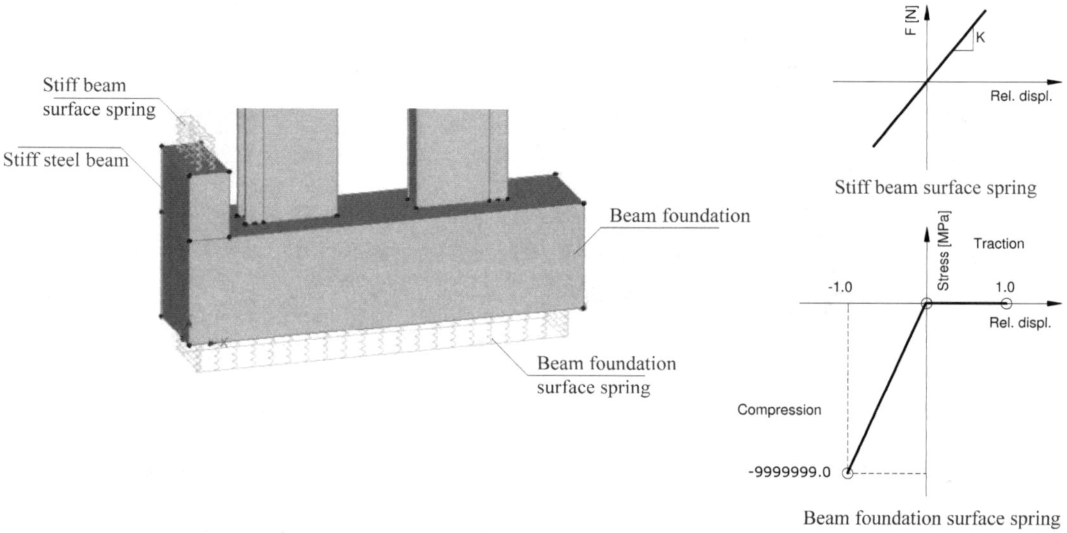


**Figure 4.** Modeling of the walls boundary conditions

The rebars embedded in concrete were modeled using truss linear discrete elements with two integration nodes, each bar being modeled individually. All the stirrups were modeled using four lines with four integration nodes, being completely closed neglecting the hooks. All the reinforcement bars were modelled with an assumption of perfect bond to concrete macroelements. The SBETA (StahlBETonAnalyse) CC3DNonLinCementitious2 material model was used to calibrate the nonlinear behavior of concrete in tension and compression, based on the recommendations from other researches [5]–[8]. All the nonlinear concrete parameters were automatically calculated by the software using the mean cubic strength of concrete determined experimentally. The 3D Steel Bilinear Von Mises material model which incorporate the Von Mises failure criterions was applied to simulate the behavior of steel in compression and tension from the steel profiles partially embedded in the edges. The following parameters were considered in the simulation: yield limit $\sigma_y$=445 MPa; hardening modulus H=2100 MPa; elastic modulus E=210000 MPa and Poisson's ratio $\mu$=0.3. In case of steel reinforcements, the CCReinforcement multilinear plastic stress-strain constitutive model was used to cover the behavior in tension and compression. The buckling of longitudinal bars in compression was not considered in this study. In figure 5 and figure 6 are presented the numerical models defined for both specimens.

IOP Conf. Series: Materials Science and Engineering 1304 (2024) 012012 doi:10.1088/1757-899X/1304/1/012012

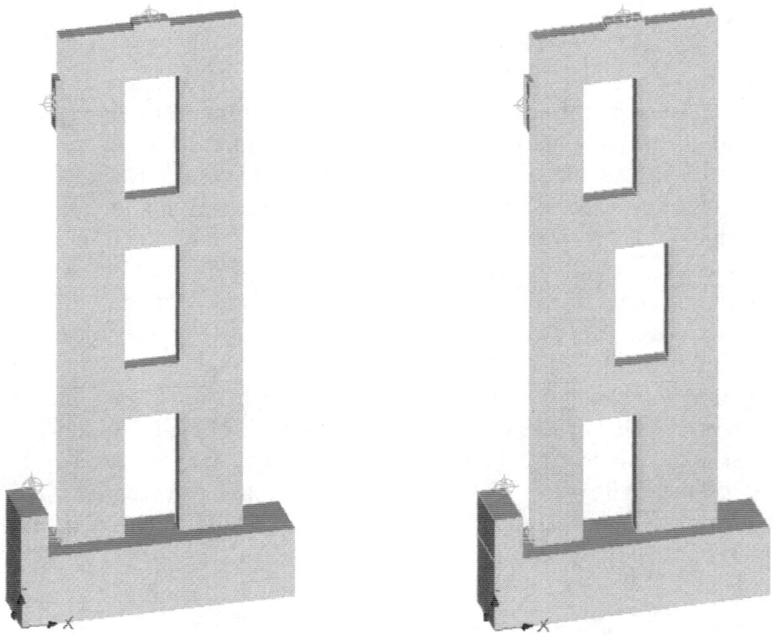

a. CSRCW-21-CO b. CSRCW-21-SO

**Figure 5.** The 3D macroelements of investigated specimens

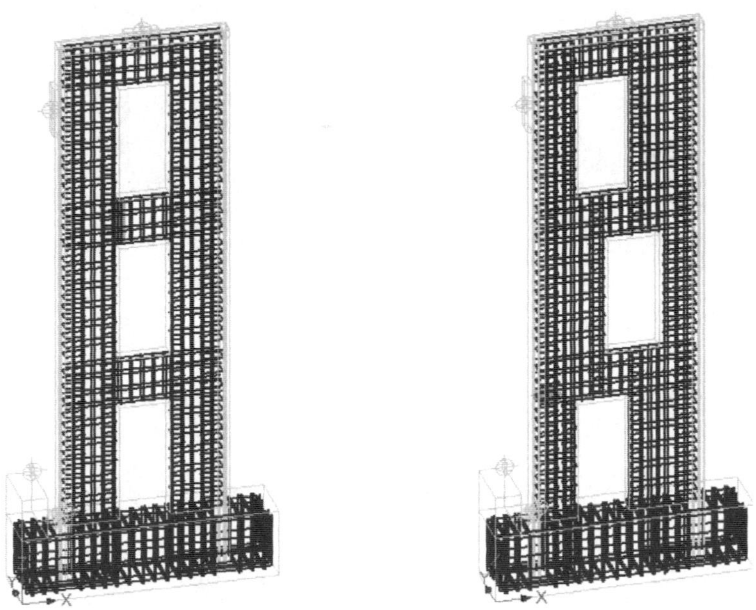

a. CSRCW-21-CO b. CSRCW-21-SO

**Figure 6.** Modeling of steel reinforcements for both specimens

## 5. Numerical results

*5.1. Envelope curves*

The numerical analyses were performed using the displacement control mode (pushover analysis). A constant vertical load of 100 kN was applied on both specimens during the numerical analysis, simulating the gravitational load. The experimental hysteretic curve of the CSRCW-21-CO specimen was used to calibrate and to compare the results of the numerical models [4]. The envelope curves of both analyzed specimens are presented in figure 7. In case of the CSRCW-21-CO specimen, the envelope curve obtained numerically is quite similar with the experimental hysteretic curve. Also, the brittle failure mode characterized by the sudden degradation of the horizontal force is recorded by the approached numerical model. The maximum load bearing capacity recorded experimentally was $F_{exp}$=266 kN while numerically $F_{num}$=263.9 kN. It can be said that the numerical approaches considered for the numerical models could be used for further investigations of such walls. By modifying the position of the openings, the seismic behavior of CSRCW-21-SO specimen is not significantly changed. The load bearing capacity of the specimen and the lateral stiffness is quite close with those recorded by the CSRCW-21-CO specimen, but the failure mode is triggered in a different way. The composite wall with staggered openings CSRCW-21-SO has an improved deformation and stresses redistribution capacity in comparison with composite wall with regular openings CSRCW-21-CO.

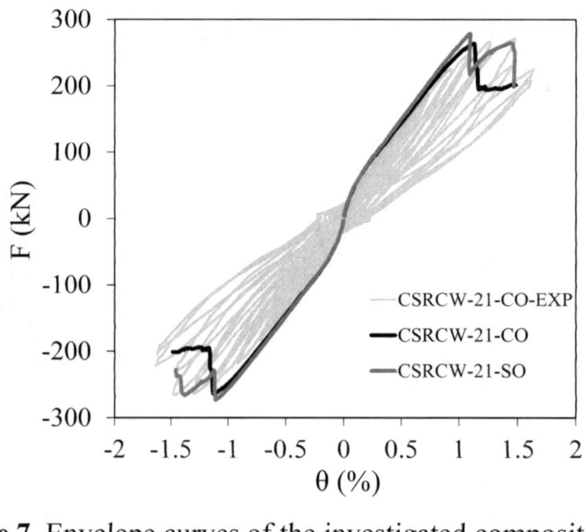


**Figure 7.** Envelope curves of the investigated composite walls

*5.2. Lateral stiffness degradation*

The lateral stiffness of the walls (figure 8) was evaluated as the ratio between the lateral force F (kN) and the top lateral displacement (Δ). The behavior of both specimens is quite similar until the failure occur, but the initial stiffnesses in the first step of loading recorded are different. The shear wall CSRCW-21-SO recorded a greater initial stiffness K=48,59 kN/mm while element with centered openings CSRCW-21-CO K=33,04 kN/mm, a difference about 30%. This aspect may be the result to the fact that, in case of the CSRCW-21-SO composite wall the coupling beams became shorter after changing of openings position and therefore the deformed length was reduced and due to a better confinement assured by the enclosed stirrups. Other aspects can be related to the vertical piers, that had an local alternate increased cross-section with an enhanced moment of inertia, that contributed on the increasing of the wall lateral stiffness.

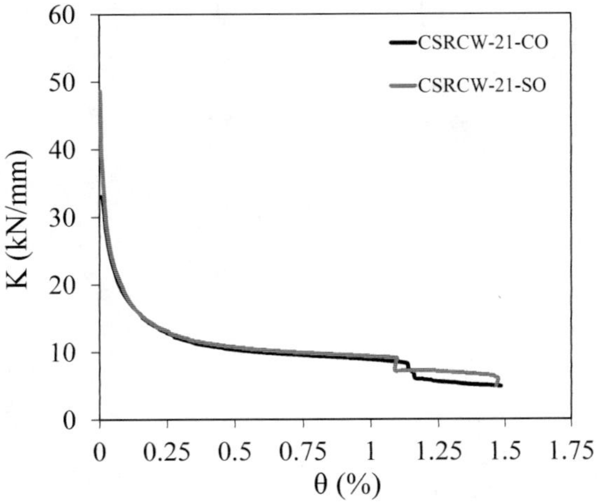


**Figure 8.** The lateral stiffness degradation of both composite walls

*5.2.1. Strain and stresses analysis*

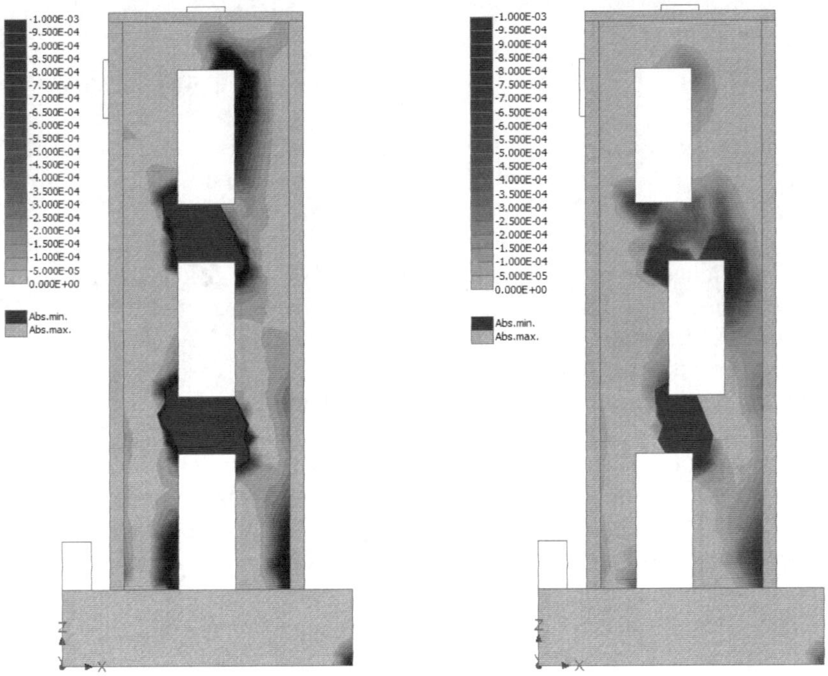


a. CSRCW-21-CO b. CSRCW-21-SO

**Figure 9.** The distribution of the compressed concrete strain

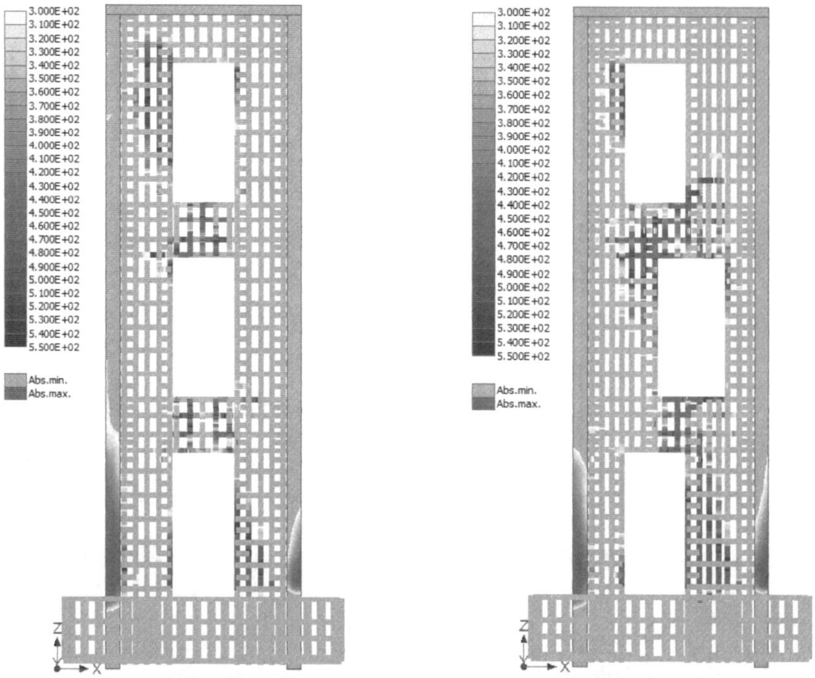

a. CSRCW-21-CO b. CSRCW-21-SO

**Figure 10.** Tensile stresses in steel profiles and reinforcements

In figure 9 are shown for comparative study the trajectories of the compressed concrete strain for both analyzed specimens. These trajectories are marked with blue color on the concrete web panel. It can be seen that for the CSRCW-21-CO wall the most damaged parts are the coupling beams. The vertical piers developed a high state of stresses just over the encasement level in the beam foundation. In case of the wall with staggered openings, this situation is slightly modified due to different openings arrangement. The compression stresses in concrete are more accentuated at the level of first coupling beam in time that for the second beam these trajectories are more widespread. Similar tensile stresses are developed in steel profiles partially embedded in the edges in case of both specimens, but the composite wall with staggered openings developed a larger state of tensile stresses in steel reinforcements as can be seen in figure 10. The composite wall CSRCW-21-SO will develop a higher capacity to dissipate the hysteretic energy due to larger plastic deformations in reinforcements and will experience a more stable seismic behavior in comparison with the composite wall with regular openings. In case of CSRCW-21-CO specimen the concrete in the coupling beams will be crushed or damaged almost simultaneously in the same time due to the uniform state of stresses recorded as is stated in figure 9.

*5.2.2. Cracks analysis*

By analyzing the cracks distribution of the composite walls (figure 11), it can be observed that the CSRCW-21-SO specimen with staggered openings presents a different crack distribution in comparison with the composite wall with regular openings. In case of CSRCW-21-CO both coupling beams are intersected by inclined shear cracks when the total drift reached only 0,25% from the total height. In case of CSRCW-21-SO specimen, these cracks occur only on the horizontal beam from the second level, in time that the first beam from the first level is intersected only by bending cracks propagated from the coupling beam edges. This mode of cracking may suggest an improved lateral stiffness of the composite wall with staggered openings.

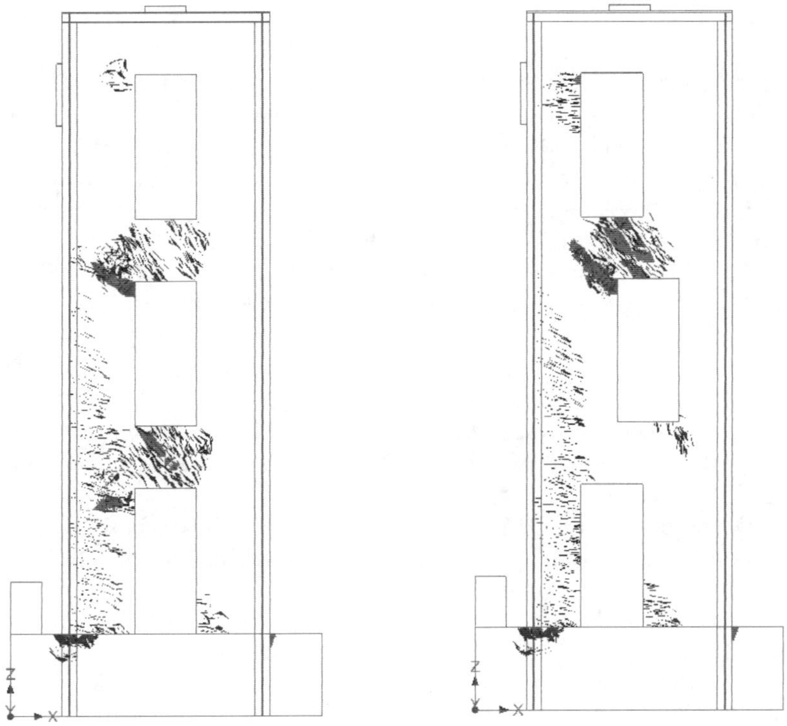

a. CSRCW-21-CO b. CSRCW-21-SO

**Figure 11.** Cracks development on the concrete web panel

## 6. Conclusions

This paper presents the results of a comparative study made on the seismic behavior of composite steel-concrete walls with rectangular openings, by performing a series of numerical analyses. Two walls with centered and staggered openings were analyzed and their seismic performance in terms of load bearing capacity, lateral stiffness degradation, deformation capacity and cracks distribution were compared and discussed during the chapters. To validate the numerical models, the experimental curve of the CSRCW-21-CO specimen was used. By analyzing the results, the following conclusions can be drawn from the present study:

- The numerical approaches considered to simulate the seismic performance of composite steel-concrete shear walls with staggered and regular openings using ATENA 3D Engineering are in a good agreement with the results recorded by the experimental tests;
- The composite wall with staggered openings CSRCW-21-SO developed a higher initial stiffness, with more than 30% in comparison with the composite wall with regular openings CSRCW-21-CO;
- The specimen CSRCW-21-SO will dissipate a higher rate of the seismic energy due to larger plastic deformations developed in steel reinforcements and will experience a more stable seismic behavior in comparison with the composite wall with regular openings CSRCW-21-CO;
- The sequence of the occurrence and cracks distribution in case of the CSRCW-21-SO specimen is quite different in comparison with the composite wall with central openings CSRCW-21-CO;

**Acknowledgments**

This paper was financially supported by the Project "Network of excellence in applied research and innovation for doctoral and postdoctoral programs" / InoHubDoc, project co-funded by the European Social Fund financing agreement no. POCU/993/6/13/153437.

## 7. References

[1] A. L. Ciutina and A. stratan, “Cyclic Performances of Shear Connectors,” in *Composite Construction in Steel and Concrete VI*, Devil’s Thumb Ranch, Tabernash, Colorado, United States: American Society of Civil Engineers, Feb. 2011, pp. 52–64. doi: 10.1061/41142(396)5.
[2] Y. Zhang, G. Yuan, Q. Shu, M. Zhu, and L. Lu, “Investigation on seismic behavior of RC shear walls with multiple post-construction openings based on experiment and simulation,” *J. Build. Eng.*, vol. 46, p. 103707, Apr. 2022, doi: 10.1016/j.jobe.2021.103707.
[3] M. Mosoarca, “Failure analysis of RC shear walls with staggered openings under seismic loads,” *Eng. Fail. Anal.*, vol. 41, pp. 48–64, Jun. 2014, doi: 10.1016/j.engfailanal.2013.07.037.
[4] V. Todea, D. Dan, S. C. Florut, and V. Stoian, “Experimental investigations on the seismic behavior of composite steel concrete coupled shear walls with central openings,” *Structures*, vol. 33, pp. 878–896, Oct. 2021, doi: 10.1016/j.istruc.2021.04.074.
[5] P. Chaimahawan, S. Suparp, P. Joyklad, and Q. Hussain, “Finite Element Analysis of Reinforced Concrete Pile Cap using ATENA,” *Lat. Am. J. Solids Struct.*, vol. 18, no. 2, p. e342, 2021, doi: 10.1590/1679-78256290.
[6] B. El-Ariss, M. Mansour, and T. El-Maaddawy, “Effects of Web Height Reduction and Skew Angle Variation on Behavior of RC Inverted T-Beams,” *Buildings*, vol. 11, no. 10, p. 451, Oct. 2021, doi: 10.3390/buildings11100451.
[7] E. Ismail, M. S. Issa, and K. Elbadry, “Finite element analysis of reinforced concrete deep beam with large opening,” *Beni-Suef Univ. J. Basic Appl. Sci.*, vol. 10, no. 1, p. 25, Dec. 2021, doi: 10.1186/s43088-021-00104-z.
[8] E.-S. I. M. Ismail, “Non-linear finite element analysis of reinforced concrete flat plates with opening adjacent to column under eccentric punching loads,” *HBRC J.*, vol. 14, no. 3, pp. 438–449, Dec. 2018, doi: 10.1016/j.hbrcj.2018.01.001.

IOP Conf. Series: Materials Science and Engineering 1304 (2024) 012013 doi:10.1088/1757-899X/1304/1/012013

# Evaluation Methods for Traditional Houses in Dobrogea

**T Şeitan[1], D I Andrieş[2], A M Grămescu[3]**
[1] Universitatea Ovidius Constanta,

E-mail: seitandoru@gmail.com

**Abstract.** Assessing the technical condition of earthen constructions is an extremely difficult task due to the fact that the materials, methods and techniques used in construction have been replaced by new materials and installation procedures. Although they have a fairly large share in the existing housing stock, earthen constructions have not benefited from the same interest in the field of construction. This paper proposes a review of the current assessment possibilities applicable to earthen constructions in relation to the current level of knowledge on a local and international level.

## 1. Introduction

The invention of Portland cement and the industrial revolution facilitated access to high-performance materials such as concrete or brick, but at the same time, the energy required to manufacture these materials is obtained from non-renewable and unsustainable sources. At the present time, the fossil fuels used in the industry are irreversible affecting the environment and there is an acute need for sustainable and ecological construction materials. A possible answer to this need can be earthen constructions, which by nature save energy right from the construction phase. The material is obtained locally, it does not need to be subjected to very high temperatures in furnaces and the installation does not require specialized labor.

Any new or existing construction must meet safety and stability requirements. For this purpose, there is a legislative basis with the help of which new structures can be designed and erected or existing buildings can be reinforced. With the help of the current regulations, the technical condition of a construction made of reinforced concrete, masonry or wood can be assessed with high accuracy, but the same regulations do not cover constructions made of earth.

## 2. Traditional houses in Dobrogea

The Dobrogea area is of particular interest in the study of earthen constructions because it is an arid steppe-like land with very few resources that can be used in constructions. The traditional Dobrogea house is made with clay adobe, often without foundations and covered with reeds thatch. The architecture is simple, the houses most often have a central room from which access is made into two larger living rooms. The main facade has a porch from where access is made to the central room and has decorations or motifs specific to the area. The traditional household is completed by annexes with the role of kitchen, bathroom or storeroom, made in similar techniques as the house, although in certain areas these rooms are under the same roof as the living quarters.

The traditional Dobrogea house has a rectangular shape and is about five meters wide by thirteen meters long with wall thickness that vary between 40 and 60 centimeters.

The case study is carried out on a typical house from a village in Constanta County, which has three roughly square rooms, the side rooms having internal dimensions of about four meters, and the central room of about three meters. The thickness of the walls is 45 centimeters, the foundations are made of stone laid with clay mortar, continuously under the walls, roof and ceiling made of wood, covered by ceramic tiles. Some of the details may be observed in figures 1, 2, 3 and 4.

**Figure 1.** Traditional house in Dobrogea. Main facade.

**Figure 2.** Traditional house in Dobrogea. Porch close up.

IOP Conf. Series: Materials Science and Engineering 1304 (2024) 012013 doi:10.1088/1757-899X/1304/1/012013

**Figure 3.** Traditional house in Dobrogea. Roof damage.

**Figure 4.** Traditional house in Dobrogea. Blind wall.

## 3. Technical evaluation of a traditional house from Dobrogea

A possible option for evaluating the technical condition of the construction would be to draw a parallel between it and a house made of brick, but using appropriate characteristic resistances for the materials used.

Taking into account the provisions of the regulations in use for the evaluation of existing masonry constructions as well as the information from the specialized literature, the values of the R1, R2 and R3 indicators can be calculated according to the P100-3/2019 standard.

For the calculation of the R1 indicator, the standard proposes an evaluation of the quality of the structural system, the quality of the structural materials, floor types, in plane configuration, the configuration in elevation and other aspects that contribute to the behavior of the structure and which will be detailed in the following lines.

The quality of the structural system takes into account the type of structural system, the efficiency of the connections between the walls, the efficiency of the connections between the walls and ceiling and the quantity of masonry for the two main directions. Regarding these criteria, the following unfavorable aspects can be noted: the structural walls are made of unreinforced masonry, without reinforced concrete columns and beams, the floors are flexible, the coupling areas are made of unreinforced masonry, without reinforced concrete beams and lintels made of wood. The crossing of the walls is made only by weaving the masonry, and the connections of the wooden beams are made by friction only.

These aspects lead to a reduced ductility and redundancy of the structural system with unfavorable behavior during dynamic loads. The coupling zones have low capacity both for axial stress and bending moment as well as from the shear force. The connection between the walls and floors is quite weak.

The distribution of masonry is approximately equal on the two main directions as well as the density of the structural walls of approx. 6.0% in both directions are favorable aspects that improve the behavior to seismic loads. The structure has vibration periods values approximately equal on the two directions.

The materials used are unfired clay bricks (adobe), clay-based mortar and wood for the superstructure and stone with clay-based mortar for the foundations. The quality of bricks and mortar is poor, having low resistance characteristics, but also low durability over time, especially when exposed directly to atmospheric factors and humidity. As a positive aspect, it can be mentioned that the clay-based mortar has superior damping capacity during seismic actions compared to cement-based mortar. The quality of the wood is poor, although it is suitable for the importance class of the construction, but with little durability over time, especially in the case of water infiltration through the frame, being sensitive to the attack of wood-eating insects and fungi, at the same being highly combustible.

The quality of the masonry execution is good and ensures a good behavior of the construction both to gravity loads and to seismic loads, and the absence of weak areas (niches or slits) reduces the risk of cracks.

The floor and the roof are made of wood without sufficient rigidity in plane to ensure the compatibility of the structural walls deformations, they cannot redistribute the loads between the structural walls from the two directions and have a diminished role in preventing their overturning due to stresses perpendicular to their plane. The weak connections (by friction) between the wood and masonry elements mean that the loads in the walls cannot be properly transmitted through the ceiling to the adjacent walls, even if the ceiling has sufficient rigidity in plan.

The regular shape, the low height regime, without withdrawals in plane or in elevation is a positive aspect because a construction with large differences between the dimensions of the two directions, as well as with significant withdrawals can have additional effects of general torsion. Other positive aspects are represented by the absence of significant vertical discontinuities in structural walls, gaps or indirect supports.

Distances less than 5.00 meters between structural walls in both main directions, their height less than 3.00 meters and the ratio between the height and thickness of the walls with a value less than 12 make the structure less prone to failure with a uniform distribution of seismic forces without major concentrations of forces in certain structural elements. High story heights can lead to significant 2nd order effects and buckling of slender vertical members.

Another negative aspect is represented by the fact that the walls are subjected to thrusts from the frame without having any elements to limit these effects and local failures may occur due to exceptional loads from gravitational or seismic loads.

The foundation soil has sufficient resistance capacity, without large swellings or contractions, and the foundations have a suitable width to transmit the loads from the superstructure. The negative aspects of the foundations are represented by the small foundation depth that does not exceed the frost depth. They are influenced by the swellings and contractions of the ground subjected to freeze-thaw cycles. More than that, the material from which the infrastructure is made does not have sufficient resistance to deformation, not being able to take over stretching efforts from seismic actions or differentiated settlements.

There are no possible interactions with the adjacent buildings because the building is at a sufficient distance from the neighboring constructions, and the building does not have decorative elements that present a risk of collapse or partition walls and claddings with inadequate fastenings. However, as a negative aspect, the degradation of the heels can be observed with a considerable risk of the roof collapsing.

Taking into account all the aforementioned aspects, the R1 indicator can be appreciated at a value of 61 points out of a maximum of 100 and the building can be included in the RsIII seismic risk class.

For the calculation of the R2 indicator, the standard proposes an evaluation of the degree of structural damage, taking into account the degree of damage of the vertical and horizontal elements. It can be appreciated that the vertical elements show moderate damage on less than 1/3 of their area (cracks) and the horizontal elements show serious damage with the affected area between 1/3 and 2/3 of the total area. By adding up these effects, a value of 65 points is obtained for the R2 indicator and the building can be included in the RsIII seismic risk class.

For the calculation of the R3 indicator, the modes of failure associated with masonry constructions must be taken into account, namely eccentric compression failure, ladder failure and horizontal sliding failure. It also takes into account the global behavior of the structures and the possibilities of redistributing the loads that appear during a dynamic action.

Starting from the evaluation of the mass of the entire structure and from the evaluation of the static loads acting on the structure, the basic seismic force can be determined. The value of the behavior factor for the unreinforced brick masonry structure has a maximum value of 1.5 according to the P100-3/2019 standard. This factor is related to structure's ability to dissipate energy and it can be considered that a building made of adobe has a lower dissipative capacity.

In specialized literature, the minimum value of the characteristic resistance to compression accepted for brick in seismic areas is 5N/mm$^2$ [4]. The studies carried out on unfired clay bricks indicate a maximum compressive strength of 3N/mm$^2$, most of the times around 1.5N/mm$^2$. The tensile strength is estimated at a value of 3% of the value of the compressive strength [1] [5] [6].

Taking into account all the above as well as the peak value of the design acceleration of the area, the R3 indicator defined as the ratio between the capacity of the structure and the value of the basic seismic force has a maximum value of 60, which means that the building falls into the RsII seismic risk class.

Using computer software (ETABS17) to model the structure we can clearly see the areas where the loads are concentrated. A common design problem of the traditional Dobrogea house is that the centre beam of the ceiling which in turn supports the roof rests on the transversal walls as seen in figure 5. One shortcoming of this approach is that the door openings are directly under the beam and the load is distributed by wooden lintels. That is why one of the most common place for cracks to appear is the transversal wall, above the door. Figures 6 and 7 show the two main vibration modes and the corresponding displacement values in millimetres.

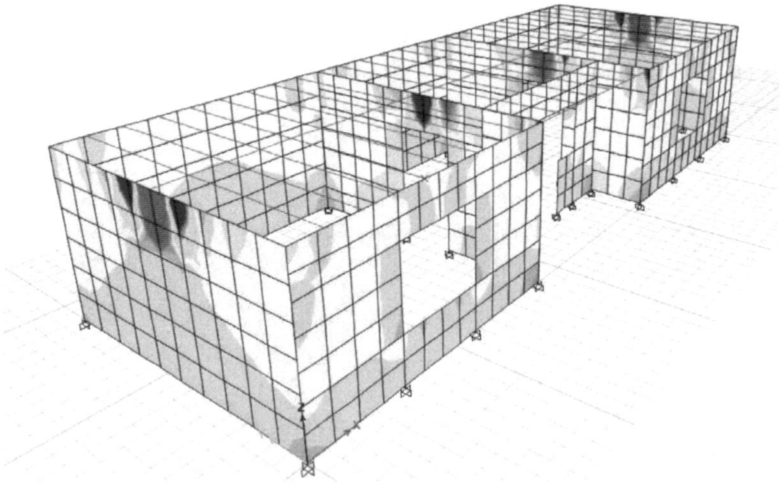

**Figure 5.** Software modelling of the structure..

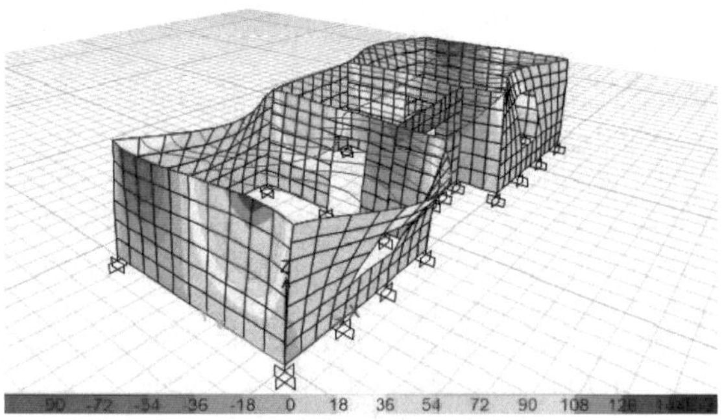


**Figure 6.** Deformed shape (mode shape 1) for vibration modes

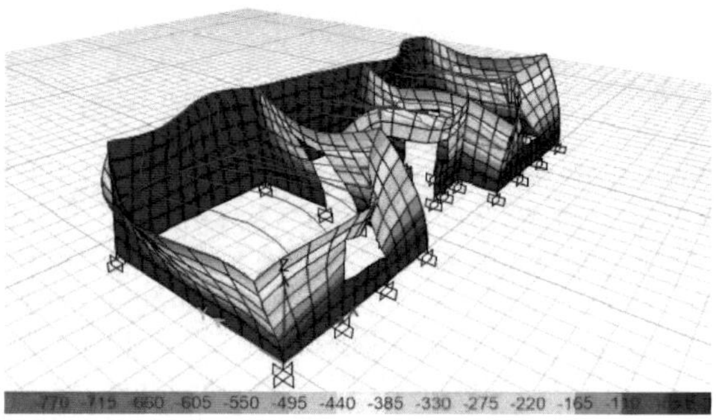


**Figure 7.** Deformed shape (mode shape 6) for vibration modes

## 4. Conclusion

The deficiencies encountered in traditional Dobrogea houses are largely influenced by the high loads placed on the transverse walls and the undersized foundations made of non-homogeneous materials. Most of them suffered some damage, sometimes major damage, but not total collapse, taking into account the major seismic events that happened in Romania, idea which is also sustained by the value of the R3 indicator.

## 5. References

[1] Minke G. 2008 *Building with Earth: Design and Technology of a Sustainable Architecture*, (Simetria)
[2] OAR *Architectural guide for the rural area*
[3] P100-3/2019 *Design code for seismic evaluation of existing buildings*
[4] CR6-2013 *Design code for masonry structures*
[5] http://www.kerpic.org/alker_technology.html
[6] Marina Rabelo Brito et. al. *Evaluation of the Properties of Adobe Blocks with Clay and Manure, Buildings* **2023**, *13, 657* https://doi.org/10.3390/buildings13030657

IOP Conf. Series: Materials Science and Engineering 1304 (2024) 012014 doi:10.1088/1757-899X/1304/1/012014

# Time history comparative analyses of a thin-walled cold-formed steel structure considering P-D effect

**G. Tăranu, M. Movilă, S.G. Luca, C. Pastia, I.O. Toma**

"Gheorghe Asachi" Technical University of Iasi, Faculty of Civil Engineering and Building Services, no. 1, Prof.dr.doc. D. Mangeron Street, Iasi, *Romania*

E-mail: george.taranu@academic.tuiasi.ro

**Abstract.** The paper presents a comparative time history analysis based on the 1D-beam Finite Element Method (FEM) for a thin-walled cold-formed steel structure previously investigated both theoretically and experimentally. The recent earthquake which occurred in Turkey with 7.8Mw and a Peak Ground Acceleration (PGA) of almost 0.5g showed numerous collapses of different types of buildings. Romania has an active seismic area located in the Vrancea region where, in 1977 a 7.5Mw earthquake also occurred. In this context, this paper presents a comparative analysis with time history accelerograms of the recorded earthquake in North – South and East - West directions. The structural model was analyzed in Robot Structural Analysis software in two scenarios of the time history accelerograms considering also the P-Δ effect. The finite elements of the structure were defined as 1-D beam elements with hinged joints and limited axial stiffness previously obtained in past studies. The results showed a substantial increase in terms of displacements and stresses.

## 1. Introduction

Researchers worldwide are continuously studying how civil engineering structures respond to seismic movements. In addition to ensuring safety and minimizing damage, buildings must also become sustainable, meaning they should achieve maximum performance with minimal material consumption. Although seismic design codes such as EN 1998-1 [1] and the Romanian code [2] provide extensive guidance on the seismic design of steel structures, they categorize cold-formed steel structures (CFSS) as having limited ability to dissipate seismic energy. As a result, further research is needed to fill the informational gap and provide accurate information on CFSS.

Some researchers [3–8] have explored different methods to enhance the seismic performance of thin-walled steel structures. For example, adding supplemental damping devices or using special shapes and configurations have been explored to improve the seismic performance of these structures. Additionally, various numerical simulations [9] and experimental studies [10–13] have been organized to better understand the behaviour of thin-walled steel structures under seismic loads and to develop reliable analytical models to predict their response to earthquakes. Overall, research in this field aims to improve the design and construction of steel structures to ensure their safety and sustainability in earthquake-prone regions.

Based on previous work [14–16] the research presented in this paper focus on the structural response of a structural skeleton made of C shape thin-walled steel profile. The structural system depicted in Figure 1 utilizes a framing approach with truss-like diagonal elements situated in the peripheral panels. The structure was tested in the past on the shaking table in two scenarios of gravitational loading. The model was numerically analysed and calibrated using experimental data. In the context of the recent catastrophic earthquake which has produced in Turkey in 2023, a time history linear and nonlinear analysis was conducted in comparison with the 1977 Vrancea earthquake that occurred in Romania. Both analyses consider an additional P-D effect so the obtained results show an important influence of the model approach.

**Figure 1.** The structural skeleton made of C shape thin-walled steel profile was experimentally tested on the shaking table 2. Materials and methodology

Table 1 provides information on the material properties of the steel sheet and connectors. The steel sheet used to create the steel profiles through cold-forming was DX51D+Z. The framing elements were connected using Self Tapping Screws (STS) measuring 4.8×22mm, while the wall frames were connected to each other using Self Drilling Screws (SDS) measuring 5.5×25mm. Additionally, diagonal elements were connected using 4.8×16mm SDS. The material properties of the steel sheet were determined experimentally by performing direct tensile tests on specimens taken from the profile web.

**Table 1**. Materials properties.

| Element | Type | Elastic modulus (GPa) | Tensile yield strength (MPa) | Tensile ultimate strength (MPa) |
|---|---|---|---|---|
| Steel sheet C profiles - fig. 4 (a) | DX51D+Z | 210 | 120 (exp) | 250 (exp) |
| SDS – for profile joints ISO 15481 class (5.6) [17] | 4.8x16 mm | 210 | 300 [18] | 500 [18] |
| SDS –for panel joints ISO 15481 class (5.6) [17] | 5.5x25 mm | 210 | 300 [18] | 500 [18] |
| STS – for profile joints ISO 7049 class (5.6) [19] | 4.8x22 mm | 210 | 300 [18] | 500 [18] |

Cold forming is a metalworking process in which steel is shaped at room temperature, rather than heated and then shaped as in traditional hot forming processes. Thin-walled steel profiles are frequently used in construction and manufacturing because of their high strength-to-weight ratio, and cold forming is often used to shape these profiles due to their efficiency and cost-effectiveness [20–22]. The process of cold-forming C shape thin-walled steel profiles typically involves feeding a coil of steel through a series of rollers that gradually shape the steel into the desired profile. This process is also known as roll forming. The rollers are often custom-designed to produce the specific profile required, and the process can be highly automated to increase efficiency and reduce labour costs [23,24].

Cold-formed steel profiles have a number of advantages over hot-rolled profiles, including better dimensional accuracy, lower material cost, and improved consistency of properties [25,26]. However, there are also some limitations to cold-formed steel, including a reduced capacity to resist lateral torsional buckling and lower ductility at low temperatures. In order to ensure the quality and high structural performance of cold-formed steel profiles, it is important to follow appropriate design and manufacturing guidelines. These guidelines may include specific tolerances for dimensions, recommendations for the use of stiffeners or other reinforcement, and guidelines for the use of appropriate fasteners and connections.

The manufacturer of these type of structures [27] can produce light steel structures made of galvanized steel. Prefabricated "U" and "C" profiles for steel structure houses, steel structure industrial buildings.

## 2. Numerical modelling of the building

There are two commonly used approaches to numerically model Cold-Formed Steel (CFS) structures that aim to replicate the actual behaviour of the structure with precision. The first approach, which is commonly referenced in scientific literature, involves employing 2D or even 3D elements with non-linear material behaviour. While this method produces very accurate results, it is costly in terms of hardware and simulation time, particularly for intricate structures. [28–34].

The second approach, as presented in design codes, considers that the joints in CFS structures are the most susceptible components. Hence, numerical investigations must account for the second-order effect and actual stiffness of the joints, which can be determined through laboratory testing. In this paper, the numerical models consist of linear and non-linear models with joints that are either infinitely rigid or semi-rigid, where the latter's rigidity may either be constant or exhibit a parabolic variation.

By utilizing previously published findings and incorporating results from cyclic tests conducted on T-joints, a finite element structural model based on 1D-beam pin-connected elements presented in Figure 2 and 3 were employed to simulate the behaviour of the structure when subjected to time history accelerograms in the Robot Structural Analysis software [35]. The connections of the beam elements are considered linear with joints defined with axial elastic stiffness $k_x$=2300 N/mm. The model rests at the base on pinned and simple supports.

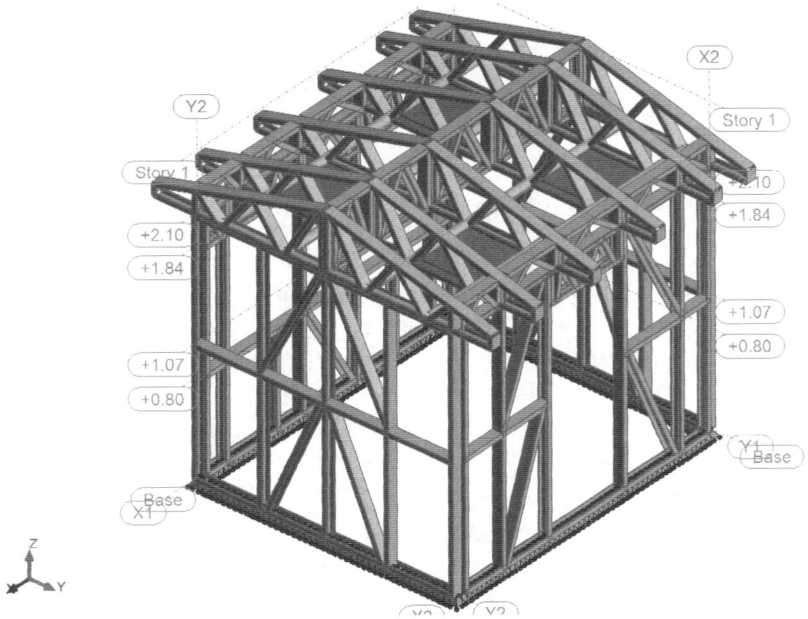


**Figure 2**. The numerical model rendered

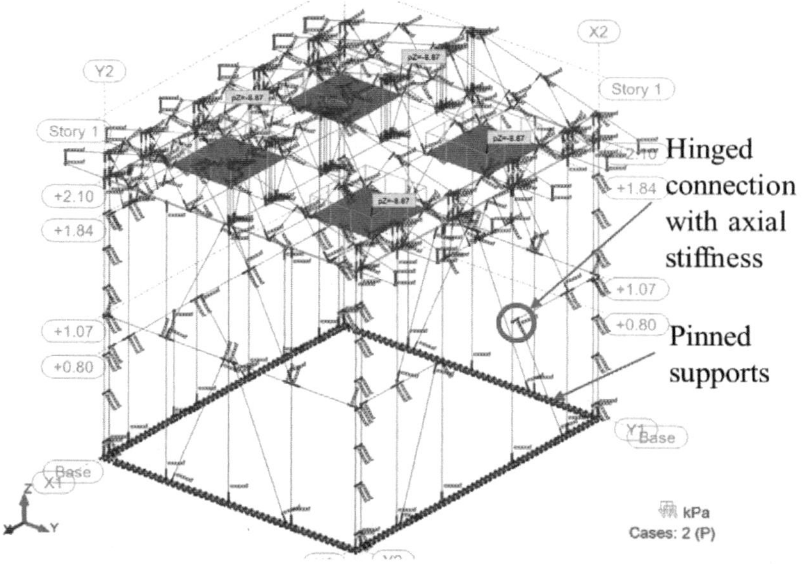


**Figure 3**. The loading schema and joints definition

The load cases and load values selected according to Romanian design codes are presented in Table 2. These were combined also according to CR 0-2012 Romanian code [36].

**Table 2.** Load cases

| Case | Name | Uniformly surface-distributed load q (kPa) | According to code |
|---|---|---|---|
| 1 | Self-weight | Automated | EC 1 [37] |
| 2 | Gravitational load | 8.87 | EC 1 [37] |

A time-history analysis is one of the methods used to evaluate the seismic response of cold-formed steel buildings. The time-history investigation involves the use of a ground motion record for a building model to determine the response of the building to seismic forces. Assessing the seismic response of cold-formed steel buildings using time-history analysis is a challenging process that encompasses various factors, such as the seismic hazard present at the site, the building's design, the steel's material properties, and the modelling techniques employed to simulate the building's behaviour. The precision of the analysis relies on the quality of the input data, including the ground motion record and building properties. Moreover, modelling techniques employed to replicate the building's behaviour should be authenticated against experimental data to verify the credibility of the results [38–45].

The input data used for the time history comparative analysis presented in Figure 4 of this paper were sourced from the Engineering Strong Motion database (ESM), a compilation of accelerometric data recorded in Europe and the Middle East since 1969 [46, 47]. The accelerograms used were defined as follows in Table 3.

**Table 3.** Time history analysis description

| Case nr. | Case name | Description |
|---|---|---|
| 3 | Modal analysis | Modal analysis |
| 8 | T-H Vrancea_77 | Y (N-S accelerogram) + X (E-W accelerogram) |
| 9 | T-H Vrancea_77_P-Δ | Y (N-S accelerogram) + X (E-W accelerogram) +P- Δ effect |
| 10 | T-H Turkey_2023 | Y (N-S accelerogram) + X (E-W accelerogram) |
| 24 | T-H Turkey_2023_P-Δ | Y (N-S accelerogram) + X (E-W accelerogram) +P- Δ effect |

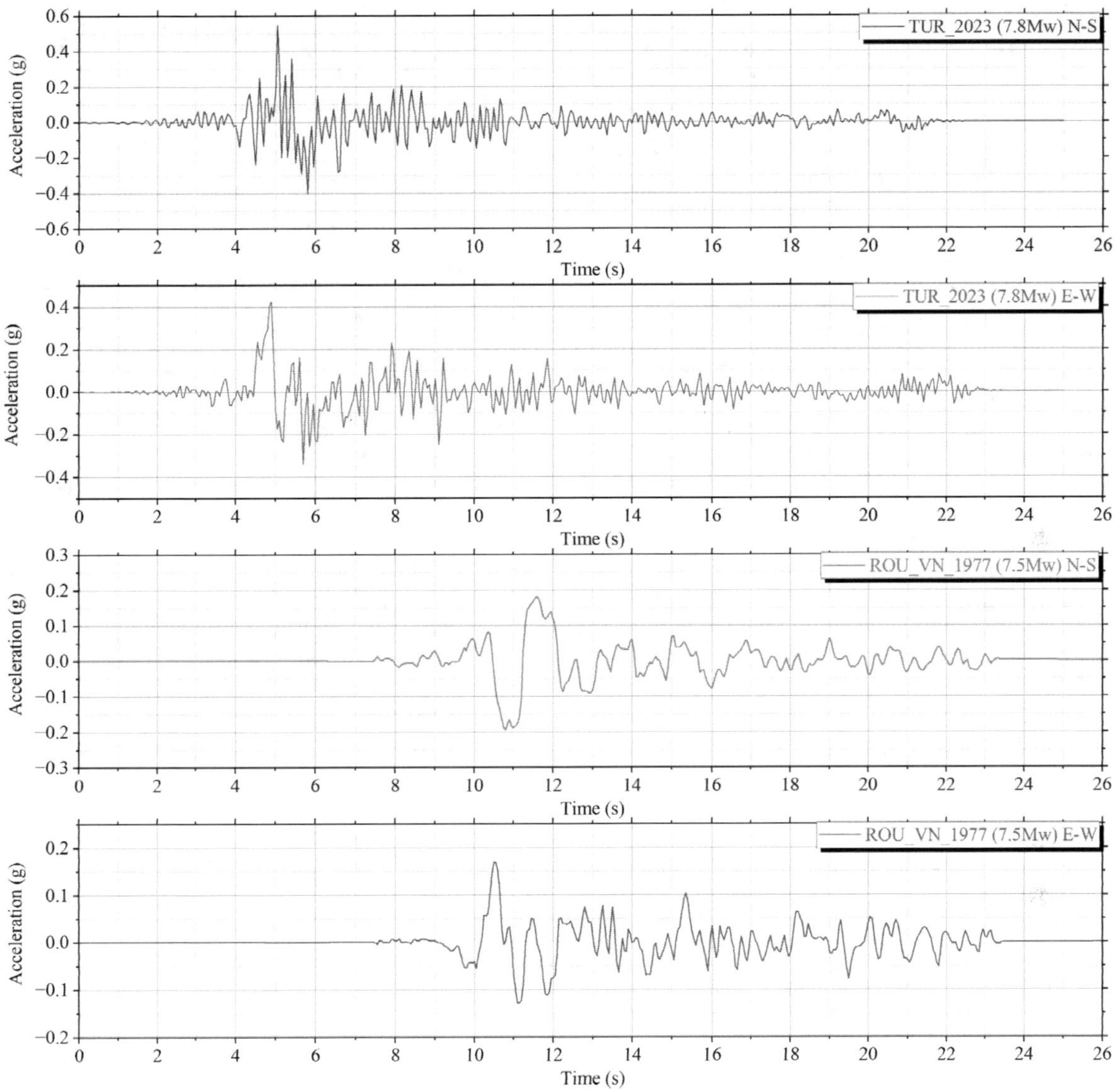


**Figure 4**. Recorded and processed accelerogram of the North – South and East – West direction of the Vrancea 7.5 Mw, Romania 1977 earthquake and 7.8 Mw Turkey 2023.

## 3. Results and discussion

From the numerical analyses, the following results were observed and compared. Table 4 presents: modal analyses result with fundamental frequencies, mass participation factor, total mass, pulsation, and damping. The first two modes of vibration are translational in X and Y directions. The deformed shapes of vibration are presented in Figure 5.

Figure 6 presents de maximum absolute displacement in the case of Vrancea 77 - time history analysis and Figure 7 the maximum absolute displacement in the case of Vrancea 77 considering the P-Δ effect. Also, the maximum absolute displacements for Turkey 2023 accelerograms are presented in Figures 8 and 9 with P-Δ effect. In terms of stress in structural elements, figures 10-13 present maximum and minimum principal stress for the mentioned time history analyses.

**Table 4.** Modal analysis results

| Mode | Frequency (Hz) | Period (sec) | X direction mass participation (%) | Y direction mass participation (%) | Total mass (kg) | Pulsation (1/sec) | Damping |
|---|---|---|---|---|---|---|---|
| 1 | 2.48 | 0.40 | 96.44 | 0.05 | 1483.63 | 15.59 | 0.02 |
| 2 | 2.92 | 0.34 | 0.06 | 96.68 | 1483.63 | 18.37 | 0.01 |

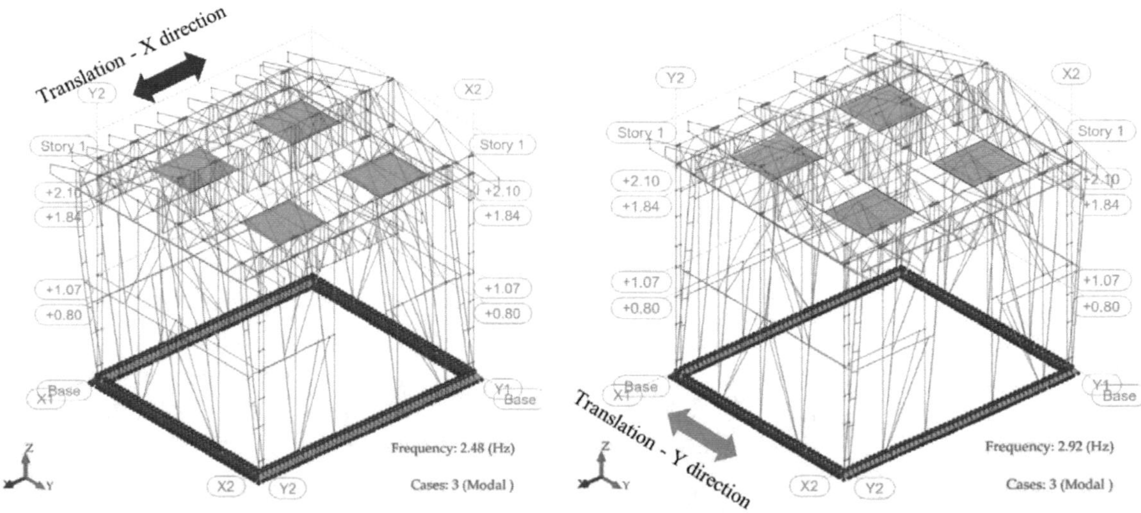


**Figure 5**. First and second vibration modes – translational in X (1st mode) and Y direction (2nd mode)

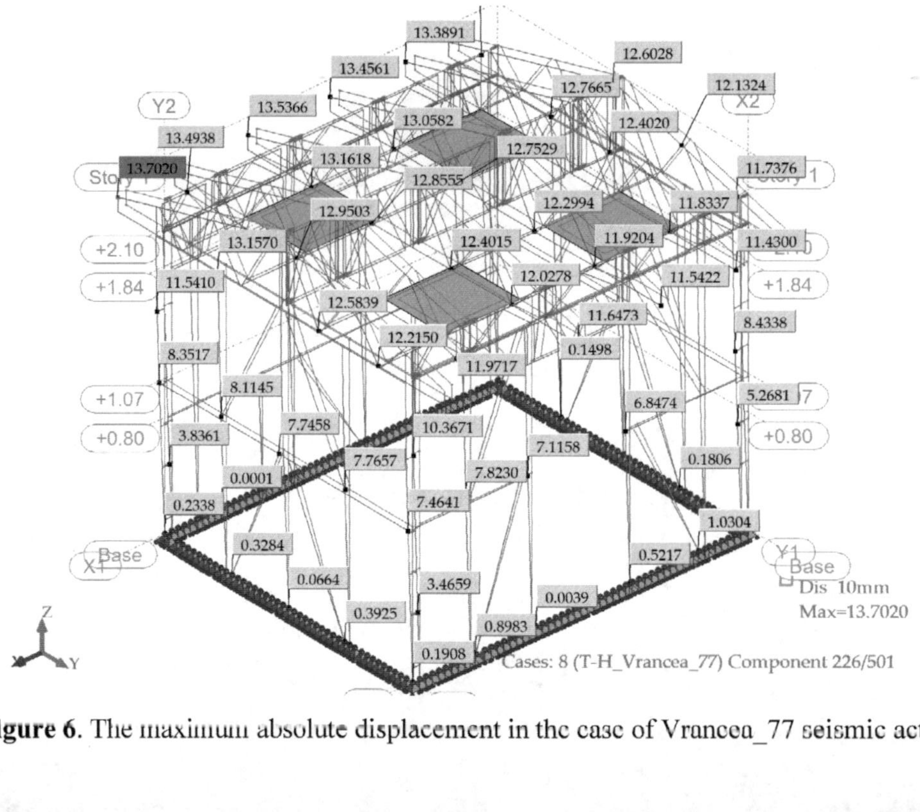


**Figure 6**. The maximum absolute displacement in the case of Vrancea_77 seismic action

IOP Conf. Series: Materials Science and Engineering 1304 (2024) 012014 doi:10.1088/1757-899X/1304/1/012014

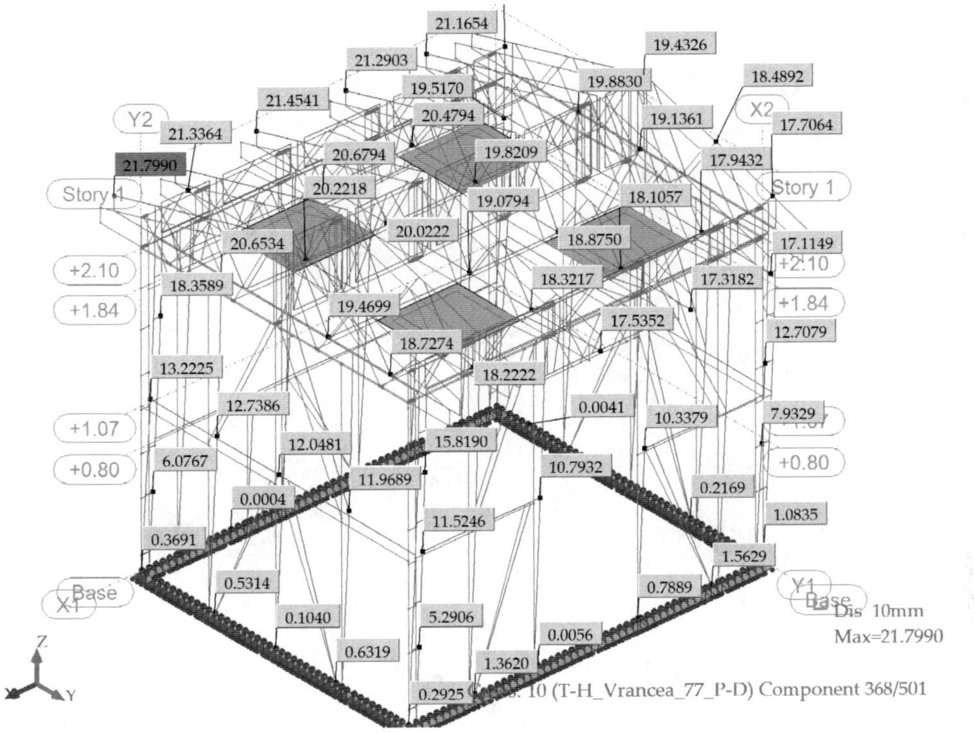


**Figure 7**. The maximum absolute displacement in the case of Vrancea_77 seismic action with P-Δ effect

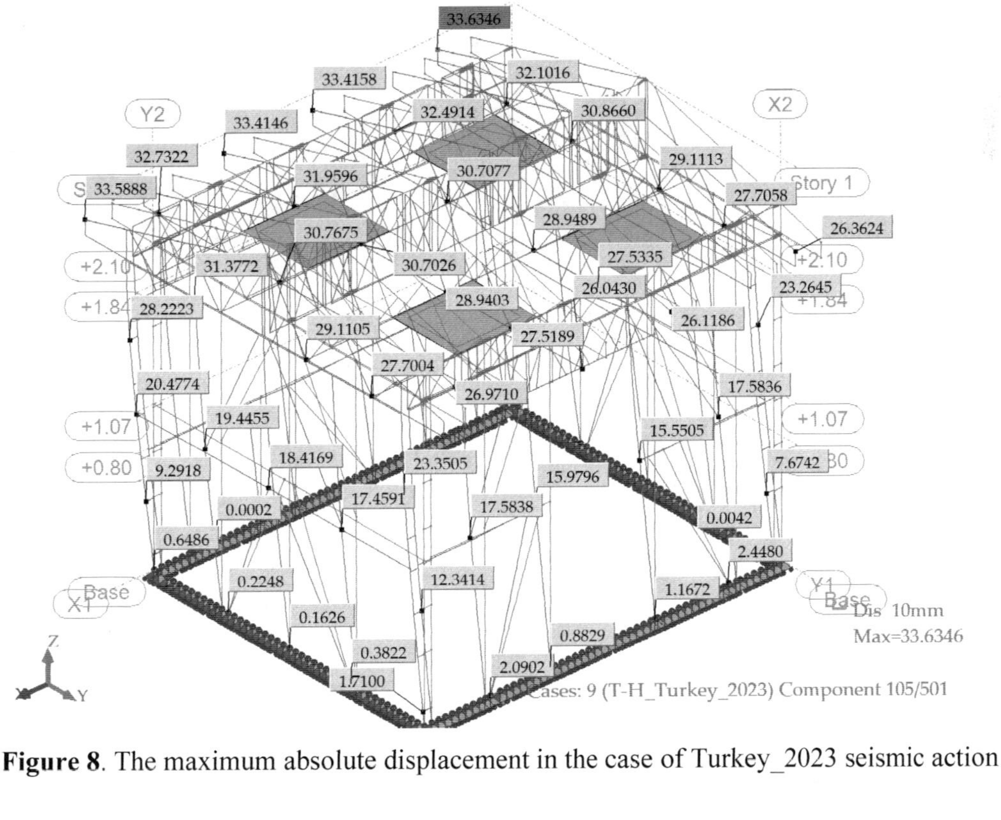


**Figure 8**. The maximum absolute displacement in the case of Turkey_2023 seismic action

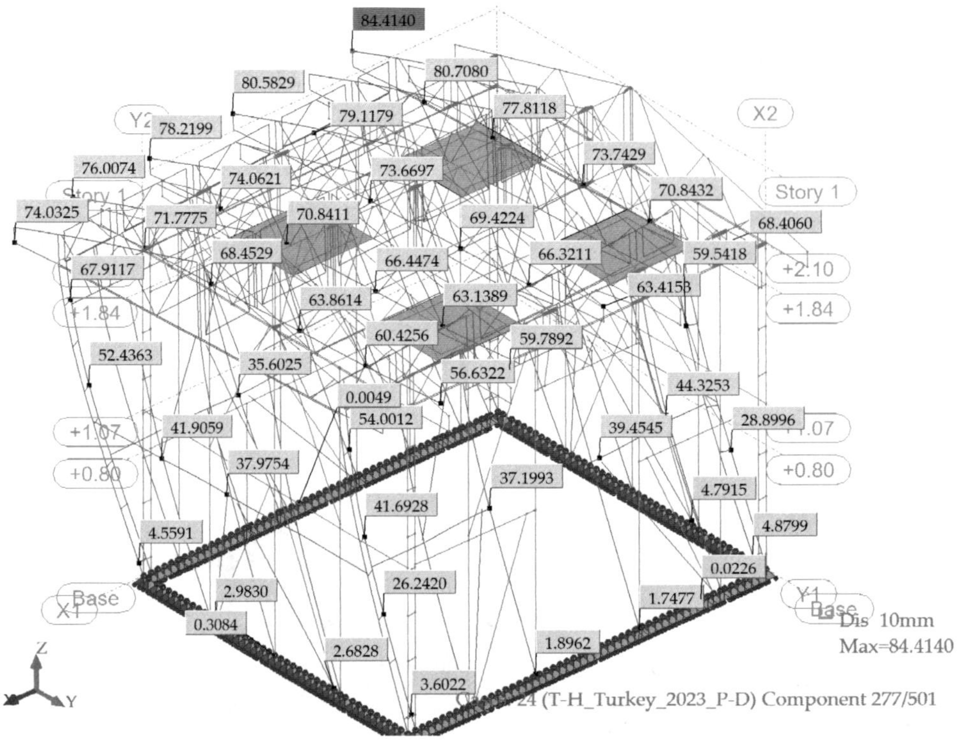


**Figure 9**. The maximum absolute displacement in the case of Turkey_2023 seismic action with P-D effect

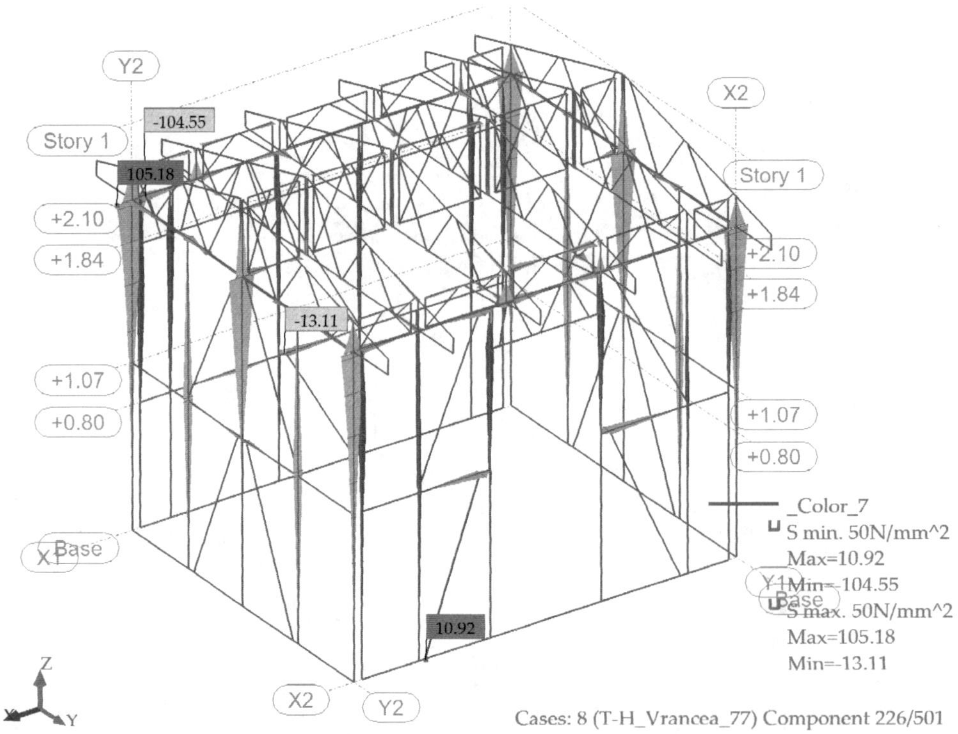


**Figure 10**. Maximum and minimum principal stress in Vrancea_77 case

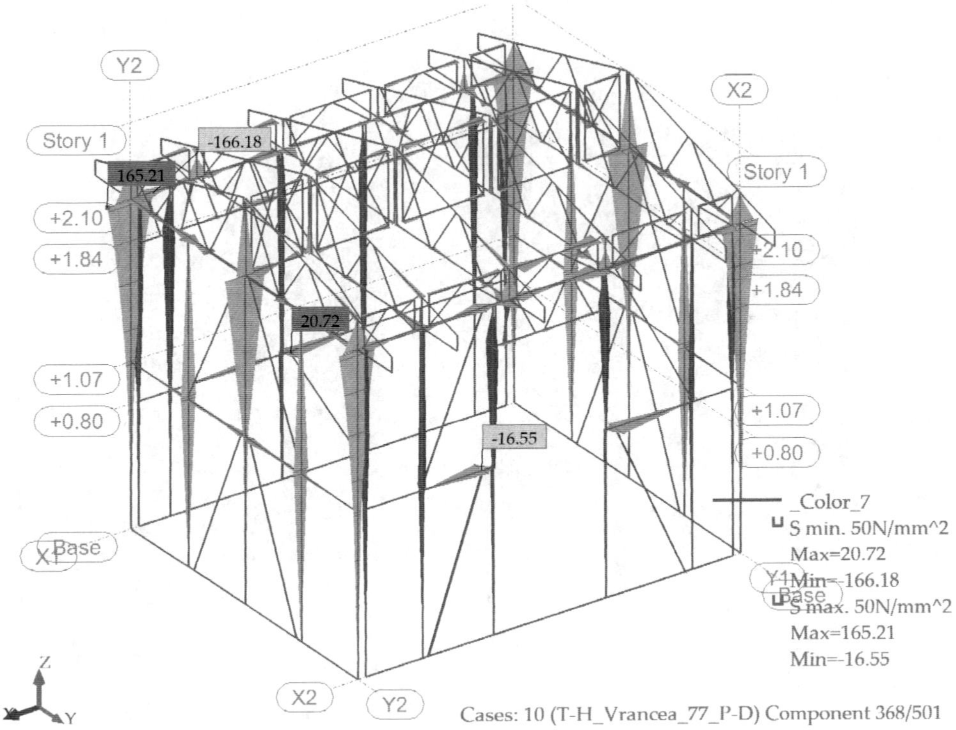


**Figure 11**. Maximum and minimum principal stress in Vrancea_77 case P-Δ effect

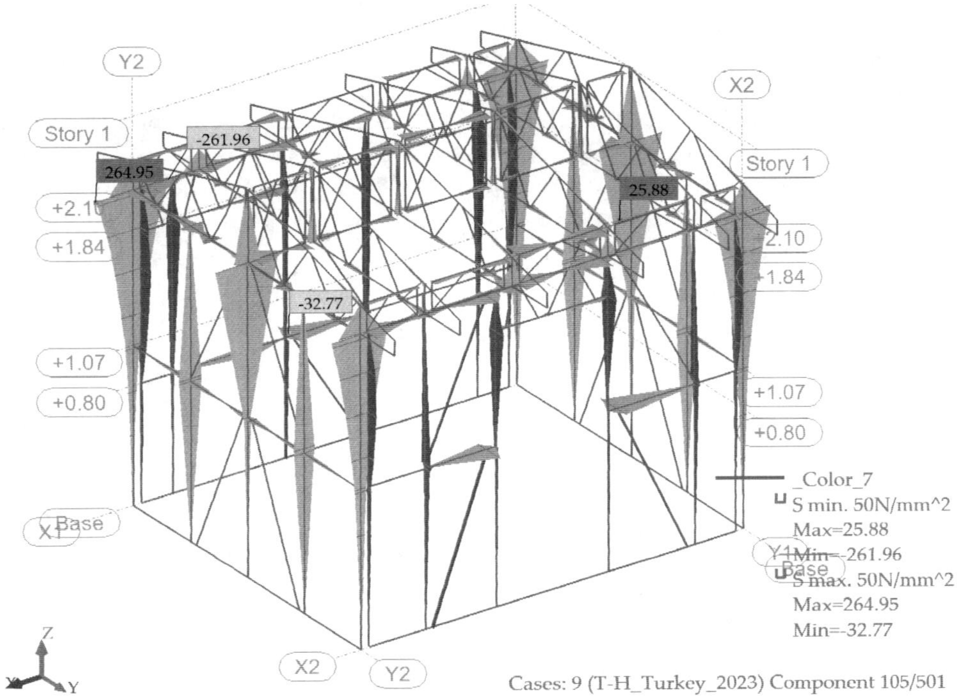


**Figure 12**. Maximum and minimum principal stress in Turkey_2023 case

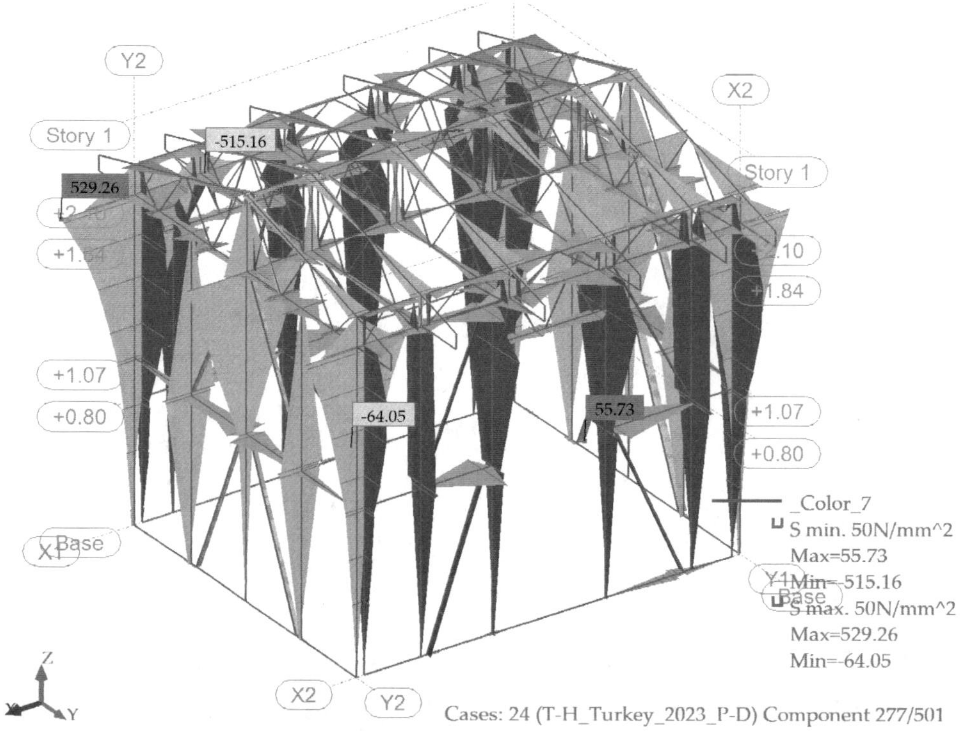


**Figure 13**. Maximum and minimum principal stress in Turkey_2023 case P-Δ effect

In figure 14 are presented the comparative graphs for the absolute maximum displacements and maximum principal stresses attained from the analyses. In the case of Vrancea _77 seismic action the maximum absolute displacement is approximately 13 mm and is very close to the maximum acceptable displacement as per seismic design codes [2]. In the case of the same action but considering the P-Δ effect, the maximum displacement was 21.79 mm which means an increase of displacement by 59%. In the case of Turkey_2023 seismic action, the maximum displacement was 33.63mm while the P-Δ effect led to an increase of 151% the maximum absolute displacement being 84.41mm.

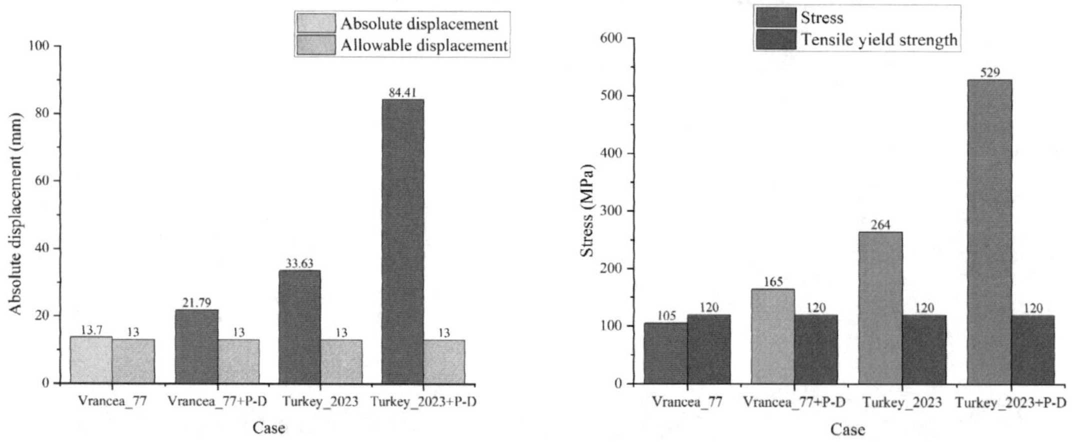


a b

**Figure 14**. Comparative graphs: a – maximum absolute displacements; b- maximum principal stress

In terms of stress, the maximum principal stress obtained from the analyses, are compared and graphically presented in figure 14 b. The comparative graphs show that the maximum principal stress have a value of 105 MPa being smaller than the yielding strength of the material which is 120 MPa. Considering the P-Δ effect for the same seismic action the maximum principal stress increase by 57% resulting in a value of 165 MPa, greater than the yielding strength of the material but smaller than the ultimate strength which is 250 MPa. In the case of Turkey_ 2023 seismic action, the maximum principal stress was located in the same element but with a value of 264 MPa, while the P-Δ effect leads to an increase of the maximum principal stress by 100% than the model without a second-order effect.

## 4. Conclusions

The paper presents a comparative Time-History numerical analysis of a structural skeleton system made of thin-walled steel profiles with DX51D+Z steel. The structure was previously tested on the shaking table and the numerical model was calibrated using experimental data. The numerical model considers the limited axial stiffness of the 1-D beam elements in joints. The recorded actions of the earthquakes were chosen as a reference for the seismic areas of Romania and Turkey.

The analyses were completed considering the P-Δ effect as separate cases. So, it was observed that in the case of Vrancea_77 seismic action combined accelerograms for N-S and E-W direction, the maximum displacement and maximum principal stress are acceptable in comparison with the design code allowable values. In the case of Turkey_2023 seismic action the maximum allowable values for displacements and strength were considerably exceeded. The P-D effect leads also to a greater value of displacements and stresses for both seismic actions Vrancea_77 and Turkey_2023.

To reduce the displacements values and stress distribution for the structural skeleton there are two possibilities that can be verified in future numerical analyses: the first one consists in adding additional diagonal and other vertical profiles, and the second one consists in adding additional connectors in joints to increase the overall stiffness.

These analyses must be followed by a future study where the structural steel profile skeleton will have also claddings that are basically made of OSB panels. The presence of the OSB panels influences both the overall rigidity and the stress distribution in the structural skeleton.

## 5. References

[1] European Comittee for Standardization 2004 *EN 1998-1: Eurocode 8: Design of structures for earthquake resistance – Part 1: General rules, seismic actions and rules for buildings*

[2] M.D.L.P.A Romania 2013 *P100-1/2013: Seismic design code - Part I: Design rules for buildings (in Romanian)*

[3] Fiorino L, Macillo V and Landolfo R 2017 Shake table tests of a full-scale two-story sheathing-braced cold-formed steel building *Eng Struct* **151** 633–47

[4] Kyvelou P, Kyprianou C, Gardner L and Nethercot D A 2019 Challenges and solutions associated with the simulation and design of cold-formed steel structural systems *Thin-Walled Structures* **141** 526–39

[5] Landolfo R, Shakeel S and Fiorino L 2022 Lightweight steel systems: Proposal and validation of seismic design rules for second generation of Eurocode 8 *Thin-Walled Structures* **172**

[6] Ye J, Mojtabaei S M and Hajirasouliha I 2019 Seismic performance of cold-formed steel bolted moment connections with bolting friction-slip mechanism *J Constr Steel Res* **156**

[7] Wang M, He H, Wang B and Yan W 2021 Aseismic performance and damage evolution of cold-formed thin-walled steel whole structure connected by self-tapping screws with lead damper *Zhendong yu Chongji/Journal of Vibration and Shock* **40**

[8] Ye J, Mojtabaei S M, Hajirasouliha I and Pilakoutas K 2020 Efficient design of cold-formed steel bolted-moment connections for earthquake resistant frames *Thin-Walled Structures* **150**

[9] Kechidi S, Bourahla N and Castro J M 2017 Seismic design procedure for cold-formed steel sheathed shear wall frames: Proposal and evaluation *J Constr Steel Res* **128** 219–32

[10] Macillo V, Fiorino L and Landolfo R 2017 Seismic response of CFS shear walls sheathed with

nailed gypsum panels: Experimental tests *Thin-Walled Structures* **120** 161–71
[11] N.Usefi, Sharafi P, Mortazavi M, Ronagh H and Samali B 2021 Structural performance and sustainability assessment of hybrid-cold formed modular steel frame *Journal of Building Engineering* **34**
[12] Mortazavi M, Sharafi P, Ronagh H, Samali B and Kildashti K 2018 Lateral behaviour of hybrid cold-formed and hot-rolled steel wall systems: Experimental investigation *J Constr Steel Res* **147** 422–32
[13] Feng R qiang, Zhu B, Xu P H and Qiu Y 2019 Seismic performance of cold-formed steel framed shear walls with steel sheathing and gypsum board *Thin-Walled Structures* **143**
[14] Ţăranu G and Toma I-O 2021 FEM analysis of a floor structural system made of thin-walled cold-formed steel profiles *IOP Conf Ser Mater Sci Eng* **1141**
[15] Taranu G, Ungureanu V, Nagy Z, Alexa-Stratulat M S, Toma I O and Luca S G 2023 Shake table test and numerical analyses of a thin-walled Cold-Formed Steel structure: Part 1 — Investigation of the structural skeleton without claddings *Thin-Walled Structures* **182**
[16] Taranu G and Toma I O 2021 Experimental investigation and numerical simulation of c-shape thin-walled steel profile joints *Buildings* **11**
[17] TR Group Self Drilling Screw *DIN 7504N ~ ISO 15481*
[18] TR Group Screws Mechanical Properties ISO 898-1 EN 20898-1
[19] TR Group Self Tapping Screw *DIN 7981C ~ISO 7049*
[20] Yu W, LaBoube R A and Chen H 2019 Cold-Formed Steel Design *Cold-Formed Steel Design*
[21] Franklin N P 2019 *Loadbearing Cold-Formed Steel for Mid-Rise Residential Buildings in Loadbearing Cold-Formed Steel for Mid-Rise Residential Buildings in Australia Australia* PhD (The University of Wollongong)
[22] American Iron and Steel Institute 2008 *AISI Manual-Cold-Formed Steel Design*
[23] AISI-American Iron and Steel Institute 2015 North American Standard for Cold-Formed Steel Structural Framing
[24] Ye J 2016 More Efficient Cold-Formed Steel Elements and Bolted Connections
[25] Mrówczyński D, Gajewski T and Garbowski T 2022 Parametric Study of the Numerical Model of a Bolted Connection of Steel Structure for Photovoltaic Panels *Materials 2022, Vol. 15, Page 6794* **15** 6794
[26] Staszak N, Gajewski T and Garbowski T 2022 Effective Stiffness of Thin-Walled Beams with Local Imperfections *Materials* **2022** 7665
[27] UNIC ROTAREX Unic Rotarex® Manufacturer of light metal structures
[28] Zhang J H and Young B 2018 Finite element analysis and design of cold-formed steel built-up closed section columns with web stiffeners *Thin-Walled Structures* **131** 223–37
[29] Sarawit A T, Kim Y, Bakker M C M and Peköz T 2003 The finite element method for thin-walled members-applications *Thin-Walled Structures* **41** 191–206
[30] Ungureanu V, Luka•ević I, Both I and Dubina D 2021 Numerical investigations on built-up cold-formed steel beams for long spans *ce/papers* **4**
[31] Lukačević I, Ungureanu V, Valčić A and Ćurković I 2021 Numerical study on bending resistance of cold-formed steel back-to-back built-up elements *ce/papers* **4**
[32] Shamim I and Rogers C A 2012 Numerical modelling and calibration of CFS framed shear walls under dynamic loading *21st International Specialty Conference on Cold-Formed Steel Structures - Recent Research and Developments in Cold-Formed Steel Design and Construction*
[33] Kechidi S, Fratamico D C, Bourahla N, Castro J M and Schafer B W 2017 Numerical study of screw fasteners in built-up CFS chord studs (Copengahagen)
[34] Taranu G, Venghiac V-M, Olteanu-Dontov I, Rotaru A and Toma I-O 2022 Sustainable Design for CFS Structures: Experimental Data and Numerical Models of Hinged Connections *Sustainability (Switzerland)* **14**
[35] Autodesk 2022 Robot Structural Analysis Professional
[36] M.D.L.P.A 2012 *CR 0-2012: Cod de proiectare. Bazele proiectarii constructiilor/ Romanian*

*design code. The basics of construction design*
[37] European Committee for Standardization 2002 Eurocode 1: Actions on structures - Part 1-1: General actions - Densities, self-weight, imposed loads for buildings EN 1991-1-1
[38] di Lorenzo G and de Martino A 2019 Earthquake response of cold-formed steel-based building systems: An overview of the current state of the art *Buildings* **9**
[39] Schafer B W, Ayhan D, Leng J, Liu P, Padilla-Llano D, Peterman K D, Stehman M, Buonopane S G, Eatherton M, Madsen R, Manley B, Moen C D, Nakata N, Rogers C and Yu C 2016 Seismic Response and Engineering of Cold-Formed Steel Framed Buildings
[40] Abhishek C and Shelke N 2019 Comparative study of time history analysis of cold formed steel frame *International Journal of Advance Research, Ideas and Innovations in Technology* **5** 236–40
[41] Leng J, Buonopane S G and Schafer B W Seismic Modeling and Incremental Dynamic Analysis of the Cold-Formed Steel Framed CFS-NEES Building
[42] Hancock G J 2016 Cold-formed steel structures: Research review 2013–2014 *http://dx.doi.org/10.1177/1369433216630145* **19** 393–408
[43] Dubina D 2008 Behavior and performance of cold-formed steel-framed houses under seismic action *J Constr Steel Res* **64** 896–913
[44] Dubina D 2008 Structural analysis and design assisted by testing of cold-formed steel structures *Thin-Walled Structures* **46** 741–64
[45] Leng J, Schafer B W and Buonopane S G Modeling the seismic response of cold-formed steel framed buildings: model development for the CFS-NEES building
[46] Puglia R, Russo E, Luzi L, D'Amico M, Felicetta C, Pacor F and Lanzano G 2018 Strong-motion processing service: a tool to access and analyse earthquakes strong-motion waveforms *Bulletin of Earthquake Engineering* **16** 2641–51
[47] Engineering Strong Motion Database ESM

# A design tool for the analysis of reinforced and prestressed concrete columns

**B Kovacs[1] and T-N Toader[1]**
[1]Technical University of Cluj-Napoca, Faculty of Civil Engineering, Department of Structures, 25 Barițiu Str., 400027, Cluj-Napoca, Romania

E-Mail: traian.toader@dst.utcluj.ro

**ABSTRACT.** This article aims to describe an automatic calculation algorithm developed for the analysis of reinforced concrete and / or prestressed concrete columns. The article describes two analysis types: sectional analysis and analysis on the element. The sectional analysis consists in the generation of three characteristic curves: bending moment - curvature (M-1/r), axial force - bending moment (N-M) and bending moment – bending stiffness diagrams. The analysis on the element provides the height – displacement (H-w) diagram, obtained by using the finite element method (FEM). For the two analysis, the calculation approach and the implementation of the algorithm is briefly described by the mathematical equations used. To ease the usage of the algorithm an application was developed in MATLAB appDesigner. For validating the algorithm, the results were compared with those obtained from experimental tests in the laboratory but also with numerical simulations using commercial computational software based on FEM. MATLAB programming language was used to implement the algorithm, while results of the analysis proved to be conservative when compared to the experimental ones. Moreover, the bending moment resistance for prestressed concrete columns using bonded tendons is better estimated with at least 10% when compared with a commercial software, based on fiber-based cross-section nonlinear analysis.

## 1. Introduction

Facing a fast urban development in Romania, the need for buildings in industrial regions increased significantly. The most widely used buildings in these regions are single-story warehouses with long slender columns fully fixed in foundation and hinged at top. The main problem of such structures is the significant lateral displacement at the roof level caused by horizontal actions such as wind or earthquake [1] [2]. This large drift can be diminished by increasing the lateral stiffness of the superstructure [3]. Such improvement can be reached by the using columns with larger cross-section or by using materials with higher modulus of elasticity. However, it has been observed that by switching from reinforced concrete to prestressed concrete we get the same effect but with significant material savings [4] [5].

The aim of this publication is to present a new design tool developed to calculate the lateral displacement of reinforced and/or prestressed concrete long columns. The algorithm is composed of two parts, first the generation of the characteristic curves on the cross-section: bending moment – curvature ($M - \phi$) and axial force – bending moment ($N - M$); and second, the variation of bending stiffness in accordance with the magnitude of the bending moment ($M - E \cdot I$), the variation of the lateral displacement along the column length ($H - w$), and generation of the force diagrams: axial loading, shear force, and bending moment. Computation on the cross-section is iterative, while the calculations on the entire element are done using the Finite Element Method (FEM). Investigations for determining the accuracy of the algorithm were done by comparing the computational results with the ones from experimental tests done by other researchers [4] [6], as well as by running in parallel numerical analysis in a FEM based commercial software, SAP2000, for the same input values.

The programming language used for writing the algorithm was MATLAB, which afterwards was transformed into an app using MATLAB appDesigner, resulting into a design tool with a friendly user interface.

## 2. Algorithm development

From a functional point of view, the algorithm can be divided in two main parts: the starting point preceding the determination of the curve and the computation of the actual curve. The input data for computation include: cross-section geometry, properties of materials, loss of prestress and material degradation parameters, followed by processing the input data: cross-section transformation and the determination of the actual prestressing force. Afterwards, the following diagrams are determined: plastic bending moment – axial force ($M_{pl}$ – N), bending moment – curvature (M – $\phi$), bending moment – flexural stiffness (M – E·I), height – displacement (H – w) and force diagrams.

### *2.1 Input*

In this section the description of the considered column is made. As a departure point, the column cross-section definition is required. The cross-section is given in two steps, first step: the height (h) and the width (b) of the column cross-section, and second step: position of the longitudinal reinforcement [2]. The reinforcement is defined by coordinates and geometrical characteristics (diameter, area). A visual representation of the input data is shown in figure 1 and in figure 2.

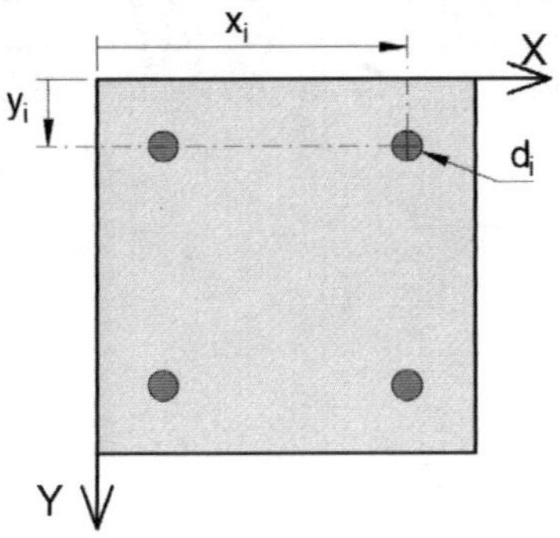


**Figure 1**: Size and position of passive reinforcement.

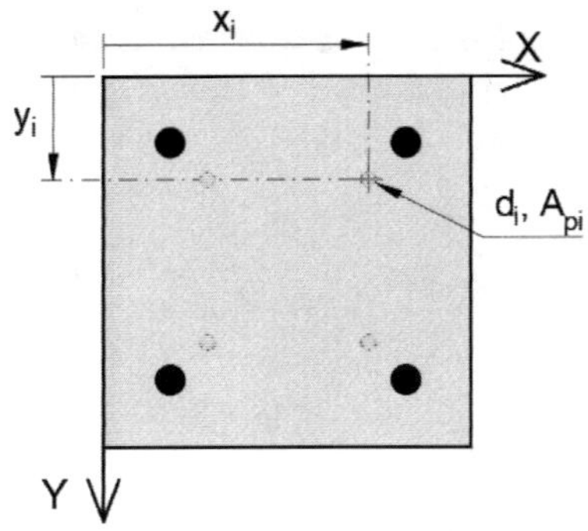


**Figure 2**: Size and position of pretensioned reinforcement.

For the rest of column`s geometry, the position of the nodes is given [7]. In this purpose a three by m matrix is used, where: m is the number of nodes defined, the first column consists of the node index, the second and the third column comprise the node coordinates. A second matrix is needed to specify which degree of freedom should be blocked for each node, therefor a four by k matrix is used where k represents the number of nodes, the first column is the index of the node which is affected by restraints and the last three columns represent the following: translation on the horizontal direction, translation on the vertical direction and rotation by the z axis. Is assigned 1 if the degree of freedom is blocked and 0 if it's free. Configuration of notations for commonly used supports are presented in table 1. The nodal loads are defined also by using matrixes.

**Table 1:** Notations for common supports.

| | Free end | Rolling support | | Hinged support | Fixed support |
|---|---|---|---|---|---|
| Translation X | 0 | 1 | 0 | 1 | 1 |
| Translation Y | 0 | 0 | 1 | 1 | 1 |
| Rotation about Z | 0 | 0 | 0 | 0 | 1 |

The elements that connect the nodes are defined in a matrix, indexing the element in the first column, mentioning the starting (i) and the end (j) nodes in the second and third column, and the fixation type in the last column. The fixation type is needed to correctly determine the stiffness matrix.

The definition of the used materials and the parameters from losses of prestress are presented in [2]. The constitutive curves are determined according to [8] and [9].

Lastly, the parameters of material degradation are needed. By determining the effect of the shrinkage and creeping of the concrete, the material degradation can be numerically quantified [10]. Each effect results in a value with which the ultimate compressive strain of the concrete is reduced, equation (1). The data required is the following: the age of concrete when loaded, $t_0$, coefficient dependent of used cement, s, the coefficients, $k_h$, , relative humidity, RH, the age of the concrete at the beginning of drying shrinkage, $t_s$.

$$\varepsilon_{cu.new}=\varepsilon_{cu}-\varepsilon_{t}-\varepsilon_{cc\_inf\_t0} \tag{1}$$

where $\varepsilon_t$ is the value with which $\varepsilon_{cu}$ is reduced due to concrete shrinkage and $\varepsilon_{cc\_inf\_t0}$ is the reduction value from the concrete creeping.

*2.2 Determining the transformed (idealized) cross-section*

After the input section is filled, the transformation of the cross-section is made [11]. The idea behind the transformation (idealization) of the cross-section is that the axial stiffness of passive and active reinforcement is taken into consideration. The reinforcement is transformed into rectangular concrete evasions which are attached to the cross-section at the corresponding coordinates in the y direction, as shown in figure 3.

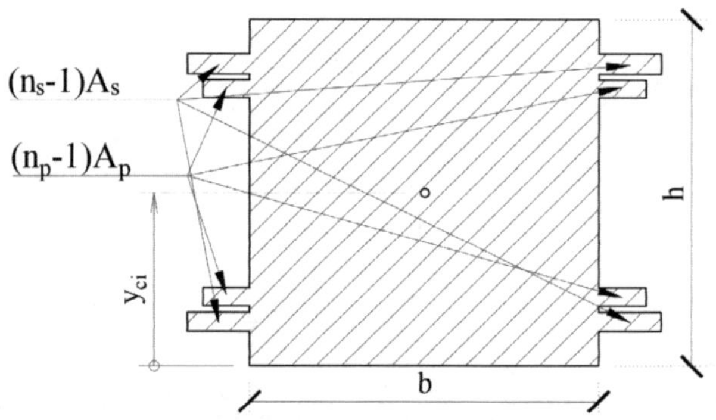


**Figure 3**: Graphical representation of the transformed cross-section.

The process in which this transformation is reached is made in the following order: the calculation of the transformation coefficient presented in equations (2), the computation of the idealized area with equation (3), the determination of the geometrical center of the idealized cross-section with equation (4) and the calculation of the moment of inertia for the transformed cross-section, using the parallel axis theorem, presented in equation (5).

$$n_s=\frac{E_s}{E_c} \qquad n_p=\frac{E_p}{E_c} \tag{2}$$

$$A_{ci}=A_c+(n_s-1)\cdot A_s+(n_p-1)\cdot A_p \tag{3}$$

$$y_{ci}=\frac{\sum(A_i\cdot y_i)}{\sum A_i} \tag{4}$$

where $\Sigma A_i$ is the sum of the equivalent concrete areas, $y_i$ is the distance from a chosen end of the cross-section to each equivalent concrete area.

$$I_{ci}=I_i+A_i\cdot(y_i-y_{ci})^2 \quad (5)$$

where $A_i$ are the equivalent concrete areas, $I_i$ are the moment of inertia of each area.

### 2.3 *Plotting the axial force – plastic bending moment interaction curve (N – $M_{pl}$)*

Succeeding the preliminary calculations, the first interaction curve determined is the axial force – bending moment (N – $M_{pl}$). To determine the curve, the behavior of the column is divided in different stages as presented in [2]. The theoretical basis for this section is presented in [12]. The step-by-step calculation is presented in [10].

An iterative calculation is realized for three out of the six stages: 2a, 2b, 2c. The reason for is that usually the behavior of the concrete columns is in these areas. The extreme points: $I_a$ and $III_b$ are determined with equations (6) and (7). The variation between these end points and the iterated segment are considered linear.

$$N^{III_b}=-A_c\cdot f_{cm}$$
$$M_{pl}^{III_b}=0 \quad (6)$$

$$N^{I_a}=A_s\cdot f_u+A_p\cdot f_{p0,1k}$$
$$M_{pl}^{I_a}=0 \quad (7)$$

At pure compression (point $III_b$), only concrete is considered to carry load. At this stage the steel reinforcement is prone to buckling and its contribution is neglected in the calculation. At pure tension (point $I_a$), only the reinforcement is considered to carry tensile stress, consequently, the concrete tensile strength is neglected throughout the computation. For prestressed concrete, the compressed active reinforcement is disregarded in calculations [13].

The three iterated segments are determined using a linear strain diagram. The strain diagram is determined as follows: for each segment the strain value in the extreme compressed concrete fiber is considered equal to $\varepsilon_{cu}$, the strain value of the tensioned reinforcement is chosen with regards to each segment's interspace, a line that connects the two values is drawn and the other fibers strain values are determined geometrically, with equation (8).

$$\varepsilon_{fib}=\phi\cdot(x-y_{fib}) \quad (8)$$

where $\phi$ is the curvature of the section, x is the compressed concrete height and $y_{fib}$ is the height of each fiber.

If the cross-section includes active reinforcement, then the iterated strain value is that of the tendon, otherwise the strain of the passive reinforcement is considered. The difference between the three iterated segments is the interspace in which the value of reinforcement strain is chosen for each calculation cycle. These interspaces are presented in figure 4.

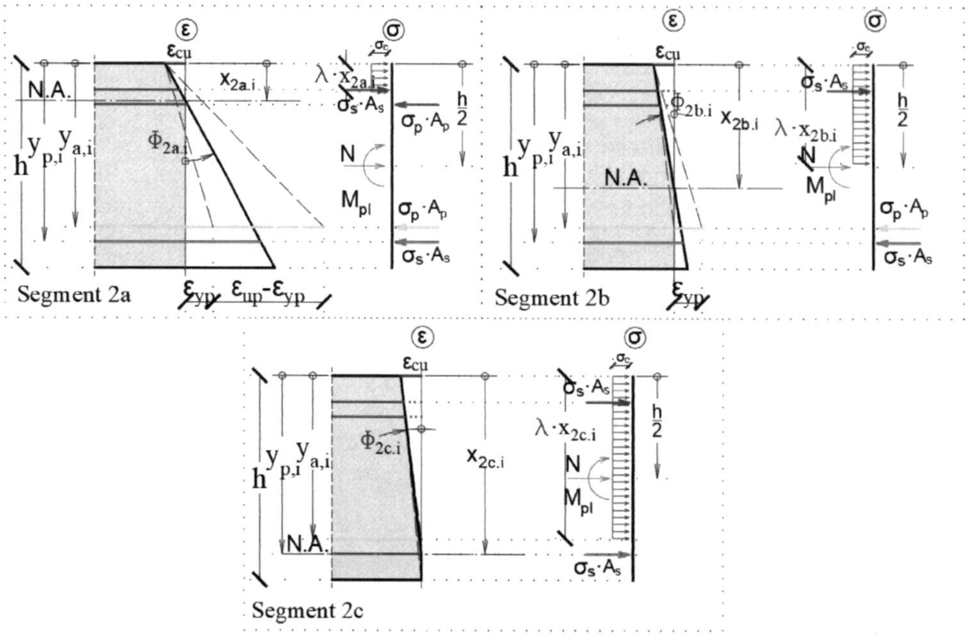


**Figure 4**: Stress state and force diagram for each segment. The reinforcement strain value interspace for each segment is the following: [$\varepsilon_{yp}$, $\varepsilon_{up}$] for 2a, [0, $\varepsilon_{yp}$] for 2b, and for 2c the neutral axis positioning interval is [h, $d_p$] from [10].

The next step is the determination of the stress diagram based on the strain diagram, using Hook`s law. For each fiber the stress value is determined by equation (9). After obtaining the stress diagram, the force value is determined for each fiber with equation (10).

$$\sigma_{fib}=E_{fib}\cdot\varepsilon_{fib} \tag{9}$$

$$F_{fib}=\sigma_{fib}\cdot A_{fib} \tag{10}$$

where $E_{fib}$ is the modulus of elasticity of each analyzed fiber, $A_{fib}$ is the area of each analyzed fiber.

Last, the axial force and the plastic bending moment are determined. The axial force is calculated by summing up the forces that are present in the fibers, while the plastic bending moment is determined by summing the moments that each force generates around the cross-section`s geometrical center. The analytical formulation is presented in equation (11).

$$N=F_{bet}+\sum \pm F_{arm}$$
$$M_{pl}=F_{bet}\cdot\frac{h-\lambda\cdot x}{2}+\sum \pm F_{arm}\cdot\left|\frac{h}{2}-y_{fib}\right| \tag{11}$$

*2.4 Plotting the bending moment – curvature interaction curve (M - Φ)*

The theoretical background for the determination of the M – ϕ interaction curve is like that of the N – $M_{pl}$ curve. The strain-stress-force diagram determination method is identical to the one presented in the previous subsection. The determination of the M – ϕ curve is done in three different stages as presented in [2]. The difference between each stage is the chosen strain that's going to be iterated. These stages are presented in figure 5. The theoretical basis is presented in [12] and [14].

IOP Conf. Series: Materials Science and Engineering 1304 (2024) 012015 doi:10.1088/1757-899X/1304/1/012015

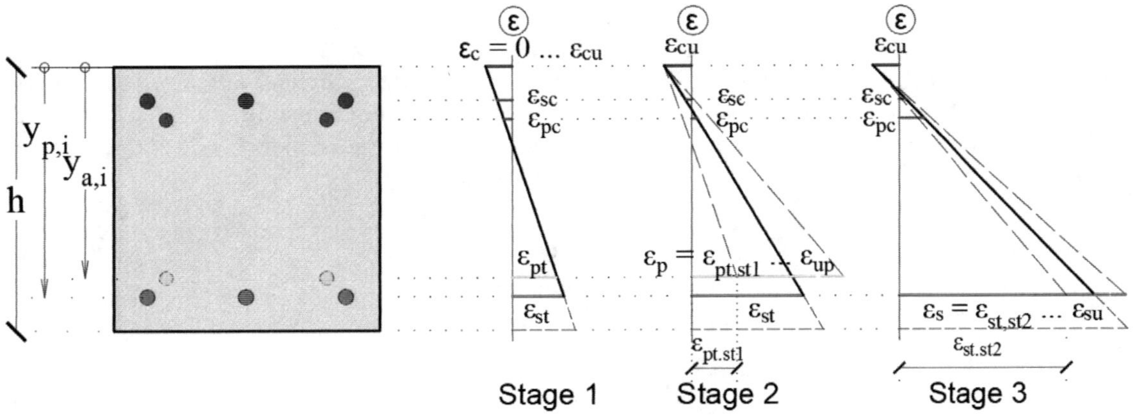


**Figure 5**: Stages of iteration for M-ϕ curve. The strain value interval for each stage is the following: $[0, \varepsilon_{cu}]$ for stage 1, $[\varepsilon_{pt.st1}, \varepsilon_{up}]$ for stage 2 and $[\varepsilon_{st.st2}, \varepsilon_{su}]$ for stage 3 from [10].

Figures 6 to 9 present an example of graphical outputs obtained with the algorithm by before described computational approach. The three stages are color coded: blue represents stage one, green stage two, and red stage three.

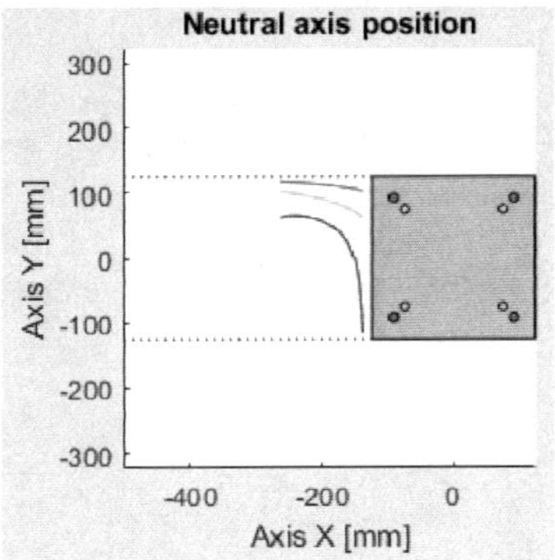


**Figure 6**: Neutral axis positioning.

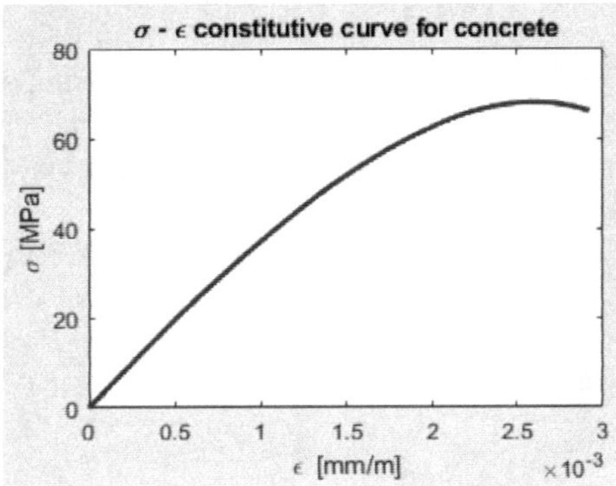


**Figure 7**: σ - ε curve for concrete in compression, generated with the algorithm.

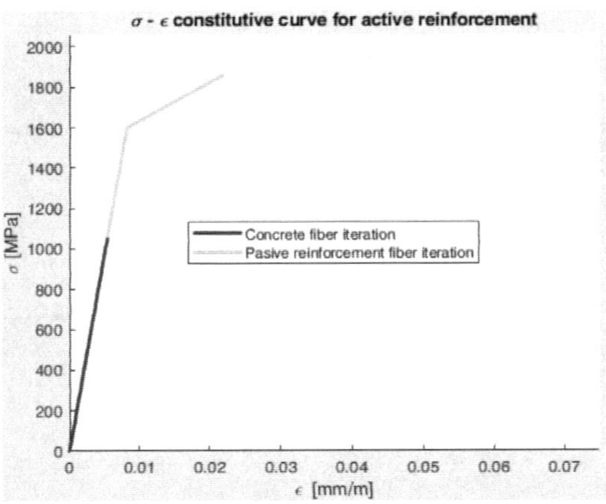


**Figure 8**: σ - ε curve for prestressing tendon in tension, generated with the algorithm.

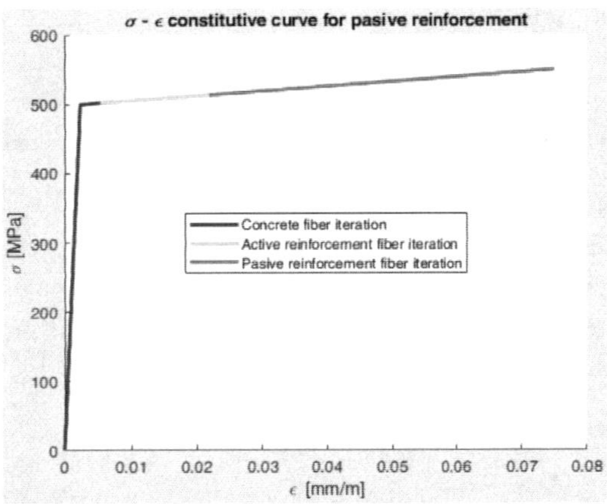


**Figure 9**: σ - ε curve for reinforcement in tension generated by the algorithm.

### *2.5 Plotting the bending stiffness – bending moment interaction curve (M – E·I)*

The bending stiffness – bending moment curve is determined from the previously obtained M – ϕ curve. The theoretical basis of the determination of this curve is presented in figure 10. For each point of the M – ϕ curve a line is drawn from the origin through that point. The angle between the horizontal axis and this theoretical line is the value of the stiffness in that point.

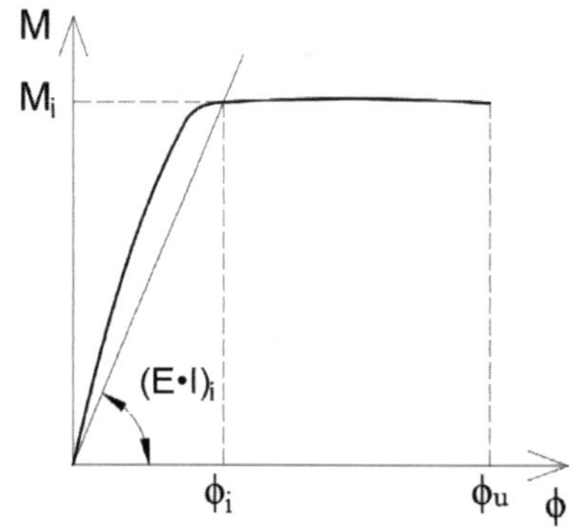


**Figure 10**: Bending stiffness value for the bending moment value at iteration „i".

For each determined plotted point of the moment – curvature diagram, the algorithm determines the stiffness value by using equation (12). Then the value obtained is capped by values obtained in equations 13 and 14. If the cross-section is uncracked then equation (13) is used as the capping value, otherwise equation (14) takes its place. The way the algorithm determines if the section is cracked or uncracked is by comparing the stress value at the extreme tensioned concrete fiber with $f_{ctk}$. If the obtained value exceeds $f_{ctk}$, then the section is considered cracked [15].

$$\phi_i=\frac{M_i}{(E\cdot I)_i}\rightarrow(E\cdot I)_i=\frac{M_i}{\phi_i} \tag{12}$$

$$E\cdot I=(E\cdot I)_{id} \tag{13}$$

$$E\cdot I=\beta\cdot E_s\cdot A_{tot}\cdot d_{tot}^{\ 2} \tag{14}$$

where $(E\cdot I)_{id}$ is the stiffness value determined for the transformed cross-section, $A_{tot}$ is the summed area of the reinforcements present in the cross-section, $d_{tot}$ represents the effective height (the distance between the center of mass of the tensioned reinforcement and the top of the section), β is a reduction coefficient.

### *2.6 Plotting the lateral displacement – height interaction curve (w – H)*

To determine the height – displacement curve the finite element method (FEM) is used, for linear elements only. The benefit of such method is that the geometry is easily customizable. The drawback of the method is that the accuracy of the results is highly dependent on the discretization used. The theoretical basis is presented in [14] and [7], while the implementation method is detailed in [10].

The implementation of the method consists in matrix calculation, having the first step the column input. This aspect was discussed in subsection 2.1. The data needed is the description of the nodes and the elements that connect the nodes. Second, the degrees of freedom for each inputted element are indexed and gathered in a matrix called 'The matrix of degrees of freedom, M_GL'. The content of the M_GL matrix is shown in expression (15).

$$M\_GL=\begin{bmatrix}\text{FE index} & u_{x_i} & u_{y_i} & w_{z_i} & u_{x_j} & u_{y_j} & w_{z_j}\\ & & & \vdots & & & \end{bmatrix} \tag{15}$$

where FE index is the index of the current element and $u_x$, $u_y$ and $w_z$ are the degrees of freedom for each node of the element.

Afterwards the 'Localization' matrix, L, is constructed, which helps collect each specific matrix element into a single matrix. The localization matrix is presented in expression (16).

$$L=\begin{bmatrix} 1 & 0 & 0 & 0 & 0 & 0 & \cdots \\ 0 & 1 & 0 & 0 & 0 & 0 & \\ 0 & 0 & 1 & 0 & 0 & 0 & \\ 0 & 0 & 0 & 1 & 0 & 0 & \\ 0 & 0 & 0 & 0 & 1 & 0 & \\ 0 & 0 & 0 & 0 & 0 & 1 & \\ 0 & 0 & 0 & 1 & 0 & 0 & \\ 0 & 0 & 0 & 0 & 1 & 0 & \\ \vdots & & & & & & \ddots \end{bmatrix} \quad (16)$$

Next a rigidity matrix is selected for each element. These matrixes are attributed to each element using the fixation type defined in the input. In addition, matrixes are defined in local coordinates for each element so a transformation matrix, R, is needed for the conversion to global coordinates, expression (17). After transformation, each rigidity matrix is gathered in a matrix using L, as in equation (18).

$$R=\begin{bmatrix} \cos(\alpha) & \sin(\alpha) & 0 & 0 & 0 & 0 \\ -\sin(\alpha) & \cos(\alpha) & 0 & 0 & 0 & 0 \\ 0 & 0 & 1 & 0 & 0 & 0 \\ 0 & 0 & 0 & \cos(\alpha) & \sin(\alpha) & 0 \\ 0 & 0 & 0 & -\sin(\alpha) & \cos(\alpha) & 0 \\ 0 & 0 & 0 & 0 & 0 & 1 \end{bmatrix} \quad (17)$$

$$K=L^T \cdot k_g \cdot L \quad (18)$$

The next step consists in the generation of the nodal load vector, P. The vector cumulates the values of the nodal loads for each node. Afterwards, the rigidity matrix and the nodal load vector are reduced with regards to the nodes that suffered losses of degrees of freedom. Then the displacement, U, and the force, S, matrix is determined. The expression that helps determine these vectors are presented in equations (19) and (20). After obtaining the displacement values in the direction of each degree of freedom, a displacement matrix is filled with the blocked displacements that were previously reduced.

$$U_{free}=K_{reduced}^{-1} \cdot P_{reduced} \quad (19)$$

where $K_{reduced}$ is the reduced rigidity matrix and $P_{reduced}$ is the reduced load vector.

$$S=K \cdot U \quad (20)$$

After calculating the value for the moment for each finite element, the stiffness is altered in the inputted data and the calculation is repeated. The stiffness value is picked using the M – E·I interaction curve. In the input, the stiffness is defined using the value for moment of inertia, I, so the expression that considers the rigidity reduction is shown in expression (21).

$$I=\frac{E \cdot I}{E_c} \quad (21)$$

## 3. App development

To make the algorithm more intuitive to use, an application was developed in MATLAB appDesigner [10]. The application interface can be seen in figure 11. The interface is divided in two main parts: input on the left-side and output on the right-side, both are further divided in subsections. For input these subsections are named as: Material, Section, Element, Prestress and time effect data, and Results, while for output three subsections are created: Geometry, Section curves, Element curves.

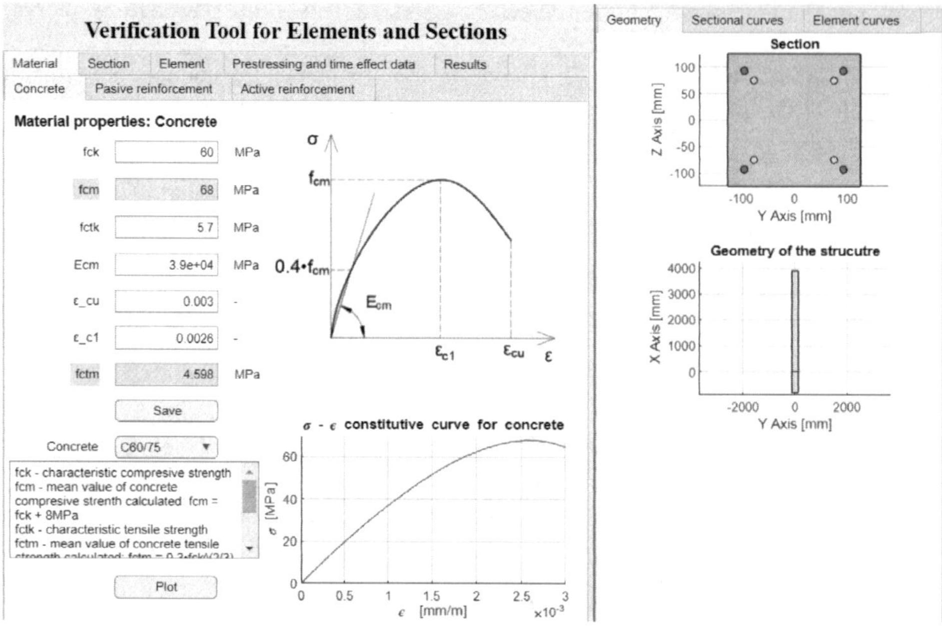


**Figure 11**: Application interface.

The data required for the input section is identical to that listed in subsection 2.1 `Input`. The 'Material' subsection has three main parts: the definition of concrete, reinforcement and tendon. The layout of each part is alike. They include the following elements: data input area (top left), informative graphic (top right), input data description (bottom left) and graph area for the added data (bottom right) activated by the `Plot` button. The layout is presented in figure 11.

The subsection 'Section' similarly to the previous one has three main parts: the definition of the concrete cross-section and the positioning of the passive and active reinforcement. The layout of the concrete cross-section definition is different compared to the reinforcement positioning. The cross-section is defined by its height (h) and with (b), values which can be added in the two input boxes (top left). By pressing the 'Save' button the application will calculate the area and the moment of inertia of the given cross-section. The layout is presented in figure 12. The reinforcement is added by mentioning the position in the cross-section and the diameter. For the active reinforcement, additionally the area of the tendon is also required. The reinforcement is added one by one, and it can be verified in the table shown at the bottom side.

The 'Element' subsection is also divided in three parts: the definition of nodes and elements, and the attribution of nodal restraints and loads. Nodes can be added by giving them an index and coordinates. The interface is presented in figure 13. Restraints and loads can be attributed to the nodes by mentioning the index of the node and the desired values, in regards of the prescriptions from the subsection 2.1 'Input'. Elements are added by defining an index, the beginning and end nodes ('i' respectively 'j') and the type of fixation. The data will be inserted to a table at the bottom of the subsection, along with the transformed area and moment of inertia, the length of the element and the angle between the horizontal axis and longitudinal axis of the element.

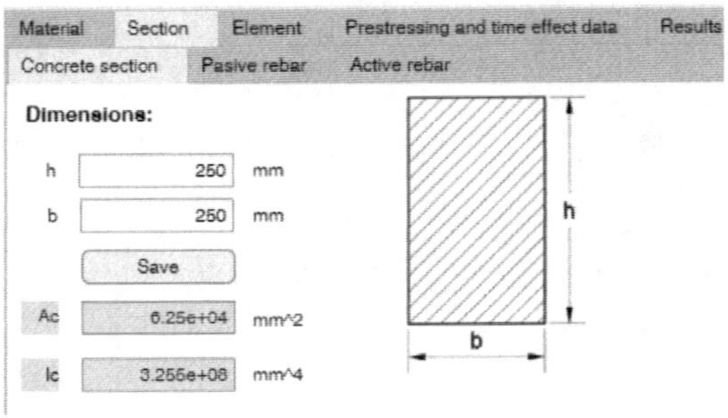


**Figure 12**: Cross-section input.

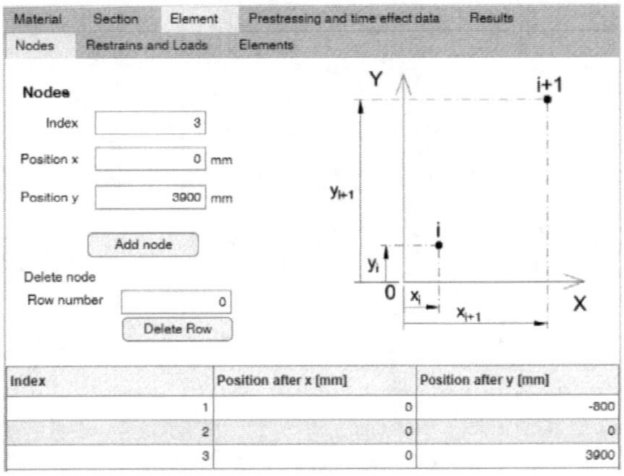


**Figure 13**: Node and element input.

The 'Prestressing and time effect data' subsection is divided in two parts. The loss of prestress calculation is made on the left-side of the subsection, and the material degradation on the right-side. In addition, each coefficient is explained in the text boxes, as in figure 14.

Through the last subsection, 'Results', the user can control de increments in which the diagrams are determined and can also obtain some numerical results as well, figure 15. This subsection is divided in four parts. The first part transforms the cross-section in an idealized one, utilizing the above-mentioned steps. If elements where defined, then this step is automatically made when the first element is added. The second part determines the number of iterations in which the N – $M_{pl}$ interaction curve is split. In this section, the coordinates of the balance point and the corresponding plastic moment to an arbitrarily chosen axial force is given. The third part determines for different values of axial force and draws the curves M – ϕ and the M – E·I, through an iterative process. The last section determines the number of finite elements in which the elements are discretized to determine the H – w interaction curve.

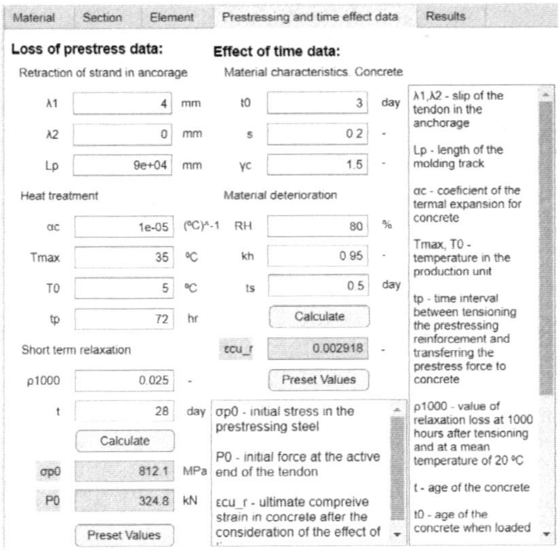


**Figure 14**: Loss of prestress and material degradation data input.

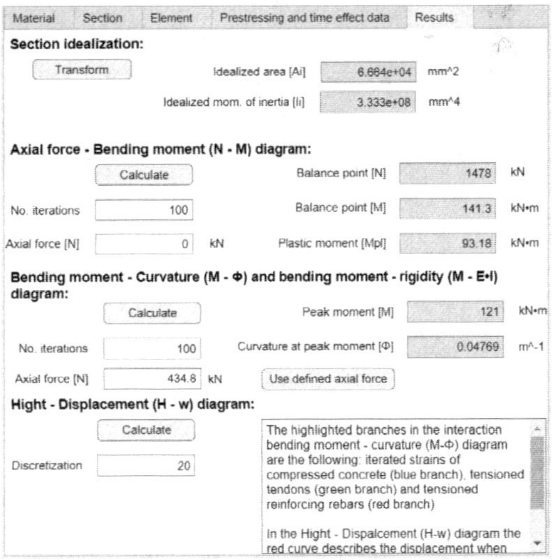


**Figure 15**: Results subsection.

The output section consists of three subsections: the geometrical representation of the input data figure 16, curves defined by sectional analysis figure 17 and diagrams determined using FEM figure 18.

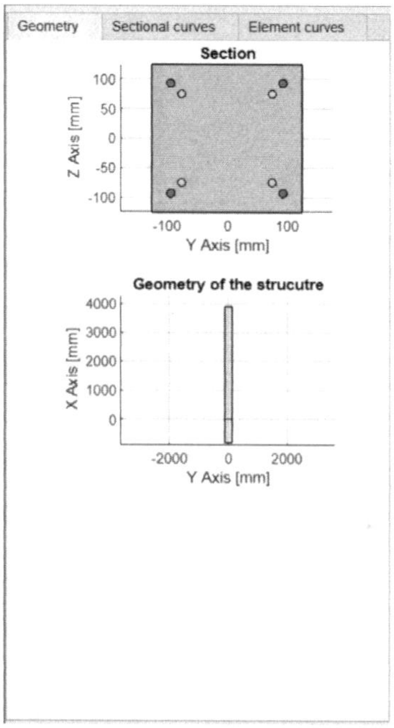


Figure 16: Graphical representation of input data.

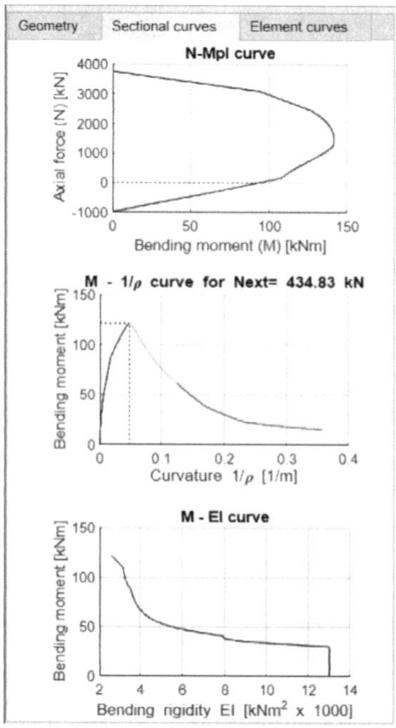


**Figure 17**: Output of cross-sectional diagrams.

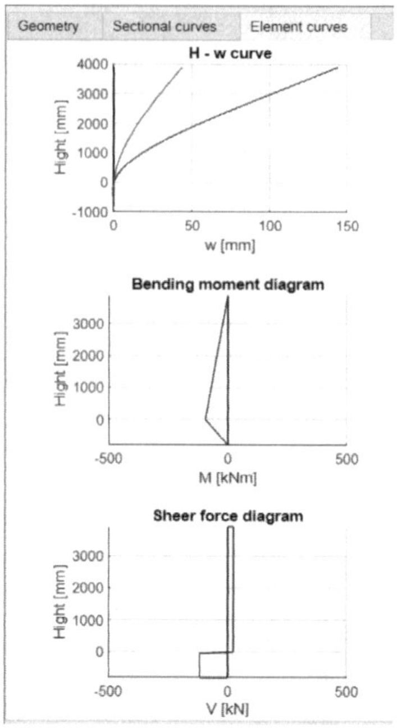


**Figure 18**: Output of displacement and force diagrams on the element.

## 4. Results

To test the accuracy of the results provided by the algorithm, eight types of cross-sections were modelled. Three cross-sections were taken from [4], figure 19, and the remaining five from [6], figure 20.

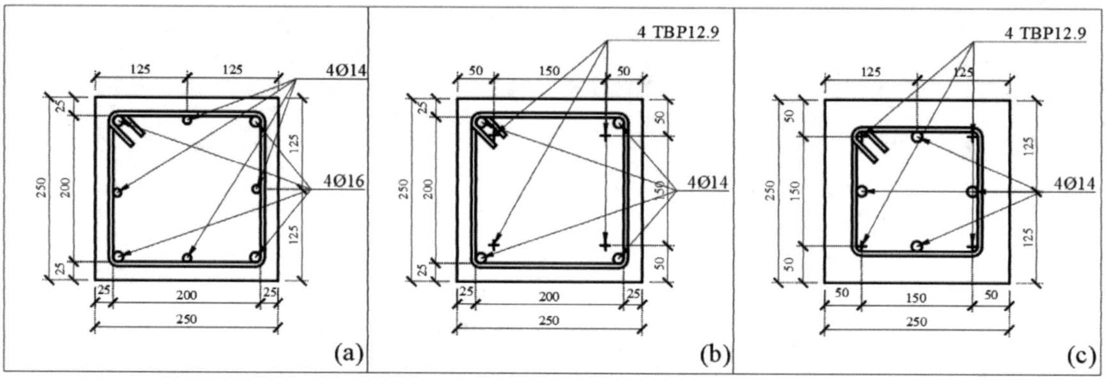


**Figure 19**: Column cross-sections: a) S01, b) S02 and c) S03, from [4].

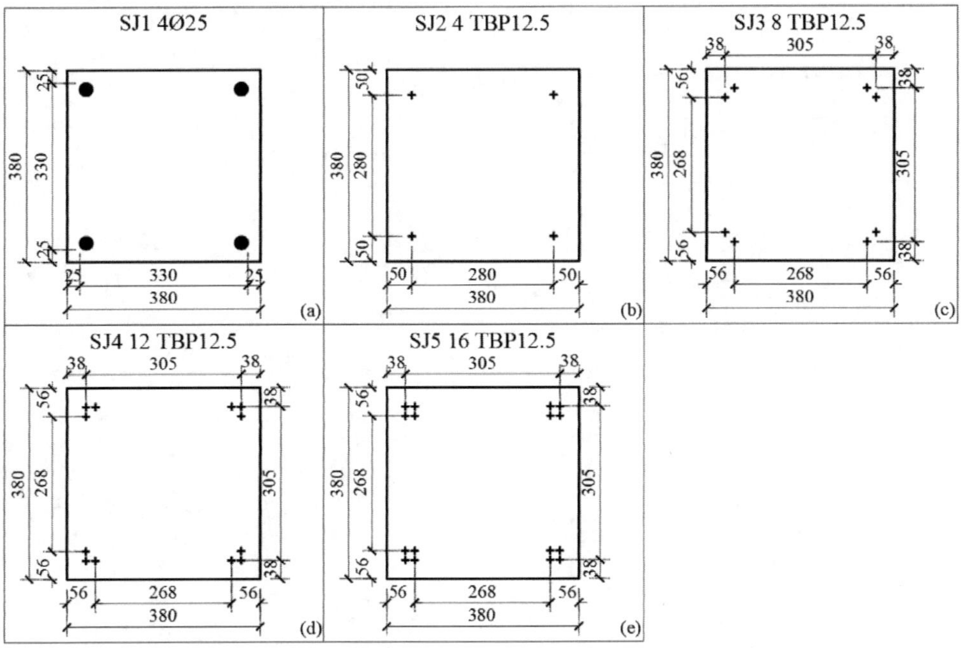


**Figure 20**: Column cross-sections: SJ1, SJ2, SJ3, SJ4 and SJ5 from [6].

The first three cross-sections were tested experimentally and modelled in a commercial software for structural analysis, SAP2000. The generated $N - M_{pl}$ and $M - \phi$ curves were compared with the ones obtained from the algorithm. The detailed results are presented in [2]. The maximum displacement was determined for the maximum lateral load and compared with the values presented in [4], table 2. The $M - E{\cdot}I$ curves are compared in figure 21.

**Table 2:** Displacement value comparison between experiment (from [4]) and the computed ones.

| | Element | | Displacement from experiment at P=Pmax | Maximum lateral force from experiment Pmax | Mean displacement | Mean lateral force | Computed displacement | Ratio between computed and experimental values |
|---|---|---|---|---|---|---|---|---|
| | - | | [mm] | [kN] | [mm] | [kN] | [mm] | - |
| **S01** | 1 | left | 294.86 | 27.12 | 285.91 | 27.24 | 373.68 | 1.31 |
| | | right | 276.95 | 27.36 | | | | |
| **S02** | 1 | left | 325.95 | 29.76 | 237.95 | 30.51 | 201.36 | 0.85 |
| | | right | 158.63 | 24.00 | | | | |
| | 2 | left | 306.39 | 32.56 | | | | |
| | | right | 262.23 | 31.14 | | | | |
| | 3 | left | 259.49 | 29.88 | | | | |
| | | right | 115.02 | 33.06 | | | | |
| **S03** | 1 | left | 206.97 | 27.28 | 273.33 | 29.505 | 257.91 | 0.94 |
| | | right | 387.25 | 30.87 | | | | |
| | 2 | left | 398.68 | 29.34 | | | | |
| | | right | 166.55 | 26.61 | | | | |
| | 3 | left | 249.05 | 30.18 | | | | |
| | | right | 231.47 | 29.67 | | | | |

IOP Conf. Series: Materials Science and Engineering 1304 (2024) 012015 doi:10.1088/1757-899X/1304/1/012015

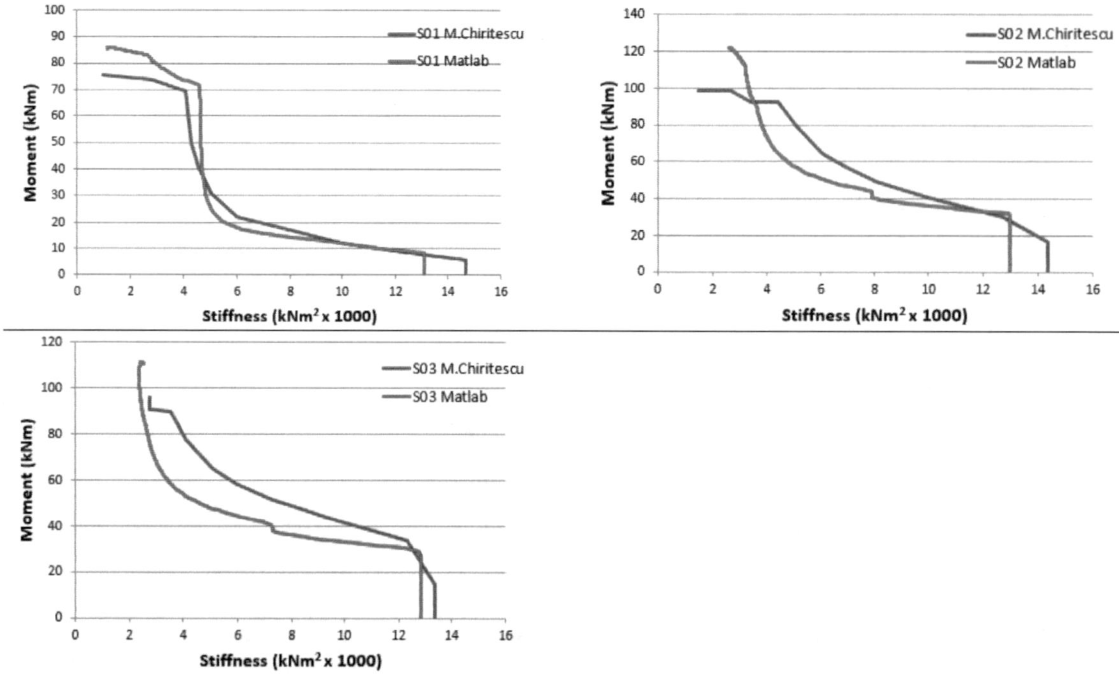


**Figure 21**: Comparison between algorithm generated M – E·I curve (red) and the one obtained by [4] (blue).

For the last five cross-sections the M – E·I curve were compared. The results are presented in figure 22.

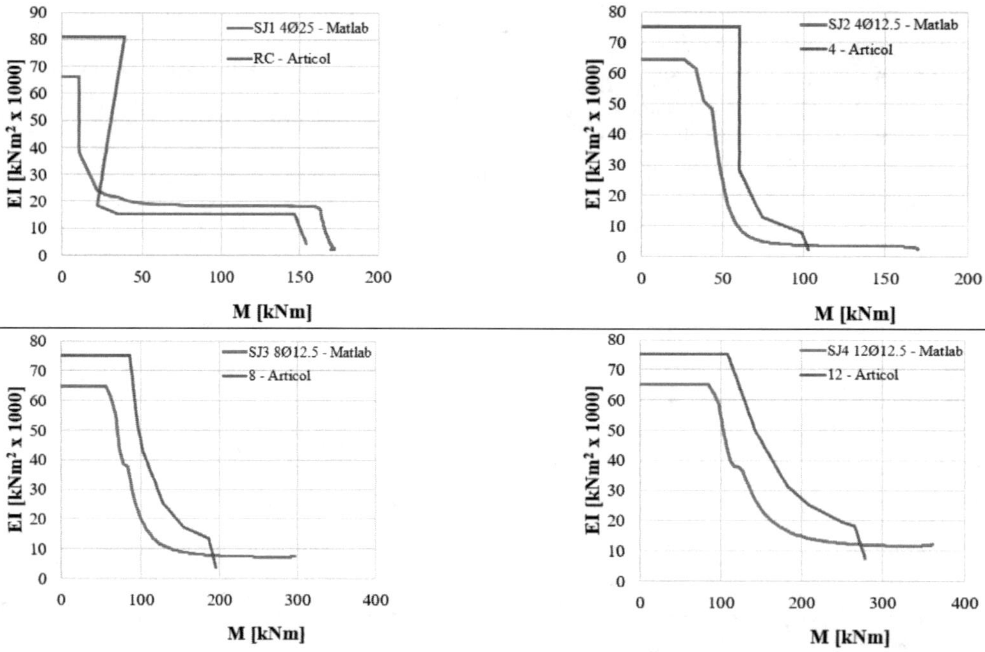

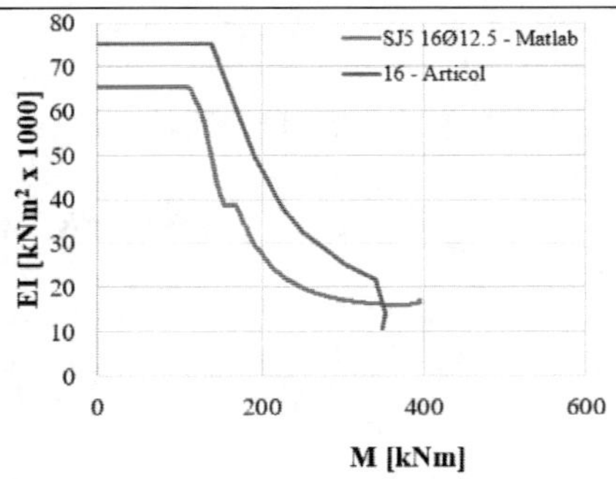


**Figure 22**: Comparison between algorithm generated M – E·I curve (red) and the one obtained by [6] (blue).

## 5. Conclusions

From the bending moment – bending stiffness (M – E·I) diagrams we can deduce the following conclusions: the axial loading has a positive effect on the stiffness of the element. For higher axial loading (for example columns S02 as well as for S03) due to prestressing the bending stiffness remains high on a larger interval of bending moments. The prestressed columns started losing stiffness after the bending moment exceeded 25kNm, while the columns that were not prestressed had a lower threshold of 10kNm. The results of the analysis when compared to the ones from experiments [4] and other published scientific papers [6], show that the approach developed initially by Faur [12] [16] [17] for calculating reinforced concrete members in normal cross-sections can be successfully further developed and applied to prestressed concrete members as well and through programming transformed into an app.

The comparison between the M – E·I diagram generated by the algorithm and the ones taken from [4] show that the value of the initial stiffness it has a lower calculated value that the one taken from references. This difference in initial stiffness can be observed in the comparison made between the algorithm and the cross-sections taken from [6].

In the case of cross-sections SJ2, SJ3, SJ4, and SJ5, we can see that the allure of the curves is similar, but the maximal bending stiffness value differs. The algorithm generates conservative bending stiffness values (between 60 and 70 $MNm^2$) when compared to the ones from [6] (between 75 and 85 $MNm^2$).

## 6. References

[1] Ostertag A, Toader T N, Bertsche B and Sobek W 2019 *System-Safety in the Application of Adaptive Load-Bearing Structures*, Conference: 2019 Annual Reliability and Maintainability Symposium, https://doi.org/10.1109/RAMS.2019.8769000

[2] Kovacs B and Toader T N 2022 *An automated algorithm for the sectional analysis of reinforced and prestressed concrete columns*, Journal of Applied Engineering Sciences, no. 1/2022, p. 10, https://doi.org/10.2478/jaes-2022-0010

[3] Toader T N and Kiss Z 2014 *Experimental tests on a precast concrete frame*, Bulletin of the Polytechnic Institute of Jassy, CONSTRUCTIONS. ARCHITECTURE Section.

[4] Chirițescu M, Teză de doctorat 2016 *Utilizarea precomprimării la elemente prefabricate lungi din beton, solicitate preponderent la compresiune cu încovoiere*, Universitatea Tehnică din Cluj-Napoca.

[5] Toader T N, Sobek W and Schmeer D 2019 *Concept for an onshore tower structure made of UHPFRC segments for wind turbines*, Acta Technica Napocensis: Civil Engineering & Architecture, vol. 62.

[6] Haavisto J, Kerokoski O and Laaksonen A 2019 *Second-order analysis of prestressed concrete columns*, FIB symposium proceedings, p. 7.

[7] Nedelcu M and Mociran H 2016 *Metoda elementelor finite* - îndrumător de laborator -, Cluj-Napoca: U.T.PRESS, ISBN 978-606-737-202-1.

[8] CEN, SR EN 1992-1-1 2004, Eurocod 2 - Proiectarea structurilor din beton, Partea 1-1: reguli generale şi reguli pentru clădiri, Bucureşti.
[9] Model Code 2010. First complete draft - Volume 1, fib Bulletin, ISBN 978-2-88394-105-2, doi.org/10.35789/fib.BULL.0065.
[10] Kovacs B 2022 *Lucrare de disertație: Un algoritm de calcul pentru stâlpii din beton armat precomprimat*, Universitatea Tehnică din Cluj-Napoca.
[11] Kovacs B, Toader T N 2022 *Algoritm de calcul automat folosind MEF pentru calculul săgeților stâlpilor din beton precomprimat cu armături aderente*, A XIX-a Sesiune Națională de Comunicări Științifice Studențești (editori: H.-L. Cucu, D.-L. Manea, N.-M. Ilieş, G. Hoda), Cluj-Napoca: UT Press, ISBN 978-606-737-532-9
[12] Faur A, Puskás A 2014 *Moment plastic, moment ultim, moment capabil*, Cluj-Napoca: U.T.PRESS
[13] Aalami O B 2014 *Post-tensioned buildings Design and Construction*, USA: PT-Structures, , chapter 12 – Section design for bending.
[14] AGENT R, BĂNUȚ V 1979 *Calculul structurilor din beton armat cu stîlpi zvelți (Design of reinforced concrete structures with slender columns)*, Bucureşti: Editura Tehnică
[15] Tertea I 1981 *Betonul Precomprimat Bazele proiectării*, Bucureşti: Editura Tehnică.
[16] Faur A, Puskás A 2016 *Plastic Moment Capacity Evaluation for Reinforced Concrete Frame Elements by Adopting the Proper Material Constitutive Laws*. II International Conference on Concrete Sustainability ICCS16, Technical University Madrid, Madrid.
[17] Faur A, Puskás A 2016 *Simplifications for considering the contribution of the reinforcement in the compression zone for designing more efficient RC frame elements*. II International Conference on Concrete Sustainability ICCS16, Technical University Madrid, Madrid.

# Predicting multi-span bridge maximum deflection: Numerical simulation and validation

**M Brinissat [1], R P Ray [2], R Kuti [3]**
[1,2] Department of Structural and Geotechnical Engineering, Széchenyi István University, 9026 Győr, Hungary.
[3] Department of Machine Design, Faculty of Mechanical Engineering, Informatics and Electrical Engineering, Széchenyi István University, Egyetem tér 1,9026 Győr, Hungary.

E-mail: brinissat.marame@hallgato.sze.hu

**Abstract.** Deflection is one of the main parameters often required to reflect and monitor the overall health state of bridge structures. Several approaches have been developed to monitor deflection accurately and economically. This research proposes a theoretical method to estimate the maximum deflection of the longest section of a seven-span continuous bridge using stress data collected through strain gauges. Firstly, initial data were collected from field tests under various controlled loadings to obtain the maximum deflections and stresses. Then, the stress-deflection regression equations are derived to reconstruct deflection from estimated gauges. Finally, the maximum deflection values are compared with analytical results using finite element simulations to verify the method's effectiveness.

## 1. Introduction

Deflection is an essential indicator of a bridge's overall health and safety. Monitoring vertical deflection provides valuable information about the bridge's condition. It helps identify potential problems: excessive component stress, fatigue cracking, or other damage [1]. There are various ways to monitor bridge conditions; one uses GPS, which is relatively expensive despite producing accurate data [2]. Less costly alternatives may also provide the necessary information to determine deflection on the bridge. A few studies adopted methods for empirically fitting curves using predefined deflection forms. Zhang et al. [3] reconstructed the estimated bridge deflection of cable-stayed bridges through inclinometer data using the finite-element model combined with the partial least-square regression method. Sun et al. [4] proposed a novel method for bridge displacement estimation based on inclinometer data by developing a hybrid monitoring algorithm that uses the finite-element model and the monitored data. In addition, placing a tiltmeter or strain gauge is also an effective method of monitoring deflection [5]. Penuntun et al. [6] proposed a new method depending on the Lagrange Interpolation Half approach to obtain deflection data from rotational data; they examined the deflection of a cable-stayed bridge using tiltmeters and approved the accuracy of the proposed method.
In order to implement the idea of hybrid monitoring identified above. Suangga et al. [7] employed a particular stress-deflection relationship to predict bridge maximum deflection by utilizing strain data of a simple span bridge. Their findings indicated that the deflection outcomes from the suggested static equation agree well with the 2D Finite Element Method (FEM) analysis results. In this study, the stress-deflection relationship will be applied to load test data of a continuous multi-span steel box girder bridge presented in Brinissat et al. [8], to predict the maximum analytical deflections. This approach will use the regression analysis method, and based on the test loadings and a 3D FEM, the maximum deflections will be correlated accurately.

## 2. Analytical approach of beam maximum deflection

A simplified analytical formulation for maximum beam deflection based on section proprieties and stress follows in this section. Consider an idealized uniform elastic beam with constant depth 'h,' breadth

IOP Conf. Series: Materials Science and Engineering 1304 (2024) 012016 doi:10.1088/1757-899X/1304/1/012016

'b,' and span 'L,' as shown in Figure 1. The beam is assumed to be loaded with a uniform distributed load 'w' over the entire span, with different support conditions, as shown in Table 1.

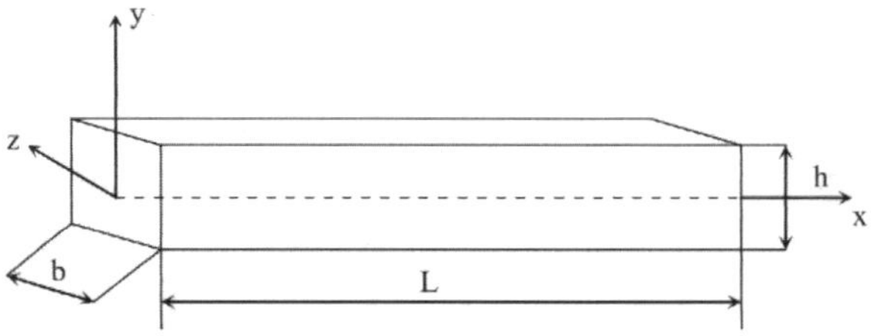


**Figure 1.** Beam model geometry.

The maximum bending stress that occurs at the center of the beam in the extreme fiber follows as:

$$\sigma_{max} = \frac{M_{max}}{I} y_{max} \tag{1}$$

Here $\sigma_{max}$ and $M_{max}$ represent the maximum stress and the maximum bending moment at the cross-section, respectively, $y_{max}$ is the distance from the neutral axis to the extreme fiber, and $I$ is the moment of Inertia of the beam. Where:

$$I = \frac{bh^3}{12} \; and \; y_{max} = \frac{h}{2} \tag{2}$$

Equation (1) can be combined with equation (2) and the maximum bending moment formula to express the maximum deflection at the center of the beam shown in Table 1 below :

**Table 1.** Maximum deflection of beams with different support systems.

| Beam configurations | Maximum deflection $\delta_{max}$ | Maximum Moment $M_{max}$ | $\delta_{max} = f(\sigma_{max})$ |
|---|---|---|---|
| **Simply supported** 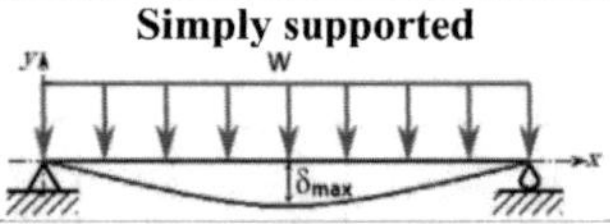 | $\delta_{max} = \frac{5wL^4}{384EI}$ | $M_{max} = \frac{wL^2}{8}$ | $\delta_{max} = \frac{5L^2}{24Eh}\sigma_{max}$ |
| **Fixed-Hinged** 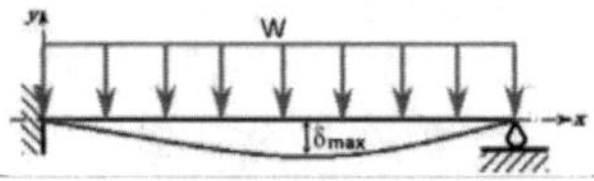 | $\delta_{max} = \frac{wL^4}{185EI}$ | $M_{max} = -\frac{wL^2}{8}$ | $\delta_{max} = \frac{16L^2}{185Eh}\sigma_{max}$ |
| **Fixed-Fixed** 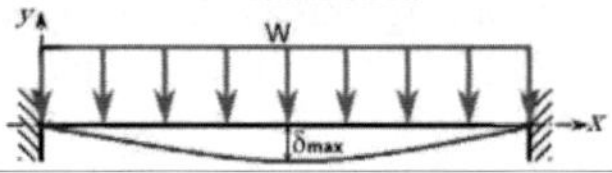 | $\delta_{max} = \frac{wL^4}{384EI}$ | $M_{max} = \frac{wL^2}{24}$ | $\delta_{max} = \frac{L^2}{8Eh}\sigma_{max}$ |

## 3. Case study: The Szapáry Bridge

The structure this research considers is the Szapáry bridge, a multi-span steel box girder bridge located in Hungary. The bridge is 758 meters long; it is composed of a single-cell steel box girder weighing 8,327 tons, orthotropic deck plates, and skewed girders with two tracks, each with two lanes, as shown in Figure 2.

IOP Conf. Series: Materials Science and Engineering 1304 (2024) 012016 doi:10.1088/1757-899X/1304/1/012016

**Figure 2.** The Szapáry Bridge.

## 4. Monitoring methods

*4.1. Monitoring parameters based on FE Model*

A Finite-Element model of the bridge superstructure was simulated in Midas Civil software to represent the condition of the right independent track of the continuous steel box girder bridge, as shown in Figure 3.

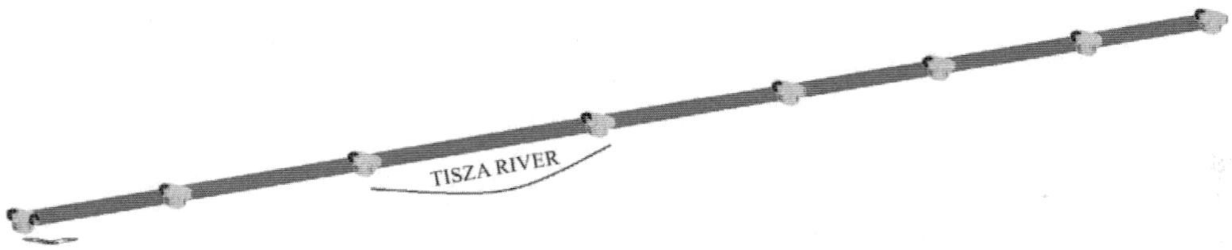


**Figure 3.** Finite element model of the bridge superstructure.

One of the significant challenges with these bridges is the substantial deflection in the midspan, which impacts bridge structure safety. Therefore, this research assessed deflections and stresses at the mid-span of the longest span (3rd span) created by numerous load cases, presented in Table 2.

**Table 2.** Arrangement of the load cases.

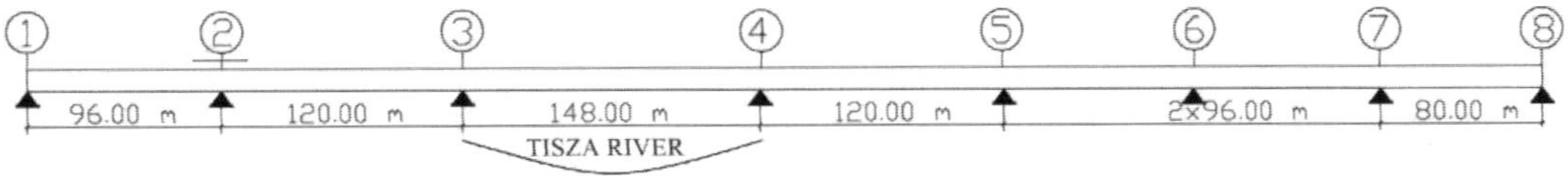


| | |
|---|---|
| 1 | 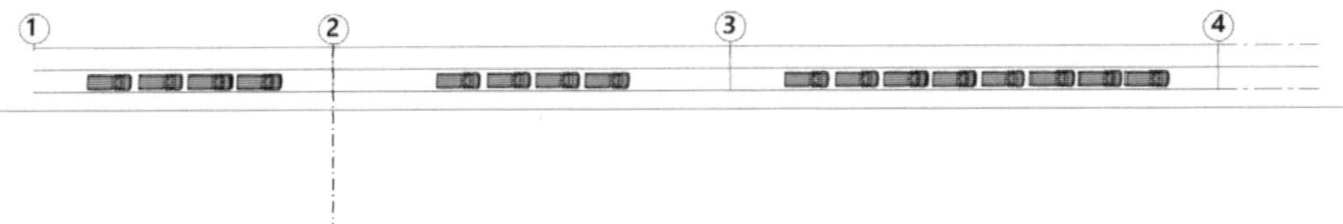 |

IOP Conf. Series: Materials Science and Engineering 1304 (2024) 012016 doi:10.1088/1757-899X/1304/1/012016

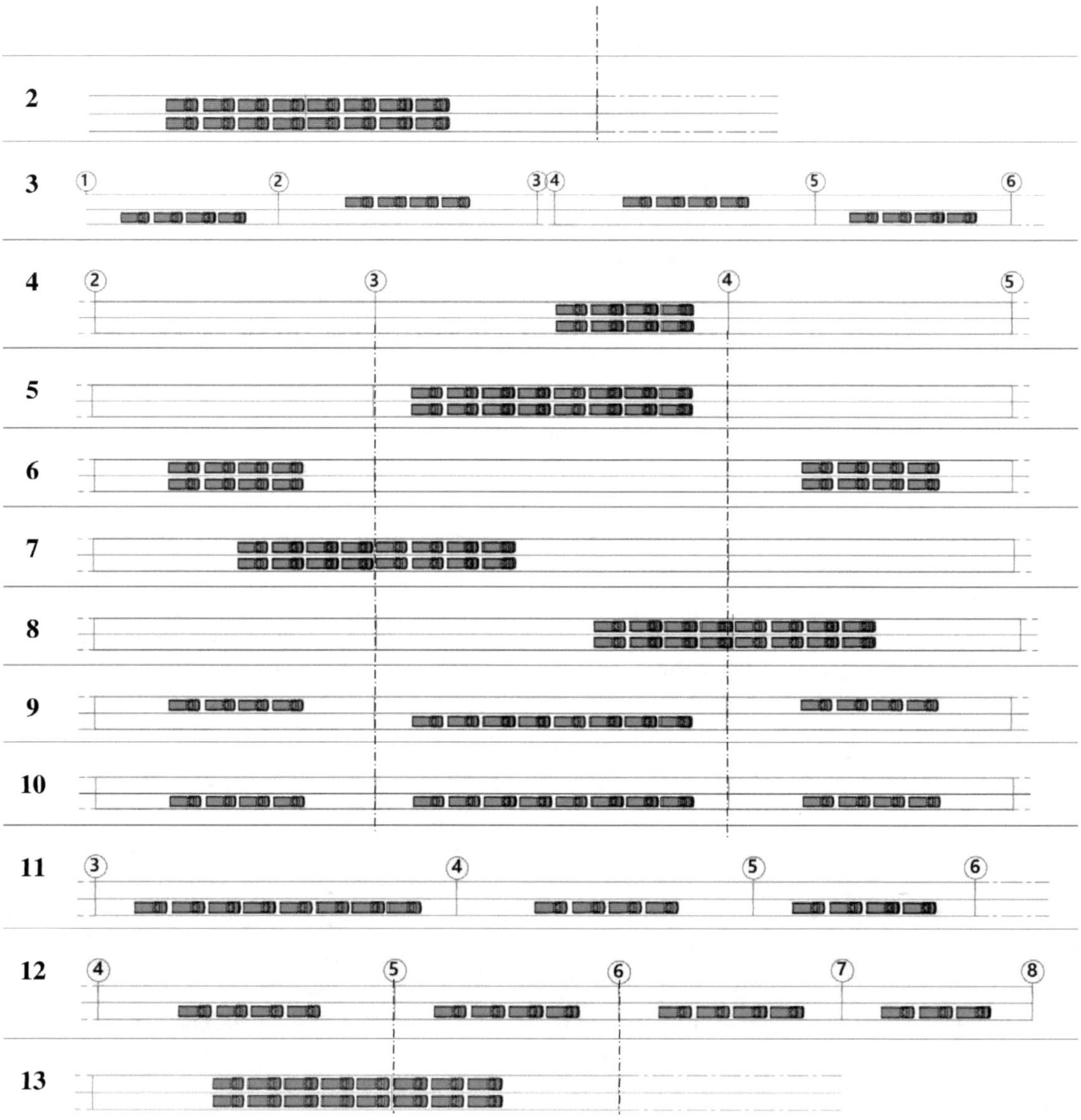


*4.2. Monitoring parameters based on in-situ test*

The deflection and the stress data were based on real-time monitoring of the critical section in the steel box girder to ensure that the bridge could be considered safe and under control.

The midspan section of the longest span provided locations for the stress measurement. Strain gauges produced data at six locations in the section, as shown in Figure 4. The top and the bottom steel plate each contained three measuring points (left, middle, and right). Gauges measured the longitudinal strain of each measuring point, and then the data were collected and analyzed. The relationship between stress and strain allowed for the conversion from strain to longitudinal stress at each measuring point. Two rotary resistance meters fixed on the deck of every span mid-section recorded the deflection measurement. The maximum deflection and stress data from the measured 13 load cases appear in Table 3.

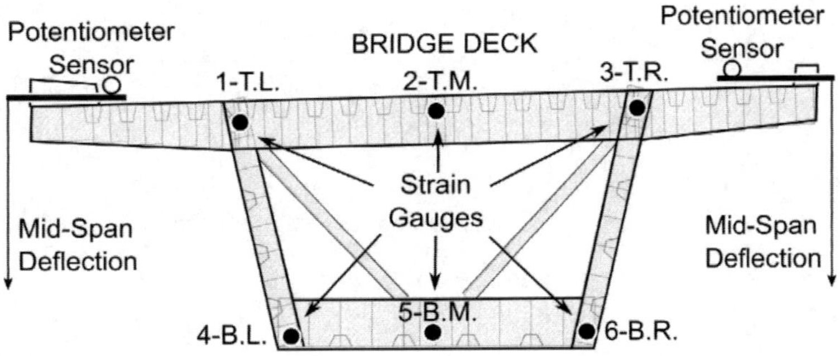


**Figure 4.** Locations of the strain gauge measurement points (FEM X-sec).

**Table 3.** Stresses and maximum deflections from recorded data.

| Load cases | $\sigma_1$ $(N/mm^2)$ 1-TL | $\sigma_2$ $(N/mm^2)$ 2-TM | $\sigma_3$ $(N/mm^2)$ 3-TR | $\sigma_4$ $(N/mm^2)$ 4-BL | $\sigma_5$ $(N/mm^2)$ 5-BM | $\sigma_6$ $(N/mm^2)$ 6-BR | Maximum deflection $\delta(mm)$ L/2 |
|---|---|---|---|---|---|---|---|
| **1** | -17.6 | -20 | -15.3 | 31.8 | 31.5 | 34.8 | -103.87 |
| **2** | 1.4 | 2 | 1.5 | -3.4 | -3.5 | -3.9 | 15.50 |
| **3** | 4.7 | 5.7 | 4.2 | -10.9 | -7.7 | -8.3 | 41.56 |
| **4** | -24.2 | -25.6 | -20.1 | 38.3 | 35.4 | 36.4 | -132.77 |
| **5** | -40.5 | -46.4 | -36.6 | 73.9 | 69.2 | 76.2 | -264.85 |
| **6** | 11.7 | 13.9 | 10.8 | -22.4 | -20.9 | -19.8 | 101.82 |
| **7** | -5 | -6.3 | -5.1 | 10.9 | 9.6 | 10.5 | -35.33 |
| **8** | -9.5 | -10.7 | -7.6 | 18.3 | 15.9 | 16.2 | -63.20 |
| **9** | -13.3 | -14.8 | -11.7 | 19.9 | 23.7 | 26 | -66.51 |
| **10** | -13.8 | -15.8 | -12.1 | 26.3 | 25.7 | 28.2 | -72.24 |
| **11** | -17.2 | -20.1 | -15.5 | 32.4 | 31.3 | 34.3 | -101.14 |
| **12** | 2.5 | 2.5 | 2.1 | -2.8 | -4.2 | -4.3 | 23.66 |
| **13** | 4.9 | 5.1 | 3.8 | -6.1 | -7.7 | -7.8 | 41.23 |

### *4.3. Monitoring parameters based on a theoretical formula*

The developed formula (deflection-stress) is based on interpolation between the recorded data using a multiple regression solver. The interpolation technique was processed to obtain the maximum deflection for each reference point, the sixth point in the middle of the largest span. Table 4 presents the equations and goodness of fit for each location using linear and quadratic regression methods.

**Table 4.** Theoretical formula of maximum deflection as a function of stress.

| Data | Linear regression Equation | $R^2$ | Quadratic regression Equation | $R^2$ |
|---|---|---|---|---|
| $(\boldsymbol{\delta_1}, \boldsymbol{\sigma_1})$ | $\delta_1 = 6.61\sigma_1 + 11.54$ | 0.9799 | $\delta_1 = 0.0001\sigma_1^2 + 6.61\sigma_1 + 11.55$ | 0.9875 |
| $(\boldsymbol{\delta_2}, \boldsymbol{\sigma_2})$ | $\delta_2 = 5.83\sigma_2 + 11.16$ | 0.9863 | $\delta_2 = 0.0001\sigma_2^2 + 5.83\sigma_2 + 11.16$ | 0.9921 |
| $(\boldsymbol{\delta_3}, \boldsymbol{\sigma_3})$ | $\delta_3 = 7.64\sigma_3 + 10.91$ | 0.9830 | $\delta_3 = 0.0041\sigma_3^2 + 7.56\sigma_3 + 10.80$ | 0.9907 |
| $(\boldsymbol{\delta_4}, \boldsymbol{\sigma_4})$ | $\delta_4 = -3.61\sigma_4 + 6.70$ | 0.9865 | $\delta_4 = 0.0001\sigma_4^2 - 3.68\sigma_4 + 9.80$ | 0.9932 |
| $(\boldsymbol{\delta_5}, \boldsymbol{\sigma_5})$ | $\delta_5 = -3.81\sigma_5 + 7.62$ | 0.9858 | $\delta_5 = 0.00015\sigma_5^2 - 3.80\sigma_5 + 9.40$ | 0.9875 |

IOP Conf. Series: Materials Science and Engineering 1304 (2024) 012016 doi:10.1088/1757-899X/1304/1/012016

| | | | | |
|---|---|---|---|---|
| $(\boldsymbol{\delta_6}, \boldsymbol{\sigma_6})$ | $\delta_6 = -3.54\sigma_6 + 9.46$ | 0.9813 | $\delta_6 = 0.0011\sigma_6^2 - 3.65\sigma_6 + 11.20$ | 0.9814 |

$\sigma$ [N/mm²] : Stress,
$\delta$ [mm] : Deflection.
$R^2$: Goodness of fit.

An example of the theoretical approach that was applied to develop the linear regression equation is presented in Figure 5.

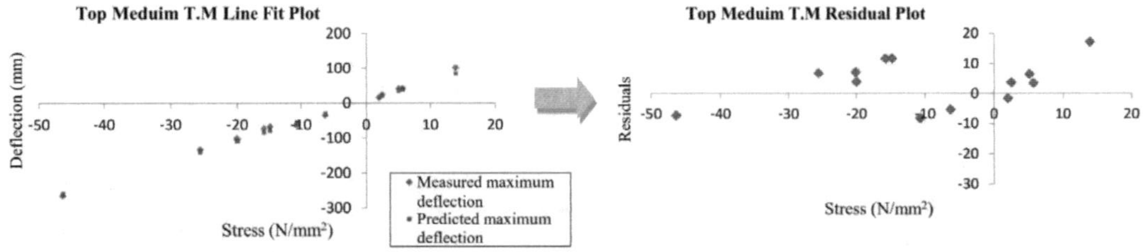


**Figure 5.** Predicted vs measured deflections from linear regression.

## 5. Validation of the theoretical formula

In order to validate the approach, the differences between the measured and the predicted deflections are computed using Equation 3 :

$$\Delta_i = \delta - \delta_i \tag{3}$$

The analysis of available data presented in Tables 5 and 6, shows that the average difference between measured deflection and deflections from regression equations is 7.61 mm for the linear regression and 7.42 mm for the quadratic regression.

**Table 5.** Comparison of maximum deflections from the field test and linear regression equations.

| Load cases | $\delta_1$ (mm) | $\Delta_1$ | $\delta_2$ (mm) | $\Delta_2$ | $\delta_3$ (mm) | $\Delta_3$ | $\delta_4$ (mm) | $\Delta_4$ | $\delta_5$ (mm) | $\Delta_5$ | $\delta_6$ (mm) | $\Delta_6$ |
|---|---|---|---|---|---|---|---|---|---|---|---|---|
| 1 | **-104.80** | 0.93 | **-105.44** | 1.57 | **-105.98** | 2.11 | **-108.17** | 4.30 | **-112.40** | 8.53 | **-113.84** | 9.97 |
| 2 | **20.79** | -5.30 | **22.82** | -7.32 | **22.37** | -6.87 | **18.98** | -3.48 | **20.95** | -5.45 | **23.28** | -7.78 |
| 3 | **42.61** | -1.05 | **44.39** | -2.84 | **43.00** | -1.44 | **46.07** | -4.51 | **36.95** | 4.60 | **38.87** | 2.69 |
| 4 | **-148.42** | 15.65 | **-138.09** | 5.31 | **-142.65** | 9.88 | **-131.64** | -1.13 | **-127.26** | -5.52 | **-119.51** | -13.27 |
| 5 | **-256.17** | -8.69 | **-259.35** | -5.50 | **-268.71** | 3.86 | **-260.23** | -4.62 | **-256.04** | -8.82 | **-260.52** | -4.33 |
| 6 | **88.88** | 12.94 | **92.20** | 9.62 | **93.42** | 8.40 | **87.60** | 14.22 | **87.25** | 14.58 | **79.61** | 22.21 |
| 7 | **-21.51** | -13.82 | **-25.57** | -9.76 | **-28.05** | -7.28 | **-32.67** | -2.66 | **-28.96** | -6.37 | **-27.74** | -7.59 |
| 8 | **-51.26** | -11.95 | **-51.22** | -11.98 | **-47.15** | -16.05 | **-59.40** | -3.80 | **-52.96** | -10.24 | **-47.94** | -15.26 |
| 9 | **-76.37** | 9.86 | **-75.12** | 8.61 | **-78.48** | 11.97 | **-65.18** | -1.33 | **-82.68** | 16.17 | **-82.66** | 16.15 |
| 10 | **-79.68** | 7.44 | **-80.95** | 8.72 | **-81.53** | 9.30 | **-88.30** | 16.06 | **-90.30** | 18.06 | **-90.45** | 18.22 |
| 11 | **-102.15** | 1.02 | **-106.02** | 4.89 | **-107.51** | 6.37 | **-110.33** | 9.20 | **-111.64** | 10.50 | **-112.06** | 10.93 |
| 12 | **28.07** | -4.40 | **25.74** | -2.07 | **26.95** | -3.29 | **16.81** | 6.85 | **23.62** | 0.04 | **24.69** | -1.03 |
| 13 | **43.93** | -2.70 | **40.89** | 0.33 | **39.94** | 1.29 | **28.73** | 12.50 | **36.95** | 4.27 | **37.10** | 4.13 |
| **Average (mm)** | | **7.36** | | **6.04** | | **6.78** | | **6.51** | | **8.70** | | **10.27** |

**Table 6.** Comparison of maximum deflections from the field test and quadratic regression equations.

| Load cases | $\delta_1$ (mm) | $\Delta_1$ | $\delta_2$ (mm) | $\Delta_2$ | $\delta_3$ (mm) | $\Delta_3$ | $\delta_4$ (mm) | $\Delta_4$ | $\delta_5$ (mm) | $\Delta_5$ (mm) | $\delta_6$ (mm) | $\Delta_6$ (mm) |
|---|---|---|---|---|---|---|---|---|---|---|---|---|
| 1 | **-104.76** | 0.89 | **-105.40** | 1.53 | **-103.91** | 0.04 | **-107.02** | 3.15 | **-107.98** | 4.11 | **-114.49** | 10.62 |
| 2 | **20.78** | -5.29 | **22.82** | -7.32 | **22.15** | -6.65 | **22.25** | -6.75 | **22.44** | -6.94 | **25.45** | -9.95 |
| 3 | **42.60** | -1.04 | **44.39** | -2.84 | **42.62** | -1.06 | **49.82** | -8.27 | **38.33** | 3.22 | **41.57** | -0.02 |
| 4 | **-148.37** | 15.60 | **-138.02** | 5.25 | **-139.50** | 6.73 | **-130.86** | -1.91 | **-122.28** | -10.49 | **-120.20** | -12.57 |
| 5 | **-256.01** | -8.84 | **-259.14** | -5.71 | **-260.41** | -4.44 | **-261.30** | -3.55 | **-244.35** | -20.50 | **-260.54** | -4.31 |
| 6 | **88.88** | 12.94 | **92.21** | 9.61 | **92.92** | 8.90 | **92.13** | 9.70 | **88.64** | 13.18 | **83.90** | 17.92 |
| 7 | **-21.52** | -13.81 | **-25.57** | -9.77 | **-27.65** | -7.68 | **-30.30** | -5.04 | **-26.81** | -8.53 | **-27.00** | -8.33 |
| 8 | **-51.26** | -11.94 | **-51.21** | -11.99 | **-46.42** | -16.78 | **-57.47** | -5.73 | **-50.30** | -12.90 | **-47.64** | -15.56 |
| 9 | **-76.37** | 9.85 | **-75.10** | 8.59 | **-77.09** | 10.58 | **-63.35** | -3.17 | **-79.23** | 12.72 | **-82.96** | 16.44 |
| 10 | **-79.67** | 7.43 | **-80.93** | 8.70 | **-80.08** | 7.84 | **-86.84** | 14.60 | **-86.62** | 14.39 | **-90.86** | 18.62 |
| 11 | **-102.13** | 1.00 | **-105.98** | 4.85 | **-105.40** | 4.26 | **-109.22** | 8.08 | **-107.24** | 6.11 | **-112.70** | 11.56 |
| 12 | **28.06** | -4.39 | **25.73** | -2.07 | **26.69** | -3.03 | **20.04** | 3.62 | **25.08** | -1.42 | **26.92** | -3.25 |
| 13 | **43.92** | -2.69 | **40.89** | 0.33 | **39.58** | 1.64 | **32.17** | 9.05 | **38.33** | 2.89 | **39.74** | 1.49 |
| **Average(mm)** | | **7.36** | | **6.04** | | **6.13** | | **6.35** | | **8.63** | | **10.05** |

Figure 6 provides a comparison chart that summarizes the measured, analyzed, and calculated maximum deflection results. By examining this chart, it is clear that the regression results demonstrate a high degree of accuracy when compared to both the full FEM model and the field test measurements.

This suggests that the generated theoretical formula proved its ability to provide adequate maximum deflection values. In light of the results, it can be considered an economical method for bridge deflection estimation.

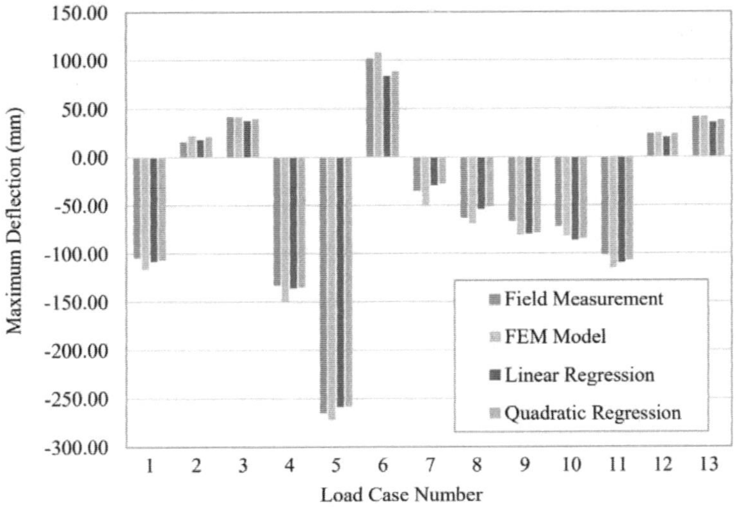


**Figure 6.** Deflections from the field, FEM, and best-fit regressions of stress.

## 6. Conclusions

Regression formulas predict the maximum deflection of a bridge deck accurately and reliably. Higher-order formulas may produce more accurate results, but the proposed method already provides a good prediction. This approach gives the engineer another method to verify field test data in a simple and rapid format for the bridge monitoring process. Using numerical simulations makes it possible to analyze further and validate the outcomes.

Nevertheless, strain gauge sensors demonstrated their high potential to measure the stress and estimate the maximum deflections of a large bridge structure. This method provides an economic estimation system in structural health monitoring.

## 7. References

[1] Nguyen D T, Nguyen X T, Ngo V M 2022 *Survey and analysis long term deflection of cantilever bridge in Vietnam* IOP Conf. Ser. Mater. Sci. Eng., vol. 1258, no. 1, p. 012063, doi: 10.1088/1757-899x/1258/1/012063.

[2] Suangga M, Haripriambodo T 2021 *Maximum deflection of three span continuous bridge using rotation data based on 3 dimensional model* IOP Conf. Ser. Earth Environ. Sci., vol. 794, no. 1, doi: 10.1088/1755-1315/794/1/012016.

[3] Zhang W, Sun L M, Sun S W 2017 *Bridge-Deflection Estimation through Inclinometer Data Considering Structural Damages* J. Bridg. Eng., vol. 22, no. 2, doi: 10.1061/(asce)be.1943-5592.0000979.

[4] Sun L, Li Y, Zhang W 2020 *Experimental Study on Continuous Bridge-Deflection Estimation through Inclination and Strain* J. Bridg. Eng., vol. 25, no. 5, pp. 1–14, doi: 10.1061/(asce)be.1943-5592.0001543.

[5] Sousa H, Cavadas F, Henriques A, Bento J, Figueiras J 2013 *Bridge deflection evaluation using strain and rotation measurements* Smart Struct. Syst., vol. 11, no. 4, pp. 365–386, doi: 10.12989/sss.2013.11.4.365.

[6] Panuntun R B, Aminullah A, Suhendro B, Wardana P K 2019 *Bridge Displacement Estimation using Tiltmeter Data* J. Civ. Eng. Forum, vol. 5, no. 2, p. 105, doi: 10.22146/jcef.43670.

[7] Suangga M, Surachmat D 2022 *Maximum deflection of simple span steel box girder bridge using stress data* IOP Conf. Ser. Earth Environ. Sci., vol. 998, doi: 10.1088/1755-1315/998/1/012017

[8] Brinissat M, Ray R P, Kuti R 2023 *Evaluation of the Szapáry Long-Span Box Girder Bridge Using Static and Dynamic Load Tests* Infrastructures, 8(5), 91, doi: 10.3390/ infrastructures8050091.

# Digital applications for preserving archaeological sites and tourism management

**D I Andrieş[1], T Şeitan [1], A M Grămescu[1], A Gemănaru[1]**

[1] Universitatea Ovidius Constanta

Email: delia.andriesh@gmail.com

**Abstract.** Cultural heritage is an important part of our life, a resource for the identity and the cohesion of communities .The paper aims to identify possible solutions for cultural heritage management of the archaeological sites and monuments from Dobrogea. The authors documented and researched above-ground archaeological structures, analysing their characteristics and suggesting solutions for conservation and promoting of archaeological heritage. The rapid development of digital technologies like augmented reality can help us finding cost-effective solutions for archaeology and cultural heritage tourism.

## 1. Introduction

Digital technologies have advanced faster than any other form of innovation in history, playing an important role in the development of science and technology. The use of computers for tedious tasks evolved and now it helps us in creating new forms of realities like virtual or augmented reality which can have multiple uses, including in the field of archaeology and cultural heritage. The digital reconstruction of artifacts or archaeological sites can be an important tool for understanding and appreciating the cultural values. At the same time, the use of this type of technology, derived from video games, can play a special role in the education of young generations, relying on the affinity that young people show for the digital world.

Augmented reality can be used as tool for scientific study, design, industrial production, training, learning or recreation. Using it in cultural tourism can improve the sensorial experience of a visitor when seeing an archaeological site [1] adding new information like images of a building, peoples or sounds. The article presents the main types of archaeological sites in Dobrogea, studying how they are conserved at the present time and how their significance is shown to the public, then reviews the traditional restoration methods used. We introduce the technologies that allowed the development of the concept of virtual heritage and the way in which augmented reality can enhance the Dobrogean sites and contribute to the development of cultural tourism, bringing as an argument similar solution used in Europe.

## 2. Touristic sites in Dobrogea

Dobrogea is an area with an important touristic potential, its geographical position and proximity to the Black Sea being an advantage. Tourists come here primarily for recreation and rest, but the multitude of archaeological sites dating back to the last 2500 years can create the premises for the development of cultural tourism.

The Greek civilization founded the first cities on the shores of the Black Sea, a place that facilitated commercial relations with the indigenous population. The expansion of the Roman Empire led to a more pronounced development of the area and to an increase in the number of settlements, many of them being fortified. The need to defend the borders of the empire against barbarian attacks led to the creation of a fortified line represented in Dobrogea by the Danube limes, formed by earth walls, defence towers and forts.

One type of archaeological site is the urban site like those found in Constanta (the ancient city of Tomis) and Mangalia (the ancient city of Callatis). Tomis and Callatis are among the first cities founded by Greek colonists in the $6^{th}$-$7^{th}$ centuries BC and they have been continuously inhabited throughout time, a fact that has advantages but also disadvantages. The advantage is that they are easily accessible, but many of the original buildings have been lost over time, the stones from their masonry were used for other buildings, the foundations remained hidden under ground or were swallowed by the Black Sea, the southern part of the of the Romanian seashore being subjected to a continuous erosion.

Archaeology, as a science, has its beginnings in the $18^{th}$ century, and only at the end of the $19^{th}$ century, the industrial revolution and the demographic growth made more people interested in it but this science was in its beginnings in Romania and not popular enough to impact people's life and way of thinking.

When these cities began their expansion at the end of the $19^{th}$ century, they used all the available space and did not pay too much attention to the historical vestiges. The streets and houses were placed according to their usefulness at the time of construction without carefully avoiding the ruins encountered. This way of thinking is changed now, recent urban planning being made in order to protect and value the archaeological remains, but part of the walls of the fortresses and houses still remain under the foundations of the new buildings, unable to be accessed by the public. In the post-war period, the two cities experienced an important development, and it is worth mentioning the protection of some of the historical remains that can be visited, like the Greek wall of the Callatis fortress, which can be visited within the premise's of a hotel. There are also vestiges hidden, at least for now, from the public, such as a Roman amphitheatre in Tomis, which was protected in the basement of a hotel.

In these two cities there are other vestiges presented to the public, some of them are currently valued, others are not. In the Archaeological Park of Constanta you can admire the entrance gate of the old Tomis, (figure 1) and in Mangalia, near the archaeology museum, the ruins of the northwest corner of the Callatis fortress (figure 2) that were recently highlighted.

**Figure 1.** The gate of the old Tomis

**Figure 2.** The northwest corner of the Callatis fortress wall

Another type of Dobrogea's sites, from cultural tourism point of view is represented by sits ready to welcome tourists, such as the Greco-Roman fortress of Histria or the Roman fort from Capidava. The archaeological excavations started in the first part of the $20^{th}$ century and were continued, with small interruptions, until today. The sites also have nearby museums where the most important discoveries are exhibited.

Other sites, where archaeological excavations are advanced, such as Ulmetum, Tropaeum Traiani, Dinogetia, Halmyris, Argamum, are more difficult to access and do not offer tourists information to facilitate understanding of the site and its component elements.

Some of the sites, part of the fortified frontier of the Roman Empire, some are completely unexplored: Sucidava, Beroe, Cius, Salsovia, others are impossible to visit, Axiopolis, or partially destroyed by the stone mining in the area: Arrubium, Capidava.

## 3. Restauration and valorisation methods used

From the very beginning of the archaeological excavations, the fragility of the uncovered ruins was noted by archaeologists who, with the help of local craftsmen, took the first primary conservation measures. And now the archaeologists are dealing with the primary conservation of the ruins, operations that consist mainly of restoring the joints with mortar and anchoring the insufficiently supported masonry elements. Compared to the beginning of the 20$^{th}$ century, a mineralogical analysis of the mortar can be done now with the aim of using materials similar to the original ones.

The restoration of the remains was done in many fortresses, using, in general, the materials found on the site or when the walls were dismantled many centuries ago, they were reconstructed with similar materials, so that the visitor can more easily imagine the greatness of the original walls. This type of restoration was applied to the walls of fortresses such as those of Tropaeum Traiani or Dinogetia.

Recently, several projects have been proposed to restore and enhance some fortresses such as Noviodunum, Histria, Carsium, Halmyris, so the expected number of tourist attractions is growing.

## 4. Use of digital technologies in virtual heritage and cultural tourism

Virtual heritage starts from existing cultural heritage and proposes the application of new digital technologies to save or reconstruct the image of a building or an object as it was originally. The first attempt to use the technology for the purpose described above was made in 1994 by reconstructing the 3d image of Dudley Castle in England in its form from 1550.

Concerns to integrate new digital technologies have materialized since 2000 in the regulations of the Krakow and London Charters, the Ename Charter and the Seville Principles that define terms such as virtual restoration, virtual reconstruction or virtual anastylosis. All these documents standardize the most suitable practices for virtual reconstructions and have entered into the common use of archaeologists and other specialists involved in the restoration process, be it physical or virtual.

There is also a concern in Romania for the digitization, mainly, of treasure objects from museum collections, objects too fragile to be exposed to the public for a longer period of time. Regarding heritage buildings, as consolidation works are carried out, 3D scanning is used more and more, and these images are used by specialists in order to identify the optimal intervention measures.

In detail digital reproduction of the surrounding world was made possible by a series of inventions and innovations which appeared in the 20$^{th}$ century.

Computer tomography, also known as CT, is an imaging technique that has exceeded the boundaries of medicine, being successfully applied in all fields. It consists in measuring the absorption of X-rays by materials and through computerized processing of the measurements, the model is digitally reconstructed in three-dimensions. Although the theoretical possibility of creating such devices has been known since the beginning of the 20$^{th}$ century, their emergence was in the early 1970s when the first thermal imaging cameras and computers powerful enough to quickly perform calculations necessary for image reconstruction appeared.

The invention of photography and laser technology allowed the development of 3D scanning and photogrammetry which together can give a realistic three-dimensional model of an object. The development of computing power has facilitated the appearance of augmented, virtual and mixed reality and the possibility to see objects in detail and even interact virtually with them.

The first steps in the digitization of tourism were achieved by creating virtual tours of museums. With the help of a PC or tablet device, the user can virtually walk through the museum, being able to follow a circuit through which they can view the exhibits. To do this, the proposed path is photographed and the images obtained are digitally combined to create an overall view with a three-dimensional appearance.

There are many possibilities offered by augmented reality or other types of mixed realities [2]. With the help of a phone equipped with a GPS and an application it is possible to visit a historical site and at the same time see on the screen a reconstructed image of the original site. The whole perception of the site will be enriched with a new dimension and the experience will be totally different.

This technology is already used in several places in Europe [3], a good example being the Kameiros site in Rhodes, Greece where virtual three-dimensional constructions are superimposed over the ruins of the old city. This experience can be accessed through the "Kameiros 3D" mobile application developed by MOPTIL.

The application's main screen presents the existing overview and a reconstruction of the site.

**Figure 3**. Satellite image of the site [4].

**Figure 4**. Reconstructed image of the site [4].

Arriving in front of a building, a comparison can be made between the image seen on the spot (figure 5 and figure 7) and the one proposed for reconstruction (figure 6 and figure 8).

**Figure 5.** Photo of the Great Stoa [5].

**Figure 6**. Reconstructed image of the Great Stoa [4].

**Figure 7.** Photo of the Great Stoa [5].

**Figure 8.** Reconstructed image of the Great Stoa [4]

This type of technology could be successfully applied to archaeological sites in the Dobrogea area. In the reconstruction, it will be possible to introduce elements from different periods and give the visitor an image both of how the fortress looked when it was full of life, during the growth periods of the empire, and how it looked later after the barbarian raids had subsided in intensity and people started to build huts in the former granaries.

## 5. Using classical methods versus a digital approach

The image of the site presented as it was at a certain moment in time can be done through digitalization, the costs required for reconstruction are greatly reduced, the only procedures that needs to be done being those of conservation and protection.

Another advantage of this method is the fact that it can easily integrate any other discoveries subsequent to the decision to reconstruct. From this point of view, the monuments and future researchers have a lot to gain, the authenticity of the ruins being ensured and their research and reinterpretation being possible at any time in the future.

A disadvantage of these techniques is the perception of the visitor, because the image obtained with the help of augmented reality is not identical to the one given by a reconstructed building. The fact that the virtual experience is more attractive compared to observing the ruins of some foundations cannot be neglected, but this technology has its drawbacks.

## 6. Conclusion

This article introduces the idea of using the augmented reality as a mean to increase the public interest for the Dobrogean archaeological sites and contribute to their conservation and valorisation. The first section highlighted the main characteristics of the sites and analysed the way in which there are presented to the public at the present moment. An overview of the main technologies related to the concept of virtual heritage was introduced, pointing out the benefits of augmented reality technologies. An example of similar approaches was then described making suggestions about the possibilities offered by augmented reality to simulate lost cities and forts.

The ideas presented in the article could represent a way to satisfy the public curiosity about the way of life of people in the past, about what a house of the ancient Romans looked like or to see how high were the walls of their fortresses. All of this can be seen standing in front of their ruins and watching what remained from the different layers of construction and what significant destructive effect the passage of time had, offering a new perspective of their importance. The idea is quite simple and can be easily put into practice using the new digital technologies, making possible the development of cultural tourism in Dobrogea with a relatively small investment.

## 7. References

[1] Stuart E 2012 *Augmenting Phenomenology: Using Augmented Reality to aid Archaeological Phenomenology in the Landscape* Preprint s10816-012-9142-7
[2] Bekele M K 2021 *Mixed Reality: A Bridge or a Fusion Between Two Worlds?* (London: Ubiquity Press)
[3] Thomopoulos C A S 2022 *SYNTHESIS: A Platform of Platforms for Integrated Management, Curation, and Visualization of Digital Cultural Experiences through VR and AR Technologies.* Heritage - New Paradigm (IntechOpen)
[4] Mobile application *Kameiros 3D*
[5] Website *https://www.google.com/maps*

# Computational assessment regarding structural rehabilitation of a reinforced concrete decanter tank

**D Covatariu[1], A Filip[2], G Covatariu[1]**
[1]Faculty of Civil Engineering and Building Services, "Gheorghe Asachi" Technical University of Iasi, Romania
[2]APAVITAL S.A. - Regional operator of water and sewerage services in Iasi County, Romania, Design Compartment

E-mail: daniel.covatariu@academic.tuiasi.ro

**Abstract**. The article describes the analysis of an industrial structure (reinforced concrete decanter tank for the drinking water's storage in the Iasi – Romania area) in order to justify the proposals for structural rehabilitation to ensure structural stability. In this regard, a computational model was created in Graitec Advanced Design, using the physico-mechanical characteristics of the materials (concrete, reinforcement) from the initial design. Subsequently, using the data obtained from the technical assessment report and the non-destructive tests carried out on the structural elements, it was found a depreciation of the concrete grade (C25/30 → C16/20), at the current stage. At this stage (the structure's service state) a modelling of the degraded structure was achieved by comparing the results obtained by Graitec Advanced Design and Axis VM, in order to realise an extended structural analysis between initial and current state. A solution for the rehabilitation of degraded elements using polymeric composite materials reinforced with carbon fibers was proposed, which was validated by the Sika CarboDur Composite systems (software program for the design of consolidation solutions).

## 1. Introduction

A civil/industrial structure has to fulfil the service, strength and stability requirements throughout the lifecycle, without any significant loss of functionality and unforeseeable maintenance work [1, 2].

In the current context, the requirements of strength, stability and safety in use must be handled with particular attention to reinforced concrete construction due to particular operating conditions [6, 7].

The operating conditions of the decanter tanks are assessed in accordance with the data collected in relation to the projected technological process taking place inside the basins, the operation of the technological equipment and the characteristics of the aggressive environment inside and outside the basins [3].

Generally, the static actions applied to the decanter tanks during operation, the long-term influence of a series of climatic factors (humidity, variations in temperature) as well as the aggressive action of the environmental agents contribute significantly to the reduction of the service life.

## 2. Particularities regarding the static behaviour of the decanter tanks

In order to store liquids and to ensure the structural stability, circular shape is mainly used for reinforced concrete tanks. The radial pressure given by the stored liquid (water) is evenly distributed in the circular sections producing axial forces.

Another factor to be considered for operational safety is represented by the dimensions of the structure, which are limited by deformation and cracking conditions. The specific deformation is independent of the decanter diameter, the risk of cracking being directly proportional to the diameter of the tank [4].

IOP Conf. Series: Materials Science and Engineering 1304 (2024) 012018 doi:10.1088/1757-899X/1304/1/012018

The site is regularly checked in regard to stability (levelling measurements on the decanter tank and the emplacement). In case of decanter tanks, the protection against infiltrations will be represented by using an effective waterproofing system.

To assure a structure's proper functionality current operating conditions have to be considered in the structural rehabilitation design stage: environmental conditions, charging / discharging of the decanter tank, pipes protection against corrosion, the utilities functionality.

## 3. Case study

### *3.1. Structural system's description*

The analysed construction is a suspension decanter with a volume of 10,000 cubic meters, semi-buried, having a circular shape, within the Chirița Treatment Plant, Iasi city (Figure 1.a, 1.b, 1.c). The decanter was put into operation in 1994, as a result of a project to extend the treatment plant.

a). Side view b). Top view

c). General view of the structural elements (including scrapping bridge)

**Figure 1.** Decanter tank no. 2 – Chirița treatment plant

The constructive structure of the decanter is in the form of a reinforced concrete basin divided into the reaction chamber and the clarifying compartment (Figure 2). The decanter's basin consists of the following structural elements: perimeter walls (1), raft foundation (2), the central structure (having the role to support the central rotor with pallets and delimiting the mixing spaces (3) and the supporting beam of the scrapping bridge (4).

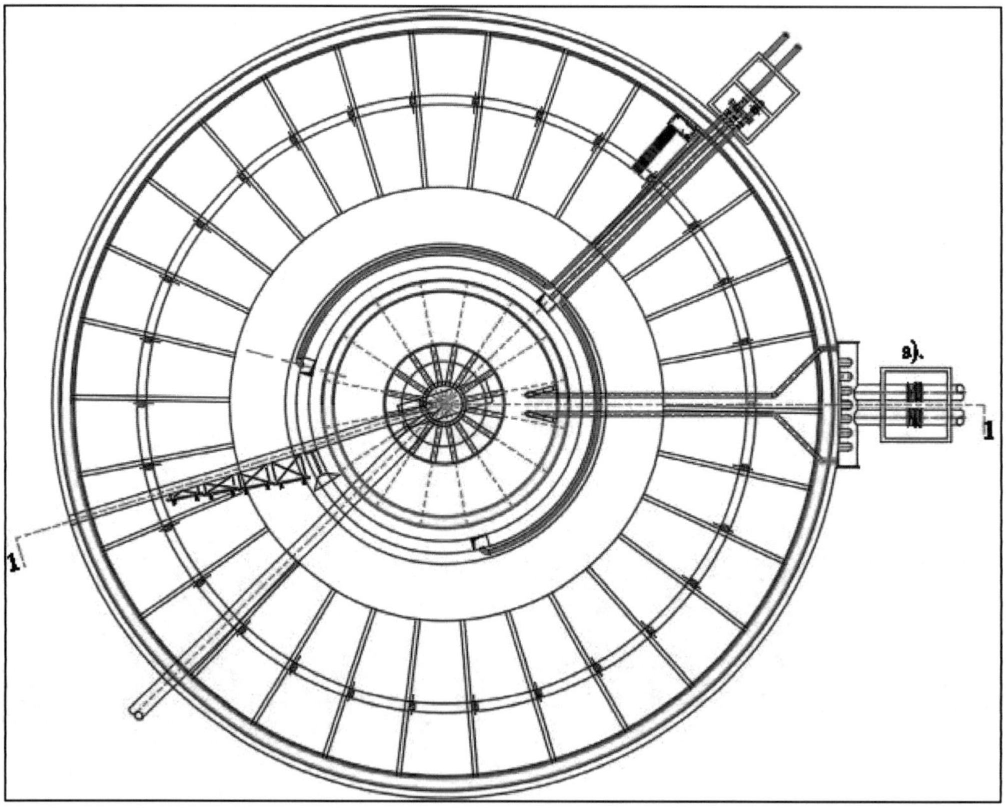


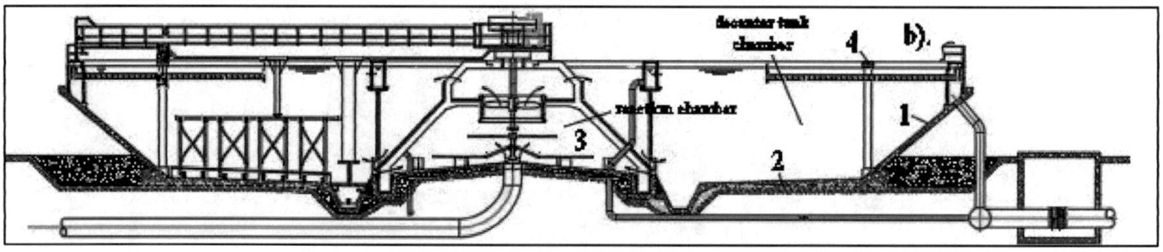


**Figure 2.** Decanter tank no. 2 – Chirița treatment plant (Iași county)
a). Top view sketch; b). Transversal section

*3.2. Numerical analysis by Finite Element Method of the decanter in initial stage*

The structures analysis (static, dynamic) in general, and of decanter tanks in particular, has been carried out using the Finite Element Method (F.E.M.). The use of F.E.M. allows the determination of stress and displacements for all load categories, considering special structural shapes [4,5].

For decanter tanks, the static analysis aims to static linear analysis, static nonlinear analysis and identification of the failure mechanism and safety factor assessment;

For the considered case study, linear static analysis was performed. At the initial stage of the analysis, the mechanical characteristics of the materials (concrete, reinforcements) were collected from the technical project (Table 1):

IOP Conf. Series: Materials Science and Engineering 1304 (2024) 012018 doi:10.1088/1757-899X/1304/1/012018

**Table 1.** Input data

| Structural element (C25/30 reinforced concrete) | Type of reinforcement | No. of longitudinal bars / stirrup arrangement, diameter and type of steel | Reinforcement label (Figure 4, 5) |
|---|---|---|---|
| Columns | - longitudinal bars | 8Ø14 PC52 | 2 |
| | - stirrups | Ø8/20 cm OB37 | 1 |
| Circular beam | - longitudinal bars | 3Ø14 PC52 | 4 |
| | - longitudinal bars | 2Ø10 PC52 | 5 |
| | - longitudinal bars | 3Ø14 PC52 | 4 |
| | - longitudinal bars | 2Ø12 PC52 | 6 |
| | - stirrups | Ø6/20 cm OB37 | 3 |

The studied structure was modelled and analysed using the Graitec Advanced Design software program, and the static linear analysis was performed to calculate the maximum internal forces according to the next structural model (Figure 3). The considered structural model was achieved by using bar-type elements. Reinforced concrete columns were considered to be embedded at the foundation's level, respectively hinged at the circular beam's level.

The circular beam (the columns supporting beam of the scrapping bridge) was calculated taking into account two hypotheses:

- continuous beam (to obtain the maximum values of the bending moment in the support);
- hinged beam (to obtain the maximum values of the bending moment in the field).

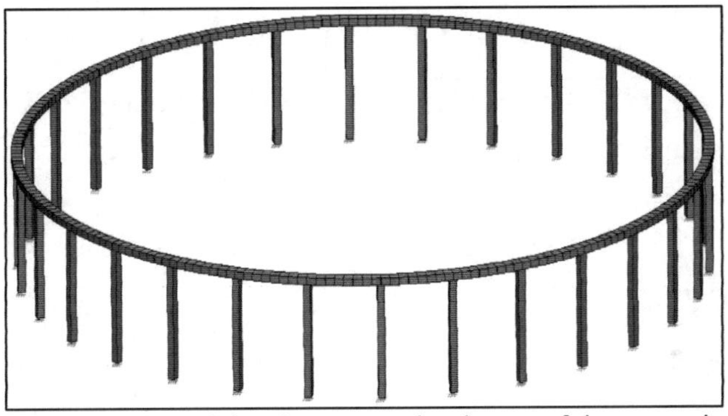

**Figure 3.** Structural model – the supporting beam of the scrapping bridge

The reinforcement of the structural elements section in the computer program using the data from the initial project was carried out for the columns (Fig. 4) and the circular beam respectively (Figure 5).

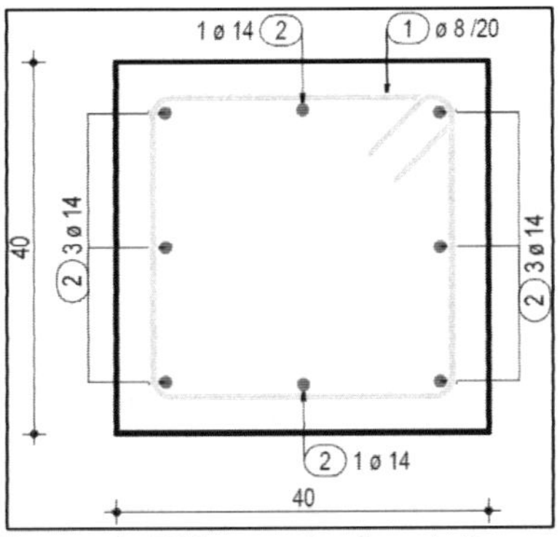


**Figure 4.** Pillar section's reinforcement

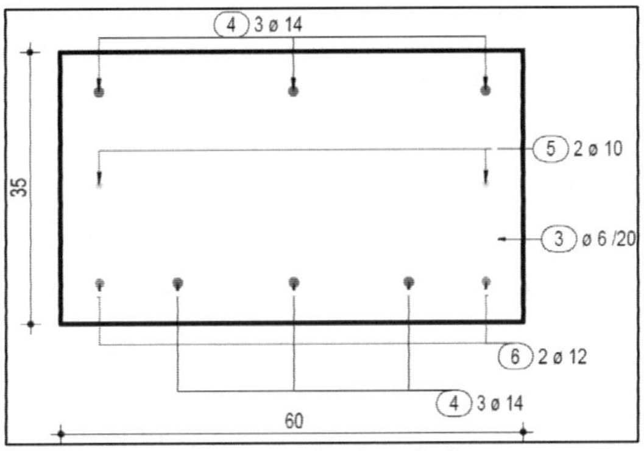


**Figure 5.** Beam section's reinforcement

IOP Conf. Series: Materials Science and Engineering 1304 (2024) 012018 doi:10.1088/1757-899X/1304/1/012018

The accuracy of the calculation depends on the considered structural model, which must be in accordance with the static scheme (nature of the links between the component elements). The loads involved in calculating the structure of the supporting beam of the scrapping bridge are:

- self-weight of the structure (circular beam, columns, perimeter wall, raft foundation, stored water);
- self-weight of pipes (supported by columns);
- self-weight of the scrapping bridge;
- seismic loading (it is considered that the scrapping bridge is stationary).

After performing the linear static analysis, the following values of the internal forces were obtained (Figure 6 ÷ 12, Table 2)

**Columns – internal forces**

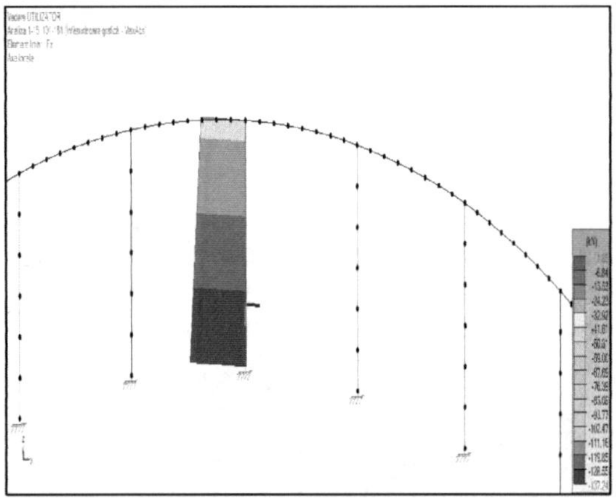

**Figure 6.** Min, max envelope – Axial force $N_x$

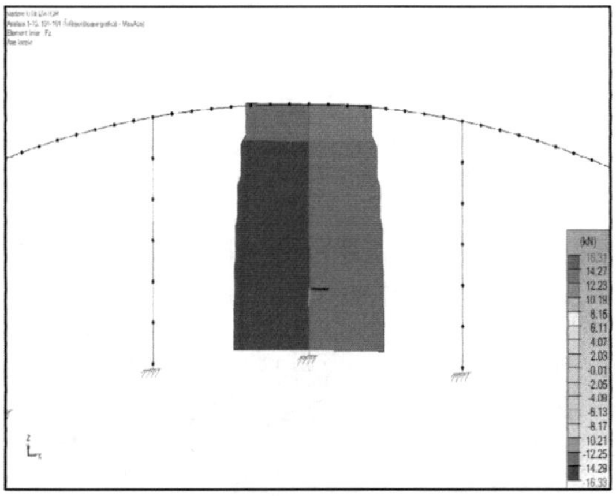

**Figure 7.** Min, max envelope – Shear force $V_y$

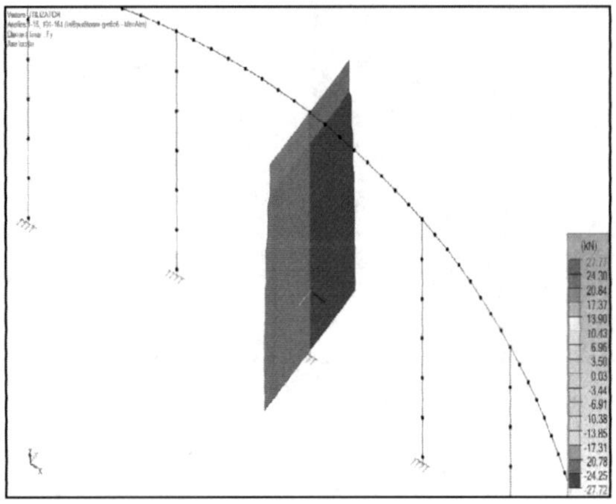

**Figure 8.** Min, max envelope - Shear force $V_z$

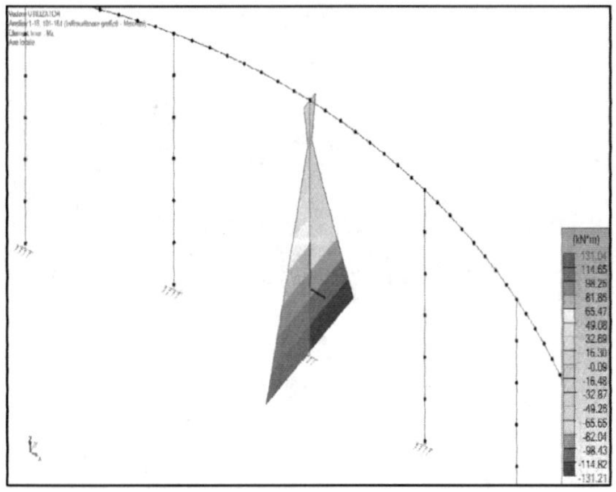

**Figure 9.** Min, max envelope – Bending moment $M_y$

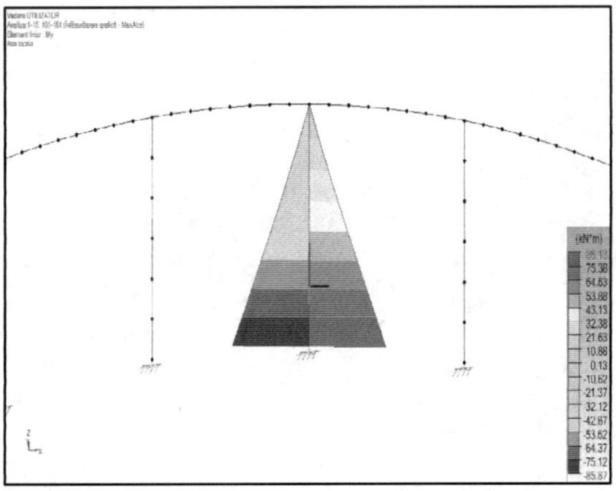

**Figure 10.** Min, max envelope – Bending moment $M_z$

**Circular beam – internal forces**

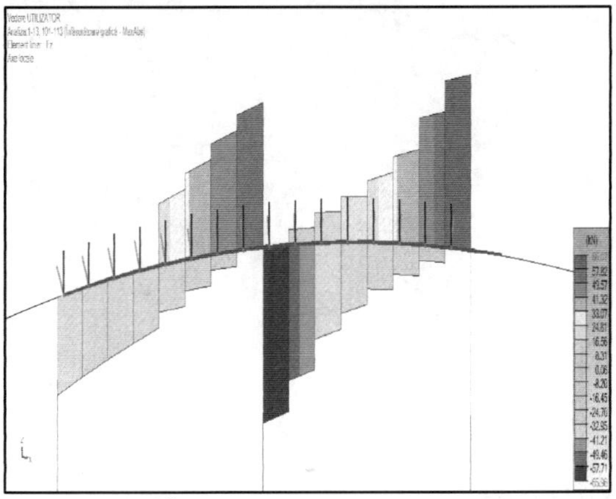

**Figure 11.** Min, max envelope – Shear force $V_z$

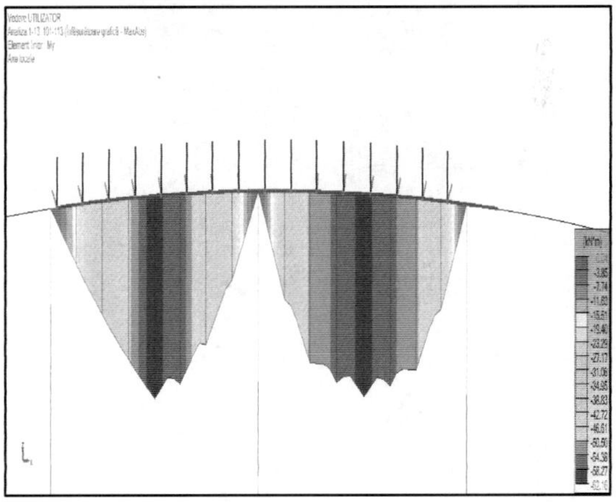

**Figure 12.** Min, max envelope – Bending moment $M_y$

**Table 2.** Values of the internal forces – initial stage

| Structural element | Internal effort | Maximum value |
|---|---|---|
| Columns | Axial force $N_x$ [KN] | 137.24 |
| | Shear force $V_y$ [KN] | 16.33 |
| | Shear force $V_z$ [KN] | 27.77 |
| | Bending moment $M_y$ [KNm] | 131.21 |
| | Bending moment $M_z$ [KNm] | 86.13 |
| Circular beam | Shear force $V_z$ [KN] | 66.07 |
| | Bending moment $M_y$ [KNm] | 62.16 |

*3.3. Visual inspection and identification of degradations*

Following the in-situ inspection, at the level of the rolling frame of the scrapping bridge, the following degradations / nonconformities of the structural elements were found:

- concrete surfaces with areas where concrete is segregated and exfoliated / delaminated (Figure 13.a);
- the presence of cracks in the supporting columns of the supporting beam on the opposite side of the running direction of the scrapping bridge; the cracks were identified next to the longitudinal and transverse reinforcements, respectively, having openings of 1-2 mm (Figure 13.b);
- delamination of the concrete coating, respectively uncovered and / or corroded reinforcement bars (Figure 13.c).

a). b). c).

**Figure 13.** Degradations to the supporting beam of the scrapping bridge

From the analysis of the degradations, it can be concluded that they are caused mainly by the action of aggressive environmental agents, water in the decanter and the efforts resulting from the braking of the scrapping bridge, as well as due to possible execution defects (reinforcements insufficient coverage, the use or inefficient waterproofing system).

In this respect, in order to ensure the requirements of mechanical strength and stability, a series of non-destructive tests (sclerometer test) were carried out on the structural elements (columns, respectively circular beam) to determine the mechanical characteristics of the supporting beam of the scrapping bridge (current stage), the resulting values for the concrete grade being corresponding to the C16/20 class (decreasing compared to the value of the concrete's class in the initial project – C25/30).

Also, due to chlorinated water actions from the interior surface of the structural elements, the concrete was degraded due to carbonation.

### *3.4. Comparative structural analysis for degraded structure*

The linear static analysis was performed according to the same structural model (Figure 14) in two software programs (Graitec Advanced Design and Axis VM) in order to realise an extended structural analysis between initial and current state. Thus, the linear analysis by two software programs could provide more accurate results.

IOP Conf. Series: Materials Science and Engineering 1304 (2024) 012018 doi:10.1088/1757-899X/1304/1/012018

**Graitec Advanced Design** **Axis VM**

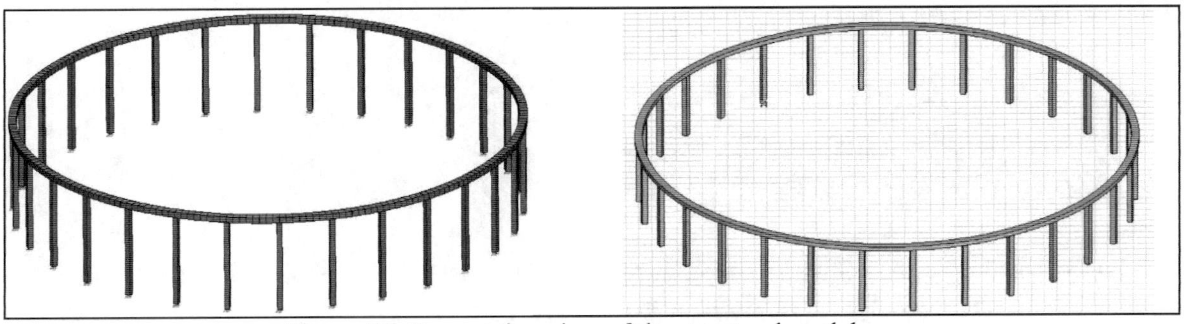

**Figure 14.** Perspective view of the structural model

**Columns – internal forces**

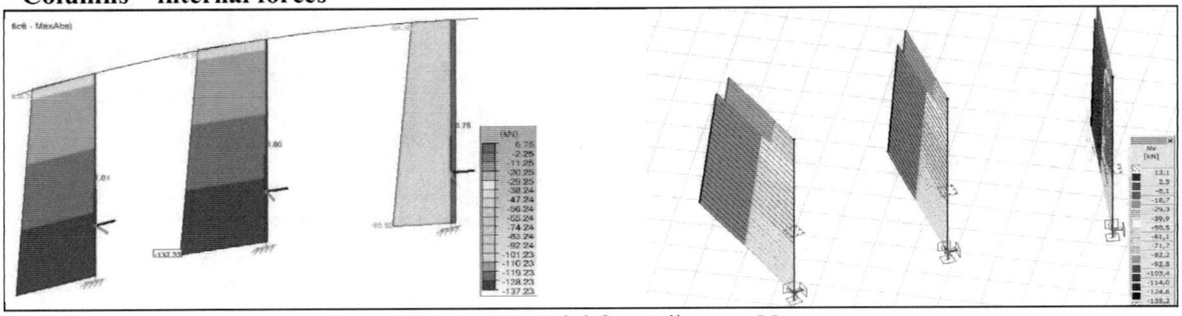

**Figure 15.** Axial force diagram $N_x$

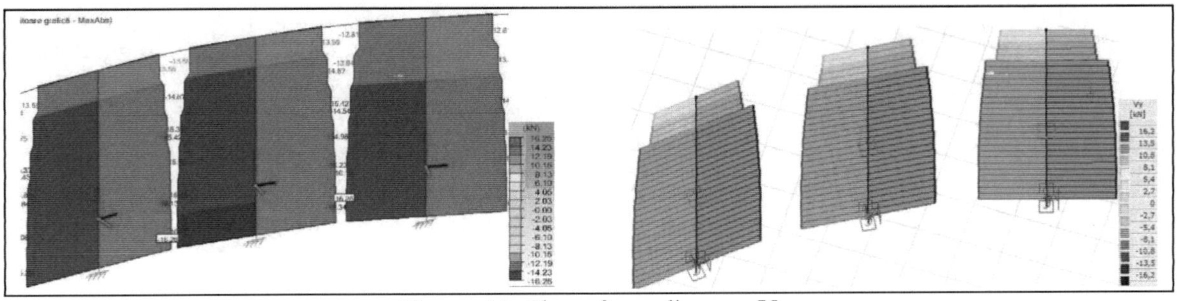

**Figure 16.** Shear force diagram $V_y$

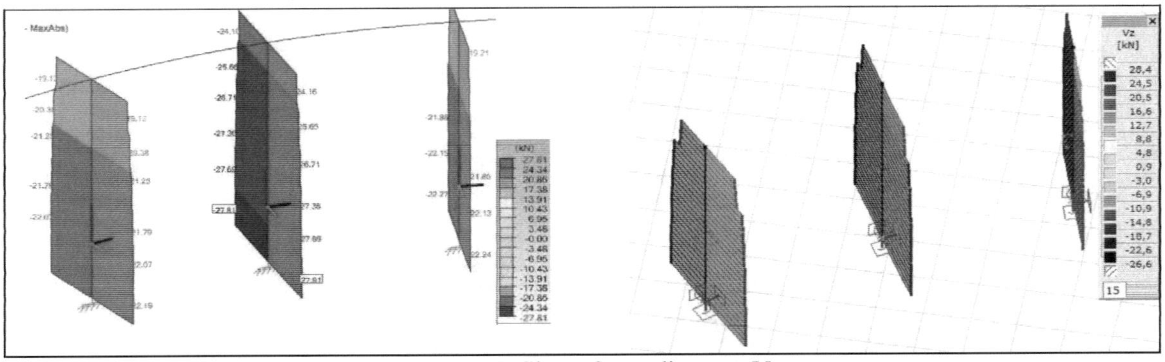

**Figure 17.** Shear force diagram $V_z$

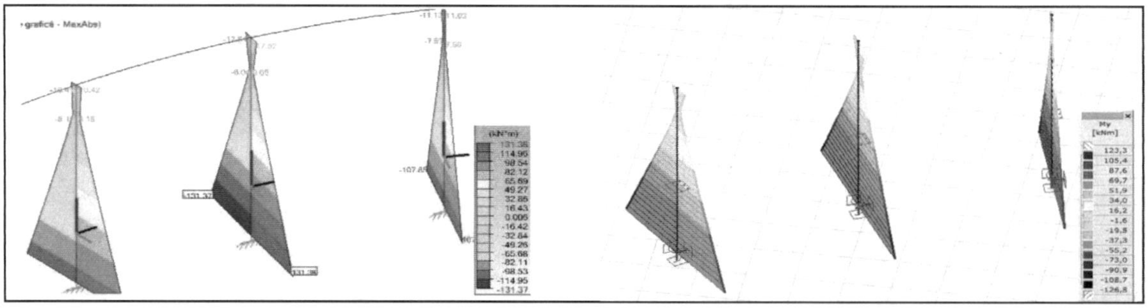

**Figure 18.** Bending moment diagram $M_y$

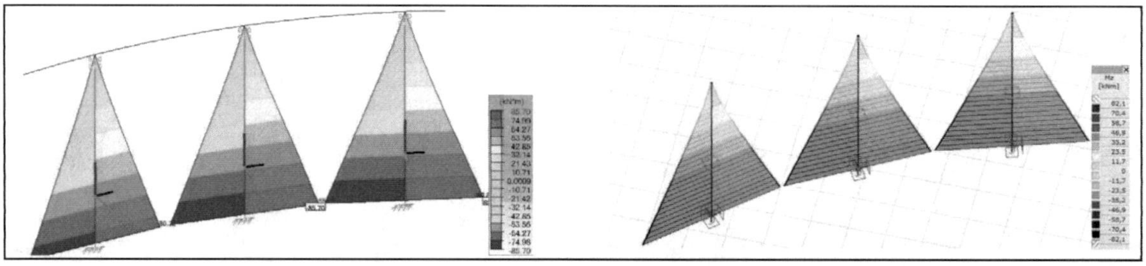

**Figure 19.** Bending moment diagram $M_z$

**Circular beam – internal forces**

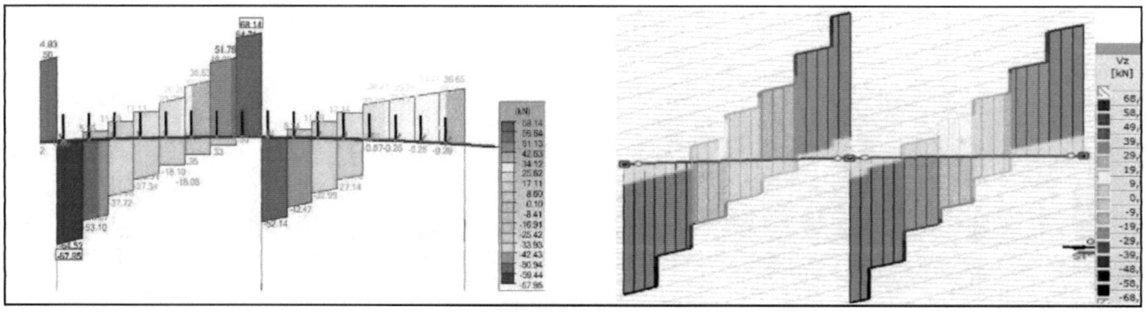

**Figure 20.** Shear force diagram $V_z$

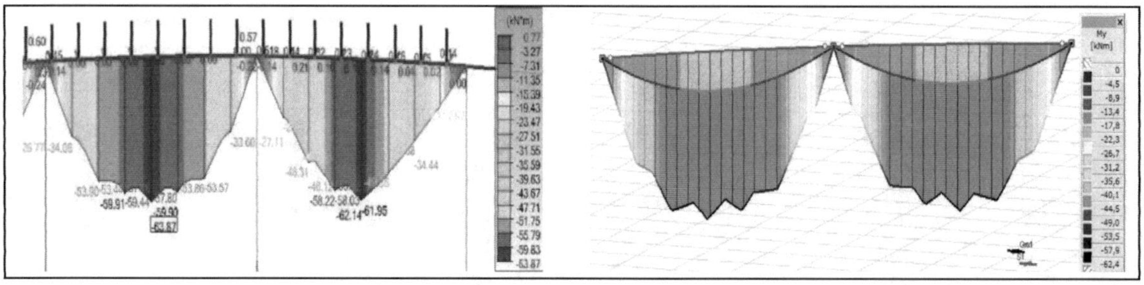

**Figure 21.** Bending moment diagram $M_y$

In terms of internal forces, the maximum values computed by both programs (Figures 15-21) are shown in Table 3.

**Table 3.** Maximum values – actual stage

| Structural element | Intern efforts | Maximum values (Graitec Advanced Design) | Maximum values (AXIS VM6) |
|---|---|---|---|
| Columns | Axial force $N_x$ [KN] | 137.23 | 135.20 |
| | Shear force $V_y$ [KN] | 16.25 | 16.20 |
| | Shear force $V_z$ [KN] | 27.81 | 28.40 |
| | Bending moment $M_y$ [KNm] | 131.38 | 123.30 |
| | Bending moment $M_z$ [KNm] | 85.70 | 82.10 |
| Circular beam | Shear force $V_z$ [KN] | 68.14 | 68.60 |
| | Bending moment $M_y$ [KNm] | 63.87 | 62.40 |

In this regard, the numerical values obtained in the actual stage (using two software programs), taking into account the analysis results, structural rehabilitation is highly recommended.

## 4. Structural rehabilitation using CFRP materials

### *4.1. General remarks regarding retrofitting solutions*

Structural rehabilitation is the main component in terms of increasing the safety in operation and the sustainability characteristics of these types of structures.

The causes that generate the state of degradation of the special water storage structures (design / execution errors, accidental actions, lack of current maintenance and repair works, the quality of the materials used, aggressive environmental agents, the long service life, etc.), can ultimately lead to the accelerated degradation of the structural elements and hence the transfer from the service state of the structure (the state of operation of the structure) to the non-compliant state.

The structural rehabilitation solutions used can be the classic ones (surfaces jacketing using special mortars - grouting) or by using modern materials of CFRP type (polymeric composite reinforced with fibers).

For reinforced concrete structures, in addition to composite materials, there are used other available methods of rehabilitation, for example:

- Shotcrete (sprayed concrete) consolidation: This method involves applying a layer of sprayed concrete onto the existing structure. This additional layer of concrete improves the load-bearing capacity and crack resistance of the structure.
- Chemical anchors: Consolidation of reinforced concrete structures using chemical anchors involves injecting a chemical material into drilled holes, which solidifies and creates a bond between the existing concrete and additional elements such as bars or plates.
- Injection of epoxy resins: This method entails injecting epoxy resins into cracks and voids in the concrete structure to enhance strength and waterproofing. The resins penetrate the pores and cracks of the concrete, solidify, and reinforce the structure, reducing the risk of further cracking and water infiltration.
- Surface restoration and protection: Various techniques can be used to repair deteriorated surfaces of the concrete structure, such as repairing with polymer mortars or applying surface protection systems like paints or waterproof coatings. These methods help restore the integrity and aesthetic appearance of the concrete and protect the structure from harsh environmental factors.

It is important to evaluate and choose the appropriate method based on the type and degree of deterioration of the reinforced concrete structure, as well as the specific needs of the rehabilitation project. Consequently, when it comes to rehabilitating reinforced concrete structures, there are several methods available beyond composite materials. Each method offers unique advantages and is suitable for different types and extents of damage. The choice of rehabilitation method should be based on a thorough assessment of the structural condition and the specific requirements of the project.

Materials such as shotcrete, chemical anchors, and epoxy resins provide effective solutions for enhancing the load-bearing capacity, crack resistance, and waterproofing properties of reinforced concrete structures. These methods address both structural integrity and durability concerns, ensuring the long-term performance of the rehabilitated structure.

Furthermore, surface restoration and protection techniques play a vital role in preserving the aesthetics and safeguarding the concrete from environmental degradation. By repairing damaged surfaces and applying protective coatings, the structure's appearance is restored, and its resistance to aggressive factors like moisture, chemicals, and weathering is improved.

Ultimately, the success of a rehabilitation project for reinforced concrete structures relies on selecting the most suitable method and implementing it with proper expertise and quality materials. This ensures the safety, and functionality of the structure, prolonging its service life and minimizing the need for future interventions.

CFRP materials are multiphase systems, obtained artificially, by the association of at least two chemically distinct materials, with a clear separation interface between the components and the resulting compound material is created in order to obtain properties that cannot be obtained by any of the components working individually [9, 10]

The use of modern consolidation systems with polymeric composite materials reinforced with fibers in the structural rehabilitation of reinforced concrete elements and / or structures has well-known and documented advantages, such as: resistance to the action of corrosive agents (chlorine ions), high durability, superior mechanical characteristics (compared to classical materials), reduced execution time, etc.

In this case, in order to comply with the requirements of mechanical strength and stability, it was chosen the consolidation systems with polymeric composite materials reinforced with carbon fibers, using the SIKA CarboDur automatic calculation program.

### *4.2. Numerical calculus using SIKA CarboDur software program*

Sika® Carbodur® FRP Design is a software program for designing consolidation solutions using Sika CarboDur Composite systems to increase the bending and shear capacity, respectively the sealing of the reinforced concrete structures and structural elements itself.

Following the model carried out by F.E.M. and the selection of the chosen structural rehabilitation method, the solution for consolidating the structural elements made of reinforced concrete (columns, circular beam) at bending and shearing by having discrete strips and continuous jacketing with CFRP was calculated. In both situations, it was opted for the use of carbon fiber fabric.

#### *4.2.1. Columns*

In the case of pillar's reinforcement (400 x 400 mm) the following calculation efforts (Table 4) were taken into account, respectively the capable efforts (Table 5) for the unconsolidated structure.

**Table 4.** Design shear forces/bending moments – unrehabilitated structure

| | |
|---|---|
| Shear forces | $V_{x,Ed}$ = 137.23 [kN] |
| | $V_{y,Ed}$ = 27.81 [kN] |
| | $V_{z,Ed}$ = 16.25 [kN] |
| Bending moments | $M_{y,Ed}$ =85.70 [kNm] |
| | $M_{z,Ed}$ =131.37 [kNm] |

**Table 5.** Shear and bending capacity – unrehabilitated structure

| | |
|---|---|
| Shear forces | $V_{y/z,Rc}$ = 41.54 [kN] |
| Bending moments | $M_{y/z,Rc}$ = 67.27 [kNm] |

IOP Conf. Series: Materials Science and Engineering 1304 (2024) 012018 doi:10.1088/1757-899X/1304/1/012018

For the pillar's consolidation, carbon fiber reinforced polymeric composite (CFRP) was used, having the following mechanical characteristics (Table 6).

**Table 6.** Mechanical characteristics of the CFRP system

| | |
|---|---|
| Thickness | $t_f = 0.172$ [mm] |
| Width | $b_f = 300$ [mm] |
| Elasticity modulus | $E_f = 242$ [kN/mm$^2$] |
| Total deformation | $\varepsilon_{f,lim} = 0.009$ (0.015) |
| Tensile strength | $\sigma_t = 2.800$ [MPa] |
| Transversal area of the fibers | $A_f = 172$ [mm$^2$] |

As a result of the calculation performed, the following numerical values for the bending check (Table 7) and the shear (Table 8) respectively resulted.

**Table 7.** Design moments

| | |
|---|---|
| Capacity moment | $M_{z,Ed} = 131.37$ [kNm] |
| Necessary area of carbon fiber | $A_{nec,f} = 72.29$ [mm$^2$] |
| Number of layers | $n = 2$ |
| Effective area of carbon fiber | $A_f = 103.2$ [mm$^2$] |
| Capacity moment after strengthening | $M_{rb,f} = 144.21$ [kNm] |

**Table 8.** Design shear force

| | |
|---|---|
| Thickness of the carbon fiber | $t_f = 0.172$ [mm] |
| Shear force capacity | $V_{fd} = 69.93$ [kN] |

To increase the bending load-bearing capacity (bending moment) two layers of unidirectional strip arranged longitudinally on each side of the pillar over the entire height of the element will be used (Figure 22 - blue colour). In case of increasing the load-bearing shear capacity (shear force), a layer of unidirectional strip with a width of 150 mm at the interaxial distance of 300 mm perpendicular to the axis of the column, arranged in the form of discrete strips, shall be used (Figure 22 - red colour);

IOP Conf. Series: Materials Science and Engineering 1304 (2024) 012018 doi:10.1088/1757-899X/1304/1/012018

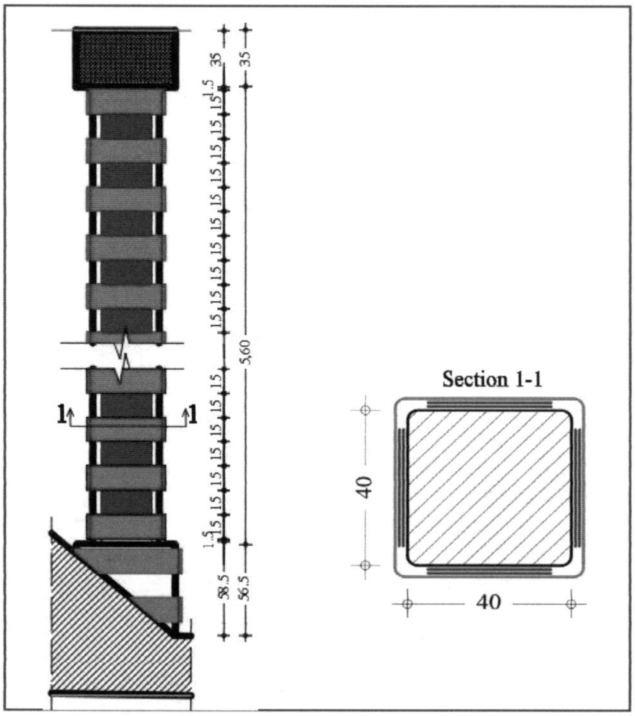


**Figure 22.** Column section – CFRP

*4.2.2. Circular beam*

In the case of the circular beam's reinforcement (600 x 350 mm), the following calculation efforts (Table 9) were taken into account, respectively the capable efforts (Table 10) for the unconsolidated structure.

**Table 9.** Design shear forces/bending moments – unrehabilitated structure

| | |
|---|---|
| Shear force | $V_{z,Ed}$ = 68.14 [kN] |
| Bending moment | $M_{y,Ed}$ = 63.87 [kNm] |

**Table 10.** Shear and bending capacity – unrehabilitated structure

| | |
|---|---|
| Shear force | $V_{z,Rb}$ = 20.12 [kN] |
| Bending moment | $M_{y/z,Rb}$ = 60.29 [kNm] (inferior) |

For columns consolidation, the following system of composite material reinforced with carbon fiber was used, having the following mechanical characteristics (Table 11).

**Table 11.** Mechanical characteristics of the CFRP system

| | |
|---|---|
| Thickness | $t_f$ = 0.172 [mm] |
| Width | $b_f$ = 300 [mm] |
| Elasticity modulus | $E_f$= 242 [kN/mm$^2$] |
| Total deformation | $\varepsilon_{f,lim}$ = 0.009 (0.015) |
| Tensile strength | $\sigma_t$ = 2.800 [MPa] |
| Transversal area of the fibers | $A_f$=172 [mm$^2$] |

As a result of the calculation performed, the following numerical values obtained for the bending consolidation (Table 12) and the shear consolidation (Table 13) respectively for both situations resulted, considering a minimum capacity increase imposed of 70% (the effective consolidation increase being of 85%).

**Table 12.** Design moments

| | |
|---|---|
| Capable moment | $M_{z,Ed}$ =109 [kNm] |
| Necesary area of carbon fiber | $A_{nec,f}$ = 43.43 [mm²] |
| Number of layers | n = 1 |
| Effective area of carbon fber | $A_f$=51.6 [mm²] |
| Capable moment after strengthening | $M_{rb,f}$=118 [kNm] |

**Table 13.** Design shear force

| | |
|---|---|
| Necessary shear force | $V_{z,nec}$ = 115.84 [kN] |
| Shear force supplement | $V_{z,sup}$ = 95.72 [kN] |
| Necessar thickness | $t_{f,nec}$ = 0.14 [mm] |
| Thickness of the carbon fiber | $t_f$ = 0.172 [mm] |
| Capable shear force after strengthening | $V_{fd}$ = 124.29 [kN] |

Based on the numerical values obtained, the consolidation of the circular beam will be carried out with polymeric composite materials on reinforced with carbon fibers as is shown in Figure 23.

In order to increase the bending load-bearing capacity (bending moment), a layer of unidirectional strip, connected longitudinally at the beam's bottom (Figure 23 – blue colour) will be used, and to increase the bearing capacity at shear (shear force), a layer of unidirectional strip will be used, connected in a transverse direction on three sides (bottom and the lateral faces of the beam) in continuous U-shape strips (Figure 23 – red colour).

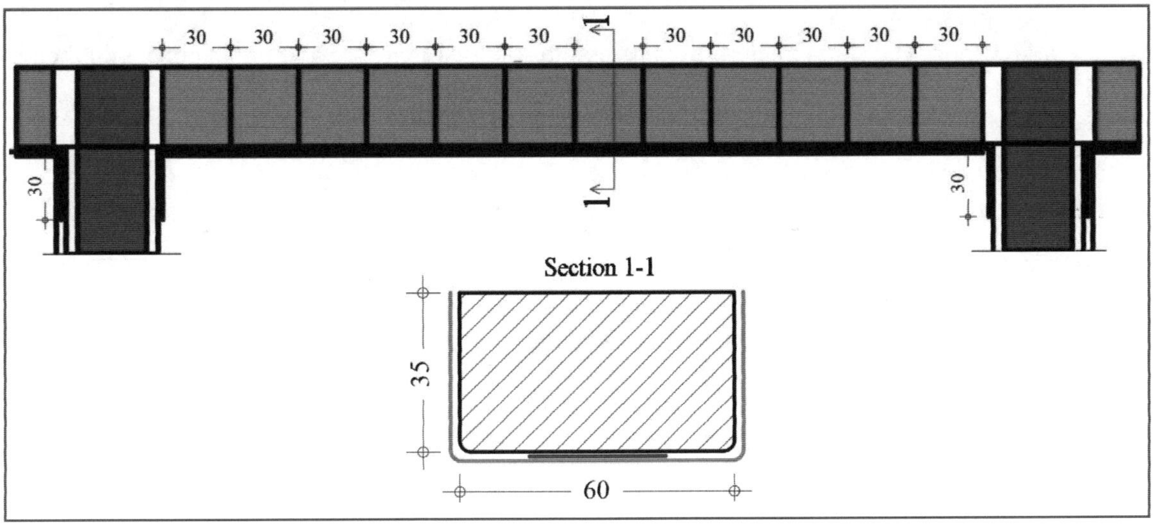


**Figure 23.** Reinforced circular beam section

## 5. Final remarks

In the current stage, the results obtained using two software programs shows that the most stressed parts of the structural elements are common in both models (Table 14), and based on the numerical values obtained, the structure have to be rehabilitated with CFRP systems.

**Table 14. The results quantification**

| Structural element | Internal effort | Design stage (Graitec) | Current stage (degraded structure) | |
|---|---|---|---|---|
| | | | Graitec | AXIS VM6 |
| Columns | Axial force $N_x$ [KN] | 137.24 | 137.23 | 135.20 |
| | Shear force $V_y$ [KN] | 16.33 | 16.25 | 16.20 |

| | | | | |
|---|---|---|---|---|
| | Shear force $V_z$ [KN] | 27.77 | 27.81 | 28.40 |
| | Bending moment $M_y$ [KNm] | 131.21 | 131.38 | 123.30 |
| | Bending moment $M_z$ [KNm] | 86.13 | 85.70 | 82.10 |
| Circular beam | Shear force $V_z$ [KN] | 66.07 | 68.14 | 68.60 |
| | Bending moment $M_y$ [KNm] | 62.16 | 63.87 | 62.40 |

Taking in consideration that each FEM software program has its proper manner to introduce the structural model corroborated with the fact there are very small differences in the resulting values (less than 5%), this research can be suitable for the linear static analysis of the reinforced concrete structures.

In the same time, obtained results shows that both computer programs are compatible in order to be used for the structural analysis of the reinforced concrete water tank, assuming that are realised from homogeneous material.

A proper numerical analysis based on conclusive data correlated with the detailed investigation of durability factors (destructive and / or non-destructive tests, in-situ and in authorised laboratories) represents an effective way to simulate the degradations level and to ensure a safe service of the structure.

## 6. References

[1] Budescu M, Ciongradi I P, Ţăranu N, Gavrilaş I, Ciupala M A, Lungu I 2001Reabilitarea construcţiilor, Editura Vesper, Iaşi, ISBN: 973-96589-9-7 (in Romanian);
[2] CR0 - 2012 – Design code. Basics of construction design (in Romanian);
[3] GT 052/2002 – Guide to investigate the technical condition of operating tanks, decanters and meta-tanks (in Romanian);
[4] Hobjilă V 2001 Calculul structurilor construcţiilor hidroedilitare – Structuri complexe din beton armat, Iaşi: Cermi Press, ISBN: 973-8000-42-4 (in Romanian)
[5] Kim Y, Quinn K, Satrom N, Garcia J, Sun W, Ghannoum W M, Jirsa J O 2012 Shear Strengthening of Reinforced and Prestressed Concrete Beams using Carbon Fiber Reinforced Polymer (CFRP) Sheets and Anchors, Center for Transportation Research;
[6] Law 10/1995 on quality in construction (in Romanian);
[7] Law 50/1991 regarding the authorization of the construction works (in Romanian);
[8] Law 458/2002 on the quality of drinking water (in Romanian);
[9] Mofidi A, Chaallal O 2011 Shear Strengthening of RC Beams with Externally Bonded FRP Composites: Effect of Strip-Width-to-Strip-Spacing Ratio, American Society of Civil Engineers.
[10] Murali G, Pannirselvam N 2011 Flexural Strengthening of Reinforced Concrete Beams using Fiber Reinforced Polymer Laminate: A Review, *ARPN Journal of Engineering and Applied Sciences*, **6**, no. 11.
[11] P100-3/2019 – Seismic design code - Part III. Stipulations for seismic evaluation of existing buildings (in Romanian);
[12] Soliman J 2018 General Behavior of T-section RC Beams Strengthened with Epoxy-Bonded Carbon Strands, *MOJ Civil Engineering*, **4**, 213-217.

# Variable diameter pile foundations – advantages and disadvantages

**O E Colţ[1], C P Neagu[1], A Motău[2], I Dmitriev[3] and L Chercheja[4]**

[1] *The "Gheorghe Asachi" Technical University of Iasi,*
[2] *Callaghan Brickwork Ltd*
[3] *S.C. Amphion proiect SRL, Iaşi*
[4]*Arcadis Excellence Center Romania*

E-mail: *oana-elena.colt@academic.tuiasi.ro*

**Abstract.** Variable diameter pile foundations can be an efficient foundation solution used for special structures, mainly for the ones with increased heights. For economic efficiency in land use, especially in urban areas, there is a tendency to build very high constructions with very small footprint related to their height, which transmit large loads to the foundation soil. Based on these aspects, it is necessary to consider pile foundations in order to ensure the desired safety and stability requirements. The technologies for piles constructions evolved a lot and a main objective was to better use the interaction between the soil and the pile. From this point a view the paper presents a comparison between the bearing capacities of a constant diameter pile and a variable diameter pile, considering different soil conditions. All the calculations were performed using finite element analyses.

## 1. Introduction

Due to the technological evolution around the world, new ideas for sustainable development are needed. The fast increase of human population leads to less available construction sites which can be solved by extending buildings in their height instead of area. Especially in highly dense urban areas, constructions are built on very small footprints related to their height, and for that transmit large loads to the foundation soil. This comes with a challenge for all types of engineers, especially for civil engineers.

In order to provide new and sustainable solutions for different types of structures, we can interfere in the construction of the infrastructures where more than 40% of the total costs of a structure are assigned. One of the purposes of the researchers is to find new technical solutions for deep foundation, aimed to reduce costs, by work technology, possibilities to increase the bearing capacity and reduction of the material consumption. Therefore, to be able to create less environmentally damaging infrastructures, the variable pile foundation, piles that have variable diameters (as increased cross section) along the length and at the base, come as a solution. Constant diameter pile foundations are the usual design solution when it comes to tall buildings. The large number of piles; the increased length of piles in order to reach the good foundation soil; the large diameters of piles, all of these related to loads and to the type of soil can become disadvantages in using such foundation solutions.

The variable diameter piles solution, allows an increase in the bearing capacity of the piles for a smaller diameter, by using increased diameters along the pile and at the base (joints), positioned in the good foundation soils. This allows increasing both the friction resistance, by increasing the pile lateral area, the resistance in compression and tension due to the created enlargements. This paper presents the theoretical results obtained by a comparative calculation, using numerical modelling, for a pile foundation in two situations: constant diameter and variable diameter for the piles considering also different soils conditions. The analysis of the results allows us to highlight the benefits of using variable diameter piles for deep foundations design.

## 2. Theoretical background of pile foundations

A pile is basically a long cylinder made of a strong material such as reinforced concrete that is pushed into the ground to act as a steady support for structures built on top of it. Pile foundations are used when there is a layer of weak soil at the surface and this layer cannot support the loadings from the building, so it is needed to bypass this layer and to be transferred to the good foundation soil or rock that is below the soft layer ore when a building has increased concentrated loads, such as in a high-rise structure, bridge, or water tanks and they have to be transfer to a soil with high bearing resistance.
One of the strongest points of a pile foundation is the increased load carrying capacity compared to spread footings.
There are two types of pile foundations which are commonly used and each of them is working in different manners.
One of them is the end bearing piles, which have the bottom end of the pile placed on a layer of special strong soil or rock. The load of the building is transferred through the pile onto the strong layer. In a sense, this pile acts like a column. The principal key is that the bottom of the pile bypasses the weak layer and the load therefore it is safely transferred to the strong layer.
The second one is the friction piles that transfers the load of the building to the soil around on the full length of the pile, by friction. Therefore, the entire surface of the pile, which is cylindrical in shape, works to transfer the forces to the soil. Figure 1 is presenting these two types of piles and their work principles.

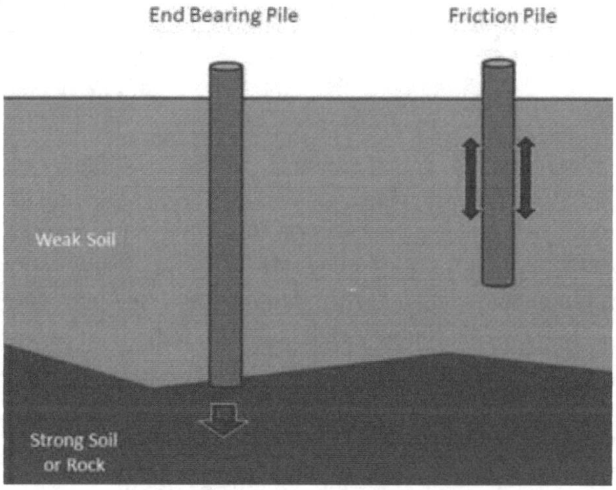


**Figure 1.** Constant diameter piles

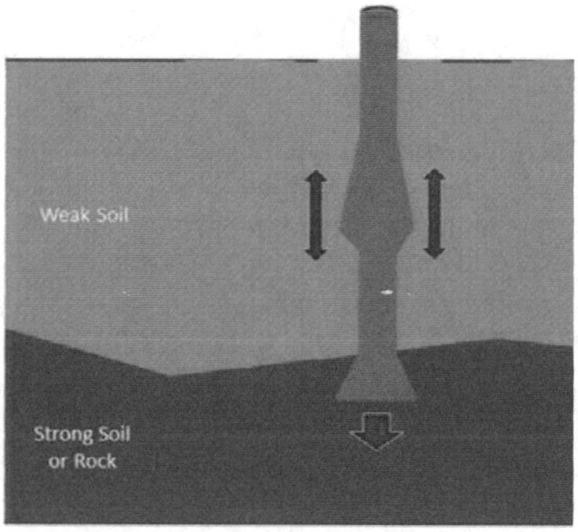


**Figure 2.** Variable diameter piles

Due to those properties of the pile foundations around the world the use of piles foundations increased rapidly and heavy multi-storied buildings were built. Starting from, this during the time, there were many improvements in piles and pile driving technology and systems.
Variable diameter piles are considered an improvement of the constant diameter pile. The variable diameter piles solution allows an increase in the bearing capacity of the piles with a smaller diameter, by using increased diameters along the length of the pile and at the base (joints), positioned in the good foundation layers. This allows an increase in both the friction resistance, through the enlarged lateral area of the pile, and resistance in compression and tension, giving the created enlargements. Figure 2 is presenting a view of the variable diameter piles.

IOP Conf. Series: Materials Science and Engineering 1304 (2024) 012019 doi:10.1088/1757-899X/1304/1/012019

## 3. Technological innovation for variable pile foundations

The construction of variable diameter piles needs a particular innovative technology. The devices used for building the variable diameter piles are more complex than the ones that are used for constant diameter piles.

In order to build variable diameters along the pile length and at the base, it was needed to improve the usual boring devices with some particular extensions to laterally expand at certain depths in the borehole. Therefore, based on these classic techniques, an innovative technique for building variable cross section piles was developed. The main steps in building either constant or variable diameter piles are drilling, reinforcing and concrete pouring.

### *3.1. Drilling process for variable diameter piles*

Drilling is a mechanical process performed with a special boring device. One of the main steps in creating a borehole is to establish the precise location of the longitudinal axis of the future pile and after that a special attention must be paid in positioning the drilling machine in accordance to the precise angle described in the design process (Figure 3).

**Figure 3.** Positioning the drilling machine (http://www.sunlyengineering.com/images/bp4.jp)

**Figure 4.** Drilling and soil extraction

In the step of drilling and soil extraction, for the constant cross section areas of the piles the technological process is the same as for regular pile construction (Figure 4). For preventing the walls from collapsing while the soil extraction is performed, bentonite mud is poured in the borehole to generate hydrostatic pressure. Care should be taken when using the bentonite mud in the borehole, as it is necessary to have special equipment to collect this material after its use in the concrete pouring stage.

For drilling and soil extraction in the variable cross section areas, to the classic drilling device a special element is attached for lateral expansion at the required depth of the position of the increased diameter. The device is made of two parallel blades which are extended on a certain angle at its sides and through a circular rotation the excavation of the bulb (increased diameter) is created. Figure 5 is presenting the described device. At the base of the pile the blade device is creating another enlargement in the same manner as the intermediary ones were drilled. After drilling the entire length of the pile bentonite mud is poured in the borehole while the drilling device is extracted.

### *3.2. Reinforcement of variable diameter piles and concrete*

The reinforcement of the piles is designed in accordance with the European design codes prescriptions. The geometrical shape of the reinforcement is the same as for constant cross section piles.

The process of concrete pouring is the same as for constant cross section piles. After the reinforcement is being placed in the borehole, the concrete is poured from bottom to the top, while the bentonite mud is extracted gradually.

**Figure 5.** Blades device for variable cross section drill (http://www.japanpile.co.jp/english/)

## 4. Finite Element Analyses

### *4.1. Analysed cases*

The attention of this study is focused on evaluating the maximum bearing capacities of constant and variable cross section piles evaluated in different types of soil. For these analyses a 20 m long pile with a constant cross section of 1 metre in diameter placed in a uniform soil layer was considered (Case 1). Three types of variable cross section piles were considered in the analyses, the piles having the same length and soil depth as in the case of constant cross sections. For the variable cross section pile cases the same constant diameter equal to 1 m was considered and, in the areas where the bulbs (enlargements) were created the diameter reached 2 m. The variable cross section cases are the following: Case 2 the pile has 1 bulb at the base of the pile, Case 3 the pile has 2 bulbs one at the base and the other one at the middle of the entire length and Case 4 the pile has 3 bulbs, one at the base and the other two equally distributed on the length. In all cases studied the piles were made of C20/25 concrete.
For each of the four studied cases 3 types of soil conditions were considered but the soil is uniform on the entire depth. The foundation soils were modelled using the Mohr-Coulomb model with the input geotechnical characteristics presented in Table 1.

**Table 1.** Geotechnical characteristics of the foundation soil

| Type of soil | Silty Sand | Sandy Clay | Gravelly clay |
|---|---|---|---|
| Natural unit weight [kN/m$^3$] | 18 | 18.5 | 19.5 |
| Friction angle [°] | 29 | 24.5 | 27 |
| Cohesion [kPa] | 5 | 14 | 10 |
| Oedometric modulus [kPa] | 10000 | 5000 | 12000 |
| Poisson ratio | 0.3 | 0.35 | 0.35 |

IOP Conf. Series: Materials Science and Engineering 1304 (2024) 012019 doi:10.1088/1757-899X/1304/1/012019

*4.2. Finite element modelling*

All the numerical simulations were performed using the finite element software Plaxis 2D. A nonlinear FEM analysis with standard settings for the iterative procedure was applied considering a gradually increasing load in a range between 0 KN to 800 KN. Due to the symmetry of the structure the models were analysed only on one half. The computation was continued until plastic failure occurred. Figures 6 to 9 are presenting the finite element models with the corresponding foundation soil for each case study.

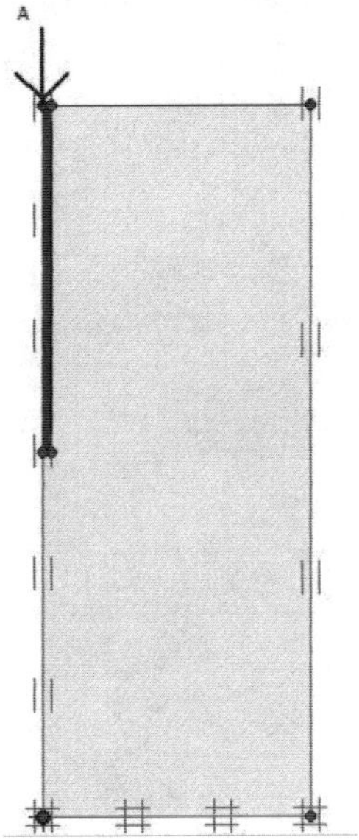


**Figure 6.** Case 1 - Model with constant cross section

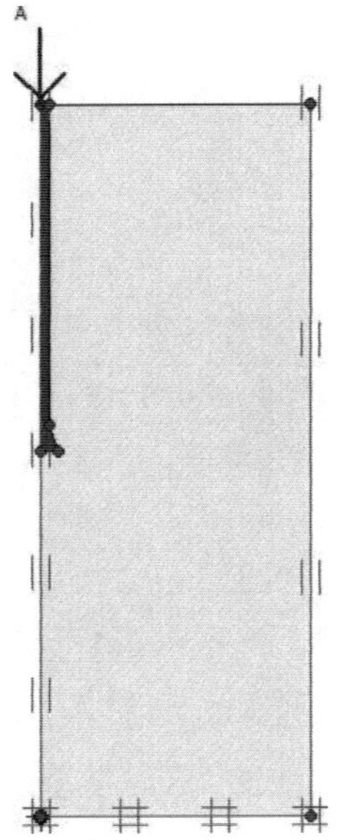


**Figure 7.** Case 2 - Model with enlarged base

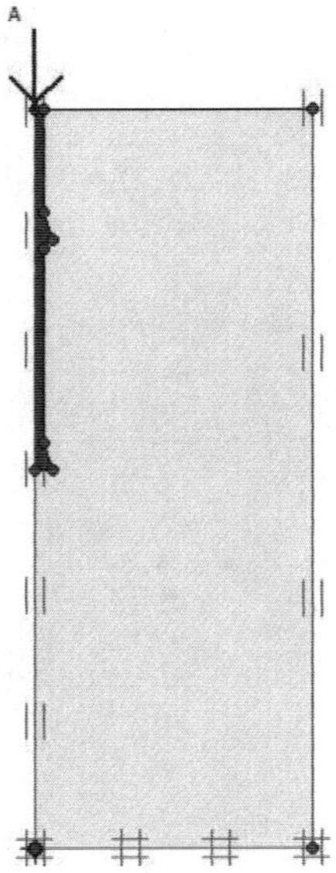


**Figure 8.** Case 3 Model with 2 enlargements

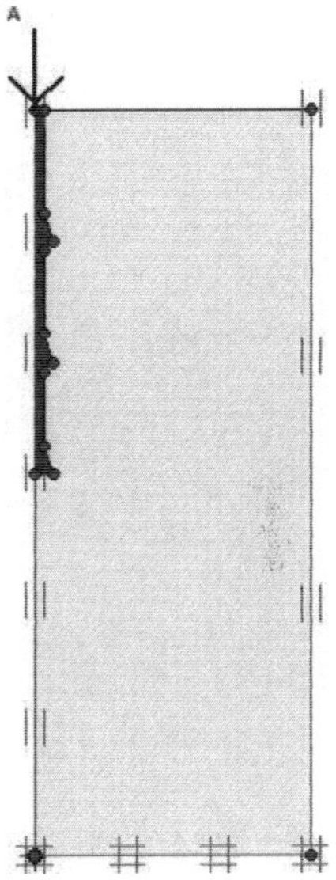


**Figure 9.** Case 4 Model with 3 bulbs

The mesh of the models was defined with a fine dense and refined step in the close surrounding of the pile, since high stresses are expected there. The mesh is further refined at the end points of the variable diameter pile foundation element representing the bulb. Interface elements are drawn beneath the piles to smooth the mesh. Figure 10 is presenting one of the model's mesh.

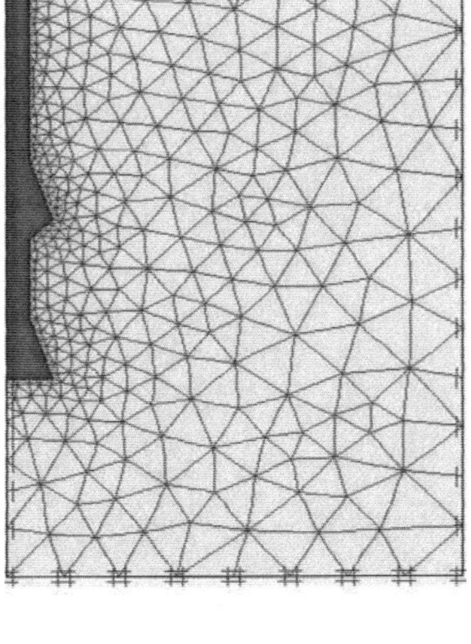

**Figure 10.** Mesh of the model

### *4.3. FEM analyses results*

Based on the FEM analyses the maximum bearing capacity was determined for each of the case studies. In table 2 the results of these analyses are presented.

**Table 2.** Bearing capacities of the case studies

| | Case 1 Constant diameter pile | Case 2 Variable diameter pile with enlarged base | Case 3 Variable diameter pile with 2 enlargements | Case 4 Variable diameter pile, 3 enlargements |
|---|---|---|---|---|
| | Uniform Silty Sand Layer, Pile length 20 m | | | |
| Fmax [KN] | 438 | 498 | 570 | 654 |
| S [cm] | 3.8 | 3,9 | 4 | 4,6 |
| | Uniform Sandy Clay Layer, Pile length 20 m | | | |
| Fmax [KN] | 229 | 251 | 275 | 303 |
| S [cm] | 5,5 | 5,6 | 5,8 | 6,2 |
| | Uniform Gravelly Clay Layer, Pile length 20 m | | | |
| Fmax [KN] | 484 | 560 | 642 | 750 |
| S [cm] | 4 | 4,1 | 4.9 | 5,5 |

From the table above a comparison in percentage of loads increasing was performed. Therefore, Table 3 is presenting a percentage increase of all case studies of this study focused on evaluating the maximum bearing capacities of constant and variable cross section piles.

**Table 3.** Comparison in load increase for all cases

| | Loads increasing | | | |
|---|---|---|---|---|
| | Case 1 vs. Case 2 | Case 2 vs. Case3 | Case 3 vs. Case 4 | Case 1 vs. Case 4 |
| | Uniform Silty Sand Layer, Pile length 20 m | | | |
| Percentage % | 11 | 14 | 15 | 49 |
| | Uniform Sandy Clay Layer, Pile length 20 m | | | |
| Percentage % | 9.6 | 9.5 | 10 | 32 |
| | Uniform Gravelly Clay Layer, Pile length 20 m | | | |
| Percentage % | 15.7 | 14.6 | 16.8 | 54.9 |

In Figure 11 a comparison graph of the bearing capacities for each of the 4 studied cases with different foundation soil is presented. From figure 11 it is noticed that apart from the differences in results influenced by the variable cross section also the foundation soil has a great impact in all analyses cases, as we all already know.

Variable diameter piles can be a solution of improving the constant diameter piles. The analysis results from computer calculation with Plaxis 8.2 shows an increase of the bearing capacity of 10 % for a pile with one bulb, 20% for a pile with 2 bulbs and 30% for the 3 bulbs pile compared to the constant cross section. Due to these results the economic efficiency is considerable. This foundation solution could reduce the number of piles needed, their length, and the material cost, without interfering with the strength of the structure. The multiple diameter-expanded parts are expected to also reduce the risk of building settlement [3].

Because of the variable diameters piles are increasing both the friction resistance, by enhanced pile lateral area, and resistance in compression and pull, due to the bulbs, they can be used also for bridge's pile and wind turbines.

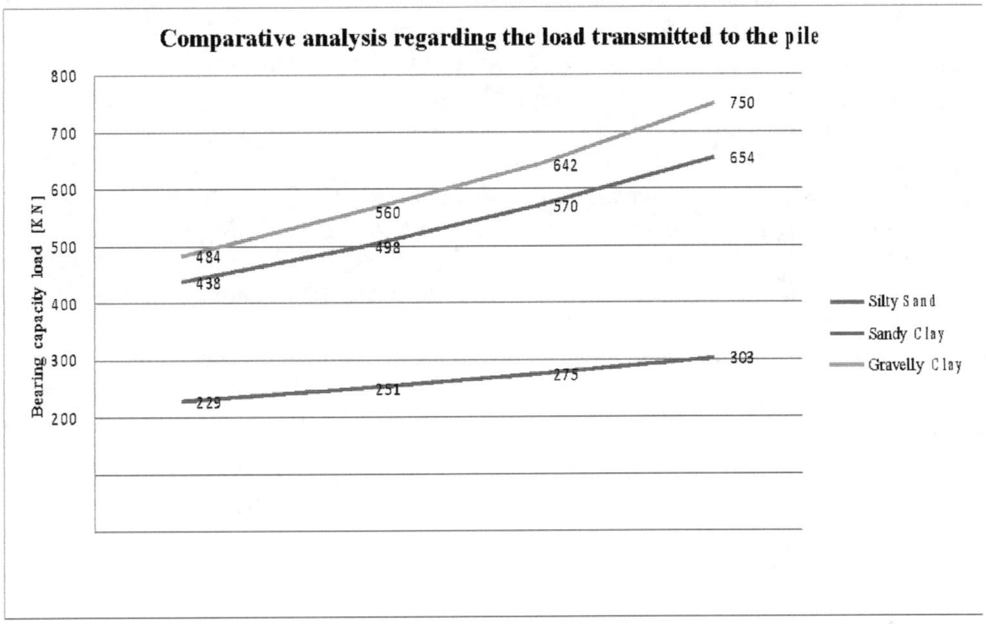


**Figure 11.** Graphical comparison of bearing capacities

## 5. Conclusion

- The field of civil engineering is one of the greatest pollutants of the world starting from the material manufacture and to the construction itself. Therefore, the sustainable development of the construction area is needed. It is necessary to realise that natural sources are limited, as foundation soils, and civil engineers have to develop better technological and ideological solutions. It is known that the cement industry is one of the most damaging ones for the environment. The variable diameter piles come as a solution for less concrete use due to the high bearing capacities which leads to a smaller number of piles on the total foundation area.
- In this paper it is studied the efficiency of using variable cross section piles compared to traditional ones. From finite element analyses it was noticed that an increase of approximately 10% for each enlargement on the pile is produced in terms of load bearing capacities. Based on this observation one can say that the technological innovation of variable cross section piles brings economical, ecological and technical improvement for a safe environment.
- From the economical point of view using this solution leads to a smaller number of piles which implies a small amount of concrete and reinforcement usage.
- From the ecological point of view, it brings a reduction of carbon emissions due to the decrease of the amount of resources and building materials. Another important aspect is the decrease of the urban expansion on sites but creating an extension on height.
- From the technical point of view it can offer an easier building process due to less boreholes created.

*References:*

[1] Ryltenius A 2011 *Fem modelling of piled raft foundations in two and three dimensions* (Dissertation thesis, Lund University)
[2] Bond A and Harris A 2008 *Geotechnical and Foundation - Decoding Eurocode 7,* Taylor and Francis
[3] Honda T, Hirai Y and Sato E 2011 *Uplift capacity of belled and multi-belled piles in dense sand.* Soils and Foundations, vol.51, No3 Jaopanese Geotechnical Society
[4] http://photos1.blogger.com/blogger/7351/2381/1600/Piling%20Markers.1
[5] http://www.understandconstruction.com/pile-foundations.html
[6] http://www.indopora.com/services/pondasi
[7] http://www.japanpile.co.jp/english/
[8] Valley M 2009 *Foundation Analysis and Design* (S.E. MEHRP)
[9] NP 112-2004 *Design code for structures with direct foundation.* Construction Bulletin (B.C.) nr. 14/2005.
[10] SR EN 1992-1:2004 Eurocode 2: Design of concrete structures — Part 1: General rules and rules for building, 2006
[11] SR EN 1997-1:2004 Eurocode 8: Geotechnical design — Part 1: General rules

*Acknowledgement:*

This paper was realised with the support of project EFECON – ECO-INNOVATIVE PRODUCTS AND TECHNOLOGIES FOR ENERGY EFFICIENCY IN CONSTRUCTION, POC/71/1/4 - Knowledge Transfer Partnership, Cod MySMIS: 105524, ID: P_40_295, Project co-financed by the European Regional Development Fund.

IOP Conf. Series: Materials Science and Engineering 1304 (2024) 012020 doi:10.1088/1757-899X/1304/1/012020

# The influence of soil specimens shape and dimensions on consolidation parameters used in foundation settlement prediction

**A V Dascălu, I Lungu, M Aniculăesi, F Bejan**
"Gheorghe Asachi" Technical University of Iași, Faculty of Civil Engineering and Building Services, Romania

E-mail: andreea-vasilica.dascalu@staff.tuiasi.ro

**Abstract.** The theory of elasticity is used in geotechnical engineering for the calculation of foundation settlement. The equation used in the evaluation of consolidation settlement uses the soil compressibility parameters which usually are determined on cylindrical soil specimens with $\phi$ 71.4 mm. Laboratory tests performed on soil specimens showed that the parameters that are used in foundation settlement analysis are influenced by the dimensions of the tested soil specimens and the shape of that specimen. The paper presents the influence of specimen size, shape and method used for the coefficient of consolidation evaluation on the final value of the foundation settlement. The analysis was performed using the Settle3 software from Rocsience, considering the case of an elastic foundation flexible square foundation (1.5 m x 1.5 m) which transmits to the ground a pressure of 100 kPa.

## 1. Introduction

The structure of a building can be divided in two parts, the superstructure which is the part of the building that develops above the ground and the substructure which is the part of the construction located under the ground surface [1]. The structural engineers in general deal with the superstructure which is made from man-made materials like concrete, steel or wood, materials for which the mechanical properties are well known. The substructure refers to foundation and the soil located inside the bulb of pressure influence beneath the foundations. In general, the substructure design is done by engineers with knowledge in soil behavior, also known as geotechnical engineers which can adopt a proper foundation based on the ground conditions encountered in that site. Geotechnical design requires to check if the soil under foundations does not reach a limit state which can be a serviceability limit state (SLS) or an ultimate limit state (ULS). The adequate behavior of the foundation to avoid the serviceability limit state implies the calculation of the total settlement and limit this value to an accepted one. The values of the settlement can induce supplementary stress in superstructure elements; therefore, it is important that the final values of the settlement to be known from the design stage in order to avoid structural damage of the building.

The most used methods for the evaluation of the shallow foundation settlements are based on the theory of elasticity. The Romanian design standard NP112/2014 presents such a method for the computation of the total settlement using the method of superposition [2], method that is based on the theory of elasticity.

The change in the stress state of the foundation soil due to construction load will lead to a rearrangement of the soil particles which in turn leads to the decrease in soil volume which results in the settlement of the structure. The rearrangement of the solid particles can be almost immediate or time dependent according to the permeability of the soil layer underneath the foundation.

In the case of the cohesive soils the rate of the settlement is time dependent. The total settlement ($S_T$) is the sum of the immediate settlement ($S_i$), from the primary consolidation ($S_c$) and from the secondary consolidation or creep ($S_s$).

$$S_T = S_i + S_c + S_s \tag{1}$$

For the cohesive soils the immediate settlements ($S_i$) represents 0.15% from the total settlement [3], [4] and the settlement from the secondary consolidation is usually small compared to the one from the primary consolidation [5]. The primary consolidation settlement can be computed using the consolidation theory developed by Terzaghi using the laboratory oedometer testing results [3]:

$$S_c = \frac{C_c \cdot H}{1+e_0} \cdot log\left(\frac{p'_0+\Delta p'}{p'_0}\right) \tag{2}$$

Equation 2 can be expressed in a different form such as:

$$S_c = \sum H \cdot m_v \cdot \Delta p' \tag{3}$$

where: $C_c$ – compression index; $e_0$ – initial void ratio; $m_v$ – coefficient of volumetric compressibility; $p'_0$ – effective overburden pressure corresponding to the layer; $\Delta p'$ - increase of effective pressure due to the foundation loading; H – layer thickness.

The rate of settlement can be calculated using the coefficient of consolidation ($c_v$). This parameter can be established in based on the laboratory testing results on undisturbed soil specimens using the one-dimensional consolidation theory. There are two methods that are the most used ones for the estimation of the consolidation coefficient value which are known as Casagrande's logarithm of time fitting method and the Taylor's square root of time method [3], [6], [7].

Both methods use the Terzaghi's consolidation theory, but the resulting values of the consolidation coefficient are different, usually the Taylor's method yields higher values compared to the values from Casagrande's method [8]. Therefore, it is very important to know which method is suitable to be used in the evaluation of the consolidation coefficient, parameter which significantly influences the consolidation settlement of a soil under the foundation.

There is limited scientific literature focused on the influence of the specimen dimension and shape on the final values of the consolidation coefficient [9]. The objective of this research is to present the influence of the soil specimen shape, dimensions and method used for evaluation of the consolidation coefficient in the prediction of the consolidation settlement of shallow foundations.

## 2. Site investigation and geotechnical findings

The soil investigated in this research was collected from the city of Iasi, Romania. Sampling and laboratory testing were performed on undisturbed soil samples. To minimize the specimen disturbance for the laboratory consolidation testing, monolith samples were collected from the lateral sides of an open excavation.

### *2.1. Geotechnical parameters of the investigated soil*

Grain size distribution tests performed on the collected samples classify the soil as silty clay and according to the Atterberg limits, the soil exhibits a medium plasticity ($I_p = (16 \div 25.1)\%$).

Oedometer consolidation testing was performed in accordance to the European standards. The tested specimens had different shapes and dimensions. Each specimen was incrementally loaded till 500 kPa and each load was maintained 24 h, time needed for the completion of the consolidation phase. The stress-strain curves of the oedometer test results obtained on saturated specimens, as corresponding values established on three identical specimens for each category of shape and dimensions, are presented in Figure 1.

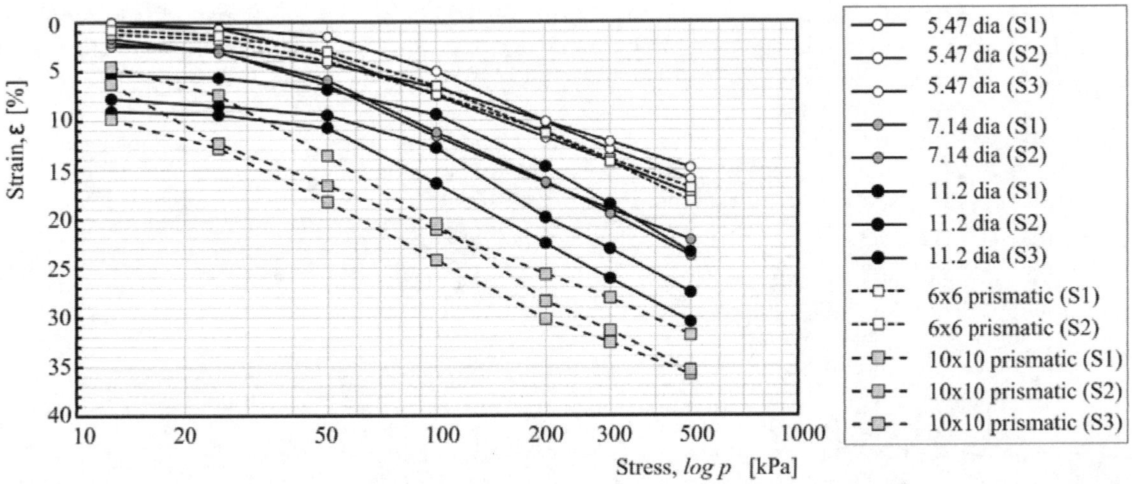


**Figure 1.** Stress strain curves for the tested soil specimens (S1, S2, S3 – for each specimen shape and corresponding dimensions).

The parameters used in the settlement evaluation and obtained from the oedometer testing as average values (based on S1, S2, S3 recordings) for specimens with circular and square cross-sections with different dimensions are summarized in table 1.

**Table 1.** Consolidation parameters for the tested soil.

| Specimens | $e_0$ (-) | $C_c$ (-) | $C_r(*)$ (-) | $\rho$ (g/cm³) | $E_{oed,1-2}$ (kN/m²) | $m_{v,1-2}$ (1/kPa) | $c_v$ (cm²/min) | |
|---|---|---|---|---|---|---|---|---|
| | | | | | | | Casagrande | Taylor |
| Circular | | | | | | | | |
| 5.47 (cm) | 1.06 | 0.234 | 0.0213 | 1.46 | 2831.91 | 0.00038 | 0.0208 | 0.1744 |
| 7.14 (cm) | 0.8 | 0.2557 | 0.0214 | 1.68 | 2692.42 | 0.00041 | 0.0212 | 0.1325 |
| 11.2 (cm) | 0.71 | 0.3845 | 0.0213 | 1.75 | 1635.88 | 0.00062 | 0.0150 | 0.1157 |
| Square | | | | | | | | |
| 6x6 (cm) | 0.765 | 0.268 | 0.0225 | 1.67 | 2338.20 | 0.00043 | 0.1963 | 0.1554 |
| 10x10 (cm) | 1.26 | 0.386 | 0.0216 | 1.71 | 1697.53 | 0.00062 | 0.0114 | 0.0929 |

(∗) Values computed using the equation proposed by Azzous [10]: $C_r = 0.003 \cdot (w/100 + 7)$

## 3. Settlement prediction

For the consolidation settlement calculation, a flexible square footing foundation (1.5 m x 1.5 m) was considered to be placed on the investigated silty clay transmitting a pressure of 100 kPa to the ground (figure 2). Settle3 software from Rocsience was used for the settlement evaluations. The loading induced stress distribution can be computed using Boussinesq method. The quarry point for the settlement was chosen in the centre of the foundation footing.

The soil behaviour can be considered as being linear or non-linear. For the linear behaviour the software requires the $m_v$ parameter for the settlement evaluation due to consolidation; for the non-linear behaviour, the modulus is not constant, being a function of the soil stress and thus, the software requires other parameters like compression index ($C_c$) and recompression index ($C_r$) resulting from the oedometer testing. The pre-consolidation pressure is equal to $p_c$ = 50 kPa.

### *3.1. The influence of the specimen dimension*

In the geotechnical laboratories the consolidation tests can be performed on specimens with different dimensions. In this study the dimensions of specimens were 54.7 mm, 71.4 mm and 112 mm as diameters for cylindrical specimens with circular cross-section and two different dimensions for prismatic specimens with square cross-section (60 mm and 100 mm). This paper refers to the influence of the used specimens' dimensions on the laboratory test values and on the final results of the primary consolidation settlement.

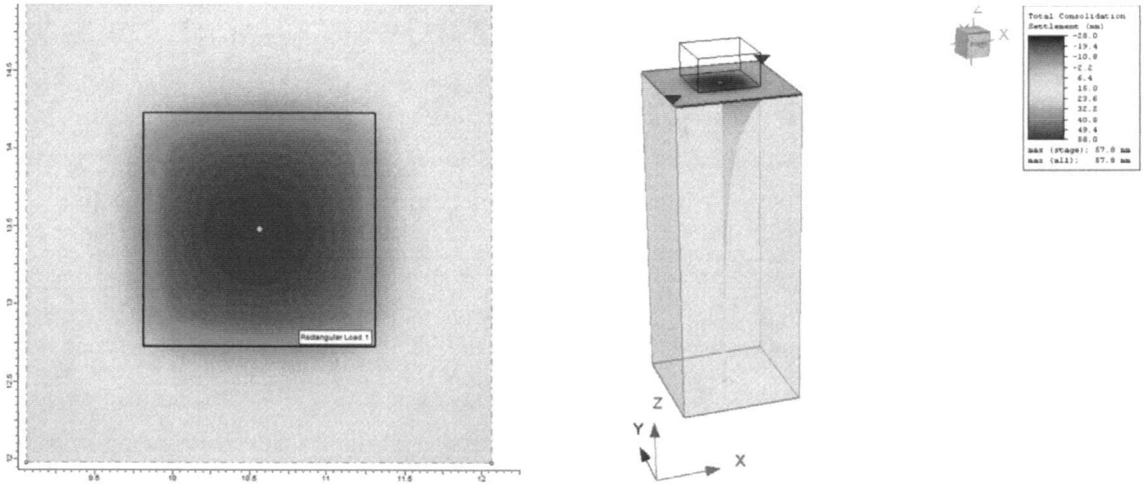


**Figure 2.** Settlement calculation model using Settle3 software.

Figure 3 shows that the specimen dimensions have an important influence on the final values of the compression index ($C_c$) and on the coefficient of volume compressibility ($m_v$). For smaller specimen dimensions these two parameters used in the settlement prediction as input parameters lead to lower settlements compared with the ones based on larger specimens' dimensions. An increase of the soil diameter from 54.7 mm to 112 mm induces an increase of approximately 40% of both soil parameters.

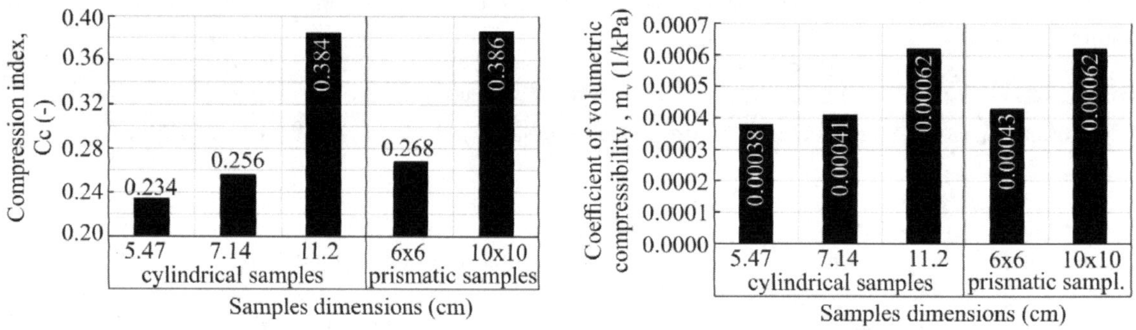


**Figure 3.** Variation of the consolidation parameters with the specimen dimensions.

In using the Settle3 software for the consolidation settlement evaluation, the following soil parameters are required from laboratory testing: the natural unit weight, saturated unit weight and depending on the analysis type, the soil behaviour can be defined as being linear, in which case it needs the coefficient of volumetric compressibility ($m_v$), or non-linear, for which the consolidation parameters need to be established ($C_c$, $C_r$, $p_c$) (Figure 4).

For the time of final consolidation, the software requires the value of the coefficient of consolidation which can be evaluated using Taylor's method or Casagrande's method.

The results of total consolidation settlement with depth after 100 years from placing the foundation loading (time at which one can be sure that the final settlement is completed) is given in Figure 5.

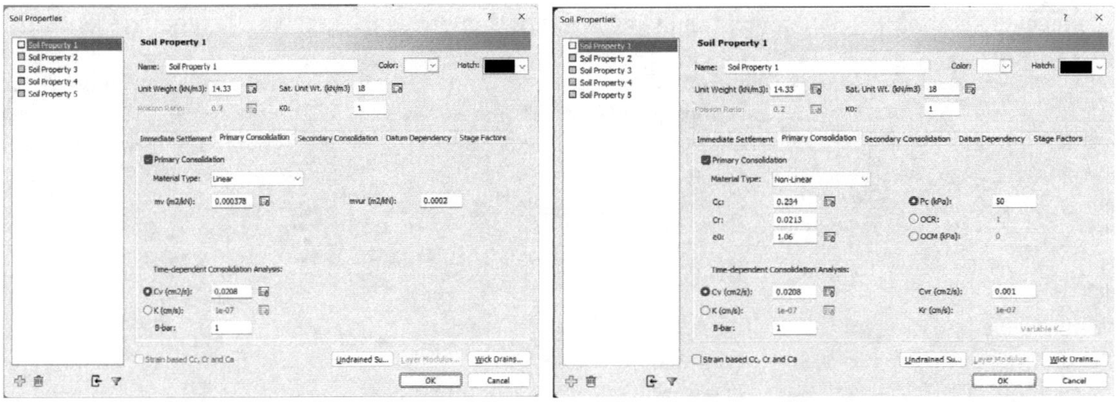


a) b)

**Figure 4.** Defining soil properties in Settle3 software for: a) linear and b) non-linear soil behaviour.

The first conclusion is that when increasing the specimen size, larger values for the final consolidation settlement will result. By approximately doubling the specimen size from 54.7 mm to 112 mm and considering the linear soil behaviour, the difference in consolidation settlement is approximately 37 mm and in the case of non-linear soil behaviour approximately 32 mm (Figure 5).

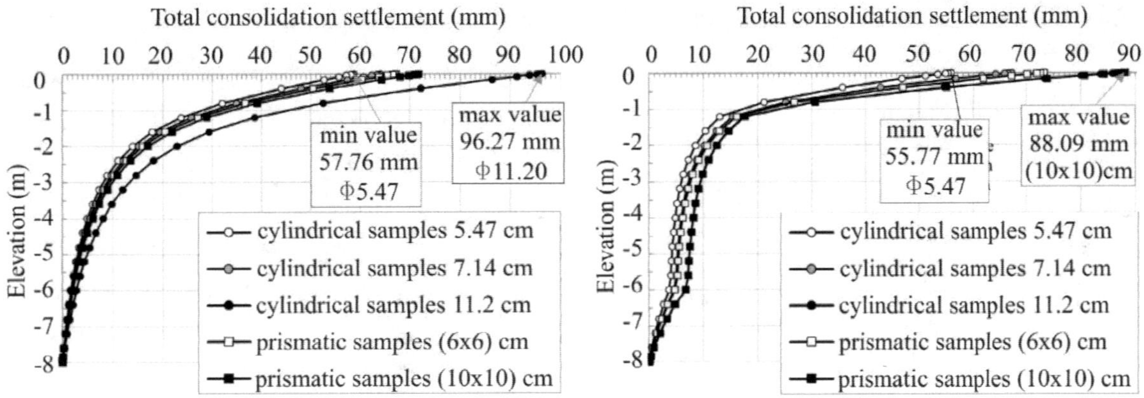


a) b)

**Figure 5.** Variation of the maximum computed value of total settlement with depth for: a) linear and b) non-linear soil behaviour.

*3.2. The influence of specimen shape*

In this research two types of specimen's shapes were used, cylindrical specimens with three different diameters (ϕ 54.7 mm, ϕ 71.4 mm and ϕ 112 mm) and prismatic specimens with square cross-sections of two different dimensions (60 x 60 mm and 100 x 100 mm).

Figure 6 presents the influence of the specimen size and shape on the final consolidation settlement under the foundation. The parameter obtained from tested prismatic specimens with square cross-sections with the dimensions of 60 x 60 mm used in the consolidation settlement evaluation led to larger value (67.31 mm) compared to the specimen that is of cylindrical shape with the diameter of ϕ 71.4 mm for which the resulting value is 64.05 mm (for linear soil behaviour, Figure 6 a). If the soil behaviour is changed to non-linear (Figure 6 b) there is the same trend in soil settlement for prismatic specimens with dimensions of 60 x 60 mm compared with cylindrical specimen with the diameter of ϕ71.4 mm; in case

IOP Conf. Series: Materials Science and Engineering 1304 (2024) 012020 doi:10.1088/1757-899X/1304/1/012020

of non-linear soil behaviour, for the larger specimens (both the cylindrical and prismatic ones) the final value of total consolidation under the foundation's centre is approximately the same (approx. 88 mm).

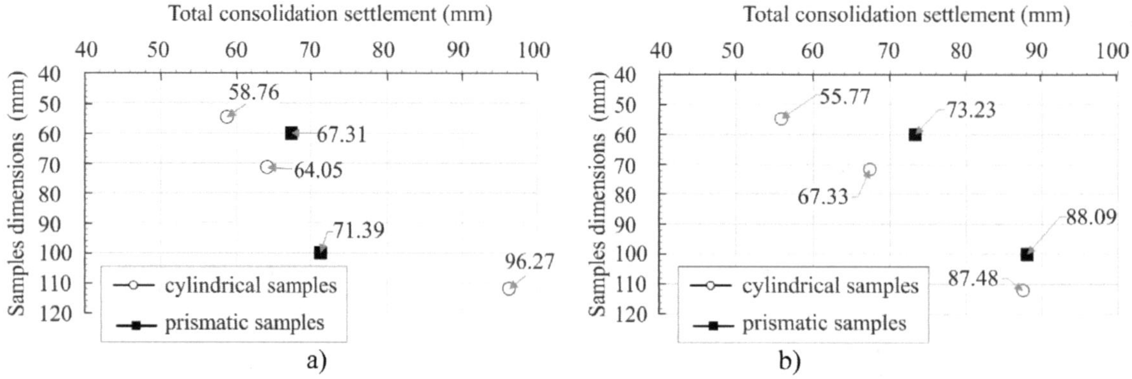


**Figure 6.** Variation of the maximum computed value of the total settlement with specimen dimension for: a) linear soil behaviour and b) non-linear soil behaviour.

### *3.3. The influence of the method used for the consolidation coefficient evaluation on the time of 90% degree of consolidation settlement*

The time needed for the final consolidation settlement can be evaluated using the coefficient of consolidation ($c_v$). The value of this parameter can be obtain using the graphical methods: Casagrande's method known as the logarithm of time fitting method and the Taylor's method also known as the square root of time method.

From Figure 7 one can see that the method used for the evaluation of the coefficient of consolidation ($c_v$) has a great influence on the parameter magnitude. For the tested soil, the Taylor's method yields higher value compared to the Casagrande's method.

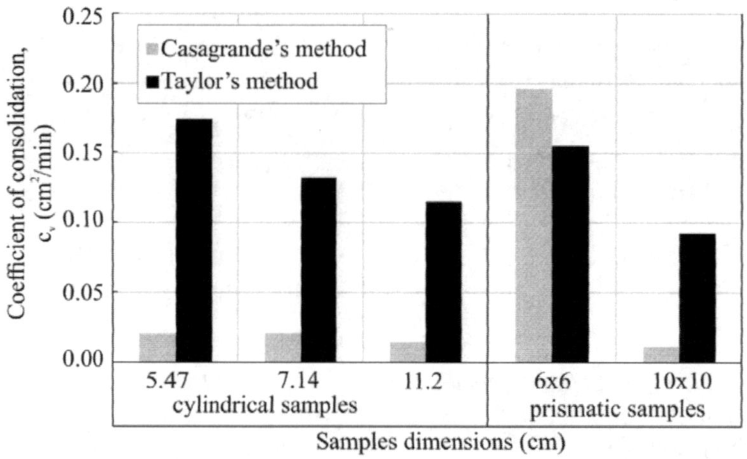


**Figure 7.** Variation of the coefficient of consolidation depending on the used method.

The evolution of the consolidation settlement in time was studied for time periods of: 0-year, 0.05-year, 0.5-year, 1.0 year, 5.0 years, 10.0 years, 50.0 years and 100.0 years. Figure 8 presents the variation of the total consolidation settlement with the log of time. The log of time scale was used to capture the time when the consolidation settlement can be considered as 90% from the final one.

Using Taylor's method, the tine needed for 90% of the total consolidation settlement is 1.0 year. The specimen size and shape does not have significant influence over the time of consolidation (Figure 8).

Using the value of the coefficient of consolidation evaluated based on Casagrande's method in the time-settlement analysis, the results indicated that the specimen dimensions influence the time for 90% of the total consolidation settlement. The consolidation parameters obtained after testing the cylindrical soil specimens with diameters of 54.7 mm and 71.4 mm led to one year time needed for reaching 90% of the total consolidation settlement. The tested specimens of 112 mm diameter indicate a time of 5 years.

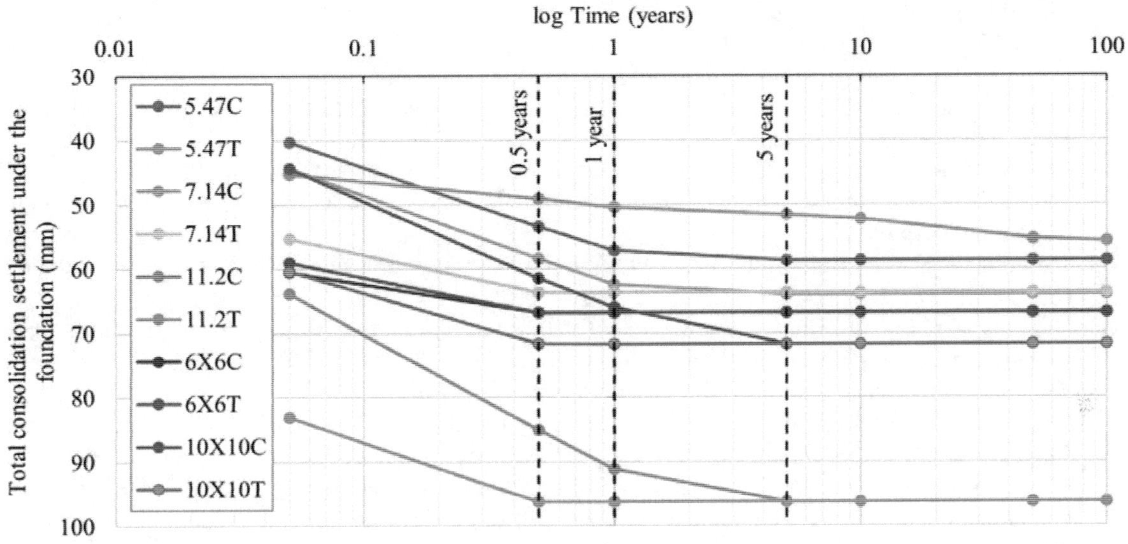


**Figure 8.** Time depending consolidation settlement (for 90% consolidation) based on the consolidation parameters evaluated using Casagrande's method (C) and Taylor's method (T)

## 4. Conclusions

The current research is focused on the analysis of the influence of the specimen dimension and shape used in consolidation tests with the application in the prediction of the final value of the foundation settlement. The final value of the foundation soil consolidation settlements is important in the geotechnical design. Usually, in geotechnical laboratories the tested specimens are cylindrical and in most of the cases with a diameter of 71.4 mm. This research presents the consolidation results obtained on three different cylindrical specimens' diameters (ϕ 54.7 mm; ϕ 71.4 mm and ϕ 112 mm) and two prismatic specimens with square cross-section (60 mm and 100 mm the side).

The consolidation settlement was computed using the Settle3 software, where the soil behavior can be defined as being linear, in which case it needs the coefficient of volumetric compressibility ($m_v$), or non-linear, for which the consolidation parameters need to be defined ($C_c$, $C_r$, $p_c$).

The specimen dimensions have an important influence of the final values of the compression index ($C_c$) and on the coefficient of volumetric compressibility ($m_v$). For the smaller soil specimen dimensions these two parameters used in the geotechnical design as input parameters lead to lower values of the settlement compared with the larger soil specimens. An increase of the soil specimen diameter from 54.7 mm to 112 mm led to an increase of both parameters ($C_c$, $m_v$) of approximately 40%.

Another conclusion is that the final value of the consolidation settlement is influenced by the specimen dimensions. Larger specimens lead to larger values of the final consolidation settlement.

The time needed for 90% of the consolidation settlement depends on the method used in the evaluation of the coefficient of consolidation ($c_v$). Using the Taylor's method, the total consolidation settlement is 1.0 year and is not influenced by the specimen size. Using the value of the coefficient of consolidation evaluated based on Casagrande's method in the time-settlement analysis, the results indicated that the specimen dimensions influence the time for 90% of the total consolidation settlement.

One year time is needed for the cylindrical soil specimens with diameters of 54.7 mm and 71.4 mm and for the rest of the larger tested specimens result a time of 5 years.

## 5. References

[1] Lin L, Hanna A, Sinha A and Tirca L 2015 Structural Response to Differential Settlement of Its Foundations, *Journal of Civil Engineering Research 2015, 5(3): 59-66, DOI: 10.5923/j.jce.20150503.02* p 59
[2] NP 112/2014 – Design Norm for shallow foundations (in Romanian)
[3] Stanciu A and Lungu I 2006 *Foundations. Soil physics and mechanics* vol I, Technical Publishing House, Bucuresti (in Romanian)
[4] Lancellotta R 1995 *Geotechnical engineering* A.A. Balkema Roterdam (Brookfield)
[5] McCarthy D F 2014 *Essentials of Soil Mechanics and Foundations: Basic Geotechnics* Pearson Education Limited
[6] Muntohar A S 2009 Reliability of the Method for Determination of Coefficient of Consolidation ($c_v$), *1st Meeting Denpasar, Bali, 5-6 November 2009* p 70
[7] Nwaiwu C M O, Anjorin J R and Mezie E O 2020 An Alternative Method for the Estimation of Coefficient of Vertical Consolidation of Coarse-Grained Lateritic Soils, *Nigerian Research Journal of Engineering and Environmental Sciences 5(1) p* 273
[8] Al-Zoubi M S 2008 Consolidation Characteristics Based on a Direct Analytical Solution of the Terzaghi Theory, *Jordan Journal of Civil Engineering, Volume 2, No. 2, p* 91
[9] Shukla S K, Sivakugan N and Das B M 2009 Methods for determination of the coefficient of consolidation and field observations of time rate of settlement — an overview, *International Journal of Geotechnical Engineering 3: (89-108), p* 89
[10] Azzouz A S, Krizek R J, and Corotis R B 1976 Regression Analysis of Soil Compressibility, *Soils and Foundations*. 16(2).

**Acknowledgments**
This paper was financially supported by the Project "Network of excellence in applied research and innovation for doctoral and postdoctoral programs / InoHubDoc", project co-funded by the European Social Fund financing agreement no. POCU/993/6/13/153437.

IOP Conf. Series: Materials Science and Engineering 1304 (2024) 012021 doi:10.1088/1757-899X/1304/1/012021

# The influence of the geometric characteristics on the behavior of braces with restrained buckling - case study of perforated steel core.

**C P Mocanu Basalic[1], S G Luca[1], I Lungu[1], P Mihai[1],**
[1]Gheorghe Asachi Technical University of Iasi, bd. D Mangeron nr. 67 Iasi, Romania

Email: cosmin-petru.mocanu-basalic@student.tuiasi.ro

**Abstract.** Buckling restrained braces are elements that equip the building with additional capacity for resistance and stability when subjected to earthquakes. As an answer to the problem of the possible loss of stability of the buckling-preventing component, the "perforated-core buckling restrained braces" (PCBRB) is a concept that involves the introduction of gaps, in the core of the bracing, to control the plasticizing mechanism of the dissipative element. The research addresses the behavior of PCBRB groups where all the elements belonging to the same group have the same area and length of the section considered active dissipative. A numerical analysis environment was developed that simulates the support conditions and material behavior. Based on the results of the numerical analyses, the influence of the geometry (b/t ratio) of the section of the dissipative zone on the behavior and the mechanism of stress strain distribution in the bracing core. It can be noticed the development of over-resistance to the compressive stress and non-uniform distribution of stresses on the surface of the elements whose b/t ratio exceeds the value of 5.5; for the elements whose b/t ratio has a value lower than 1.75 local instability problems in the dissipative zone occurred.

## 1. Introduction

Buckling restrained braces (BRB) are a type of dissipative elements, with a hysteretic behavior whose theoretical concept appeared for the first time at the end of the 80s. In 1988, the team of researchers cited in bibliographic sources [1-3] developed and tested the first fully functional bracing model with prevented buckling. Based on the promising results obtained experimentally, this type of control elements of the displacements of structures has been adopted in several countries since the end of the 90s.

The functionality of these elements is based on the deformation of the metal core as the dissipative component, and on the ability of the buckling prevention component to block the movements in the transversal directions of the core, when subjected to compression. Thus, the control component of the core movements must, at the same time, ensure the stability under the compressive stress of the dissipative component (the BRB core), while allowing at the same time the free, unobstructed movement of the core under the action of the tensile stress.

This behavior requires the introduction of a theoretical layer, first in the design and numerical analysis stage and later applied in the manufacturing stage, as a physical layer, with well-defined elastic characteristics. It allows the movement of the steel core inside the control component along the direction of the axial force. The initial concept proposed a ductile steel core as a dissipative component, and a sleeve composed of a steel pipe filled with concrete as a displacement control element - figure 1.

Relatively recent studies on these movement control elements have highlighted a series of inconveniences, as well as some territories to be further investigated in order to improve the concept.

In specialized literature, certain studies published in 2010 [4], there is highlighted the phenomenon of the local stability loss of the buckling prevention element. Under the effect of the seismic action, the displacements and deformations of the BRB core produce lateral thrusts on the elements that ensure the stability under the axial forces of the core, a fact that, under certain conditions, can lead to the premature failure of the buckling prevention mechanism.

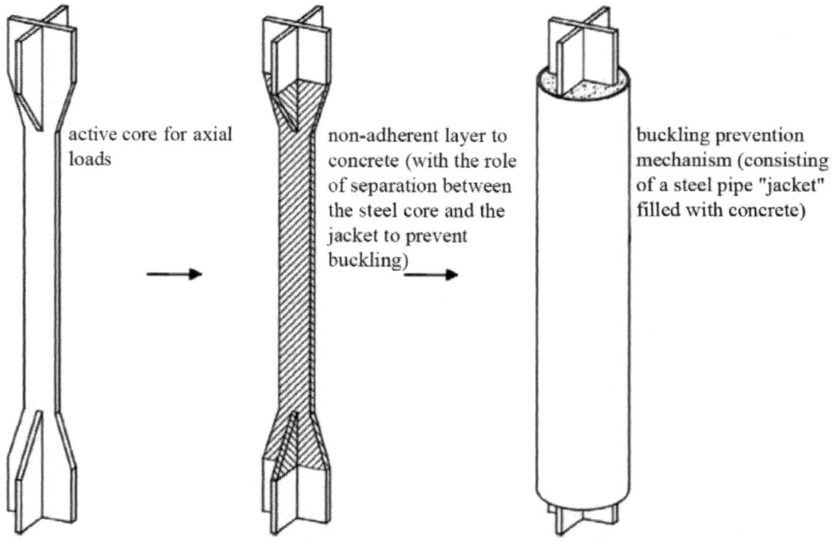


**Figure 1.** Conceptual system of the BRB [6].

The manufacturing process of BRB revealed the dependency of the implementation of the dissipative elements on the maturation time of concretes and high-strength mortars based on cement binders. The maturation time of concrete and mortar based on cement binders is considered at least 28 days.

The limited predictability of seismic actions means that the implementation/activation of the BRB is not recommended before reaching the characteristic resistances of the prevention mechanism.

After implementation, the inspection of the dissipative component (of the BRB core) becomes impossible due to its constructive nature. Due to the same closed box feature of the blocking mechanism, the replacement of the damaged elements for various reasons (accidental loads, seismic events, etc.) becomes impossible.

The above led to the need to investigate new methods of controlling the deformation mode of the BRB core as well as to furthering the implementation of new constructive concepts of buckling control mechanisms. The first requirement indicated above is addressed by putting into practice the concept of perforated BRB core (PCBRB).

Practically, in order to limit the axial stress within the elements of the buckling prevention component, a series of gaps are imposed along the core. In this way, in this way, as presented and concluded in previous research works [5-8], the axial deformations of the dissipative core are directed inside this gap, the stress induced on the hindering elements being diminished.

IOP Conf. Series: Materials Science and Engineering 1304 (2024) 012021 doi:10.1088/1757-899X/1304/1/012021

Other proposal [7 ] suggests that the buckling control elements should be made of thick sheet metal plates assembled with screws (figure 2), that would significantly reduce the manufacturing and implementation time of the BRB and would allow a periodic inspection of all the component elements.

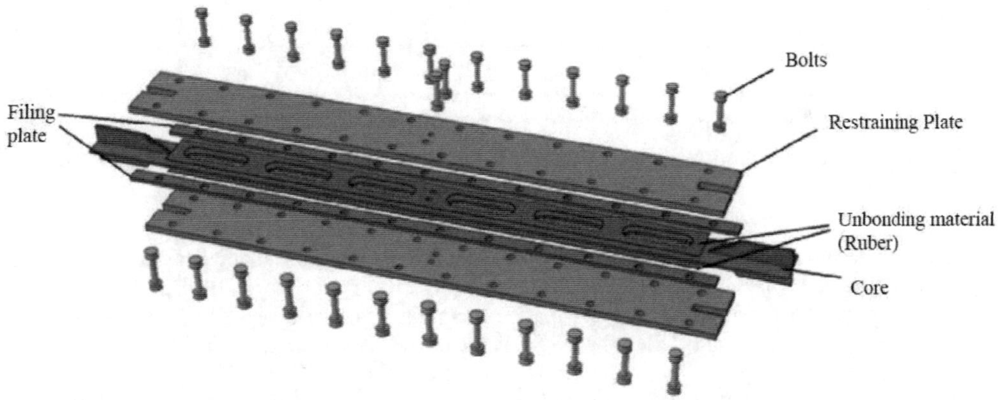


**Figure 2.** Conceptual system of "dry" BRB [7].

## 2. The research objective

The computation relationships of the classical stability factors (Euhler's theory) have limited applicability in this specific situation. The Bernoulli's hypothesis this theory is based upon (that refers to the flat and perpendicular sections to the initial axis of the element remaining the same after deformation occurred), lead to the approximate behavior of the materials. Also, this calculation hypothesis fails to capture the variable nature of the longitudinal modulus of elasticity (E) under cyclic actions, in the situation where the value of deformation or stress exceeds the limit of the material's elastic behavior. However, one cannot ignore the importance of the geometric characteristics that are the basis of the calculation principle of the stability factor, in particular the moment of inertia and the buckling length of the analyzed element.

As highlighted as presented and concluded in previous research works [5-8], the slenderness ratio can provide a preliminary prediction on the stability of PCBRB; this influence needs further investigation. The study developed in this paper focuses on the influence of the length of the elastic zone that delimits the gaps and at the same time ensures the stability of the plastic elements, upon the formation of the isostatic lines of 1st and 2nd ranks.

Starting from Navier's relationship that results in the computation of the normal stresses $\sigma$, and taking into account the fact that the area of the section (which in the studied case undergoes variations) directly influences the value of the stresses, the excessive presence of an area with a larger section, implicitly leads to a stress reduction on the respective area, thus reducing the deformable potential of the respective area. Figure 3 shows the conceptual model of PCBRB, the core areas and the interspaced gaps, highlighting the following geometrical characteristics:

- Lt – the total length of the PCBRB;
- Lea - the elastic length of the anchorage - the end zone of the PCBRB, that must be sized to display throughout the load/deformation cycle an elastic behavior;
- Lp - the plastic deformation length - the zone that is taken into account as a dissipative component;
- Let - the elastic length of the zone that ensures the stability of the median strips that sum up the dissipative component of the core;

- Lem – the length of the elastic-median zone - the zone that will also ensure the physical connection between the dissipative component and the buckling control component of the BRB.

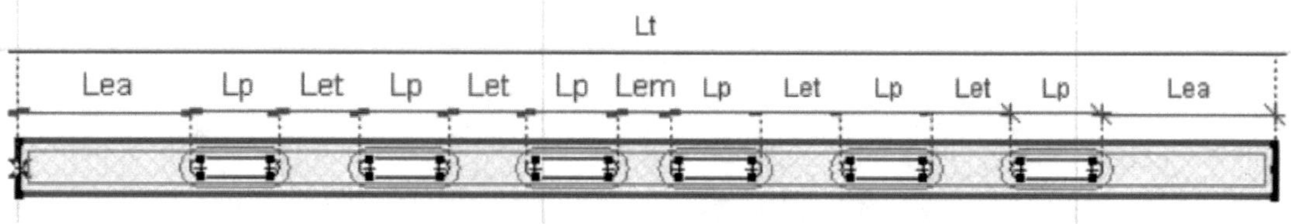


**Figure 3.** The PCBRB model and the delimitation of areas.

A final objective focuses on the distribution and values of the reactions that will be actually delivered by the PCBRB core to the buckling prevention elements. The Romanian Norm P100/2013 acknowledges the buckling-restrained bracing as „... dissipative elements that are designed to develop significant plastic deformations when subjected to the design value of the seismic action. The braces are made from a steel core inserted in a system that prevents the buckling of the core" [9].

The axial deformations of the BRB core will produce lateral thrusts towards the buckling prevention elements. In 2017, Stratan et al. [10] proposes a safety factor with a value of 3 between the critical buckling force of the buckling prevention elements and the maximum plastic capable force of the core.

Summarizing the above-mentioned facts, this study focuses on the followings:

- the influence of the slenderness ratio, along the two orthogonal directions of the PCBRB section, on stability against buckling;
- the influence of the buckling length of the plastic zone on the behavior of the PCBRB (figure 4).

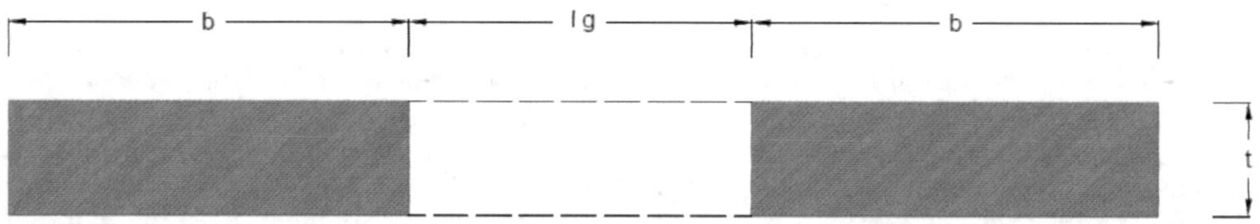


**Figure 4.** The geometrical characteristics of the plastic section: *b* – the section width, *lg* – the gap width, *t* – the core thickness.

### 3. Input parameters for the numerical simulations

S235JRZ15 steel type was chosen as the PCBRB material for the next series of experimental numerical tests; the corresponding mechanical characteristics are presented in table 1.

**Table 1.** Mechanical characteristics

| Material | $f_y$ [N/mm$^2$] | $f_u$ [N/mm$^2$] | $f_u/f_y$ [-] | ε [%] |
|---|---|---|---|---|
| S235JRZ15 | 235 | 360 | 1.532 | 15 |

One of the material feature is the high value of the self-consolidation factor as the ratio between the value of the ultimate stress and the value of the yielding stress. In order to highlight the influence of the geometric characteristics on the stability of the behavior of PCBRB subjected to cyclic loads, a series of groups of PCBRB elements with the following characteristics were defined:

- all models in the same group have the same plastic section area;
- all analyzed models have the same total length;
- all models in the same group have the same Lp;
- the sectional characteristic *t* is varied (as presented in figure 4) for the elements from the same group, taking into account the delivery availability of suppliers on the European market.

In table 2 and in figure 5, the variation of the characteristics and geometries of the elements in the test group with area equal to 700 mm$^2$ is presented.

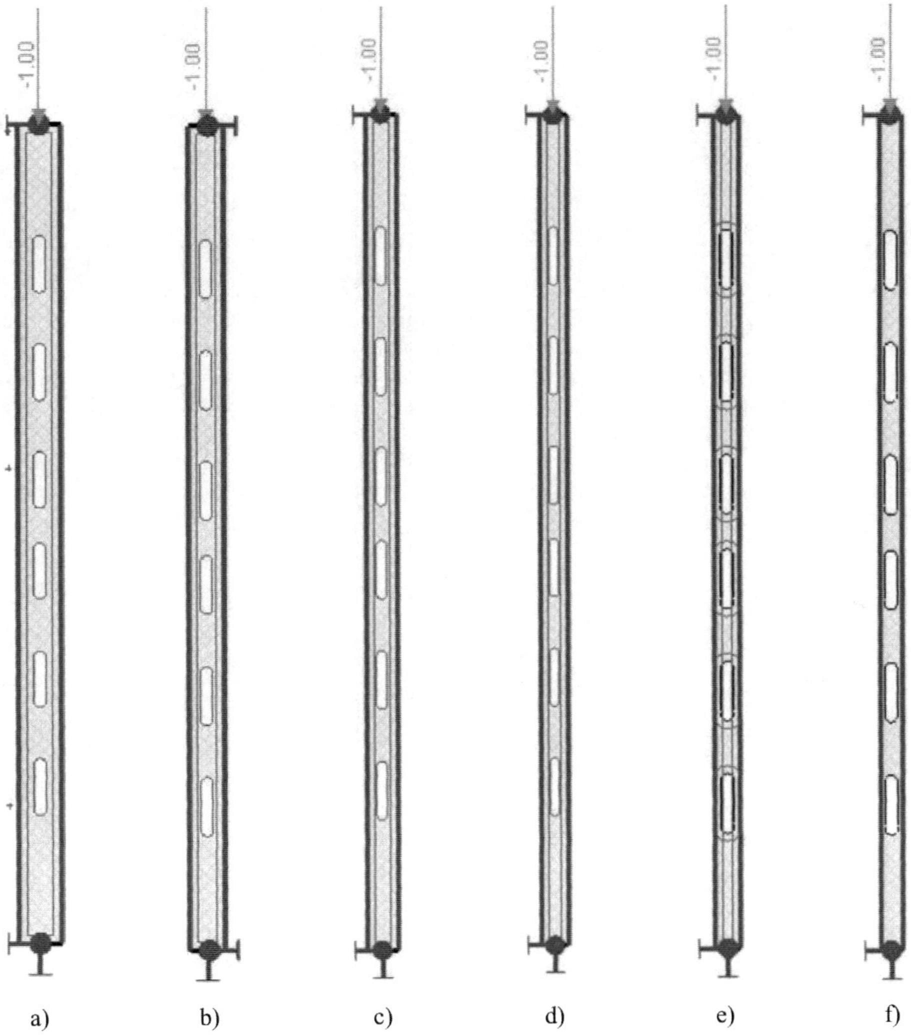


**Figure 5.** Geometry of the PCBRB group with area 700 mm$^2$: a) PCBRB 8*43, b) PCBRB 10*35, c) PCBRB 12*29, d) PCBRB 15*23, e) PCBRB 20*17, f) PCBRB 20*14.

**Table 2.** Sectional characteristics for the groups with the area of 700 $mm^2$

| Geometrical characteristic | PCBRB 8*43 | PCBRB 10*35 | PCBRB 12*29 | PCBRB 15*23 | PCBRB 20*17 | PCBRB 25*15 |
|---|---|---|---|---|---|---|
| t [mm] | 8 | 10 | 12 | 15 | 20 | 25 |
| b [mm] | 43.75 | 35 | 29.167 | 23.33 | 17.5 | 15 |
| $I_y$ [$mm^4$] | 55826.82 | 35729.17 | 24811.92 | 15879.63 | 8932.29 | 5716.67 |
| $I_x$ [$mm^4$] | 1866.67 | 2916.67 | 4200 | 6562.50 | 11666.67 | 18229.17 |
| $i_y$[$mm^2$] | 8.93 | 7.14 | 5.95 | 4.76 | 3.57 | 2.86 |
| $i_x$ [$mm^2$] | 1.63 | 2.04 | 2.45 | 3.06 | 4.08 | 5.10 |
| $\lambda_y$ | 9.52 | 11.89 | 14.27 | 17.85 | 23.76 | 29.74 |
| $\lambda_x$ | 52.05 | 41.64 | 34.70 | 27.76 | 20.82 | 16.65 |
| b/t | 5.47 | 3.5 | 2.43 | 1.55 | 0.87 | 0.56 |
| $\lambda_x/ \lambda_y$ | 5.47 | 3.5 | 2.43 | 1.55 | 0.87 | 0.56 |
| $L_p$ [mm] | 170 | 170 | 170 | 170 | 170 | 170 |

The geometries of the elements that are the subject of this study were modeled using the FEM AxisVM6 program. For all analyzed PCBRB models, the same support conditions, material behavior and cyclic loading were considered. The material behavior was modeled using von Mises' isotropic plasticization model, based on a multipoint characteristic curve of the material figure 6a. In defining the characteristic curve in the program, it was aimed to capture as accurately as possible all the inflection points considered relevant. A cyclic loading program whose characteristics correspond to the model described by European regulations [11] (figure 6b) was applied. All the models with the same type of finite element were discretized, with a maximum size of 10 mm to capture the stress state developed in the entire material mass (figure 6c).

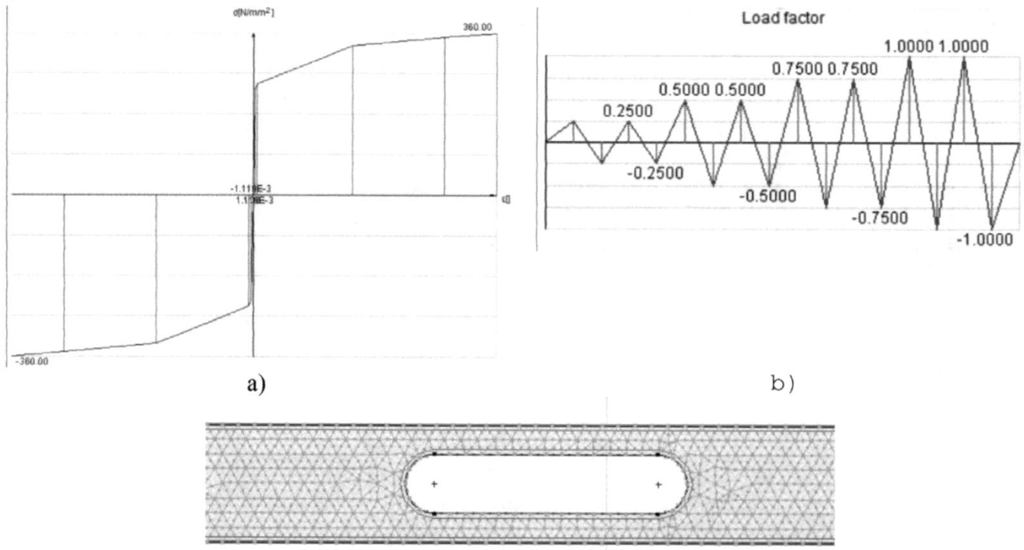


c)

**Figure 6.** Characteristics of the PCBRB model: a) multipoint characteristic curve, b) cyclic load curve, c) discretization in finite elements.

**4. Numerical analysis and interpretation of the results**

As mentioned in the previous chapters, the aim is to monitor the influence of the sectional characteristics on the stability and ductility of the analyzed PCBRB elements. As presented in table 2, the ratio between the two sides of the wing sections that form the active plastic length of the core is equal to the inverse ratio of the slenderness ratios, which derives from the calculation relationships of the sectional characteristics.

$$\lambda_y = \frac{l_f}{i_y} \tag{1}$$

where: $\lambda_y$ – slenderness ratio according to the local axis y, $l_f$ – the buckling length of the element, considered as half of Lp, $i_y$ – radius of gyration of the section along the local axis y.

$$i_y = \sqrt{\frac{I_y}{A}} \tag{2}$$

where: Iy – the geometric moment of inertia along the local axis y, A – the cross-sectional area of a strip.

$$I_y = \frac{t * b^3}{12} \tag{3}$$

where: t- section thickness, b – section width

$$\frac{t}{b} = \frac{\lambda x}{\lambda y} \tag{4}$$

As the ratio between the two dimensions of the cross section equals the ratio between the slenderness along the two local axes of the section, the behavior dependency of the PCBRB to the slenderness ratio the b/t ratio will be further used. All the analyzed models presented the same deformation mode: the displacement of the lateral wings towards the hollow part under compressive stress (figure 7). Analyzing the stress distribution within all analyzed elements, the following conclusions can be formulated:

- the length of the gap does not influence the final failure criterion of the PCBRB element, but it directly influences the distribution and values of the normal stress within the element;
- the ratio of the two geometric characteristics b and t directly influences the normal stress distribution mode on the surface of the element.

The elements with a b/t higher than 4.5 show an unstable behavior during cyclic loads, the strains developed in the solid part of the element being up to 19% higher for stresses of tension than in compression (figure 7).

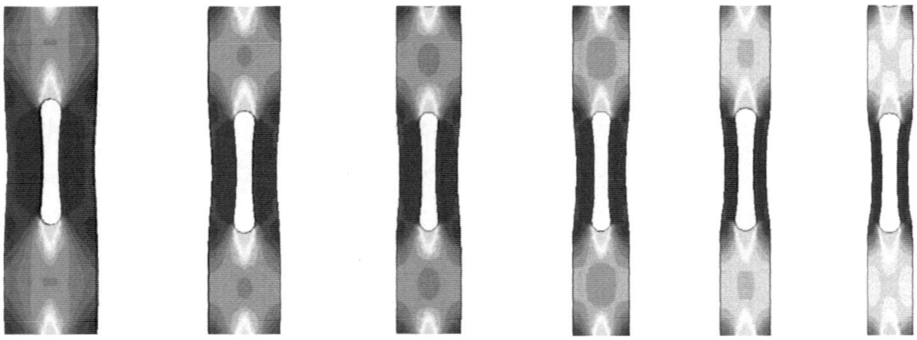

a) b) c) d) e) f)

**Figure 7**. Plasticization and stress distribution in the analyzed elements: a) PCBRB 8*43, b) PCBRB 10*35, c) PCBRB 12*29, d) PCBRB 15*23, e) PCBRB 20*17, f) PCBRB 25*14.

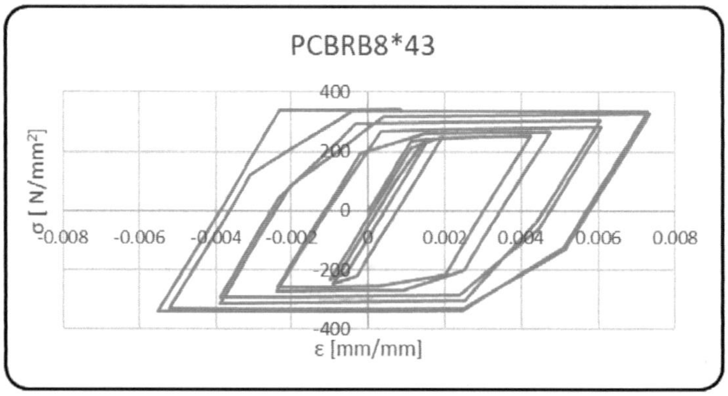


**PCBRB8*43**
b/t 5.46
σ Cmax 340.31
ε Cmax 0.00553
σ Tmax 337.13
ε Tmax 0.0074

**Figure 8.** Stress-strain curve for PCBRB 8*43

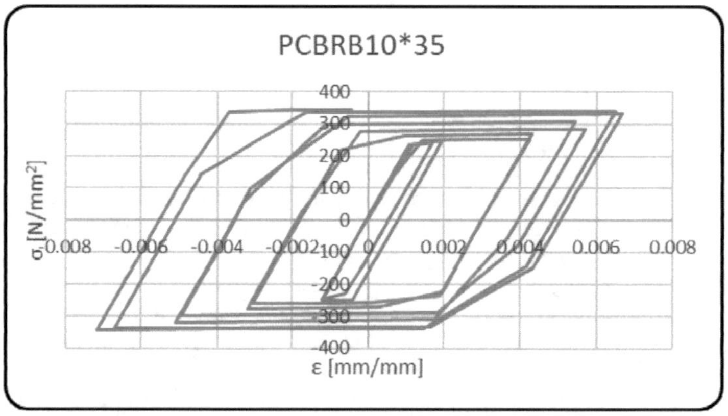


**PCBRB 10*35**
b/t 3.5
σ Cmax 342.48
ε Cmax 0.00715
σ Tmax 332.15
ε Tmax 0.0067

**Figure 9.** Stress-strain curve for PCBRB 10*35

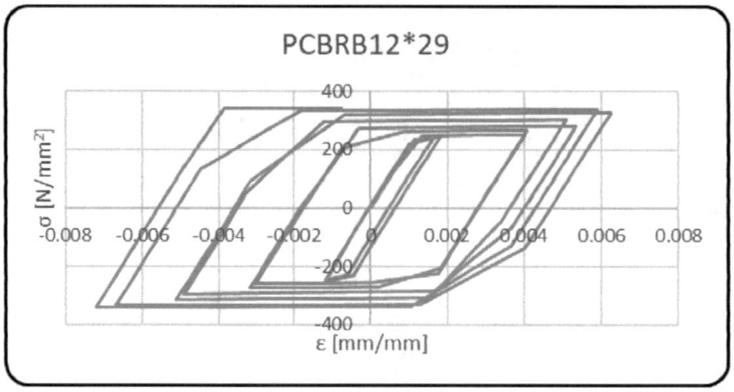


**PCBRB 12*29**
b/t 2.43
σ Cmax 341.17
ε Cmax 0.0072
σ Tmax 328.7
ε Tmax 0.0063

**Figure 10.** Stress-strain curve for PCBRB 12*29

IOP Conf. Series: Materials Science and Engineering 1304 (2024) 012021 doi:10.1088/1757-899X/1304/1/012021

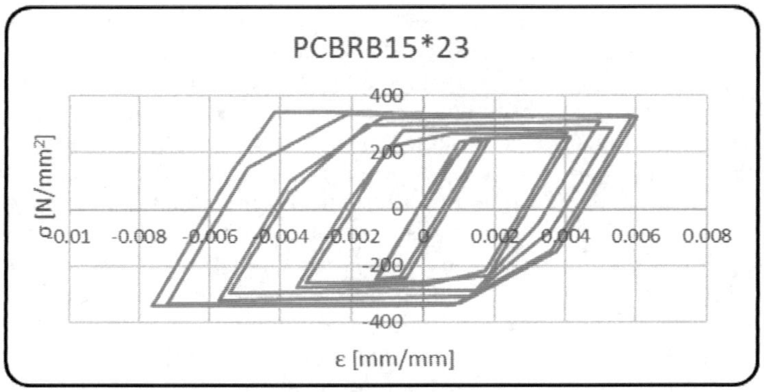


**PCBRB 15*23**
b/t 1.55
σ Cmax 342.9
ε Cmax 0.0076
σ Tmax 327.18
ε Tmax 0.0062

**Figure 11.** Stress-strain curve for PCBRB 15*23

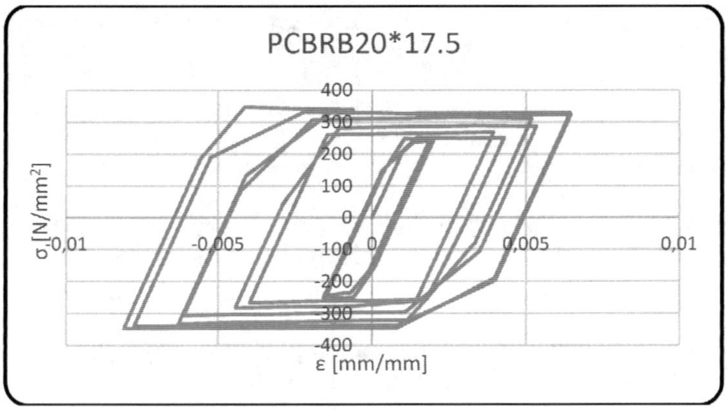


**PCBRB 20*17**
b/t 0.875
σ Cmax 347.04
ε Cmax 0.0081
σ Tmax 323.12
ε Tmax 0.0065

**Figure 12.** Stress-strain curve for PCBRB 20*17.5

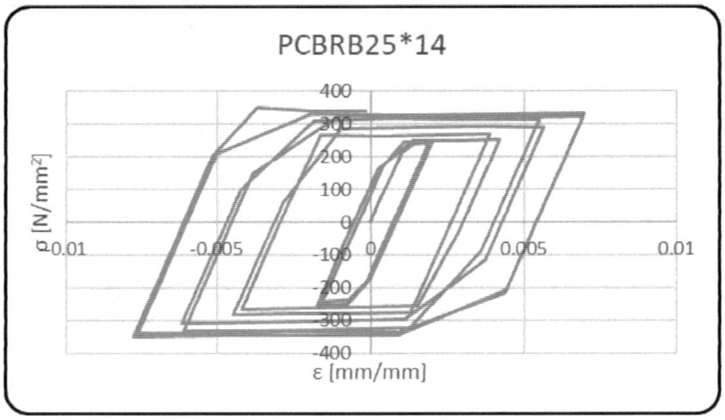


**PCBRB 25*14**
b/t 0.56
σ Cmax 348.63
ε Cmax 0.0078
σ Tmax 331.7
ε Tmax 0.0069

**Figure 13.** Stress-strain curve for PCBRB 25*14

However, it should be mentioned that in all groups, elements whose ratio b/t is less than 1.75, the software reported convergence problems between loading steps 80 and 85. This is due to the excessive deformation of the lateral bands that forms the plastic zone of the PCBRB core, under the action of compression stress. Consequently, further investigation will consider values of this ratio greater than 1.75. A similar effect is observed in the elements that present values of the b/t ratio greater than 4.5. These elements present two particularities: the initiation of the phenomenon of over-resistance to compression, and the non-uniform distribution of the normal stress on the entire plastic zone of the PCBRB. At the same time, for all the analyzed elements that are the subject of this discussion, a non-uniform stress distribution can be observed within the plastic zone. As it can be seen in figure 14 and in figure 15 for all these elements, the maximum values of the normal stress developed inside the solid of the plastic zone are directed towards the inner sides of the plastic zones. It can also be distinguished in the same figure the influence of Lp (of the length of the gap that defines the area of plastic deformation of the PCBRB) on the solid stress distribution. Thus, the length of the gap itself, which defines the area of plastic deformation of the PCBRB, cannot influence the behavior of the PCBRB, if the geometric characteristics of the section (b and t) are not taken into account. This phenomenon reflects the influence that the geometry of the cross-section has on the stability of the ductile behavior of PCBRB subjected to cyclic stretching-compression actions. However, in all groups, for the elements whose b/t ratio takes a value greater than 5.5, the analyzed elements tend to acquire a non-ductile behavior. The punching effect of the core fades, in close connection with the increase in the value of this ratio. This phenomenon can easily be attributed to the increase in the characteristic Iy of the sections (geometric moment of inertia along the local axis y of the section).

a)

b)

c)

**Figure 14.** Stress distribution inside PCBRBs: a) PCBRB 8*50; b) PCBRB 8*56; c) PCBRB 8*625

From the results of the numerical analyses, regarding the reactions transmitted by PCBRB, the following conclusions can be mentioned:

- The displacement and deformation values directly influence the reaction values. The elements that show over-resistance to compression transmit lower reaction values, and the elements that develop greater deformations transmit reactions with higher values.
- The median areas of the active part, areas that, due to the induced deformations, lose contact with the support, do not transmit reactions to the support.

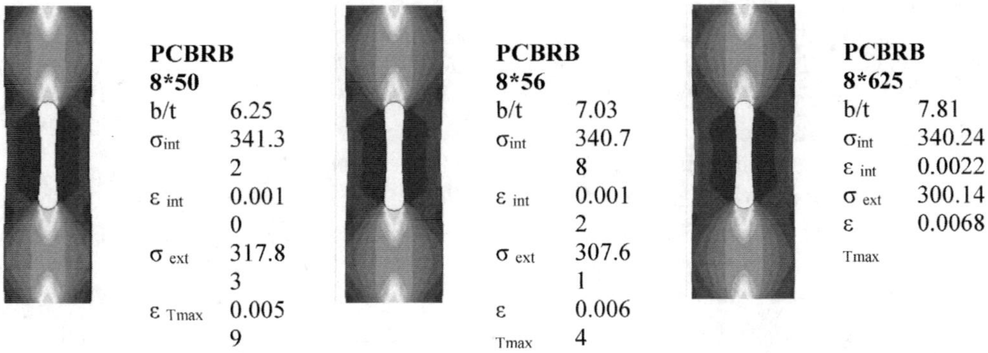


a) b) c)

**Figure 15.** Stress distribution in PCBRBs: a) PCBRB 8*50; b) PCBRB 8*56; c) PCBRB 8*625

## 5. Conclusions

Based on the present research and taking into consideration the objectives established in the second section of this paper, and the following initially imposed initial conditions:

- all analyzed elements, which are part of the same group, have the same sectional characteristics regarding the plastic active sectional area;
- all analyzed elements, regardless of the group they belong to, have the same active plastic length;
- all analyzed elements, regardless of the group they belong to, have the same maximum allowed displacement conditioned by the active plastic length;

one can conclude the followings:

- The geometric conformation of the section (b/t ratio) directly influences the character of PCBRB behavior under cyclic actions.
- The stiffness increase of the side wings perpendicular to the direction of action during the loading-unloading cycle and at the same time on the parallel direction to the direction of the degrees of freedom released by the presence of the gap leads to a non-ductile behavior, but up to a certain limit.
- The stiffness decrease in the same directions leads to a weighted increase in the ductile character of the behavior, but with the decrease in stiffness, the more significant occurrence of the local stability loss is observed for all the analyzed elements.
- The plastic length, or better said the length of the gap, becomes irrelevant outside of the values mentioned in the paragraphs above. For values of the b/t ratio greater than 5.5, a significant decrease in the ductile behavior is observed, while for values lower than 1.75, an increase in the same ductile behavior can be noticed, but local stability problems arise.
- The ratio (b/t) between the two slenderness ratios ($L_y/L_x$) assimilated with the two dimensions of the lateral wing geometry could be used in possible PCBRB pre-sizing stages.
- The value of the reactions transmitted by the PCBRB to the supports is directly influenced by the deformations and displacements, and the deformed areas, which lose contact with the supports, do not transmit reactions to the supports.

## 6. References

[1] Fujimoto M, Wada A, Saeki E, Watanabe A and Hitomi Y. A Study on the unbonded brace encased in buckling-restraining concrete and steel tubes. *J Struct Constr Eng (Trans AIJ)* 1988;34B:249–58 *(in Japanese).*

[2] Fujimoto M, Wada A, Saeki E, Watanabe A and Hitomi Y. A Study on brace enclosed in buckling-restrained mortar and steel Tubes (Part 1). *Annual Research Meeting Architectural Institute of Japan, Kanto Area, Japan, October;* 1988. *(in Japanese).*

[3] Fujimoto M, Wada A, Saeki E, Watanabe A and Hitomi Y. A Study on brace enclosed in buckling-restrained mortar and steel tubes *(Part 2). Annual Research Meeting Architectural Institute of Japan, Kanto Area, Japan, October;* 1988. *(in Japanese).*

[4] Takeuchi T, Hajjar J F, Matsui R, Nishimoto K and Aiken I D, Local buckling restraint condition for core plates in buckling restrained braces, *J. Constr. Steel Res.* **66** (2) (2010) 139–149.

[5] D. Piedrafita, X. Cahis, E. Simon and J. Comas A new perforated core buckling restrained brace, *Eng. Struct.* **85** (2015) 118–126.

[6] Y. Zhou, H. Shao, Y. Cao and E.M. Lui Application of buckling-restrained braces to earthquake-resistant design of buildings: a review, *Eng. Struct.* **246** (2021) 112991.

[7] Zhou Yun, Yongsheng Cao, Jiro Takagi, Genquan Zhong and Zhiming He Experimental and numerical investigation of a novel all-steel assembled core-perforated buckling-restrained brace *J. Constr. Steel Res.* **193** (2022)107288.

[8] Zhou Y, Chen Gong, Junxian Zhao, Genquan Zhong and Shiyu Tian Strong axial stability and seismic performance of perforated core plate buckling-restrained braces, *Thin-Walled Structujres* **156** (2020) 106997.

[9] Seismic design code - part 1 - design provisions for buildings - indicative P100-1/2013 *(2019).*

[10] Stratan A, Zub C, Dogariu A, Florea Dinu, Dubina D, Voica T-F, Ganea M-A, Marcu A-D, Coman M and Badea I-C Design recommendations for frames with buckling restrained braces, *University Horizons Publishing House (2017).*

[11] Romanian Standard SREN15129 – Anti-seismic devices (2010).

IOP Conf. Series: Materials Science and Engineering 1304 (2024) 012022 doi:10.1088/1757-899X/1304/1/012022

# Topographic Monitoring of the Operational Behaviour of the Longest Cantilever Viaduct in Romania

**Gh. Boacă, M.C. Scutaru**
"Gheorghe Asachi" Technical University of Iasi, Romania, Faculty of Civil Engineering and Building Services, Department of Transportation Infrastructure and Foundations, Blvd. Mangeron, No. 1, 700050, Iasi, *Romania*

E-mail: gheorghita.boaca@academic.tuiasi.ro

**Abstract.** Over time, bridges have been built to fulfil different roles. With the development of industry came the need to build new transit routes linking industrial areas to cities or freight routes. The peak period of industrial development in Romania was in the 1960s, when a large number of bridges were built and put into operation. One of these bridges is the Sidex viaduct, located in Galați. The construction has a total length of 1,070 m, with 15 spans of different lengths (47.40 m + 13 * 75.00 m + 47.40 m). The resistance structure is made of 7 prestressed concrete frames linked together by joints located at the level of the cantilevers, the structure being the largest cantilever viaduct in operation in our country. The present article focuses on the presentation of the implementation of a special monitoring program of the viaduct's in-service performance. For this purpose, special technologies have been used, such as topographic monitoring of the evolution of structural displacements at the most representative points of the structure, an implementation that we develop in detail.

## 1. Introduction

When in use, buildings in general are constantly subjected to the aggressive action of both the environment and the conditions of use [1]. Over time, these actions directly affect the physico-chemical properties of the materials from which the structure is made, leading to changes in structural response that can even impair or endanger the safety and comfort of users.

Bridges are also among the most affected by environmental action. They are also subjected to very high loads, both from their own considerable weight and from the action of cars. Given their specific nature, bridges must be capable of supporting the traffic they serve for a service period of around 120 years, a challenge accepted by managers.

In order to ensure optimal operating conditions [2] but also to be able to intervene in the shortest possible time in the event of substantial changes in structural behavior, various special construction monitoring technologies have been developed over time [3].

Special monitoring is the activity of monitoring the operational behavior of a structure by periodically measuring, recording, processing and interpreting the values of the parameters that define the extent to which the bridge maintains over time the strength, stability and durability requirements necessary to ensure safe traffic [4].

There is a very wide range of bridge technologies on the market for monitoring the most important parameters [5]. One of the most commonly used bridge tracking technologies is topographic tracking. This allows regular monitoring of the evaluation of vertical and horizontal movements of various characteristic points, which can be done at a distance from the structure without hindering traffic in the area.

This technology has been successfully used for the Sidex viaduct, located in the industrial area of Galați, which provides the link between the city, the industrial platform of the SIDEX S.A. steel complex and the country road DJ 252. The structure was chosen for research and for the present study due to its particular static scheme, the viaduct being the longest cantilever bridge in operation in our country.

Noting the high degree of degradation and advanced age, in 2019, the City Hall of Galati has commissioned a technical expertise to facilitate the process of starting the design and execution of rehabilitation and modernization of the passage. Analysing the structure and knowing the significant time periods required for an administrator to start the execution of the works, but also due to the specificity of the recommended extensive works, works that impacted the resistance structure, an extensive topographic follow-up program was implemented. This programme was carried out over an extended period of time, making records before, during and after the execution of the recommended works.

The article aims to present the application of a modern special tracking system for the SIDEX Viaduct. The technology used in the case under analysis monitors the evolution of the degradation of the structure by means of topographic measurements at specified time intervals, continuously comparing the new data captured with a set of reference data. Any change in the readings compared to the standard values was analysed and different threshold values of displacements were set, the exceeding of which triggers a strong alarm signal regarding structural safety.

## 2. Technology for topographic tracking of the operational behavior of a structure

Structural deformation monitoring of a bridge is part of the broad field of engineering condition monitoring (referred to as SHM in the literature), a field that is in full development thanks to the latest technological developments.

According to [6], [7] and [8], SHM includes all systems mounted on any construction, whether it is a building, a bridge, a wind turbine or an oil platform, whose main purpose is to inform the responsible personnel about the occurrence of any changes between the parameters recorded for the structure in good or very good working condition and the actual condition.

One such technology is the tracking of how a structure responds to permanent and live loads through topographic means. These methods [9] consist of the periodic measurement of points materialised on the structure by means of landmarks arranged on the visible faces of the component elements. There are two types of landmarks:

- Movable landmarks, which were placed on the surface of the elements, i.e. on the heart of the beams and on surfaces that were not touched by the rehabilitation works before the works started. At a later stage, additional fixings were installed on the parapet beam, on the kerbs and in the longitudinal axis of the structure
- Fixed landmarks installed on the ground, outside the area of influence of the construction, i.e. at the side of the structure for monitoring landmarks on the heart of the beam, and immediately after the access ramps, on an area that is stable in terms of terrain and from which landmarks installed at path level can be viewed

A total station located at points considered fixed landmarks and targeting the moving landmarks is used to track the position of the moving landmarks. The height at which the mobile landmark is located is read and automatically recorded in the database. This reading is compared with a reading recorded when the monitoring starts, considered as a reference reading, the difference between them representing the change in position of the element or the whole structure. If these changes reach or exceed the limits imposed by the expert, the results of the stability and strength calculations are taken as soon as possible, and the necessary measures are taken, which may go as far as prohibiting traffic on the structure.

## 3. Sidex viaduct structure

The Sidex viaduct connects the centre of Galati to the SIDEX S.A. industrial platform, ensuring the continuity of the county road DJ 252. This viaduct was a first for the field of bridge construction when it was inaugurated in 1969, being the longest cantilevered road viaduct in Romania.

The viaduct is built from a prestressed concrete structure, with the superstructure forming the cantilever frames together with the infrastructure. This resulted in 15 spans of different lengths (47.40 m + 13 * 75.00 m + 47.40 m), the total length measured at the level of the splice covers on the slabs being 1,070 m (Figure 1). In order to allow the trams to travel in optimal conditions, the longitudinal slope of the passage is 0.8 ‰.

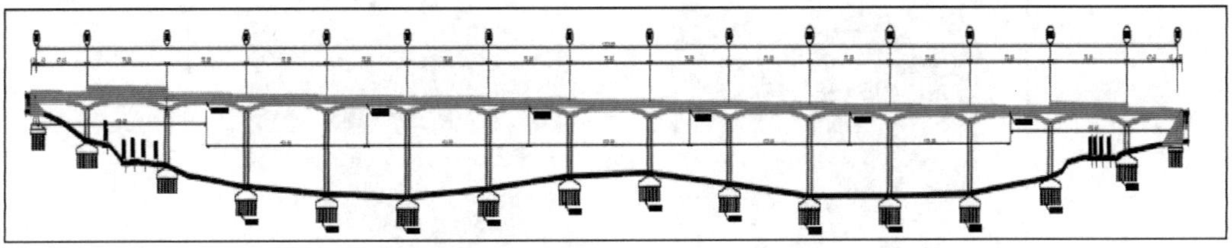

**Figure 1.** Side view of the viaduct.

The viaduct's resistance structure consists of seven frames made of prestressed concrete, joined at the top by joints. Given the specific traffic characteristics of the county road and the road serving the Iron and Steel Works, in cross-section, the roadway on the bridge has a total width of 18.70 m, consisting of a carriageway serving two tram lines and two car lanes totalling 14.00 m, plus the width of the two pavements, i.e. 1.90 m each (Figure 2), a considerable width given the location of the viaduct within the town.

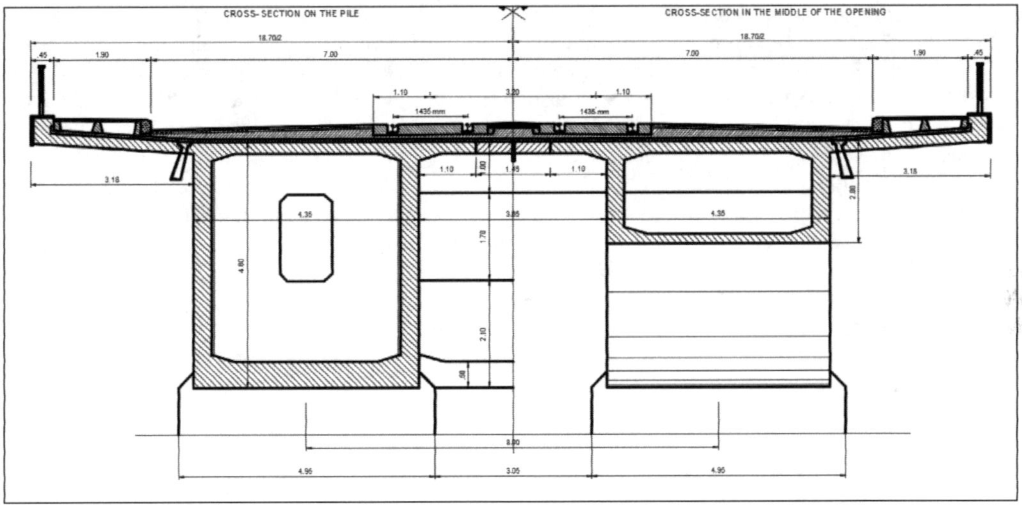

**Figure 2.** View of the path on the bridge.

The superstructure consists of two boxes (Figure 3) with variable heights, with a minimum height of 2.00 m in the middle of the openings, which increases towards the reservoir area, reaching up to 4.80 m at the piers.

In cross-section, the two boxes are 4.35 m wide and positioned 8.00 m apart. The walls of the boxes are 35 cm thick in the central section of the openings, rising to 40 cm at the bases in order to take up the cutting force properly. For strength reasons, the bottom plate of the cassette also varies in thickness, from 18 cm in the centre of the opening to 50 cm in the areas where the ruler joins the frame posts.

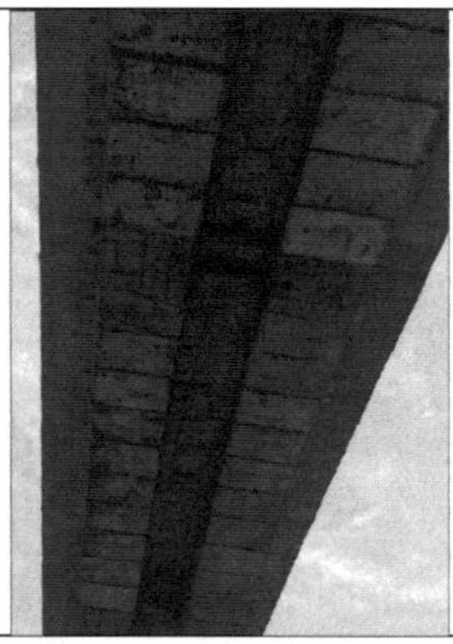

**Figure 3.** View of the box beams of the superstructure

**Figure 4.** View of the P4 pile

A particular feature of the structure was the way it was built, considered revolutionary for the time. Thus, the frame girders were made of 3.00 m long cassette sections, built of reinforced concrete in a special polygon located in the immediate vicinity of the site. The assembly of the sections was the main challenge of the execution, being cantilevered and fixed with pre-stressed cables. The two boxes are connected transversely to each other by means of a massive reinforced concrete slab. In addition, given the large spans, it was found necessary to lay out two other trusses in the field. In the longitudinal direction, the two boxes are also connected by means of a reinforced concrete slab, pre-compressed after curing.

The viaduct superstructure transmits loads to the foundation via the infrastructure. These consist of 2 reinforced concrete piers and 14 piers with massive elevations. The piers consist of two reinforced concrete columns with a box section (Figure 4), with a variable shape and height due to the specific terrain of the site, starting at 14.40 m for pile P14 and reaching a maximum height of 39.42 m for pile P4.

The foundation of the infrastructures is indirect, the good foundation ground in which the piles are embedded being at depths of between 11.50 m and 14.50 m. The piles are of the Franki type with a diameter of 600 mm, with an impressive 1,516 piles being built for all 14 piles. At the top, the piles are encased in reinforced concrete rafts between 3.50 and 4.50 m thick, varying for each pile depending on the plan dimensions and number of piles.

The bridge abutments are of the 4-post, sunken type, the bank with turned walls and guard walls being constructed of reinforced concrete. The foundations of the bridge abutments are made of 3 rows of 16 Franki piles embedded at 12.24 m (bridge abutment towards the combined) and 14.90 m (bridge abutment towards the Galati municipality) in the good foundation ground, where the loads are discharged.

**4. The need and the purpose of monitoring**

The viaduct was designed in 1966 and commissioned in early 1969. Analysing the design standards of the time and the specific traffic served at that time, it appears that the bridge was designed for Class I loading convoys (A30 truck convoy and V80 special wheeled vehicle), to which electric tram convoys were added. Over time, the payloads underwent changes, greatly increasing the values of the stresses they gave, which affected the strength structure that was already "weakened" by the occurrence of degradations.

These degradations occurred during the operating period and were mainly caused by the action of the environment. They were not dealt with in good time, and it is easy to see that maintenance work was not carried out at the intervals recommended by the standards.

In addition, during the period of operation, the structure was subjected to the action of several earthquakes with an intensity greater than 6.1 on the Richter scale, which seriously damaged the resistance structure. The damage caused by the earthquakes includes:

- severe damage to infrastructure elevations and expansion joints, the most severe being due to the 7.2 Richter earthquake of 4 March 1977
- piers damage resulting from the cumulative stresses of the August 1986 and May 1990 earthquakes

Following these earthquakes, the administrator intervened by reinforcing the infrastructure, carrying out exterior pile lining, caulking and crack injection.

In recent years, following a technical expertise which drew attention to the large number of degradation processes observed and their degree of extension, but also due to the particular constructive composition, it was decided to implement an extensive programme of special monitoring, aimed at achieving the following objectives:

- ensuring the operational safety and durability of the structure by early detection of dangerous degradation processes and affected areas
- monitoring the evolution of identified degradation processes
- recording of operational behaviour and identification of ranges corresponding to normal operation for various parameters considered representative

One of these parameters we will focus on in the article is the monitoring of superstructure displacements. By analysing their evolution, a programme has been developed to warn the administrator of the imminence of degradations with an increased impact on traffic safety.

**Table 1.** Alarm limit of reference parameters.

| Reference parameters | | Values of | | |
|---|---|---|---|---|
| | | **Caution (blue)** | **Waring (yellow)** | **Alarm (red)** |
| For the arrow measured in the middle section of the viaduct spans | | 10 mm | 20 mm | 30 mm |
| For the cross-section measured on the viaduct piers | | 10 mm | 15 mm | 20 mm |
| For expansion joint covers | Horizontal movement | ± 30 mm | ± 40 mm | ± 50 mm |
| | Vertical movement | ± 30 mm | ± 50 mm | ± 60 mm |
| For tilting the piles elevation | | ± 15 mm | ± 20 mm | ± 30 mm |

Depending on the displacement values identified and the structural element affected, the monitoring team created 3 warning thresholds, colour-coded. The administrator already knows how to intervene according to each threshold, thus minimising the reaction time needed to implement the necessary remedial measures. Table 1 gives an overview of the reference parameters used for monitoring the operational behaviour with topographic means and the warning values according to the 3 chosen thresholds.

**5. Installation of topographic tracking devices and measurements**

The equipment used to monitor the viaduct consisted of a LEICA TCA 1201 high-precision total station (Figure 5) and a LEICA DNA 03 high-precision level (Figure 6). These devices have the most appropriate characteristics for the specific works and can be easily used by the administrator's staff during the current maintenance of the works.

IOP Conf. Series: Materials Science and Engineering 1304 (2024) 012022 doi:10.1088/1757-899X/1304/1/012022

**Figure 5.** LEICA TCA 1201 total station [10]

**Figure 6.** LEICA DNA 03 level [11]

The measurements also required the installation of tracking targets made of reflective sheeting (Figure 7), so that they could be easily seen from convenient distances. These markers were mounted in characteristic sections, i.e., in the central section and in the area of the bearing apparatus for each opening.

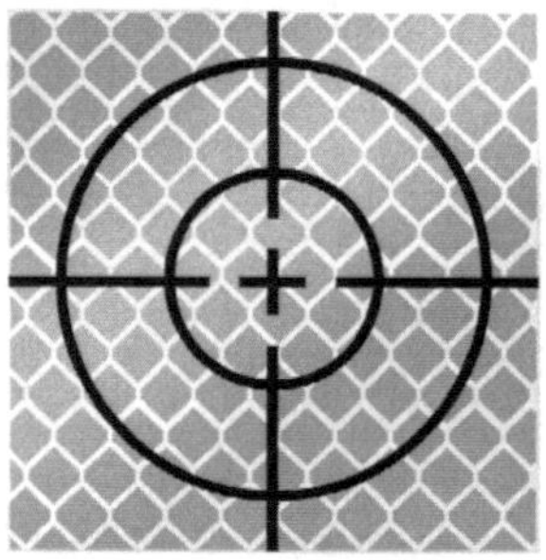

**Figure 7.** Reflective topographic target [12]

In order to be able to follow the behaviour in operation, the most representative points on the bridge were chosen, points whose displacement characterises the whole structure. At the same time, account was also taken of the period of perspective for which the topographic monitoring system is to be used.

The installation of the whole monitoring system is uncomplicated and easy to install, which helps managers who want to carry out monitoring at minimum cost.

Therefore, as a first step, tracking targets were installed on the right side of the beam heart, in the central sections of the spans and in the resemination sections. Given that rehabilitation and upgrading works were started during the period set for monitoring, and at the request of the administrator who wished to be able to easily continue the current monitoring of the structure, additional topographic benchmarks will be installed at the parapet beam level, on the kerbs and in the longitudinal axis of the viaduct upon completion of the rehabilitation works. Topographic markers have also been installed at the expansion joint covers for spans 3, 5, 9 and 13 (where the structure's joints are located) to monitor their behaviour under the action of road traffic or in the event of an earthquake.

The fixed reference points, where the total station or level is installed during the measurements, have been chosen so that all coordinates of these points are known, and all measuring points on the structure can be targeted.

After completion of the rehabilitation works and definition of the new measuring points, it was found necessary to modify the fixed landmarks as well. Thus, other points of known coordinates were identified, this time in the immediate vicinity of the access ramps, points that allow the visualization of the characteristic points on the structure.

In order to achieve the most accurate monitoring possible, particular attention has been paid to how measurements are taken. Thus, at the time periods recommended in the special monitoring project, the team of surveyors carried out repeated measurements on all points in the network. This was calculated in a local projection system and determined as a constrained grid. The offsets required for the reading were carried out using the indirect measurement method, i.e. the least squares method. The methods used were chosen because of the small degree of error in the determination of the points that can occur, an error that was quoted towards the minimum of the range recommended by the specific quality requirements of the work.

Also, with the LEICA TCA 1201 total station at hand, using the laser beam function provided, combined with the erasure method, measurement targets were targeted from at least two known coordinate monuments, reducing the risk of errors to near "0". For verification it was recommended that for at least one third of the targets the measurements should be taken from three different fixed monuments.

## 6. Evolution of the in-time structural behaviour

Following a current review, in 2015 structural deterioration was found to have progressed, endangering the safety and comfort of users. At that time, the administrator commissioned the preparation of Design Documentation in order to obtain the European funds necessary to start the rehabilitation works. One of the stages of the documentation included determining the elevations of the most important points of the structure. At that stage, it was concluded that the uncovered weak point was located between joint 4 and joint 5, where 8 points were chosen for reading the elevations as shown in Figures 8 and 9. The absolute coordinates, determined on 20.06.2015, are shown in Table 2.

When the administrator decided to start a new monitoring program in early 2021, the responsible technical staff considered it appropriate to use the data measured in 2015 as baseline data. Then, from 26.03.2021, regular readings of the most important points were taken, which were reported to the initial reading.

Following the first set of measurements, a subsidence of the analysed area could be observed, which accumulated in more than 4 years, an average of about 7.7 cm, the breakdown of subsidence for each analysed point can be seen in Table 3 below. The responsible staff, together with the technical expert, decided to continue monitoring for about 1 year.

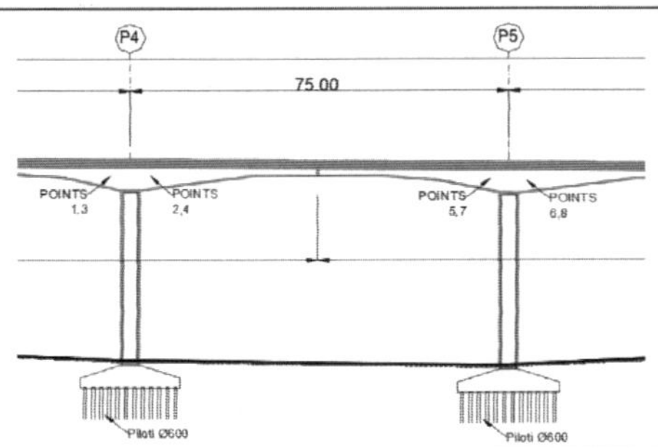


Figure 8. Location of measuring points in side view

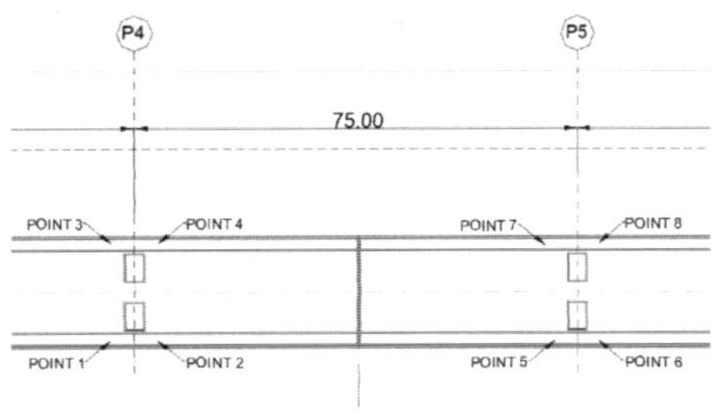


Figure 9. Location of measuring points in plan view

**Table 2.** Absolute heights determined in 20.06.2015.

| Point | Absolute height |
|---|---|
| 1. | 49.456.31 |
| 2. | 49.461.97 |
| 3. | 49.397.62 |
| 4. | 49.398.92 |
| 5. | 51.287.14 |
| 6. | 51.247.17 |
| 7. | 51.213.74 |
| 8. | 51.218.54 |

**Table 3.** Settlement compared to previous reading, in mm.

| | 1 | 2 | 3 | 4 | 5 | 6 | 7 | 8 |
|---|---|---|---|---|---|---|---|---|
| 26.03.2021 | -8,18 | -7,25 | -8,17 | -8,09 | -9,50 | -9,20 | -10,25 | -10,04 |
| 28.06.2021 | -6,00 | -5,40 | -5,02 | -4,40 | -6,25 | -6,12 | -7,10 | -7,20 |
| 24.08.2021 | -4,10 | -4,45 | -4,98 | -5,05 | -5,20 | -5,05 | -4,90 | -4,15 |
| 30.09.2021 | -5,50 | -5,10 | -5,00 | -5,05 | -4,05 | -5,62 | -5,10 | -5,85 |
| 28.10.2021 | -4,20 | -4,10 | -4,25 | -4,85 | -4,95 | -5,38 | -5,00 | -5,15 |
| 26.11.2021 | -3,20 | -3,25 | -3,70 | -3,15 | -4,05 | -4,25 | -4,91 | -4,34 |
| 27.01.2022 | -4,20 | -3,33 | -3,95 | -3,82 | -3,02 | -3,69 | -4,10 | -4,20 |
| 15.03.2022 | -5,87 | -6,10 | -5,05 | -5,08 | -4,50 | -4,55 | -5,00 | -6,02 |
| 19.05.2022 | -4,75 | -4,00 | -5,05 | - | -5,07 | -4,58 | -5,99 | -5,92 |

Thus, a consolidation of heaps was observed during the warm season (Figure 10), but also an increase with the arrival of autumn. This dynamic is due to the site's specific soil and bedrock characteristics.

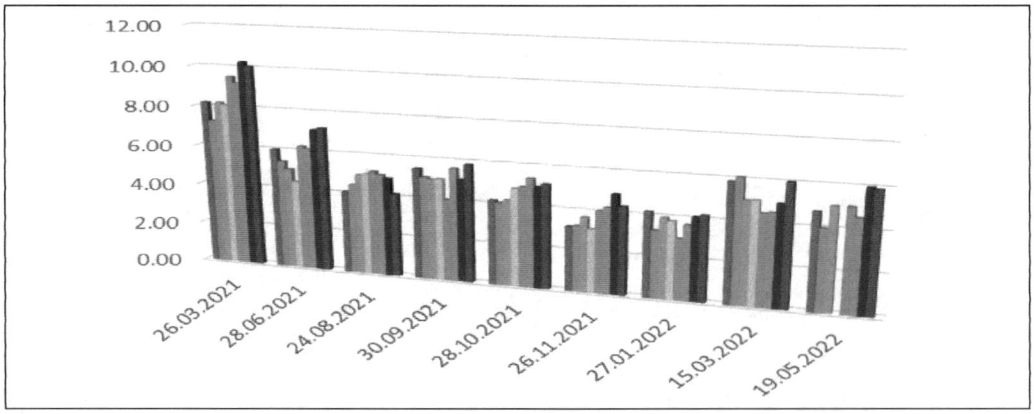


**Figure 10.** The settlement compared to the previous reading, in mm.

Also, if we compare the measurements of the 4 sections of the analysed area, we can observe that the maximum displacements are found in the right of the points 7 and 8 of measurement. This observation draws a strong alarm signal regarding a possible occurrence of local subsidence phenomenon only in the area of the pier P4, a phenomenon that may endanger the structural safety.

It should be noted that during the measurements on 14.04.2022 it was found that the materialisation element of measurement point 4 had been destroyed, which made it impossible to continue measuring it.

Analysing the graph of cumulative landings (Figure 11 and Table 4) we can see a relatively linear increase over time. By extrapolating this evolution assuming that the rate of increase is maintained in the coming years, the administrator can estimate the time remaining before the alarm values shown in Table 1 are reached. This can help the administrator to underline the need for rehabilitation works, which is one of the strongest arguments needed to obtain European funds.

**Table 4.** Cumulative settlement, in mm.

| | 1 | 2 | 3 | 4 | 5 | 6 | 7 | 8 |
|---|---|---|---|---|---|---|---|---|
| 26.03.2021 | -75,21 | -69,22 | -71,59 | -74,92 | -72,91 | -75,17 | -86,95 | -86,54 |
| 28.06.2021 | -92,31 | -86,47 | -87,62 | -89,32 | -92,14 | -93,29 | -107,74 | -108,34 |
| 24.08.2021 | -100,41 | -95,97 | -97,62 | -98,87 | -103,09 | -104,17 | -118,64 | -118,54 |
| 30.09.2021 | -105,91 | -101,07 | -102,62 | -103,92 | -107,14 | -109,79 | -123,74 | -124,39 |
| 28.10.2021 | -110,11 | -105,17 | -106,87 | -108,77 | -112,09 | -115,17 | -128,74 | -129,54 |
| 26.11.2021 | -113,31 | -108,42 | -110,57 | -111,92 | -116,14 | -119,42 | -133,65 | -133,88 |
| 27.01.2022 | -120,44 | -114,88 | -117,57 | -118,89 | -123,14 | -127,17 | -141,84 | -142,52 |
| 15.03.2022 | -126,31 | -120,98 | -122,62 | -123,97 | -127,64 | -131,72 | -146,84 | -148,54 |
| 19.05.2022 | -137,31 | -131,97 | -134,47 | - | -139,76 | -141,75 | -159,70 | -161,48 |

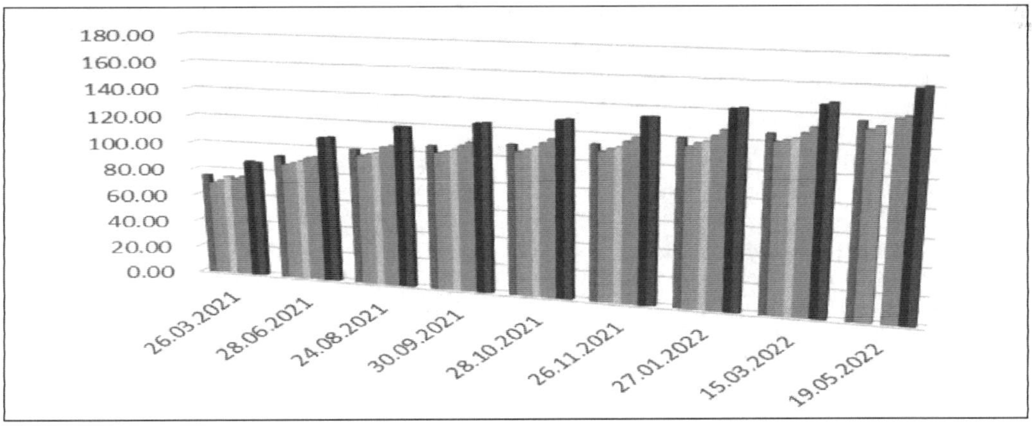


**Figure 11.** Cumulative settlement, in mm.

## 7. Conclusions

The widespread implementation of modern systems for monitoring the operational performance of bridges is an important step towards knowing the state of viability of the entire road network in operation. But this represents a major challenge for management and research in the field. This is why small steps are being taken, such as the search for technologies that can be most easily adapted to the physical and financial needs of any type of administration, be it local or national.

One such technology is the topographic tracking of the operational behaviour of bridges. It can be successfully implemented especially in the case of structures with subsidence or changes in the position of component elements, and is one of the most affordable technologies in the field.

In the present work, the research team has focused its attention on the implementation of a programme for monitoring the operational behaviour of the largest cantilever viaduct in our country by topographic means. Thus, it was possible to observe the importance of implementing such a technology at the structural level, and the collective also developed a warning scale concerning the imminence of a situation that would endanger the safety of users. This scale makes the administrator's work easier, reducing to just a few hours the reaction time needed to bring the structure to the optimal parameters for safe operation.

This work is part of an extensive research program developed by the Faculty of Construction and Building Services, "Gheorghe Asachi" Technical University of Iasi. The research is focused on the study and development of programs for automatic in-time monitoring of the bridges behaviour, programs that are easy to implement by any administrator, and that do not involve impressive costs.

## 8. References

[1] Dabous SA and Feroz S, 2020 Condition monitoring of bridges with non-contact testing technologies, *Automation in Construction* 116 103224

[2] Gimiga GR and Iriciuc SC, 2012 Workability of storable asphalt mixtures, measuring devices, *Bul. Polyt. Inst. Jassy* **58** 175-80

[3] Pachón P, Castro R., García-Macías E, Compan V. and Puertas E, 2018 E. Torroja's bridge: Tailored experimental setup for SHM of a historical bridge with a reduced number of sensors, *Engineering Structures* **162** 11-21

[4] Gómez-Martínez R, Sánchez-García R, Escobar-Sáachez JA, Arenas-García LM, Mendoza-Salas MA and Rosales-González ON, 2021 Monitoring two cable-stayed bridges during load tests with fiber optics, *Structures* **33** 4344-4358

[5] Scutaru MC, Ţăranu N., Comisu CC, Boacă G and Ungureanu D, 2018 Sensors for bridge structural health monitoring, *Bul. Polyt. Inst. Jassy* **64(68)** 25-39

[6] Beshr AAA, 2015 Structural deformation monitoring and analysis of highway bridge using accurate geodetic techniques, *Eng.* **7** 488-98

[7] Scutaru MC, 2019 Bridge monitoring using modern methods, *Roads and Bridges.* **187(256)** 18-22

[8] Boacă G and Ionescu I, 2010 Bridge instrumentation, *Intersections.* **7** 94-100

[9] Hann FEI 2012 *In situ behavior of constructions and their suitability for exploitation (*București), pp. 105

[10] https://www.topocadvest.ro/wp-content/uploads/2021/10/tcrp.jpg - Bing

[11] Hejduková JB 2017 *High-accuracy local positioning network for the alignment of the MU2E experiment (*Prague), pp. 32

[12] https://survey-solutions.ro/Tinte-reflectorizante-6x6-cm

IOP Conf. Series: Materials Science and Engineering 1304 (2024) 012023 doi:10.1088/1757-899X/1304/1/012023

# The Use of Artificial Neural Network for Predicting the Thermal Conductivity of Cement Based Mortar with Natural Zeolites

**G Covatariu[1], I O Toma[1], S M Alexa-Stratulat[1], D Covatariu[1], G Bunea[1] and I Olteanu[1]**

[1] The "Gheorghe Asachi" Technical University of Iasi, Faculty of Civil Engineering and Building Services, no. 1, Prof.dr.doc. D. Mangeron Street, Iasi, *Romania*

E-mail: ionut.ovidiu.toma@tuiasi.ro

**Abstract.** Zeolites, in their natural state, have been used in construction materials since ancient times. The pozzolanic activity of zeolites and their use as supplementary cementitious materials has been investigated over the past three decades. The indoor comfort provided by a modern life style comes at a cost of consuming enormous amounts of energy. Materials with improved insulation properties are continuously researched and the use of zeolites showing encouraging results. The paper presents experimental results obtained on different cement-based mortar mixes incorporating natural zeolites. The main parameters of the research were: substitution of cement and sand by natural zeolites, three different replacement percentages for cement and sand (10%, 20% and 30%) and the curing age of mortar specimens (14 days, 21 days and 28 days). An artificial neural network (ANN) was developed to predict the values of the thermal conductivity by taking into account parameters such as density, humidity and surface temperature of the mortar samples. The ANN was able to accurately predict the experimental results for the thermal conductivity of cement-based mortars with natural zeolites.

## 1. Introduction

Zeolites, in their natural state, have been used in construction materials since ancient times. Zeolites are hydrated alumino-silicates and, therefore, contain large amounts of reactive $SiO_2$ and $Al_2O_3$. These high concentrations give them pozzolanic properties similar to fly-ash, silica fume and metakaolin [1]. The pozzolanic activity of zeolites and their use as supplementary cementitious materials has been investigated over the past three decades with results in terms of durability and mechanical properties of cement-based materials varying over a wide range mainly due to zeolite source and replacement percentages [2,3]. Zeolites showed promising results in mitigating the autogenous shrinkage of high-performance concrete due to their ability to retain water and gradually release it in time [4,5] which helps to significantly reduce the shrinkage induced cracking. Moreover, recent studies showed the possibility of using zeolites in self-healing cement-based mortars, acting as a bacterial carrier for the self-healing process [6].

The indoor comfort provided by a modern life style comes at a cost of consuming enormous amounts of energy which, inevitably, leads to large $CO_2$ emissions. Although, recently, an increasing quantity of required energy comes from renewable sources, there is still a stringent need to develop new materials [7] with improved thermal insulation properties and to correctly assess their performance. Even in this aspect, zeolites showed promising results and research is still on-going [8].

The field of computational intelligence expanded significantly during the last decade, with leaps in the development of artificial intelligence methods. These methods gradually found their applications in a variety of scientific fields, civil engineering being one of them. Various mechanical, durability, cleaning properties of construction materials are now able to be accurately predicted by means of evolutionary approaches (neural networks) as a subset of computational intelligence. As such, artificial neural networks (ANN) are used in a wide variety of civil engineering problems, mainly related to concrete properties considering the highly heterogeneity of the material [9,10]. The development of a good ANN architecture could be beneficial in terms of long-term prediction of sought material parameters and, therefore, significantly reduce the waiting time. Of course, some long-term material characteristics could be determined by means of accelerated tests, e.g. durability issues, but those too require specialized equipment and a correct set-up of the experiment.

The paper presents experimental results obtained on different cement-based mortar mixes incorporating natural zeolites either as a substitute for cement or as a substitute for sand. As such, the main parameters of the research were: substitution of cement and sand by natural zeolites, three different replacement percentages for cement and sand (10%, 20% and 30%) and the curing age of mortar specimens (14 days, 21 days and 28 days). The thermal conductivity of all mixes was measured on specially cast cylindrical specimens. An ANN was developed, trained and used to predict the values of the thermal conductivity by taking into account parameters such as density, humidity and surface temperature of the mortar samples. The ANN was able to accurately predict the experimental results for the thermal conductivity of cement-based mortars with natural zeolites.

## 2. Materials and methods

### *2.1. Mix proportions*

The mortar mix consisted in a 1:4:0.6 mix, by volume, of constituent parts: CEM II/B-M 42.5R (rapid hardening composite cement), natural sand with a maximum particle size diameter of 4 mm and water. An initial water/binder ratio of 0.6 was considered but, depending on the consistency of the mortar, it was adjusted to obtain a mortar that could be easily cast in 70 × 30 mm (diameter × thickness/height) cylindrical steel molds. This was deemed necessary especially for mixes where cement was replaced by natural zeolites.

The natural zeolite used in this study was based on Clinoptilolite. It had a mean particle distribution of 29 μm, when used to replace cement, and a maximum particle size of 4 mm, when used as sand replacement. The mix proportions, expressed as mass quantities, are presented in [11], for cement replacement, and in [12], for sand replacement.

### *2.2. Methods*

A total of nine specimens were cast for each mix proportion, resulting in a total of 63 cylindrical specimens. As previously stated, the main parameters of the research were: cement or sand replacement by natural zeolites, replacement percentage and curing age. There were seven mix proportions considered: a control mix, without any zeolites, three mix proportions where cement was replaced by micronized zeolite (10%, 20% and 30% replacement, by volume) and three mix proportions where sand was replaced by natural zeolites (10%, 20% and 30% replacement, by volume). The mortar specimens were demolded after 24 hours from casting and placed in tap water for curing up to 28 days.

The thermal conductivity coefficient was measures by means of an ISOMET 2114 equipment (Figure 1) at the ages of 14, 21 and 28 days for all 7 mix proportions. Six individual measurements/readings were taken for each mortar sample, for a total of 1134 data entries. The cylindrical specimens were taken out of water, wiped by a cloth to remove the excess surface water and kept at room temperature for 12 hours before the thermal conductivity coefficient was measured. This would ensure that a surface dry condition of the specimens was obtained, without influencing the hydration process of cement inside the specimen, so that the obtained results would not be biased by the presence of water.

The specimens were measured, using a digital calliper with a precision of 0.1 mm, and weighted, with a precision of 0.1 grams, before each measurement of the thermal coefficient. This would help obtain of the bulk density of each specimen. The surface moisture content and temperature of the mortar sample were measured by means of a Merlin EVO CC equipment (Figure 2). Density, moisture content and temperature are influencing parameters for the thermal coefficient.

Using the Matlab and the Machine Learning Tool, the experimental data were used for creating a neural network (NN) feed-forward architecture. The data, 1134 values, was divided into 70% for training (851 values), 15% for testing (141 values) and 15% for validation (142 values). Several algorithms and configurations were considered for the learning process of the NN but the best solution found was a network with 12 neurons on the hidden layer and trained with the Levenberg-Marquardt algorithm. After learning the performance of the network were: Mean Squared Error MSE = 0.02 and Mu=1e-05. The schematic layout of the NN is presented in Figure 3. The MSE method has been frequently used and generally accepted as a method for assessing the performance of ANNs in different application fields [13,14]. The MSE is a metric that tells the distance, or how far apart, are the predicted values from the observed, in this case experimentally determined, values in a dataset. According to earlier studies the number of neurons in the hidden layer is an essential parameter influencing the accuracy of the ANN [15]. A too large number of neurons in the hidden layer may lead to overfitting while a too low number may result in underfitting issues [16].

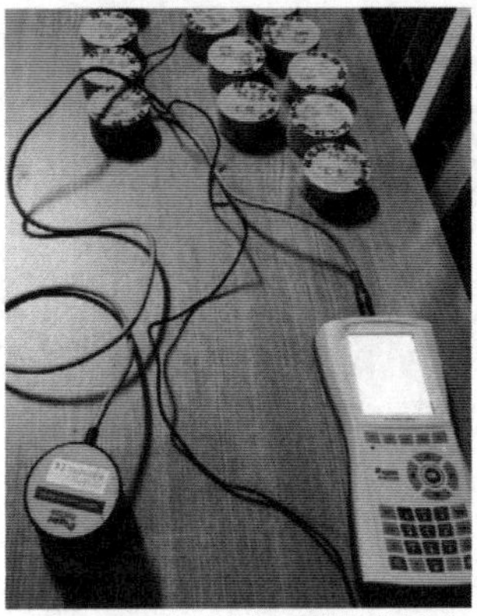

**Figure 1.** ISOMET 2114

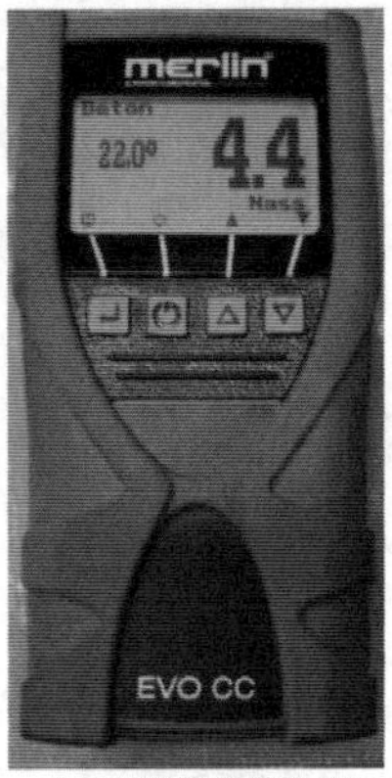


**Figure 2.** MERLIN EVO CC

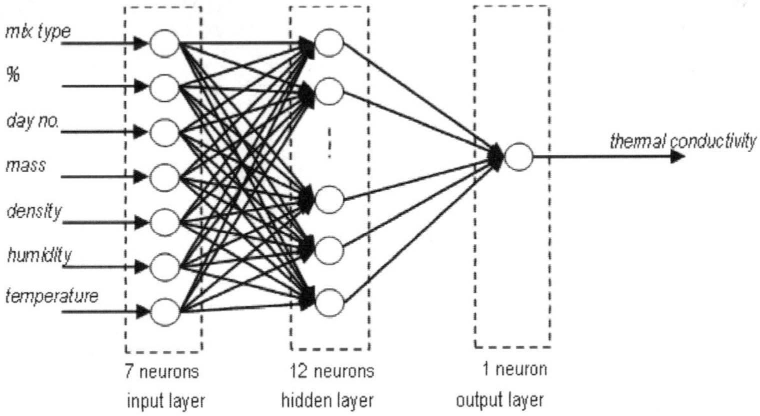


**Figure 3.** Schematic layout of the considered neural network

## 3. Results

### *3.1. Bulk density*

The bulk density of all considered mixes at different curing ages are presented in Figure 4. The values were obtained by averaging the measurements from all 9 samples belonging to each mix proportion. In Figure 4 ZP stands for Zeolite Powder, used to replace the cement and having a mean particle size of 29 μm, and ZA stands for Zeolite Aggregate. The numbers following each notation represents the replacement percentage, by volume.

It can be observed that increasing the replacement percentage, of either cement or aggregate, resulted in a decrease in the value of bulk density. This decrease was more significant in case of replacing sand (ZA series). None of the considered mix proportions exhibited large significant variation in the values of bulk density from one curing age to another. Similar trends were previously reported [12].

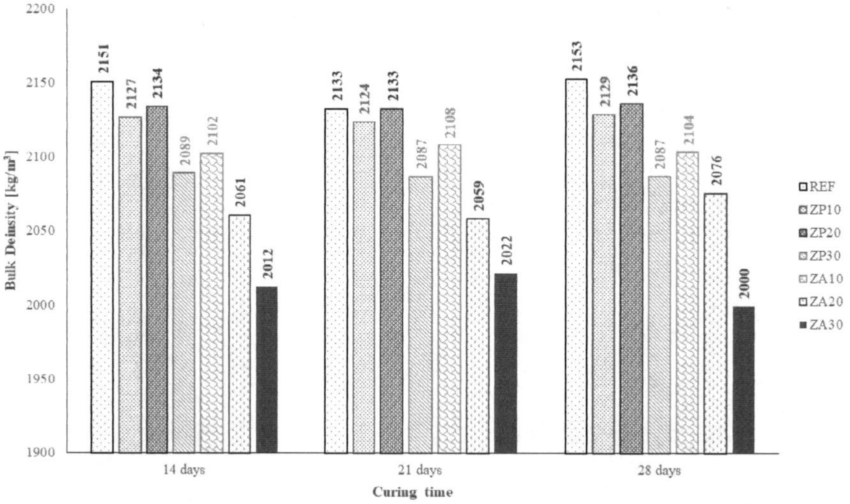


**Figure 4.** Variation of bulk density considering curing time and mix proportion

### *3.2. Surface moisture*

Figure 5 presents the surface moisture of each mix proportion as an average of nine individual measurements. It can be seen that ZA series exhibits the highest values of the surface moisture. Taking into account the porous structure of the zeolite aggregates and the fact that is absorbs the water during the mixing process to gradually release it over time, the obtained results are in line with previously reported data [17]. The ZP series exhibited the lowest values of surface moisture. The higher the replacement percentage of cement by natural micronized zeolite, the lower the surface moisture. This trend is opposite to the one recorded for the ZA series.

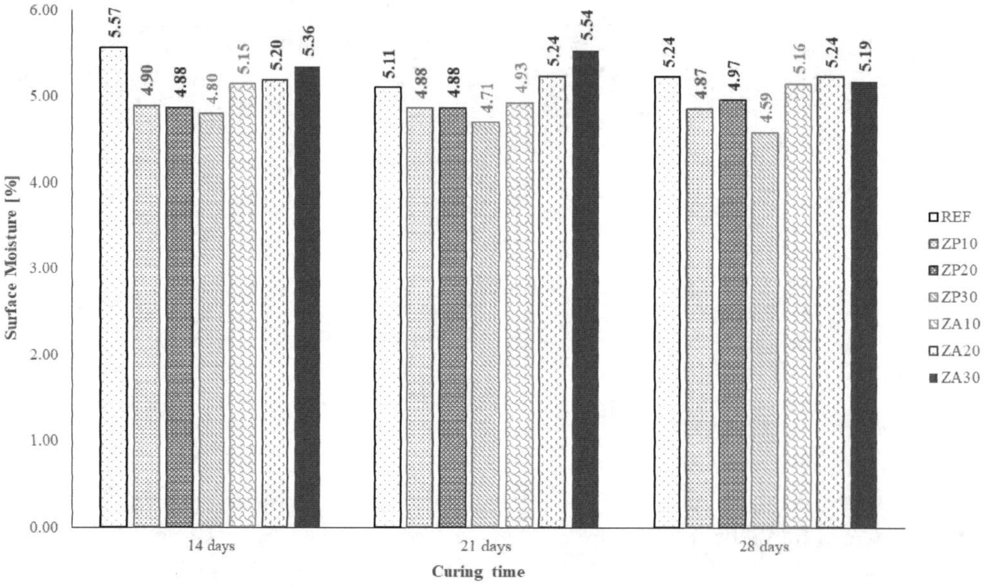


**Figure 5.** Variation of surface moisture considering curing time and mix proportion

### *3.3. Artificial neural network (ANN)*

In order to correctly design the structure of the ANN, a statistical analysis was run on the influencing parameters upon the values of the thermal conductivity. According to the generally agreed knowledge, the thermal conductivity of a material depends on its density, moisture content and temperature. Taking into account that all specimens were kept in identical laboratory conditions, both during curing and measuring procedures, the effect of temperature could be neglected at this stage of the research. Hence, a regression analysis was conducted to assess the influence of bulk density and surface moisture on the obtained values for the thermal conductivity. Both parameters are highly dependent on the zeolite content of the mix proportion.

Figure 6a presents the spread of the obtained values of the thermal conductivity as function of bulk density for all 63 specimens. A sixth-degree polynomial function was used for regression analysis resulting in a value of $R^2 = 0.677$. A similar plot is presented in Figure 6b in terms of surface moisture. The obtained value of $R^2$ was 0.094.

It was, therefore, concluded that the moisture content did not influence the measured values of the thermal conductivity. The moisture content would affect the internal structure of the material in the long term but it did not show significant influence at the age of 28 days. This conclusion was also supported by the fact that all specimens were kept in water, therefore they were subjected to the same curing conditions.

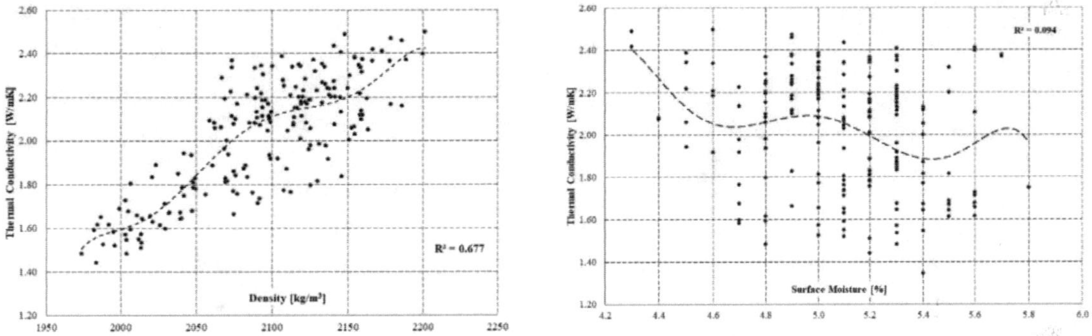


**a.** Influence of density **b.** Influence of moisture

**Figure 6.** Regression analyses on the influence of density and moisture on the thermal conductivity

Therefore, the ANN configuration presented in Figure 3 was considered. The best validation performance of the ANN is presented in Figure 7, whereas the obtained error histogram is shown in Figure 8. From the two graphs it could be concluded that the error spread was limited, indicating a good accuracy of the results provided by the ANN. A similar trend was observed during the testing and validation stages of the ANN by using the assigned sets of data.

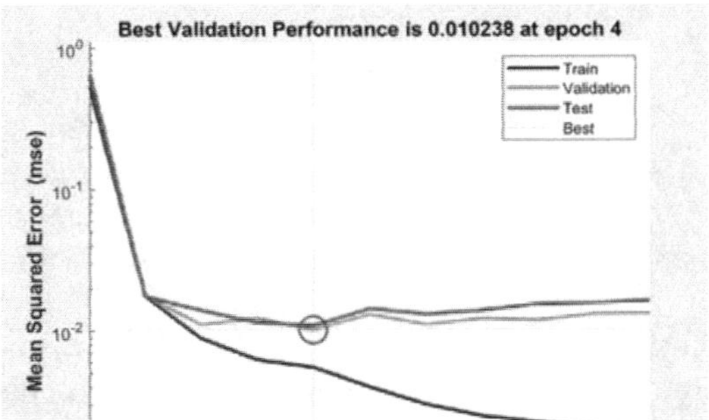


**Figure 7.** Best validation performance of the ANN

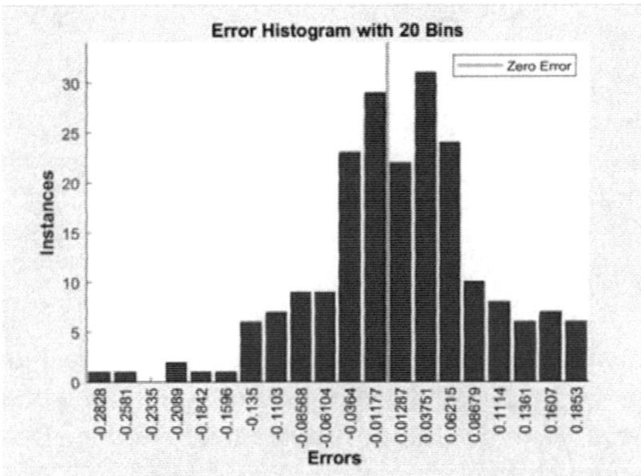


**Figure 8.** Error histogram

The overall performance of the ANN in predicting the values of the thermal conductivity of mortar with natural zeolite, at the age of 28 days, is presented in Figure 9. It can be seen that during the learning process, the output values of the neural network were close to the target values. During all stages of learning (training, validation and test), the training data showed a good fit to the experimental results. The correlation coefficient had a value of approximately 0.95. It was used to assess the accuracy of the predicted data versus the experimentally obtained data, for a linear regression model, during the training, validation and testing stages of the ANN. The best validation performance was found in fourth epoch and the program ran six more epochs, without overtraining [18].

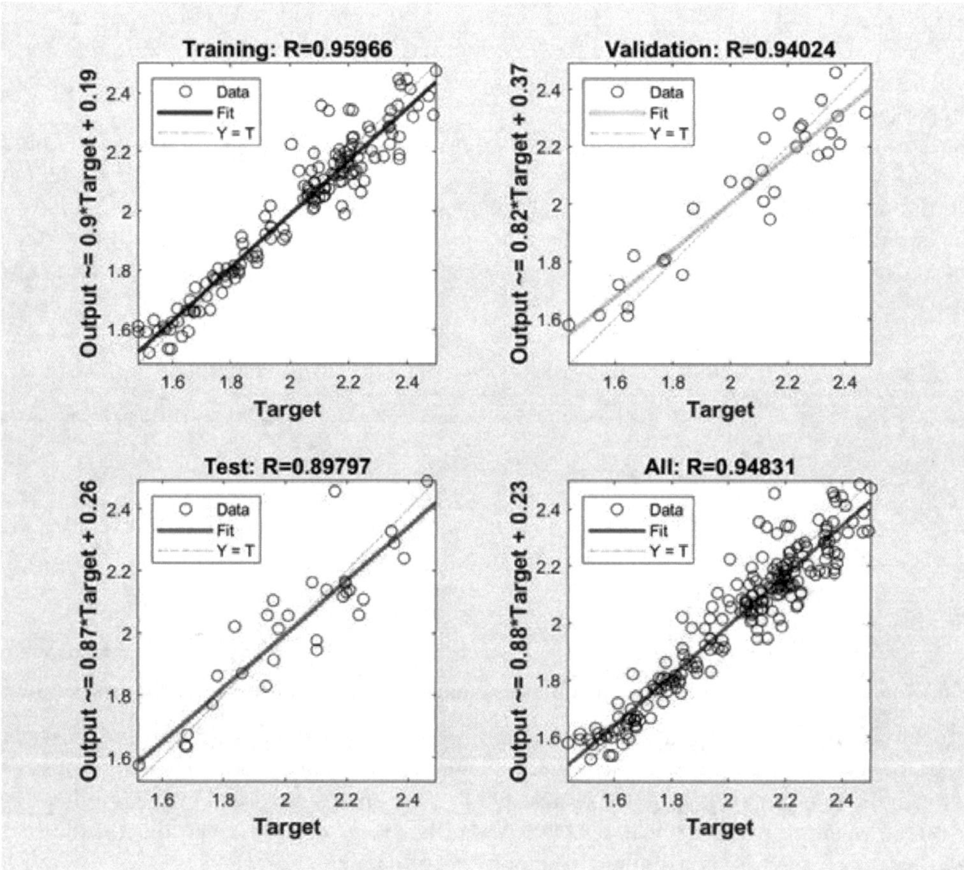

**Figure 9.** Overall performance of the ANN

## 4. Conclusions

The paper presents experimental results in terms of thermal conductivity obtained on different cement-based mortar mixes incorporating natural zeolites either as a substitute for cement or as a substitute for sand and the use of ANN to predict the obtained values. The parameters of the research work are: the replacement percentage, the curing age and the component of the mortar to be replaced by the natural zeolites.

Replacing cement by micronized zeolite results in bulk density values similar to the reference mix. At the same time, the moisture content is smaller owing to a denser structure of the obtained mortar. When sand is replaced by natural zeolite the bulk density decreases with the increase in the replacement percentage but the moisture content increases. This can be explained by the porous structure of the zeolite which leads to the formation of void inside the mortar and facilitates the absorption of water.

The developed ANN is able to accurately predict the experimental data in terms of values of the thermal conductivity. Based on regression analysis it is concluded that the bulk density plays a significant role in the outcome of the results whereas moisture content and temperature are less significant. Taking into account the fact that all specimens are subjected to identical curing conditions, the moisture content and temperature are expected to have smaller influence on the thermal conductivity. During all stages of learning (training, validation and test), the data obtained from the ANN shows a good fit to the experimental results. The correlation coefficient has a value of approximately 0.95. This means that the proposed structure and algorithm can be further expanded and used to predict the values of the thermal conductivity at later ages.

## 5. References

[1] Ranjbar M M, Madandoust R, Mousavi S Y, Yosefi S 2013 *Effects of natural zeolite on the fresh and hardened properties of self-compacted concrete* Constr. Build. Mater. 47 806–13
[2] Valipour M, Pargar F, Shekarchi M, Khani S 2013 *Comparing a natural pozzolan, zeolite, to metakaolin and silica fume in terms of their effect on the durability characteristics of concrete: A laboratory study* Constr. Build. Mater. 41 879–88
[3] Burris L E, Juenger M C G 2020 *Effect of calcination on the reactivity of natural clinoptilolite zeolites used as supplementary cementitious materials* Constr. Build. Mater. 258 119988
[4] Thang N C, Tuan N Van, Yang K-H, Phung Q T 2020 *Effect of Zeolite on Shrinkage and Crack Resistance of High-Performance Cement-Based Concrete* Materials (Basel). 13 3773
[5] Wang Q, Zhang J, Ho J C M 2020 *Zeolite to improve strength-shrinkage performance of high-strength engineered cementitious composite* Constr. Build. Mater. 234 117335
[6] Yuan H, Zhang Q, Hu X, Wu M, Zhao Y, Feng Y, Shen D 2022 *Application of zeolite as a bacterial carrier in the self-healing of cement mortar cracks* Constr. Build. Mater. 331 127324
[7] Yoo D H, Jeon I K, Woo B H, Kim H G 2021 *Performance of energy storage system containing cement mortar and PCM/epoxy/SiC composite fine aggregate* Appl. Therm. Eng. 198 117445
[8] Giosuè C, Pierpaoli M, Mobili A, Ruello M L, Tittarelli F 2020 *Multifunctional Lightweight Mortars for Indoor Applications to Improve Comfort and Health of Occupants: Thermal Properties and Photocatalytic Efficiency* Front. Mater. 7 1–10
[9] Golafshani E M, Behnood A 2021 *Predicting the mechanical properties of sustainable concrete containing waste foundry sand using multi-objective ANN approach* Constr. Build. Mater. 291 123314
[10] Shahmansouri A A, Yazdani M, Hosseini M, Akbarzadeh Bengar H, Farrokh Ghatte H 2022 *The prediction analysis of compressive strength and electrical resistivity of environmentally friendly concrete incorporating natural zeolite using artificial neural network* Constr. Build. Mater. 317 125876
[11] Toma I-O, Alexa-Stratulat S-M, Mihai P, Toma A-M, Venghiac V-M 2022 *Influence of elevated temperature on the early age properties of cement mortar with zeolite powder* Proceedings of 22nd International Multidisciplinary Scientific GeoConference SGEM 2022 pp 429–36
[12] Toma I-O, Olteanu I, Taranu G, Banu O-M, Alexa-Stratulat S-M 2022 *Early age mechanical properties of cement based mortar with zeolite aggregates* Proceedings of 22nd International Multidisciplinary Scientific GeoConference SGEM 2022 pp 405–12
[13] Guresen E, Kayakutlu G, Daim T U 2011 *Using artificial neural network models in stock market index prediction* Expert Syst. Appl. 38 10389–97
[14] Abiodun O I, Kiru M U, Jantan A, Omolara A E, Dada K V, Umar A M, Linus O U, Arshad H, Kazaure A A, Gana U 2019 *Comprehensive Review of Artificial Neural Network Applications to Pattern Recognition* IEEE Access 7 158820–46
[15] Huang X, Zhang J, Sresakoolchai J, Kaewunruen S 2021 *Machine Learning Aided Design and Prediction of Environmentally Friendly Rubberised Concrete* Sustainability 13 1691
[16] Yoon J Y, Kim H, Lee Y-J, Sim S-H 2019 *Prediction Model for Mechanical Properties of*

*Lightweight Aggregate Concrete Using Artificial Neural Network* Materials (Basel). 12 2678

[17] Kazemian M, Shafei B 2022 *Internal curing capabilities of natural zeolite to improve the hydration of ultra-high performance concrete* Constr. Build. Mater. 340 127452

[18] Tetko I V, Villa A E P 1997 A*n Enhancement of Generalization Ability in Cascade Correlation Algorithm by Avoidance of Overfitting/Overtraining Problem* Neural Process. Lett. 6 43–50

**Acknowledgement**

The research presented in this paper was conducted as part of the PN-III-P2-2.1-PED-2021-0677 research grant "Sustainable Concrete for Energy Efficient Buildings" funded by Executive Agency for Higher Education, Research, Development and Innovation Funding – UEFISCDI.

IOP Conf. Series: Materials Science and Engineering 1304 (2024) 012024 doi:10.1088/1757-899X/1304/1/012024

# Exploring the connection between adaptive architecture and artificial intelligence

**I Dohotariu**[1]
[1] "Gheorghe Asachi" Technical University of Iasi, "G.M. Cantacuzino" Faculty of Architecture, 3, Prof. Dr. docent Dimitrie Mangeron Blvd., Iasi, 700050, Romania.

E-mail: ionut.dohotariu@academic.tuiasi.ro

**Abstract.** Interactive adaptive architecture refers to the design of systems that can adapt to changes in their operating environment through interaction with users or other agents in the system. Artificial intelligence, on the other hand, refers to the ability of computer systems to perform tasks that would otherwise require human intelligence. Interactive adaptive architecture and artificial intelligence are two interconnected concepts that have evolved with the development of cybernetics. The development of interactive architecture based on the use of artificial intelligence will not mean the disappearance of architecture, as we know it, but rather a change in the way it is perceived. In the future, architecture will be participatory, with an active role of inhabitants in shaping and transforming architectural habitats, but there will always be a need for a moderator, the architect, to design and set the rules of the game for humans.

## 1. Introduction

The continuous and accelerated state of change, reshaping and adaptation of the contemporary ethos creates an environment conducive to the emergence of forms of architecture based on the absorption of research and technological innovations from other fields. These constant and accelerating changes also directly imply the transformation of some aspects of how we think about the design and use of contemporary architectural objects. The speed of technological change is increasing; therefore, observation and analysis of the phenomenon are necessary in order to direct the effects, in a controlled way, to an area where they benefit man. The evolution of contemporary lifestyles therefore implies a rethinking of the paradigm by which the individual, architecture, technology and artificial intelligence relate to each other.

Today, buildings are designed to perform certain functions, but they are not always perfectly adapted to the needs of the users. They are designed to function in a certain way, which cannot easily be changed once the construction work is completed. These inconveniences therefore equate to excessive consumption of energy and resources, as well as reduced potential user comfort. The possibility of defining an architecture responsible for mediating interactions (dialogue) between humans and computers, i.e. automating certain processes necessary for reconfiguring the interior space and the way in which it relates to the exterior space, are relevant arguments for deepening the relationship between artificial intelligence and the built environment. Adaptive architecture aims to address these issues by creating buildings that can adapt to changes from the environment and to the needs of users, offering them a higher level of comfort and satisfaction. Adaptive architecture denotes a form of architecture capable of permanent self-determination in the face of changes from the environment.

## 2. Evolution history of the concept

The design of architectural space is constantly changing. In the past, this approach focused mainly on aesthetic, functional and ergonomic aspects, on meeting functional needs and adhering strictly to standards for creating quality architectural spaces, designed and built with care and skill. However, buildings were conceived as static structures, without taking into account the possible changes that

IOP Conf. Series: Materials Science and Engineering 1304 (2024) 012024 doi:10.1088/1757-899X/1304/1/012024

may arise from their use, in contrast to adaptive architecture which focuses on the implementation of flexible principles of design and use that allow the building to adapt to changes induced by the user or the environment. This new approach is based on the use of technology and intelligent systems to update the functional layout and interior climate according to the needs and requirements of the user. The history of this trend dates back to experimental projects in the 1960s and 1970s that explored the possibility of creating flexible, adaptable and reprogrammable spaces according to the needs and requirements of users. This new direction is based on the emergence of a literature of anticipation (science fiction), such as the writings [1] of J G Ballard, or on the development of systematic and critical research on the problems related to possible future developments of the society-technology binomial (futurology).

Indications of the imminence and magnitude of the transition to ubiquitous technologies were first presented by the American futurologist and writer A Toffler, best known for his book Future Shock (1970). He published a trilogy in which he warns of the negative psychological implications of accelerating technological change, described as "the future shock" or "the disease of change" [2].

In his essay "The Question of Technology", M Heidegger undertakes a search for the essence of technology, how it has affected man. M Heidegger believes that the instrumental definition of technology does not accurately present the truth and therefore cannot fully explain the complexity of the phenomenon. In his view, modern technology becomes an exploitation of nature, and thus a pure transformation of nature into a stock of raw materials subject to human orders, in contrast to traditional craftsmanship, in which the essence of technology [3] is closely linked to the rhythm of nature and the principles of art. The philosophical approach to contemporary technology provides an analysis of how it influences our perception of the world, and by implication how it shapes our context.

But the scientific community's concerns about the relationship between man and technology go back much further. In an attempt to find new ways of examining the complexity of intelligent behaviour in humans, animals and machines, the philosopher and natural historian G Canguilhem illustrated in his 1947 lecture "Machine et Organisme" [4] (first published in 1952) a different view of the living world, hitherto explained only on the basis of mechanistic philosophy and biology. It presents a new biological understanding of technology and machines as extensions of the human organism, explaining their construction in terms different from the simple application of scientific theorems or processes of industrialization and typification. Thus, G Canguilhem proposes an understanding of machines by referring to the structure and functions of organisms, which, unlike the quantifiable, predictable, calculable and periodic movements of mechanical systems, are characterised by self-regulation, self-maintenance, reproduction or polyvalence. This new vision is embodied in an emerging discipline of that period in the USA - bionics. The ideas presented above anticipated the research agenda proposed by the new science of cybernetics, founded by N Wiener in the 1940s. Just one year after (1948) G Canguilhem, Wiener presented the idea of informational feedback and self-regulation of systems as the basis for understanding biological, mechanical and social processes, suggesting the possibility of creating a new type of machine [5] capable of responding to the environment similar to living organisms.

The possibility of building machines more complex than the human body was considered a pipe dream until the mid-20th century. The 1940s and 1950s saw a series of attempts by scientists to find new ways of exploring the complexity of intelligent behaviour in humans, animals and machines by looking at message exchange - feedback-based communication - in both biological and technological systems. Mathematicians, physiologists and engineers such as J von Neumann, N Wiener, C Shannon, W McCulloch and M Mead synthesised information theory, communication theory and machine control theory in a series of conferences (Macy [6]) that ran from 1946 to 1953. They established the principles of cybernetics, marking the beginning of an intellectual revolution supported by the emergence of new fields of research such as artificial intelligence[7] (1956), cognitive science[8] and information and communication networks[9]. The aim of cyberneticists was not to bridge the gap between organic and inorganic or natural and artificial scientifically; they did not propose a prototype

IOP Conf. Series: Materials Science and Engineering 1304 (2024) 012024 doi:10.1088/1757-899X/1304/1/012024

simulating a living organism, but a new type of machine, capable at a theoretical level of moving beyond the opposition between the two poles. The adaptability of this new type of machine is given by the implementation of the "feedback loops" needed to ensure the predictability of the system's behaviour. Their role is to control the entropy of the system through feedback - the future behaviour of the system will be governed by its previous states. Thus, it has the ability to learn from previous experiences: at a theoretical level, this type of machine is equipped with sensors that record both external conditions and operating history, feeding this information back as input to the next self-regulating loop.

Self-regulation characterised systems in the first cybernetic wave (1945-1960). In the biological world, this term was recognised as a property of homeostasis - the ability of living organisms to maintain an equilibrium by reducing energy losses in relation to changes in the environment. Researchers therefore believed that the homeostatic (first-order) cybernetic system was capable of reducing the loss of information from the system, while enhancing its control and stability through 'negative feedback'. Thus, a homeostatic machine will resist the tendency to disorder induced by informational noise in the environment (disturbances), maintaining the stability of its operating state by using feedback loops; in principle, their purpose is to correct all possible destabilising activities.

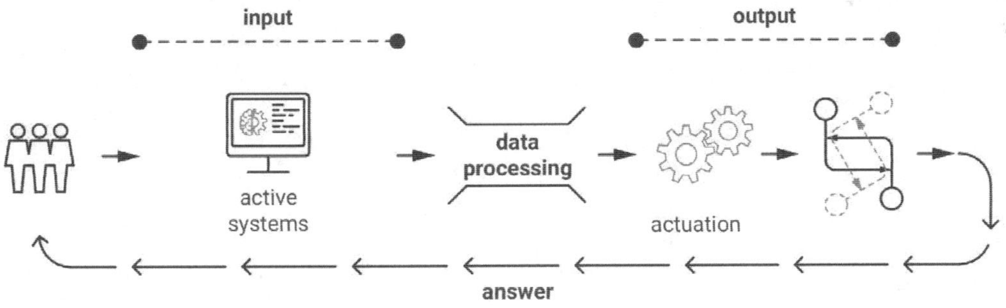


**Figure 1.** A feedback loop: improving the operating parameters of the adaptive architecture by adding new datasets.

Attempts to infuse architectural space with elements of cybernetics have resulted in defining a new potential quality of architectural space - adaptability. The phenomenon of architectural adaptability can materialise in many ways (through the use of removable elements, mechanical modules such as folding and sliding systems), but its character is not necessarily interactive, but rather static. From an architectural perspective, the first references to the phenomenon of adaptation through cybernetics or digital technologies can be found in the works of A Rabeneck, G Pask and C Eastman [10].

In 1969, A Rabeneck [11] proposed the incorporation of cybernetic technologies in order to create buildings that were as adaptable as possible to the future needs of their occupants; the expected result was to support both architects and future occupants by extending the life of a building. That same year, cybernetics pioneer G Pask outlined in his article "The Architectural Relevance of Cybernetics" [12] how cybernetics could induce evolutionary behaviour that responds and adapts to users' intentions and goals. Building on the earlier ideas of G Pask and A Rabeneck, C Eastman outlines the concept of "Adaptive-Conditional Architecture" [13], namely buildings capable of adapting according to feedback extracted from the environment (homeostatic model - first-order cybernetic system). It suggests that adaptability could be conceived as a self-adjusting feedback-based system, capable of dynamically managing spatial adjustments suited to the user's needs, similar to the way a boat constantly corrects its stability disturbed by changes in the environment (waves) [13]. In 1975, N Negroponte exemplified this principle by means of an experimental greenhouse [14] equipped with glazing operated by thermostat-type mechanisms (of the cognitive node type), capable of controlling, decoding and processing information about the comfort level of plants. The operating scheme of a cognitive node is based on storing (in a database) and processing the identified requirements/needs of individuals. This database is accessed by a learning engine and a decision engine (made by repeated comparisons with already existing data). The recombination of information from the two leads to actions. The cycle repeats itself once the system receives feedback from the environment/user and rewrites certain information in the database.

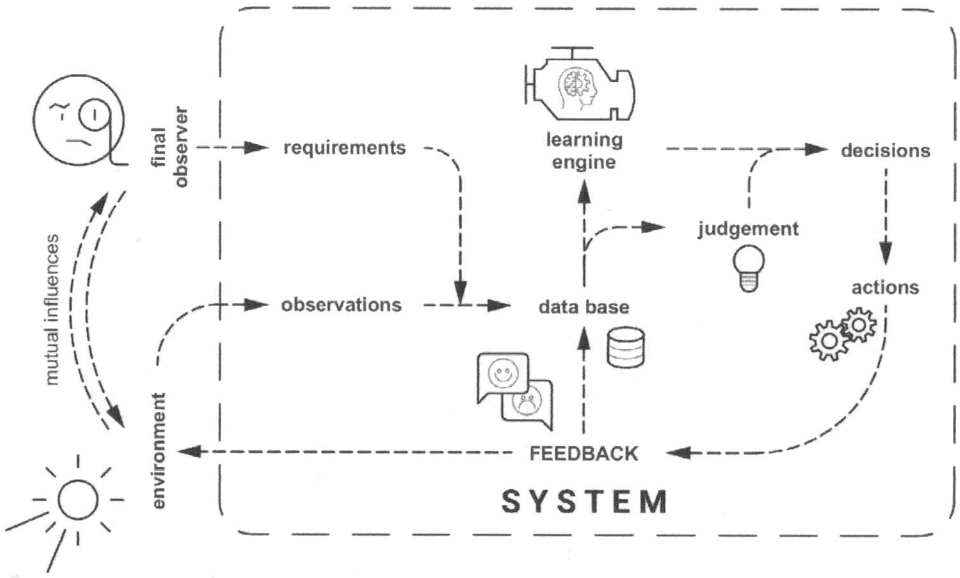


**Figure 2.** Conceptual architecture of a cognitive node

## 3. Building interaction

The Extrapolating the phrase "to live" we deduce that a living architecture is made up of animated buildings that actively react to external stimuli, are capable of reproduction (but most importantly, self-sustaining). A 'living' architecture is one that is actively involved in larger ecosystems in which it interacts with other architectures, but also with other 'living' natural or artificial entities. Imagining such an architecture may at first evoke dystopian visions.

Popular examples of science fiction culture (such as the books/movies "2001: A Space Odyssey"[15], "I, Robot" [16], "Ex Machina" [17] or "Blade Runner" [18] ) bring into the focus of consciousness architectural spaces controlled by artificial autonomous technological entities. In countless such scenarios, protagonists are immersed in environments saturated with interacting "smart" devices (from flying machines and robots to ubiquitous interactive displays, sensors and control systems) - a future that, in contemporary society dominated by smartphones, drones, cloud-computing, satellite constellations [19] and highly sophisticated military technology, does not seem too far away. Narratives presented in science fiction culture often involve the loss of human control over an artificial environment.

However, the influx of digital technology into architecture must be directed to the benefit of individuals and society as a whole. Cities 'populated' with 'smart buildings' that communicate with each other, able to modify and change their usage patterns, can generate constant feedback of information; thus 'smart and responsive' streets, walkways and public spaces could, through light, heat or cold, create microclimatic zones that attract and encourage certain human activities. The metabolizations behind these changes create different densities of activities that aim to redirect movement patterns. This 'intimacy' between the built environment, occupation of space and human movement can extend into the emergence of relationships between the urban/architectural landscape and society, energy and material resources. M Fox and M Kemp argue in Interactive Architecture [20] that interaction is a circular system - "'interacting', not just 'reacting'".

A clear definition of interactivity is fundamental. Even though the term 'interactive' is often overused in contemporary culture to define almost any IT product or service, it retains some ambiguity in common language. Interactivity is normally understood as the chain of interdependent bi- or multi-directional actions of two or more subjects.

The degree of interaction between subjects is directly proportional to the amount of information exchanged; inherently, such a relationship between interacting subjects is reciprocal, and they are altered in one way or another during the course of the interaction. Underlining the distinction that must exist between interactive and two-way or merely reactive communication, S Rafaeli proposes an interpretation of the notion of interactivity; according to him, interactivity is "the expression of the extent to which in a given series, any third (or subsequent) transmission (message) is related to the degree to which previous transmissions have referred" [21].

This seemingly complex description can easily be reduced to a classification of all types of communication that can occur between two or more subjects:

- when messages are sent in two ways, but are not linked, *bi-directional communication* occurs;
- if each of the messages sent only refers to a previously received message, we are dealing with *reactive communication*;
- if this response is more complex and each message sent refers to several messages received in the past by the respondent, we are dealing with an *interactive communication*; thus, interactivity can be defined as a measurable, positive variable (>0) only when *a communicator is able to relate his response* to the information received before the one that triggers his present response. The value of interactivity increases with the number of (previously received) messages used to formulate responses.

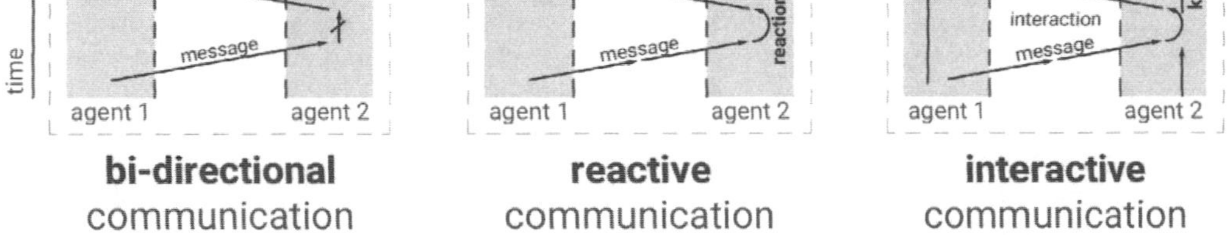


**Figure 3.** The information circuit - based on S. Rafaeli's diagram in "Interactivity: From New Media to Communication.", Sage Annual Review of Communication Research, 1988, 110-34

It is also important to stress the distinction between interactivity and interaction. If interaction is any action that takes place when two or more subjects interact, interactivity can be seen as the build-up of multiple interactions. Therefore, the notion of interactive (adaptive) architecture can be provisionally defined by buildings and constructed spaces capable of supporting an active dialogue with users and their environment - as T Jaskiewicz notes, the word "dialogue" is used metaphorically[22], as a reference to G Pask's Conversational Theory [23].

In the case of adaptive-interactive architecture, a dynamic action of a component could be seen not only as a reaction, but also as part of the overall interactivity scenario, therefore the distinction between interaction and reaction (i.e. the response of a system) is not clearly defined. Interactive architecture is a compelling alternative to simple building automation, with the potential to radically change the way buildings operate, are used and maintained. Applying the definition given by S Rafaeli, we deduce that automated spaces function purely reactively. Certainly, the automation of architecture has been an intermediate step in achieving adaptability through individual-built environment interaction, but at least two weaknesses can be identified in the automation of contemporary buildings from an interactivity perspective:

- Automation systems are rarely integrated into the architecture of buildings, despite the more than obvious benefits they bring (economic, ecological, aesthetic etc.).[24]
- Embedded systems have a linear logic, being built on chains of conditional statements ("if X, then Y, otherwise Z"). Although building automation is by definition flexible, it cannot respond favourably to a wide range of dynamic factors as the system can only develop a response based on procedures anticipated by the system designers. They work optimally as long as the building is used exactly as it was designed to be used; in short, the ability to adapt to unforeseen conditions is poor or non-existent. However, the frequency of local adaptations to specific predictable factors is high. Replacing the linear logic guiding the behaviour of these systems with the ability to reason and learn autonomously (via Artificial Intelligence) can lead to interactivity.

## 4. (r)evolution of self- discovery

Artificial Intelligence (AI) as a discipline has already penetrated countless fields. However, the integration of AI into architecture is still in its infancy, but the results are promising and represent a potential reshaping of the discipline. Architecture is undergoing a truly profound (r)evolution, and technology is certainly one of the factors that will have a strong impact on architecture[25]. Building design is already in a process of slow transformation both through the exploitation of new building technologies (e.g. 3D printing, laser scanning, the use of physico-chemical properties of natural materials - wood - or synthetic materials - shape memory polymers etc.) and through sustained efforts to develop appropriate software or the implications of artificial intelligence in the built environment.

The phrase "artificial intelligence" was coined by J McCarthy in 1955; it was needed to define the study of machines that exhibit and simulate "cognitive functions" (learning) or "problem solving" and the manifestation of intelligent behaviors[26]. J McCarthy even believes that any notion in the sphere of learning or intelligence can be described precisely enough for a machine to simulate it. In contemporary times, artificial intelligence is considered the ability of a system to correctly interpret external data and learn from it to achieve specific goals through flexible adaptation".

One of the common definitions of artificial intelligence embedded in the building fabric can be seen as the ability of architectural space to exhibit emergent autonomous behaviour that emerges through computational processing of large data sets - in this context, any architectural object that develops its own ability to know its context (through sensors) and to interact dynamically and actively with the world can be considered intelligent [20].

The applications of Artificial Intelligence (AI) and Machine Learning (ML) in architecture are based on principles from cognitive science , which is why it is sometimes also called "cognitive architecture". "Artificial intelligent architecture" / "cognitive architecture" is a system capable of performing a range of cognitive tasks, including problem solving, as well as learning about aspects of its own performance. Currently, the demand for artificial intelligence and its influence in the building and architecture sector is low compared to other industries. Therefore, logically, the implementation of AI in architecture will transfer proven knowledge from other fields such as telecommunications, finance and commerce[28].

Artificial intelligence can be used both to optimise the performance and efficiency of buildings and to provide new ways of interacting with users and the environment; however, artificial intelligence embedded in adaptive architecture will not replace classical architecture and its paradigm, but will create a new discipline that will evolve alongside classical architecture. This may lead to a more adaptive and intelligent architecture. Classical architecture, based on aesthetic and cultural principles deeply rooted in the history and tradition of architecture, will continue to evolve and adapt to new technologies and trends.

The ubiquity of interactive adaptive architectural systems based on artificial intelligence has a significant impact on the cultural structure of society and on the individual in terms of psychological impact. In terms of society as a whole, this architecture may lead to an increase in individualism and a decrease in direct interpersonal interactions. Artificial intelligence-driven technology can give people a lot of options and possibilities, but it can also give them less motivation and a reduced ability to connect with others authentically. In addition, this can lead to increased feelings of discomfort and stress among people who feel overwhelmed by the amount of information and options available, which can lead to anxiety and social isolation.

As far as the individual is concerned, interactive adaptive architecture based on artificial intelligence can lead to a number of psychological impacts. On the one hand, it can lead to increased self-esteem and sense of control over life. For example, using AI-based technology to manage and optimise individual health can lead to an increase in self-esteem and confidence in one's ability to control own health and well-being. On the other hand, AI-based interactive adaptive architecture can lead to decreased ability to make decisions and adapt to unexpected changes. Due to the fact that these technologies are designed to provide suggestions and recommendations, the individual may become dependent to them and lose the ability to make autonomous decisions.

Thus, complex interactive adaptive architecture based on artificial intelligence has a significant impact on the cultural structure of society and on the individual in terms of psychological impact. It is important to consider both the social and individual impact of using AI-based technology and to take measures to minimise its negative effects and maximise its positive effects.

## 5. Conclusions

Interactive adaptive architecture and artificial intelligence are two interconnected concepts that have evolved in parallel with the evolution of cybernetics. Interactive adaptive architecture refers to the design of systems that can adapt to changes in their operating environment through interaction with users or other agents in the system. This approach has been developed mainly in the development of control systems and has been influenced by ideas from cybernetics, an interdisciplinary field concerned with the study of control and communication systems in living organisms and machines. Artificial intelligence, on the other hand, refers to the ability of computer systems to perform tasks that would otherwise require human intelligence, such as image recognition or natural language processing. This approach has been influenced by ideas from cybernetics and information theory, as well as developments in psychology and neuroscience.

The evolution of cybernetics has also influenced the evolution of interactive adaptive architecture and artificial intelligence through the development of models and algorithms inspired by biology and the study of complex systems. Thus, models of artificial neural networks and machine learning algorithms have been developed that allow systems to learn and adapt to changes in their operating environment. Natural language processing and image recognition technologies have also been developed, allowing systems to interact with users and interpret information from their operating environment.

Adaptive architecture and artificial intelligence present both considerable limitations and opportunities. Implementing artificial intelligence in architecture involves generating systems capable of learning and adapting to new situations, similar to the way the human brain develops and adapts. However, the whole phenomenon has a number of potential weaknesses:

- Computing power: adaptive architecture that bases its operation on the use of artificial intelligence requires considerable computing power to process large volumes of data and perform complex calculations;
- Training data: to obtain accurate and efficient results, artificial intelligence requires large amounts of diverse data to be trained. In certain domains or for more specific tasks, it can be difficult to create a sufficiently large, high-quality database.

However, we can also point to a number of promising predictions about the future evolution of the connections between adaptive architecture and artificial intelligence:

- Increasing performance: technological advances and increasing computing power imply increasing efficiency and performance of adaptive architecture. Systems with integrated AI will be able to better process and understand complex information sets;
- Increased interpretability: the evolution of AI and adaptive architecture will lead to the development of methods and techniques to improve the interpretability of AI. This will allow better understanding and justification of decisions made by AI - essential in areas such as health, security or housing;
- Integration of other technologies: adaptive architecture and artificial intelligence can be combined with other technologies such as blockchain, IoT or virtual/augmented reality.

The development of interactive architecture is not equivalent to the disappearance of architecture as we know it. To some extent, much of today's architecture is already dynamic, interactive and adaptive. However, the speed and efficiency of current transformations are low. The development of interactive adaptive architecture is therefore equivalent to increasing the tempo of transformation. The revolution is therefore not in classifying adaptive architecture as a new type of architecture, but in changing the perception of any architecture as complex and adaptive. At the same time, the above statement does not imply that rapid spatial adaptation will necessarily become ubiquitous. On the contrary, scenarios can be imagined in which rapid architectural transformation is not and will never be necessary. Nevertheless, the complex adaptive view of architecture retains its validity and relevance. Adaptive architecture does not equate to the dystopian image in which future cities are full of dynamically changing or moving buildings. On the contrary, this development can equally guide us towards the development of seemingly static but actively optimising urban settlements only when these changes are explicitly desired by the city's inhabitants.

In conclusion, interactive adaptive architecture and artificial intelligence have been influenced by developments in cybernetics and information theory, as well as the study of complex systems in biology and other fields. The main common feature of all interactive spaces is the active role of inhabitants in shaping and transforming architectural habitats. Beyond any doubt, the future of architecture is a participatory one, focused on the ability of people to shape and improve the spaces in which they live their lives; but we should not assume that these can be created by users alone - there will always be a need for a moderator to design and set the rules of the game for people, while ensuring the development of enabling scenarios. It is up to the architect to take on this role.

**ACKNOWLEDGEMENTS**
This paper was realized with the support of project EFECON – ECO-INNOVATIVE PRODUCTS AND TECHNOLOGIES FOR ENERGY EFFICIENCY IN CONSTRUCTION, POC/71/1/4 - Knowledge Transfer Partnership, Cod MySMIS: 105524, ID: P_40_295, Project co-financed by the European Regional Development Fund.

## 6. References

[1] Ballard J G 2009 *The Complete Stories of J.G. Ballard* (New York: W.W. Norton & Co) pp 305–20
[2] Toffler A 1973 *Şocul Viitorului* (Bucharest: Editura Politică)
[3] Heidegger M 2011 *Originea Operei de Artă* ed T Kleininger and G Liiceanu (Bucharest: Humanitas) pp 120-69
[4] Canguilhem G 1952 *La Connaissance de la Vie* (Paris: Hachette) pp 125-59
[5] Wiener N 2007 *Cybernetics or Control and Communication in the Animal and the Machine.* (Cambridge: MIT Press) p 212
[6] Stoud J *et al.* 2003 *Cybernetics - the Macy Conferences 1946-1953: the Complete Transactions* ed C Pias (Zürich Berlin: Diaphanes)
[7] Myers A *Stanford's John McCarthy, seminal figure of artificial intelligence, dies at 84* [Internet]. 2011 [cited 2023 Apr 1]. Available from: https://web.archive.org/web/20150517012355/http://news.stanford.edu/news/2011/october/john-mccarthy-obit-102511.html
[8] Thagard P *Cognitive science* [Internet]. Stanford University; 2007 [cited 2023 Apr 1]. Available from: https://plato.stanford.edu/archives/fall2008/entries/cognitive-science/
[9] *A short history of the internet* [Internet]. 2020 [cited 2023 Apr 2]. Available from: https://www.scienceandmediamuseum.org.uk/objects-and-stories/short-history-internet
[10] Fox M 2010 *Catching up with the past: a small contribution to a long history of interactive environments* Digitally-Driven Architecture 6 14 https://doi.org/10.7480/footprint.4.1.716
[11] Rabeneck A 1969 *Cybermation: A Useful Dream* Architectural Design 9 497-500
[12] Pask G 1969 *The architectural relevance of cybernetics* Archit Des. 7/6 494-6 https://cmusyntheticecologies.files.wordpress.com/2014/01/pask_gordon_the-architectural-relevance-of-cybernetics.pdf
[13] Eastman C 1971 *Adaptive conditional architecture* (Pittsburgh: Carnegie-Mellon University)
[14] Negroponte N 1975 *Soft architecture machines* (Cambridge: MIT Press) pp 126-9
[15] Kubrick S 1970 *2001: A Space Odyssey* (Metro-Goldwyn-Mayer: Stanley Kubrick Productions)
[16] Proyas A 2004 *I, Robot* (Twentieth Century Fox, Mediastream Vierte Film GmbH & Co. Vermarktungs KG, Davis Entertainment)
[17] Garland A 2014 *Ex Machina* (Universal Pictures, Film4)
[18] Scott R 1982 *Blade Runner* (The Ladd Company, Shaw Brothers, Warner Bros.)
[19] McDowell J 2020 *The Low Earth Orbit Satellite Population and Impacts of the SpaceX Starlink Constellation* Astrophys J. 892(2):L36.
[20] Fox M and Kemp M 2009 *Interactive architecture* (New York: Princeton Architectural Press) p 256
[21] Rafaeli S 1988 *Interactivity: From New Media to Communication* Sage Annu. Rev. Commun. Res. pp 110-31
[23] Jaskiewicz T 2013 *Towards a methodology for complex adaptive interactive architecture* (Delft: TUDelft)
[24] Pask G 1976 *Conversation theory: Applications in education and epistemology* (Amsterdam; Oxford; New York: Elsevier)
[25] Davila Delgado J *et al.* 2019 *Robotics and automated systems in construction: Understanding industry-specific challenges for adoption* J Build. Eng. 26 100868.

IOP Conf. Series: Materials Science and Engineering 1304 (2024) 012024 doi:10.1088/1757-899X/1304/1/012024

[26] Chaillou S 2019 *AI + Architecture. Towards a New Approach* [Internet]. Available from: http://stanislaschaillou.com/articles.html
[27] Bishop C 2016 *Pattern recognition and machine learning* (New York: Springer)
[28] Blanco J, Fuchs S, Parsons M and Ribeirinho M 2018 *Artificial intelligence: Construction technology's next frontier* [Internet]. Available from: https://www.mckinsey.com/capabilities/operations/our-insights/artificial-intelligence-construction-technologys-next-frontier

# Fiber reinforced polymers in pervious concrete - a state of the art review

**R D Chiriac[1] , Z I Kiss[2]**
[1]PhD Student, Technical University of Cluj-Napoca, Cluj-Napoca, Romania
[2]Prof.Dr.Eng, Technical University of Cluj-Napoca, Cluj-Napoca, Romania

E-mail: raul.chiriac@dst.utcluj.ro

**Abstract.** Porous concrete, entitled also as permeable or pervious concrete, is a type of concrete that has interconnected voids that allow water and also air to go across it. It is widely used in the construction of pavements, parking lots, and sidewalks, among other applications, because of its ability to reduce stormwater drainage, better quality of the water, and decrease urban heat island effects. The addition of fibers to porous concrete has been the subject of numerous studies in recent years, with researchers examining the effects of diversified types and amounts of fibers on the material's properties. Fiber reinforcement can enhance the strength, durability, and ductility of porous concrete, making it more suitable for use in structural applications. Polymer fibers, in particular, have shown promise in improving the properties of porous concrete. They can increase the material's resistance to cracking, improve its flexural strength, and enhance its heat-absorbing property. However, the characteristics of polymer-reinforced porous concrete are not yet fully understood, and more research is needed to determine the optimal types and amounts of fibers for specific applications. Despite the potential benefits of fiber-reinforced porous concrete, there are also some challenges associated with its use. The addition of fibers can increase the material's cost and complexity of production, and it may also affect its permeability and drainage capacity.

Fiber-reinforced porous concrete represents an exciting area of research and development in the field of civil engineering. While the addition of fibers can enhance the material's properties, more research is needed to determine the optimal types and amounts of fibers for specific applications. By addressing these challenges and continuing to refine the material's properties, fiber-reinforced porous concrete has the potential to expand its use in a large spectrum of applications, including structural engineering, transportation infrastructure, and environmental protection.

Porous concrete is a type of concrete that has high porosity, which allows water to go across it with ease. It is also known as permeable concrete or pervious concrete [1]. It was created in the 80's, in Japan, as an environmental friendly material, and then it spread around the world, including Europe [1]. One of today's main general research objective is to reduce pollution, waste of materials and an environmental friendly material process industry [13]. This type of concrete can be used instead of classic concrete and it allows us to recycle aggregates and to diminish energy consumption [2].
One of the challenges of using porous concrete is its poor workability, which makes it difficult to produce precast products and apply it without the use of vibration equipment. To overcome this challenge, a ordinary permeable concrete can be produced by using a uniform size of aggregate and a low water-cement ratio. This helps to improve workability and compaction [1]. If we add too much water, the concrete mix will segregate, while if we have too little amount of water, the mixture will form balls and will not mix properly. So, the mechanical properties of porous concrete depend on the mix ratios: water-cement, aggregate size and distribution, type and dosage of admixtures [7]. Porous concrete typically has low compressive strength and stiffness compared to conventional concrete, but it can still be used for his structural performances in some applications [7].

## 1. Applications examples

### *1.1. Retaining walls*
Retaining walls are a common feature in urban development projects, mostly in areas where we have uneven land and slopes. They are design to hold back soil and prevent it from sliding down, as well as to support, if necessary, a structure built on top of them [5]. Traditionally retaining walls are constructed using masonry, concrete, stones and bricks, and can be designed also to satisfy the aesthetic needs of the surrounding landscape [5].

### *1.2. Permeable pavement*

Green infrastructure involves the use of natural or engineered systems to manage storm-water its source. Cases of green infrastructure include vegetative roofs, construction elements that collects rain water, permeable pavements, and bio swales. These systems help to capture, store, and infiltrate storm-water on site, thereby reducing the amount of runoff that enters the urban drainage. In to these solutions, urban planning and design can also participates in decreasing the impact of impermeable surfaces on the water cycle. Strategies such as low-impact development, which emphasizes preserving natural features and minimizing the disturbance of existing landscapes, can help to reduce the amount of impermeable surface area in urban areas and promote more sustainable water management practice [10].

### *1.3. Parking lots and driveways*
Pervious concrete allows rainwater to infiltrate into the ground, reducing stormwater runoff. This helps to replenish groundwater supplies and reduces the strain on drainage systems.
Since water passes through pervious concrete, it helps prevent the accumulation of surface water on parking lots and driveways. This can improve safety by reducing the risk of hydroplaning and ice formation.

### *1.4. Landscaping projects and gardens*
Allowing water to infiltrate through its porous structure is particularly beneficial in garden walkways, as it helps prevent water runoff and allows rainwater to reach the soil beneath. This promotes healthy plant growth by providing direct access to water and reducing the need for additional irrigation. By enabling water to penetrate the surface and reach the underlying soil, pervious concrete helps prevent soil erosion. Traditional impervious surfaces, such as solid concrete or asphalt, can contribute to erosion by channeling rainwater away from the soil. Pervious concrete acts as a natural erosion control measure, retaining water and reducing the risk of soil loss.

### *1.5. Greenhouse floors and foundations*
By using pervious concrete for greenhouse floors and foundations, we can contribute to sustainable water management and minimize the strain on local drainage systems. Designed and installed correctly, it can provide sufficient structural support for greenhouse foundations and floors. It has the capacity to withstand moderate loads and offers stability for the greenhouse structure, equipment, and plant containers. However, it's essential to ensure proper engineering and design to meet the specific load-bearing requirements of the greenhouse.

### *1.6. Artificial reefs*
Pervious concrete has been considered for use in artificial reefs due to its potential benefits in marine environments. While traditional concrete structures are often used for reef construction, porous concrete offers some advantages in certain scenarios. It provides a rough and textured surface that can serve as an attachment point for marine organisms such as corals, sponges, and other invertebrates.

The porosity of the concrete allows water to flow through, which can create favorable conditions for the settlement and growth of marine life, enhancing the ecological value of the reef.

## 2. Porous Concrete Properties

If we increase the quantity of reused concrete aggregate in porous concrete can have an effect on its workability, from medium to low, which is a measure of its ease of placement, consolidation, and finishing. The workability of porous concrete is influenced by many elements, such as the type and quality of the RCA, the mix design of the concrete, and the method of placement and compaction [2].

Reinforcing fibers in pervious concrete can help enhance its mechanical properties, reduce cracking, improve durability, and increase overall structural performance. However, it's crucial to consult with concrete experts or engineers experienced in pervious concrete to determine the most suitable fiber type and dosage for your specific application.

Porous concrete reinforced with polypropylene fibers is a common approach to enhance the performance and durability of the material. [11] Polypropylene fibers are synthetic fibers that are often used as reinforcement in pervious concrete due to their improved properties. These fibers help control cracking in porous concrete. The fibers act as micro-reinforcement throughout the material, distributing stress and reducing the formation and propagation of cracks. This improved crack control helps maintain the structural integrity of the pervious concrete and minimizes potential damage caused by shrinkage or external loads.

The addition of polypropylene fibers can improve the durability of porous concrete. The fibers enhance the material's resistance to weathering, abrasion, and impact. Polypropylene fibers are non-corrosive, making them particularly suitable for applications where the pervious concrete may be exposed to corrosive environments or chemical exposure. Unlike steel fibers, polypropylene fibers do not rust or deteriorate over time, ensuring long-term performance and reducing the risk of structural degradation.

Using steel fibers in pervious concrete is another approach to reinforce the material and enhance its mechanical properties. Steel fibers provide additional strength, durability, and also crack resistance to pervious concrete. This type of fibers enhance the tensile strength of pervious concrete, making it more resistant to cracking and improving its overall structural integrity. The fibers act as reinforcement throughout the material, distributing stress and resisting tensile forces that can lead to crack formation and propagation. [13]

Incorporating steel fibers in pervious concrete increases its load-bearing capacity. The fibers help distribute loads and enhance the flexural and compressive strength of the material. This can be particularly beneficial in applications where the concrete may be subjected to heavy loads or vehicular traffic, such as in parking lots or driveways.

Glass fibers can also be used to reinforce porous concrete. They are lightweight compared to other types of fibers, such as steel fibers. This lightweight characteristic makes them suitable for applications where weight reduction is desired, such as in elevated or lightweight structures. Also they have good thermal and electrical insulation properties. It can help reduce thermal conductivity in porous concrete, providing insulation against heat transfer. This can be advantageous in applications where temperature control is essential, such as in buildings or infrastructure.

Glass fibers can also enhance the visual appeal of porous concrete. They can create a unique and attractive appearance when exposed at the surface, providing a decorative element to the material.

Natural fibers offer environmental advantages as they are renewable, biodegradable, and derived from sustainable sources. Using natural fibers in pervious concrete aligns with sustainable construction practices and reduces the reliance on non-renewable resources. We can provide some degree of thermal insulation to pervious concrete. They help reduce thermal conductivity and can contribute to temperature regulation in structures or surfaces made with pervious concrete.

Porosity, permeability, and compression resistance are important variables that can influence the performance of porous concrete. Porosity refers to the amount of void space or pores in the concrete, while permeability refers to the ease with which water can pass through the material [3]. Taking into consideration both compression resistance and permeability, a specific requirement regarding the porosity can be selected for porous concrete. Based on the analysis presented, the optimal porosity range appears to be between 20% and 25%. Within this range, the compression resistance should be in the range of 10 to 20 MPa, and the permeability should be from 8 to 10 mm/s. This range can help to ensure that the porous concrete has the necessary strength and water permeability properties for its intended application [3].

The characteristics of cement paste and the properties of porous concrete were studied through experimental tests [20]. The results showed that the properties of the cement paste depend on the A/C ratio, the additives and the time of mixing the paste. Cement paste with high viscosity and consistency for the preparation of pervious concrete can be produced from an A/C ratio between 0.20 – 0.25, by adding 1% superplasticizer additive and sufficient mixing. Good quality porous concrete with porosity 15-25% and compressive strength of 22-39 MPa is obtained using cement paste with a consistency of 150-230 mm and by vibrating the surface for 1 second with an energy of 90 $kNm/m^2$.

The same authors study how the resistance of the binder and the size of the aggregates influence the compressive strength and porosity of the porous concrete [21]. To study how it influences the resistance of the binder, porous concrete test tubes were made in which 4 types of binders were used, having resistors between 30-130 MPa. The test tubes subjected to experimental tests had a porosity between 15- 35%. To study how they influence the size of the aggregates, porous concrete test tubes were made in which aggregates with the following nominal dimensions were used: 5.13 and 20 mm. The results showed that for the same type of aggregate the rates of decrease of mechanical resistances due to the increase of porosity were the same regardless of the strength of the binder used. Experimental tests showed that the mechanical strength of pervious concrete is influenced by the size of the aggregates. The rate of decrease of the mechanical resistances of porous concrete with small aggregates is higher than that of concrete with large aggregates. For the same porosity, the mechanical resistances of porous concrete with large aggregates have values higher than those of concrete with small aggregates.

In the paper "Evaluation of high-performance porous concrete properties" [15], the main purpose of the experimental tests was to evaluate the mechanical properties of high-strength porous concrete 3 sorts of large aggregates were used in the optimal porous concrete mixtures along with a suitable quantity of water and additives that increase the viscosity and thickness, and the porous concrete analyzed did not require vibration and special treatments. Experimental tests performed: compaction, consistency, porosity, permeability and mechanical tests such as bending strength and maximum strength. It was proposed to perform a mechanical test to study the effect of water reducing and thickening additives on the compaction of high-strength porous concrete, and the purpose of the proposed test was to evaluate the properties of porous concrete reinforced from the point of view of practical applications. The rate of maximum resistance has been assessed at 1, 3, 7, 14 and, respectively, 28 days since the implementation, at the standard temperature of 20 degrees Celsius and the relative humidity of 60%. Following the tests performed, the high-strength porous concrete has good workability, has not segregated or bled/ exudate. High-strength porous concrete has reached its maximum mechanical strength at a higher speed than classic porous concrete.

The capacities of porous concrete regarding the compressive resistance and the water purification capacity were highlighted by Sung-Bum Park and Mang Tia in the work "An experimental study on the water-purification properties of porous concrete" [16]. Two varieties of large aggregates were utilized in the production of porous concrete for test tubes: 5-10 mm and 10-20 mm, and 3 ratios were used in the preparation of the mixture A/C-aggregate, for each gravels, as follows: 30%, 40% and 50% respectively. The capacity of porous concrete to purify water is analyzed by the amount of organisms attached to its surface, as well as by the amount of phosphorus and nitrogen removed. The amount of organisms is analyzed indirectly by the amount of dissolved oxygen consumed by them. Experimental tests found that the compressive strength of porous concrete is higher when the size of the aggregates and the A/C -aggregate ratio are small. Chemical analysis noted that porous concrete made from gravels with small and high porosity aggregates has a superior ability to remove phosphorus and nitrogen from water. This is due to the large specific surface of the porous concrete in question.

How the consistency of the cement and the type of aggregate influence the mechanical and acoustic characteristics of the porous concrete has been studied by experimental tests [17]. Porous cement concretes were prepared with the following consistency levels in turn: 80%, 110% and 140% respectively. Aggregates from the following gravels were used in the porous concrete mixture: 8-13 mm and 13-19 mm respectively and light aggregates from the following gravels: 4-8 mm, 8-12 mm and, respectively 12-19 mm. How additives affect the physical and sound-related properties of porous concrete has also been studied. The acoustic characteristics of porous concrete were analyzed by studying the configuration of the layers that make up the test tubes, those test pieces being made of a single layer and 2 layers, respectively. The parameters used to evaluate the test tubes were: porosity, compressive strength and sound absorption coefficient. Based on the results of experimental studies, a porous concrete is created that absorbs noises with an absorption coefficient of approximately 1 with the minimum value of the sound absorption coefficient for 2-layer test tubes greater than 0.60 at a frequency of 400 Hz or more and taking into account the required tolerances [17].

A research project has been carried out to create a special porous concrete that is fragmented into small pieces in shock actions, but which retains a resistance reserve. This type of porous concrete is intended for protective walls and buildings that were designed to enhance safety [22]. Porous concrete with high compression resistance has been prepared by modifying recipes and compaction techniques. The test tubes made from the proposed mixture were analyzed by mechanical tests on a meso and macro scale, by CT, by microscopic analysis and by X-ray analysis. When making this type of porous concrete, in addition to the large mechanical resistors, the high porosity is essential for the formation of multiple cracks that lead to the fragmentation of the concrete. Experimental tests showed that the degree of compaction and the dimensions of the aggregates have a predominant influence on the mechanical properties of the special porous concrete. The influence of silica dust on the mechanical properties of the proposed porous concrete does not was significant [22].

How porosity affects compressive strength was also analyzed by creating a mathematical model that would characterize it. The mathematical model involves the analysis of empirical results and theoretical derivatives [18]. The mathematical model for porous concrete is derived from Griffith's Theory. And the results are similar to the results of the experimental tests. The paper showed that the proposed mathematical model can determine the compressive strength of porous concrete depending on its porosity. The paper showed that the proposed mathematical model can determine the compressive strength of porous concrete depending on its porosity.

Also through a mathematical model, the phenomenon of carbonation of porous concrete was described. [19]. It is based on physic-chemical processes in which the transfer of ions through the carbon-moisteal- calcium dioxide route is modeled. The rate of dissolution of calcium hydroxide and mass equations are modified to could describe more precisely the chemical process. A method of solving the system of nonlinear parabolic equations is proposed, which is called the finite volume method. The paper presents numerical simulations of the porous concrete carbonation process, carbonation is presented at normal speed and higher speed, respectively.

In the paper "Investigation of the permeability of porous concrete reconstructed using probabilistic description methods" [23], the authors present the analysis of the spatial distribution of gaps in porous concrete that influence their physical properties such as mechanical resistors or permeability. This is done by C.T. and using lower probability functions. With the help of lower probability functions, different porous concrete test tubes are rebuilt with a variety of ways of distributing the gaps. It is analyzed whether the real test tubes and the reconstructed test tubes show the same behavior. The results of the study confirm that the test tubes rebuilt using the probabilistic optimization process proposed by the authors have, statistically, properties identical to the original models and show material behavior similar to them. Reconstructed test tubes can be used in numerical tests to reduce the number of real experimental tests that consume time and material resources.

In the paper "Effect of CO2 on porous concreteness" [24], the author presents the influence of carbon dioxide on porous concrete. The study is performed by experimental tests, with porous concrete test tubes stored for 30 days in rooms with a concentration of 30% and 10% carbon dioxide, respectively. Other test tubes are kept in the air, where the carbon dioxide concentration is 0.03%. The results of the experimental tests showed that the lowest values of the compressive resistance were obtained in test tubes kept in rooms with a carbon dioxide concentration of 30% and 10%, respectively, for 30 days. In test tubes kept in rooms with high concentrations of carbon dioxide for 1 year, the values of the compressive resistance decreased by less than 10%. The primary element that affects the contraction of porous concrete is the concentration of carbon dioxide followed by air humidity. Porous concrete exposed to air for 1 year at a relative humidity of 50, 75 and 100%, respectively, increased its insignificant volume.
The influence of additives on used to optimise the mechanical properties and durability of porous concrete used in paving was also studied [25]. The influence of different types of additives on the mechanical strength and anti-exfoliation properties of porous concrete is analyzed. The effects of the additives on the characteristics of the pores, on the permeability, on the mechanical resistors and on the anti-exfoliation properties of the porous concrete were analyzed using C.T. The relationships between permeability, mechanical strength, anti-exfoliation properties and pore characteristics were analyzed. The additives used to prepare porous concrete test tubes were: superplasticizer additive, silica smoke, micro silica, slag, nano silica, graphene oxide and carbon nanotubes.
The porosity of pervious concrete varied significantly based on the type of additives used during its preparation. This indicates that the density value of the concrete is not sufficient to characterize its porosity when the cement mortar utilised in its production is altered. The permeability of porous concrete prepared with different additives was more than 131%. The use of suitable additives may improve the mechanical strength of porous concrete, and the effect of the same additive on improving the resistance to compression and bending has not been completely consistent. There was no obvious relationship between the mechanical resistances of porous concrete with different additives and the characteristics of pores, which shows that cement mortar has a greater impact on mechanical strength than on pores. The use of additives can improve the anti-exfoliation properties of porous concrete. The anti-exfoliation properties of slag concrete were relatively good, while in concrete with superplasticizing additive they were very weak. The anti-exfoliation properties of porous concrete prepared with different additives had little to do with the characteristics of the pores and its mechanical strength. Based on the observations made after the experimental tests, the authors recommend the following procedures to improve the anti- exfoliation properties and mechanical strength of porous pavement concrete: choice of suitable additives, reduction of porosity to meet the permeability conditions, choice of water-cement ratio depending on the type of additive [25].
In the paper "In-situ pore size investigations of loaded porous concrete with non-destructive methods" [26], the authors present non-destructive tests on porous concrete from the manufacturer Ytong. The volume of pores and their size distribution are analyzed. The following non-destructive methods are used: nuclear magnetic resonance analysis and X-ray computed tomography analysis. The test tubes are subjected to gradually increasing pressures. The results obtained from the 2 non-destructive methods were comparable and complementary. In the case of both non-destructive methods it was observed that the pores in the test tubes decreased by a radius of 0.1 mm when increasing applied pressure. In both non-destructive methods it was observed that small pores increased by a radius of approximately 0.0001 mm when the pressure increased. The results of the CT analysis showed that at high pressure some pores they collapse, and the growth of small pores is not the primary cause of element fracture, but by collapsing pores and remnants of material.

In the paper "Studies on the sound absorption characteristics of porous concrete based on the content of recycled aggregate and target void ratio" [27], the authors present the analysis of the physical and mechanical properties and the sound attenuation capacity of the porous concrete, this analysis being based on the target porosity and the content of recycled aggregates. The aim of the study is to reduce the noise caused by road traffic, rail traffic and activity in residential areas and central areas of cities, respectively, the purpose of the study is to prepare concrete with recycled aggregates from waste from demolished concrete structures. The compressive strength of porous concrete decreased rapidly when the target porosity exceeded the value by 25%, and the content of recycled aggregates was exceeded by 50%. The results of the study showed that the difference between the target porosity and the real porosity of the porous concrete is 1.7%. In the scenario of pervious concrete prepared with reprocessed materials, the noise absorption coefficient has optimal values for porosity of 25%. The amount of recycled aggregates had little influence on the value of the noise absorption coefficient. A porous concrete prepared with recycled aggregates that absorb noise efficiently has the following optimal characteristics: the porosity is 25% and the content of recycled aggregates is 50%.

## 3. Modified Porous Concrete

Polymer-modified porous concretes are becoming popular in construction due to their improved mechanical properties, durability and sustainability, compared to traditional concrete. Total void ratio, permeability and compressive strength are three important properties to consider [8]. The use of polymer latex and polymer powder can improve the properties of the porous concrete. When added to the mix, it can enhance its adhesion, flexibility and workability. These concretes also have higher compressive, tensile and flexural strengths than traditional porous concretes due to the presence of the polymer [8].

The possibility of making a porous concrete using aggregates from recycled crushed concrete waste was studied [28]. Porous concrete from recycled aggregates has permeability and acceptable mechanical resistors. The optimal mixtures of porous concrete used in the manufacture of test tubes were made, in turn, from ordinary aggregates and recycled aggregates. The following experimental tests were performed: porosity, determination of the permeability coefficient, mechanical tests to determine bending and compressive strength. The effect of the reused materials on the porousness, mechanical strength and permeability of the proposed porous concrete was also analyzed. To improve the mechanical properties of the analyzed porous concrete, polymer powders (butadiene styrene) and latex were introduced into the mixtures. Porous concrete made with reused materials exhibited lower compressive strength when compared to porous concrete made with regular aggregates. The conclusions of the study show that the porosity of concrete with recycled aggregates was higher than that of concrete with ordinary aggregates. The addition of polymers in the concrete mixture led to a small increase in the porosity of the concrete regardless of the type of aggregate used. Also, the addition of polymers led to a significant increase in the compressive strength of porous concrete with ordinary aggregates by 57% and porous concrete with recycled aggregates by 79%, respectively. The preparation of porous concrete with recycled aggregates and the addition of polymers leads to the realization of an acceptable porous concrete with porosity and acceptable mechanical resistors.

"Investigation on polymer–rubber aggregate modified porous concrete" [29] has a porous concrete prepared with polymer-based aggregates from rubber waste, which is intended for pavements with non- slip material. Concrete made with polymer-based aggregates has been subjected to experimental tests to study its mechanical strength, material curves (stress-deformation), abrasion resistance and impact resistance. The results of the experimental tests showed that concrete with aggregates based on rubber polymers, having the optimal ratio between aggregates based on polymers and mineral aggregates, has a bending strength and a compressive strength greater than porous concrete with ordinary polymer-based aggregates. The impact energy dissipation capacity of porous concrete with materials based on rubber polymers is higher. Also ductility, abrasion resistance and impact resistance of this type of concrete are improved. The porosity and permeability of this type of concrete meet the OGFC standards (Open Graded Fraction Course).

Through experimental tests, the study "Thermal conductivity and mechanical properties of porous concrete materials" [30] aims to analyze the influence of the water-binder ratio, replacement of cement with flying ash and incorporation of water-reducing additives on thermal insulation capacity and mechanical strength in porous concrete with a density of 600 kg/m3. It was observed that porous concrete prepared with a water-binder ratio of 0.32, containing 30% fly ash and 0.65% water reducing additive achieved optimal mechanical and thermal performance. Due to this optimal mixture with which the analyzed porous concrete was prepared, the following properties were obtained: the compressive strength measured at 28 days is 4.37 MPa, the water absorption rate is 12% and the thermal conductivity coefficient is 0.116 W/mK. Generally, porous concrete with a density of 600 kg/m3 and cast in a 150 mm layer meets the requirements of mechanical strength and thermal insulation in order to be used as a thermal insulation layer on the circulating terrace roofs. The porous concrete studied is efficient in terms of production costs and can be used in various fields.

The effects of the use of pumice as an option for ordinary aggregates were analyzed to improve the mechanical characteristics (compressive, bending, tensile strengths, hardness, roughness) and physical ones (density, different types of conductivity, specific heat) of pervious concrete, by experimental tests. The physical and mechanical properties of porous pumice stone are compared with the results of other experimental studies and with the A.C.I. The usage of pumice stone in the mixing of porous concrete leads to their porosity and to their decrease modulus of elasticity. By using pumice stone, porous concretes with tensile strength greater than those prescribed by the A.C.I. [31].

“An experimental study on the hazard assessment and mechanical properties of porous concrete utilizing coal bottom ash coarse aggregate in Korea” [32] analyzes the physical properties and toxicity of aggregates based on coal ash. Also, the mechanical characteristics of pervious concrete with a varied aggregate content based on coal ash are analyzed. In porous concrete prepared with aggregates based on coal ash, their porosity and permeability changed slightly, but the resistance to compression and bending decreased. For an aggregate based on coal ash content of more than 40%, the compressive strength of porous concrete decreased sharply by 11.7-27.1%, and bending strength decreased by 26.4%. As the aggregate based on coal ash content increased, the test tubes were predominantly destroyed by crushing the aggregates and less by crushing the binder or by breaking the surfaces. Improvements such as the incorporation of reinforcement materials or tougher aggregates based on coal ash are needed to use in the preparation of porous concrete. Following laboratory tests, aggregates based on coal ash meet the characteristics of solidity and abrasion resistance imposed by specific rules. Following the leaching test, it turned out that both aggregates based on coal ash and porous concrete prepared with them meet environmental standards.

A study that makes a new type of ecological porous concrete that is prepared mainly with industrial by- products was treated in “A novel eco-friendly porous concrete fabricated with coal ash and geopolymeric binder: Heavy metal leaching characteristics and compressive strength” [33]. When preparing the new type of porous concrete, bottom ash from coal was used as an aggregate, and as a geopolymer binder. The new type of porous concrete has been subjected to experimental tests, analysing the compressive strength and disposal capacity of heavy metals (heavy metal leaching). Experimental results showed that the concentrations of heavy metals removed from the ash in porous concrete were below the values established as evaluation criteria. Geopolymers used as binders prevented the diffusion of heavy metals from ash. Based on the experimental results, an empirical formula was created by which the relationship between the porosity of the new type of concrete and its compressive resistance is expressed. The study concludes that porous concrete made with ash and geopolymer is able to efficiently retain heavy metals as solidified / stabilized products.

Light porous concrete can be prepared using only solid residues. The purpose of this type of concrete is to improve the compressive strength in the first days after commissioning and to reduce the consumption of resources during the preparation of the mixture as well, can be put into operation in a short time [34]. This type of mixture was obtained by sintering (process of gluing metal, ceramic powders, etc. following their heating and pressing) the following substances: aluminum powder, flue gas desulphurization gypsum, carbide slag and red mud. The result was a cement material, which was then mixed with other industrial residues, such as fly ash, thus obtaining the paste for light porous concrete that can be put into operation quickly. The raw material used in the preparation of this type of porous concrete comes only from industrial residues. The results of the experimental tests showed that light porous concrete from industrial residues has a minimum thermal conductivity of 0.15 W/mK.

The compressive strength measured at 28 days is 3.57 MPa for light porous concrete from industrial residues with a density of approximately 600 kg/m$^3$. The compressive strength measured at 4 hours is 1.31 MPa, and the one measured at 1 day is 1.57 MPa. These values are higher than the compressive strength at 28 days of ordinary light porous concrete. In conclusion, lightweight porous concrete prepared from industrial residues has a smaller impact on the environment than slightly obscured porous concrete [34].
The paper [35] presents a prototype of bio-receptive porous concrete covered with vegetation that is subjected to heat transfer, ventilation and air conditioning tests, using water as an energy transfer medium. The coils for energy transfer were evenly distributed in a cylindrical formwork. At the hardening of the porous concrete poured into the respective formwork, a plant of the species Lolium perenne was sown, which was hydroponic irrigated and increased for 60 days. The cylindrical porous concrete test tubes covered with vegetation was connected to an air duct, air from 3 centrifugal fans. Experimental tests were performed on the test piece in which it was heated, cooled, humidified and dehumidified to determine its heat transfer capacity. The maximum heat transfer capacity of the test piece for heating was 238.8±4.1 W, and for cooling was −116.65±7.99 W. The maximum humidification capacity was $3.42\pm0.34\ g\cdot kg^{-1}$, and the dehumidification was $-9.04 \pm 0.9\ g\cdot kg^{-1}$. Porous concrete with vegetal layer showed a passive dehumidification capacity of $-\ 4.57 \pm 0.22\ g\cdot kg^{-1}$ without the need for other energy sources. The energy-free dehumidification capacity of porous concrete with vegetal layer shows the tendency of the vegetation layer to neutralize extreme atmospheric conditions. Experimental results showed that there is a new scope of green infrastructure, demonstrating the ability of porous concrete with a vegetal layer to control environmental conditions. Also, porous concrete with vegetal layer can be applied to the hydroponic growth of plants.

## 4. Conclusions

The use of porous concrete reinforced with polymers can potentially contribute to the research objective of reducing pollution and waste in the construction industry. Porous concrete allows for water to infiltrate the surface, which can help to reduce stormwater runoff and prevent flooding. This can help to protect nearby bodies of water from pollution and reduce the demand for traditional drainage systems. Additionally, reinforcing the porous concrete with polymers can increase its durability and strength, which can reduce the need for frequent repairs or replacements. This can help to reduce the amount of materials that are wasted during construction and minimize the environmental impact of construction activities.
Polymer-modified porous concrete is becoming increasingly popular in construction due to its improved properties and sustainability. The addition of different types of polymers to the pervious concrete mix can significantly enhance its mechanical properties, including its compressive strength, permeability, and total void ratio. Polymer-modified porous concrete has a higher compressive, tensile, and flexural strength compared to traditional porous concrete due to the presence of the polymer. The polymer optimizes the adhesion, flexibility, and workability of the concrete, which can contribute to the overall durability of the structure.
Synthetic fibers, steel fibers, glass fibers, and natural fibers are commonly used and have been tested for reinforcing pervious concrete. These fiber types have been extensively studied and applied to enhance the mechanical properties and performance of pervious concrete in various applications. Extensive research and testing have been conducted on the performance of these fiber types in pervious concrete, considering factors such as fiber dosage, length, and distribution. The selection of the most suitable fiber type depends on the specific project requirements, desired performance objectives, and environmental considerations.
The aim of reducing noise caused by traffic and activity in residential areas and city centers is an important research objective, as excessive noise can have negative effects on human health and well-being. Using recycled aggregates from waste from demolished concrete structures to prepare new concrete can contribute to this objective in several ways.

However, it's important to note that the use of polymers may have its own environmental impact, depending on the type and source of the polymer. Therefore, it's important for researchers to carefully consider the environmental implications of any new materials or processes being developed, and to ensure that they are truly environmentally friendly and sustainable.

## 5. References

[1] Malani S R & Pandey R 2023 Optimizing the Properties of Porous Concrete for Practical Applications. **14**. 163-178.

[2] Muda M & Legese A & Urgessa G 2023 Strength, Porosity and Permeability Properties of Porous Concrete Made from Recycled Concrete Aggregates. *Construction Materials*. **3**. 10.3390/constrmater3010006.

[3] Sanchez-Mendieta C & Galán J & Lage I 2021 Physical and Hydraulic Properties of Porous Concrete. *Sustainability*. **13**. 10562. 10.3390/su131910562.

[4] Mazur B & Kotwa A 2019 Porous Concrete as an Anti–Aquaplaning Building Material. IOP Conference Series: Materials Science and Engineering. **471**. 032027. 10.1088/1757-899X/471/3/032027.

[5] Suryo E & Arifi E & Zaika Y 2021 Study on sliding stability of porous-concrete retaining wall model test subjected to rainfall infiltration. IOP Conference Series: Earth and Environmental Science. **930**. 012101. 10.1088/1755-1315/930/1/012101.

[6] Cahya E & Haribowo R & Arifi E 2021 Inclined porous concrete surface impact on infiltration using recycled concrete aggregate. IOP Conference Series: Earth and Environmental Science. **930**. 012100. 10.1088/1755-1315/930/1/012100.

[7] Yehia S & Fawzy M 2017 The impact of using Polymer Impregnated Porous Concrete in Structural Engineering Applications. *International Journal of Current Engineering and Technology*. **07**. 634-640.

[8] Ariffin NF & Md Jaafar M F & Abdul Shukor Lim NH & Bhutta Aamer & Hussin M 2018. Mechanical properties of polymer-modified porous concrete. IOP Conference Series: Materials Science and Engineering. **342**. 012081. 10.1088/1757- 899X/342/1/012081.

[9] Zaman Md & Islam G M 2014 Prospects of Porous Concrete in Civil Engineering Applications.

[10] Ginting A & Adi P & Prasetyo Y 2019 Porous Concrete using White Cement as Binding Agent. 16-20. 10.5220/0009200900160020.

[11] Mukesh TS & Sampathkumar K & Muthuselvi C & Ranjith M. & Kumar S. 2021 Experimental Investigation on Enhancing the Strength of Porous Concrete by adding Polypropylene Fiber. IOP Conference Series: Materials Science and Engineering. **1055**. 012013. 10.1088/1757-899X/1055/1/012013.

[12] Krishnappa, S 2018 Porous concrete & its characteristics with mineral admixtures. 5.

[13] Valcuende M & Lliso Ferrando J & Roig Flores M & Gandía-Romero J 2021. Porous Structure of Ultra-High-Performance Fibre-Reinforced Concretes. Materials. 14. 1637. 10.3390/ma14071637.

[14] Ariffin, NF & Md Jaafar MF & Abdul Shukor Lim N & Bhutta A & Hussin, M 2018. Mechanical properties of polymer-modified porous concrete. IOP Conference Series: Materials Science and Engineering. **342**. 012081. 10.1088/1757- 899X/342/1/012081.

[15] Bhutta M A R, Tsuruta K, Mirza J 2012 Evaluation of high-performance porous concrete properties, *Construction and Building Materials*, **31**, 67-73

[16] Park S-B, Tia Mang 2004 An experimental study on the water-purification properties of porous concrete, *Cement and Concrete Research*, **34**, 2, 177-184

[17] Kim HK, Lee HK 2010 Influence of cement flow and aggregate type on the mechanical and acoustic characteristics of porous concrete, *Applied Acoustics*, **71**, 7, 607-615

[18] Lian C, Zhuge Y, Beecham S 2011 The relationship between porosity and strength for porous concrete, *Construction and Building Materials*, **25**, 11, 4294-4298

[19] Zhang Q 2016 Mathematical modeling and numerical study of carbonation in porous concrete materials, *Applied Mathematics and Computation*, **281**, 16-27

[20] Chindaprasirt P, Hatanaka S, Chareerat T, Mishima N, Yuasa Y 2008 Cement paste characteristics and porous concrete properties, *Construction and Building Materials*, **22**, 5, 894-901

[21] Chindaprasirt P, Hatanaka S, Mishima N, Yuasa Y, Chareerat T 2009 Effects of binder strength and aggregate size on the compressive strength and void ratio of porous concrete, *International Journal of Minerals, Metallurgy and Materials*, **16**, 6, 714- 719

[22] Agar-Ozbek A, Weerheijm J, Schlangen E, Breugel K 2013 Investigating porous concrete with improved strength: Testing at different scales, *Construction and Building Materials*, **41**, 480-490

[23] Chung S-Y, Han T-S, Kim S-Y, Lee T-H 2014 Investigation of the permeability of porous concrete reconstructed using probabilistic description methods, *Construction and Building Materials*, **66**, 760-770

[24] Šauman Z 1972 Effect of CO2 on porous concrete, *Cement and Concrete Research*, **2**, 5, 541-549

[25] Zhang Y, Li H, Lu Q, Yang J, Wang T 2022 Effect of Different Admixtures on Pore Characteristics, Permeability, Strength, and Anti-Stripping Property of Porous Concrete. *Buildings*. 2022; 12(7):1020. https://doi.org/10.3390/buildings12071020

[26] Rifai H, Staude A, Meinel D, Illerhaus B, Bruno G 2018 In-situ pore size investigations of loaded porous concrete with non-destructive methods, *Cement and Concrete Research*, **111**, 72-80, https://doi.org/10.1016/j.cemconres.2018.06.008

[27] Park S B, Seo D S, Lee J 2005 Studies on the sound absorption characteristics of porous concrete based on the content of recycled aggregate and target void ratio, *Cement and Concrete Research*, **35**, 9, 1846-1854, https://doi.org/10.1016/j.cemconres.2004.12.009

[28] Bhutta M A R, Hasanah N, Farhayu N, Hussin M W, Md Tahir M bin, Mirza J 2013 Properties of porous concrete from waste crushed concrete (recycled aggregate), *Construction and Building Materials*, **47**, 1243-1248, https://doi.org/10.1016/j.conbuildmat.2013.06.022

[29] Shen W, Shan L, Zhang T, Ma H, Cai Z, Shi H 2013 Investigation on polymer– rubber aggregate modified porous concrete, *Construction and Building Materials*, **38**, 667-674, https://doi.org/10.1016/j.conbuildmat.2012.09.006

[30] Sun Y, Gao P, Geng F, Li H, Zhang L, Liu H 2017 Thermal conductivity and mechanical properties of porous concrete materials, *Materials Letters*, **209**, 349-352, https://doi.org/10.1016/j.matlet.2017.08.046

[31] Hariyadi H, Tamai H 2015 Enhancing the Performance of Porous Concrete by Utilizing the Pumice Aggregate, *Procedia Engineering*, **125**, 732-738, https://doi.org/10.1016/j.proeng.2015.11.116

[32] Park S, Jang Y, Lee J, Lee B-J 2009 An experimental study on the hazard assessment and mechanical properties of porous concrete utilizing coal bottom ash coarse aggregate in Korea, *Journal of Hazardous Materials*, **166**, 1, 348-355, https://doi.org/10.1016/j.jhazmat.2008.11.054

[33] Jang J G, Ahn Y B, Souri H, Lee H K 2015 A novel eco-friendly porous concrete fabricated with coal ash and geopolymeric binder: Heavy metal leaching characteristics and compressive strength, *Construction and Building Materials*, **79**, 173-181, https://doi.org/10.1016/j.conbuildmat.2015.01.058

[34] Yao X, Wang W, Liu M, Yao Y, Wu S 2019 Synergistic use of industrial solid waste mixtures to prepare ready-to-use lightweight porous concrete, *Journal of Cleaner Production*, **211**, 1034-1043, https://doi.org/10.1016/j.jclepro.2018.11.252

[35] Rahman Md S, MacPherson S, Akbarzadeh A, Guerini A, Chapelat J, Lefsrud M 2022 A study on heat and mass transfer through vegetated porous concrete for environmental control, *Journal of Cleaner Production*, **366**, 132984, https://doi.org/10.1016/j.jclepro.2022.132984

# Environmental and Economic Sustainability Indicators for External Walls

**P C Miltiade[1], I Baran, L Dumitrescu[1], I S Entuc[1] and R A Pescaru[1]**
[1] "Gheorghe Asachi" Technical University of Iași, Bd. Dimitrie Mangeron, nr. 67, 700050, Romania

E-mail: irina.baran9@gmail.com

**Abstract.** The current context of the intensification of climate changes, on one hand, and the sharp increase in prices in the buildings sector, on the other, impose the need to find solutions that considers both economic and environmental protection aspects.

This paper proposes a method of evaluating the solutions most often used today in Romania for the exterior walls in single-family residential buildings based on a global index that takes into account both environmental performances (adjusted thermal resistance, carbon footprint) as well as the economic ones (costs).

## 1. Introduction

The improvement of living conditions due to technological advances in recent years has led to a better life for mankind, but most of these advances are based on technologies that do not take into account the possibility of environmental degradation during their use, a fact that has led to adverse climate change. Buildings and infrastructure are major contributors to global warming. The construction sector is responsible for 40% of the total energy consumption and 36% of the greenhouse gases emitted at the level of the European Union [1].

The heat loss through the external walls represents a very significant percentage of the total heat loss of a building, namely 35% for buildings with ground floor level and 65% for buildings with several levels [2]. That is why it was considered relevant to investigate the effect of adopting different types of solutions for the opaque part of building facades.

Thus, the following types of walls were considered, from the point of view of economic and environmental sustainability, for a single-family residential building with ground floor:

- external closing walls with a metallic frame type structure;
- external closing walls with frame-type structure made of softwood;
- external closing walls with a confined masonry structure made of ceramic hollow bricks (with vertical cores);
- external closing walls with a confined masonry structure made of autoclaved aerated concrete (AAC).

The elements taken into account from the point of view of environmental sustainability were:

- unidirectional specific thermal resistance R [$m^2K/W$];
- mean adjusted thermal resistance R' [$m^2K/W$];
- carbon footprint for the production of the construction materials used [kg $CO_2$ eq/$m^2$].

Considering economic sustainability, a price was calculated for a representative area of approximately fifteen square meters of wall.

Then, a global index that takes into account both environmental and economic sustainability performances was developed and tested. Previously conducted research focused on evaluating building materials [3] – [5] or even entire buildings [6], [7] using combined sustainability criteria to obtain a more holistic approach.

## 2. Methods

The study consists in the analysis, from the point of view of sustainability in terms of environmental protection and financial aspects, of some solutions of external closing walls for different structure types used on a large scale in Romania, for dwellings with ground floor.

Thus, a simple dwelling plan (figure 1) was considered (to make it easier to follow the purpose of the work by enabling simpler calculations), but which would offer at least the minimum comfort level required by the NP57/2002 Normative. The level height was considered 2.80 m.

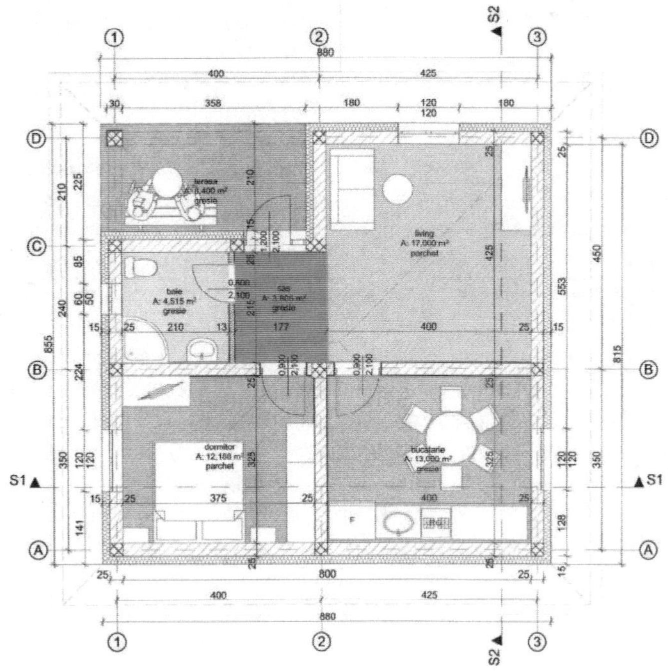

**Figure 1.** The ground floor plan for the analysed building.

The four solutions of the composition of the external walls analysed in this study were:

a. Metallic frames

In the case of the solution with a frame-type structure made of metallic profiles, a calculation was made to determine the types of profiles required, taking into account the provisions of the P100-1/2013 and Eurocode 3 regulations. For the calculation of the seismic force, a site characterized by the design ground acceleration, ag=0.25g and a corner period Tc=0.7s was considered. Thus, a HEA 180 metallic profile was established for columns.

A detail was considered for an exterior closing and interior partitioning wall to allow a specific finishing system for dwellings (washable lime on the inside and decorative plaster on the outside). The use of closing elements generally used for structures with metallic profiles, namely prefabricated sandwich panels, specific to industrial units, was avoided. It was adopted a system in which on the columns are welded metallic profiles TP30x30x3 type, which will represent the support for the oriented strand boards (OSB) boards on the outside and for the gypsum-cardboard boards on the inside (figure 2). A thermosystem can be applied over the OSB boards on the outside, which can be finished with decorative plaster, and the plasterboard boards on the inside can be finished with washable lime.

In the case of metallic frames, the openings were bordered by TP 30x30x3 profiles. These details were taken into account in the economic evaluation.

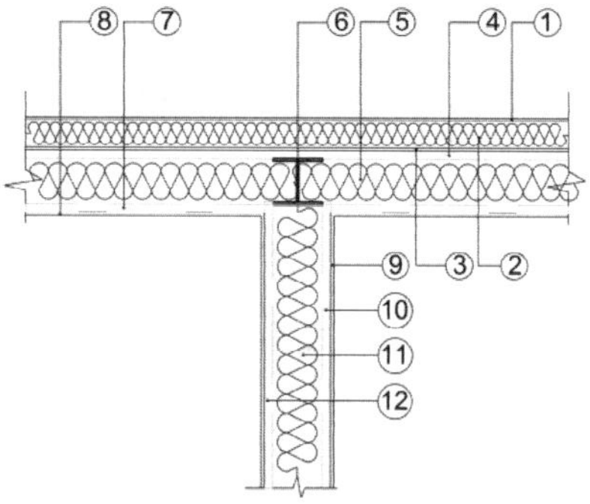


1 - decorative plaster 10 mm
2 - expanded polystyrene 100 mm
3 - OSB board 10 mm
4 - metallic profile TP 30x30x3 / 1000 mm
5 - mineral wool 170 mm
6 - metallic profile HEA 180
7 - metallic profile TP 30x30x3 / 1000 mm
8 - gypsum cardboard 12 mm
9 - gypsum cardboard 12 mm
10-metallic profile TP 30x30x3 / 1000 mm
11-mineral wool 180 mm
12-metallic profile TP 30x30x3 / 1000 mm

**Figure 2.** Detail of the exterior and interior wall – metal frames solution.

b. Softwood timber frames

The AxisVM software [8] was used to determine the necessary sections of the elements for the softwood frames, in compliance with the prescriptions of NP 005-2022 and SR EN 1995-1-1:2004. It was found that a section of the wooden columns approximately equal to that of the columns of metal profiles would not have been enough, so it was decided to increase the number of frames in this structural solution (where the architectural solution allowed this) so that the thickness of the walls results about the same. Thus, it was possible to use a 170 mm thermal insulation on the inside of the wall - the same thickness as in the case of the metallic frame structure.

The aim was to provide the two solutions with roughly the same unidirectional thermal resistance, with thermal bridges of comparable size, so that the results of this study would be as accurate as possible.

In order to determine the sequence of layers inside the wall, it was tried a similarity between this solution and the one with metallic profile frames. The softwood frames will be plated externally with 10 mm OSB boards and internally with 12 mm plasterboard boards (figure 3). The plane rigidity of the wall will be ensured both by the OSB boards on the outside and by the arrangement of bracing at the intersections between the columns and beams, respectively columns and fixing base slab, a fact taken into account in the financial evaluation. For plating with OSB plates or plaster-cardboard, a grid of metallic profiles type CW 60x0.45 or UV 30x0.45 was used as a support.

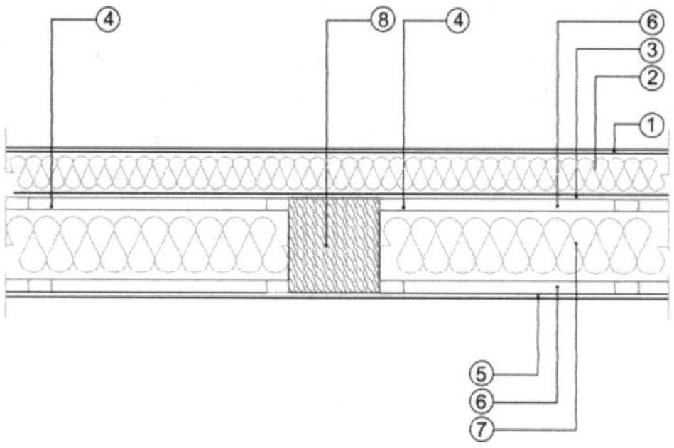


1-decorative plaster 10 mm
2-expanded polystyrene 100 mm
3-OSB boards 10 mm
4-metallic profile CD 60x0.45 mm
5-gypsum cardboard 12 mm
6-unventilated air (in the area of the perimetral metallic profile UD30x0.45 mm)
7-mineral wool 170 mm
8-softwood timber column C22 230x230 mm

**Figure 3.** Detail of the exterior wall – wood frames solution.

c. Hollow brick confined masonry

In this case, the structural system was made taking into account the minimum constructive details imposed by P100-1/2013 and CR6-2013, for a confined masonry that ensures a takeover of gravitational and seismic loads. The relatively small spans (maximum interaxial dimension of 4.5 m) and the height level (ground floor) were taken into consideration.

For masonry solution, a widely used brick, namely ceramic brick with vertical holes with plan dimensions of 240x290x188mm and a mortar of M5 class, was considered. The C20/25 class reinforced concrete cores are provided at all wall intersections and for the reinforced concrete wall beams. Reinforced concrete lintels, C20/25 concrete class, were provided over the openings. The lintel height is equal to the brick height and the length exceeds the opening by 45 cm on each side (figure 4).

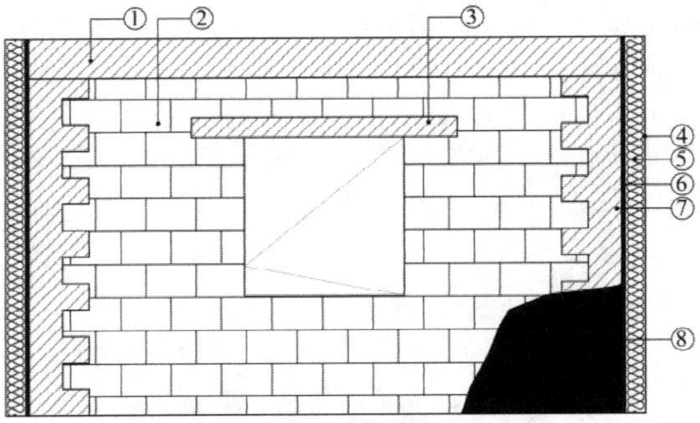


**Figure 4.** Detail of the exterior wall – confined masonry solution.

d. Autoclaved aerated concrete (AAC) block confined masonry

For this solution, the same constructive details were adopted as in the case of masonry with ceramic brick with vertical holes.

According to the provisions imposed by P100-1/2013 and CR6-2013, for the structural walls of the analysed building (dwelling on the ground floor, located in an area characterized by an acceleration of the ground ag≤0.25g) can be used AAC block with standardized unit compressive strength $f_b$=3.5N/mm$^2$ and $f_b$=5N/mm$^2$ and M5 brand mortar.

Table 1 shows the calculated thermotechnical characteristics of the used materials, according to Normative C107/1-2005, annex A.

**Table 1.** Thermal characteristics of materials.

| Material | Apparent density ρ [kg/m$^3$] | Thermal conductivity λ [W/mK] | Mass caloric capacity c [J/kgK] |
|---|---|---|---|
| Metal | 7850 | 58 | 480 |
| Decorative plaster | 1600 | 0.87 | 840 |
| Reinforced concrete | 2500 | 1.74 | 840 |
| Mortar | 1700 | 0.87 | 840 |
| Brickwork | 1700 | 0.75 | 870 |
| Expanded polystyrene | 20 | 0.04 | 1460 |
| OSB boards | 600 | 0.17 | 2510 |
| Mineral wool | 60 | 0.04 | 750 |
| Gypsum cardboard | 1000 | 0.37 | 840 |
| Pine, fir wood | 550 | 0.17 | 2510 |
| AAC block masonry | 750 | 0.28 | 840 |

Based on the thicknesses values of the material layers presented in figures 2-4 and the standard thermal conductivities in table 1, the unidirectional specific thermal resistances were calculated for all four external wall solutions considered.

For the calculation of the adjusted specific thermal resistances, values of the linear thermal transmittances determined after modelling the thermal field with the help of the RDM7 calculation program were used [9]. According to C107/3-2005, the calculation relationship is:

$$\psi = \frac{\Phi}{\Delta T} = -\frac{B}{R} \quad (1)$$

where: $\psi$ is the linear thermal transmittance, [W/mK];
$\Phi$ is the thermal flow, [W/m];
$\Delta T$ is the temperature difference between indoor and outdoor, [K];
B is the width of the analyzed domain, [m];
R is the specific unidirectional thermal resistance, [$m^2K/W$].

In figure 5 is exemplified the allure of the heat flow density in the case of thermal bridges of the "external corner" and "intersection exterior wall - interior wall" type for solution 4 (AAC block confined masonry).

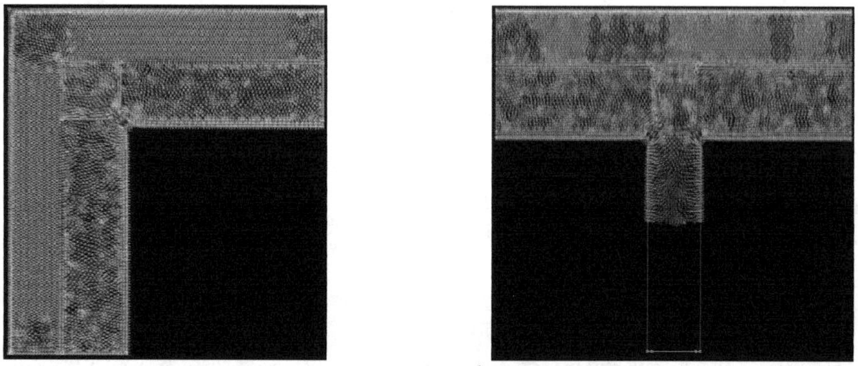

a b

**Figure 5.** Specific heat flows resulted from thermal bridge modelling: a) external corner; b) T-joint intersection.

There were considered the thermal transmittance values for the following types of thermal bridges: external corner, internal corner, T-joint intersection, intersection of the external wall with the concrete slab on the ground, intersection of the external wall with the upper floor, window and door frame joint. The lengths of the thermal bridges were established according to the information from figure 1. Thus, the adjusted specific thermal resistances related to the four opaque external walls (Northern, Southern, Eastern, and Western façade, respectively) were determined and then the mean adjusted specific thermal resistance of all external walls was calculated.

The carbon footprint expressed in kg $CO_2$ eq./$m^2$ was calculated using the UBAKUS software [10], based on the thicknesses of the material layers shown in figures 2-4 and the characteristics of the materials shown in table 1. Figure 6 shows the result obtained in the solution 4 (AAC block confined masonry). Both the embodied primary energy (from non-renewable sources) and greenhouse gas emissions are highlighted, total values and for each individual component material.

For the financial evaluation, the wall in axis D between axes 2 and 3 was considered to be representative, it having an area of 14 $m^2$, which represents approximately 15% of the total area of the external walls and also containing a carpentry opening for which, depending on each structural system analyzed, additional structural compensation measures must be provided. These constructive solutions were described for each solution separately and were included in the calculation of the quantities that formed the basis of the economic evaluation. The values presented are in RON (Romanian leu), do not include VAT, and the prices are at the level of February-March 2022.

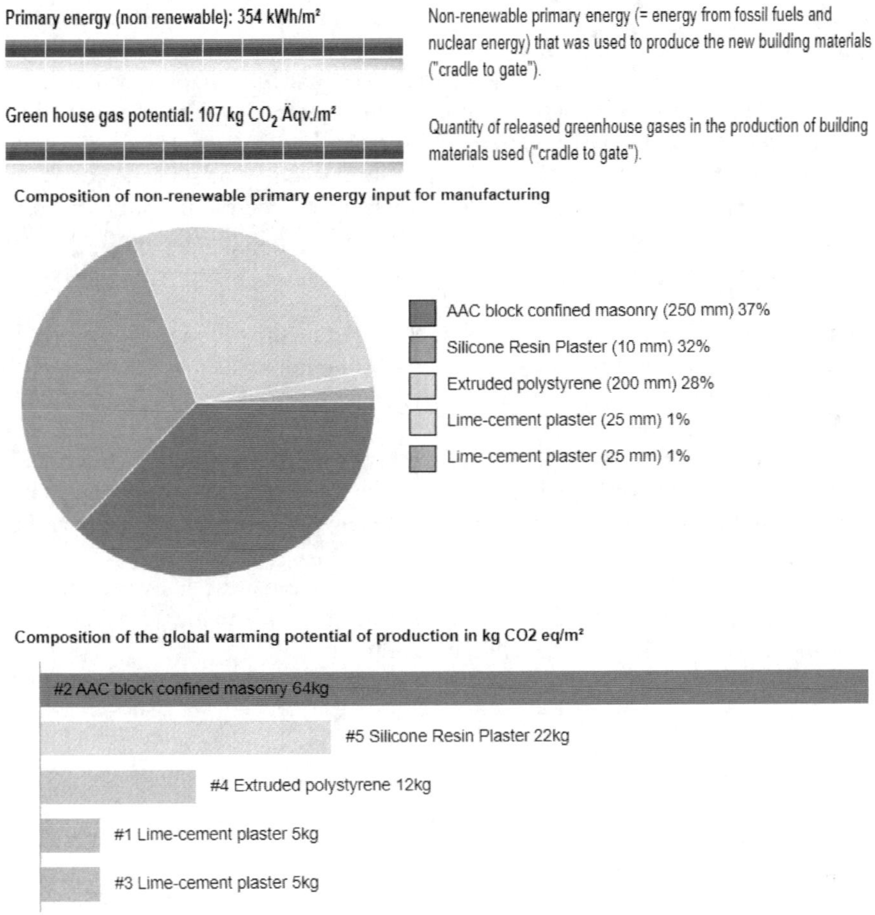


**Figure 6.** Embodied primary energy and carbon footprint for solution 4.

## 3. Results and discussions

The results of the calculation of the environmental performance indicators (adjusted specific thermal resistance, embodied primary energy, carbon footprint) and economic performance (costs) are presented in table 2, for all four solutions adopted for the external walls of the studied building.

**Table 2.** Environmental and economic performance indicators for external walls.

| Solution | 1-Metallic frames | 2-Wood frames | 3-Hollow brick masonry | 4-AAC block masonry |
|---|---|---|---|---|
| R [$m^2K/W$] | 7.385 | 6.872 | 5.569 | 6.128 |
| $R'_{North}$ [$m^2K/W$] | 3.239 | 3.152 | 2.896 | 3.069 |
| $R'_{South}$ [$m^2K/W$] | 3.805 | 3.686 | 3.341 | 3.572 |
| $R'_{East}$ [$m^2K/W$] | 3.491 | 3.390 | 3.094 | 3.291 |
| $R'_{West}$ [$m^2K/W$] | 3.384 | 3.288 | 3.009 | 3.195 |
| $R'_{mean}$ [$m^2K/W$] | 3.48 | 3.38 | 3.09 | 3.28 |
| Primary energy [$kWh/m^2$] | 480.1 | 396 | 387 | 354 |
| Carbon footprint [kg $CO_2$ eq./$m^2$] | 109.35 | 84 | 80 | 107 |
| Costs [RON] | 7548 | 6823 | 5220 | 5135 |

As can be seen from the values given in table 2, the highest value of the unidirectional specific thermal resistance R is in the case of the metallic profile structure, and the lowest is in the case of the hollow brick masonry. A 170 mm layer of mineral wool was placed inside the walls, in the case of the metallic structure and in the case of the wooden frame structure, due to the width of wall related to the dimensioning of the wooden columns. Supplementary, a layer of 10 cm expanded polystyrene was placed on the outside part as an external thermal insulation system. The presence of the two layers of thermal insulation, internal and external, leads to positioning the mentioned solutions on the top two. Taking this into account, in order to increase the accuracy of the study results, for the two solutions with load-bearing masonry type structure was considered a thermal system with a 20 cm of expanded polystyrene on the outside.

It can be noticed that following the calculation that takes into account the influence of thermal bridges, the "ranking" remains unchanged compared to the one in the case of the unidirectional specific thermal resistance R, but the relative difference between the most efficient solution and the least efficient solution, decreased by about 2 times.

Regarding the carbon footprint, the best results are given by the solution with wooden frames and the solution with confined brickwork made of fired clay ceramic brick. Also, a rather large difference can be noticed between the solutions with metallic profiles and AAC block masonry and the other two solutions taken into account.

Concerning the economic performance, it is observed that the most expensive solution is the metallic frames solution, while the solutions based on load-bearing masonry made of AAC block or ceramic brick prove to be more efficient from a financial point of view.

An explanation of this situation would be given by the recent exponential increase in the prices of metal products used in construction, while for ceramic brick, the raw material being more accessible (clay quarries generally located at a short distance from the factory), price increases were much lower.

In order to establish a ranking of the debated solutions taking into account both environmental and economic performance, the following method was proposed:

- the values of the environmental and economic indicators related to each solution were compared with the reference values, resulting in a certain degree of compliance of the criteria taken into account;

- this degree of compliance of the criteria was associated with an importance coefficient, thus resulting in some values that led to the possibility of creating a ranking of the analysed solutions.

Thus, the results obtained for the indicators related to mean adjusted specific thermal resistance, carbon footprint and costs were reported to a series of reference values, as follows:

- Reference $R'_{mean} = 4.00$ $m^2K/W$;
- Reference carbon footprint = 75 kg $CO_2$ eq./$m^2$;
- Reference cost = 5,000 RON.

The degree of compliance of the imposed requirements was represented graphically in figures 7-9.

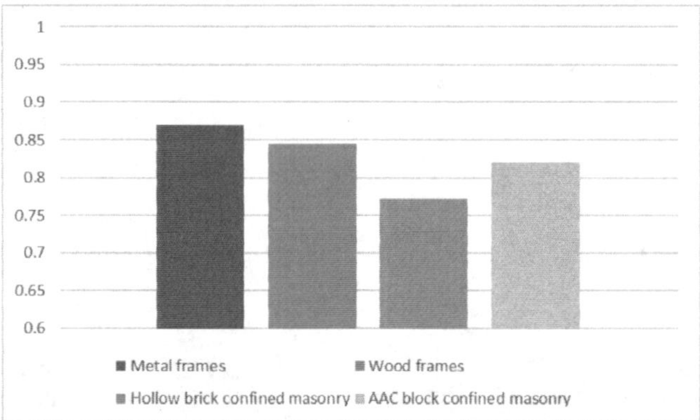


**Figure 7.** The degree of compliance with the reference $R'_{mean}$ value [$m^2K/W$].

IOP Conf. Series: Materials Science and Engineering 1304 (2024) 012026 doi:10.1088/1757-899X/1304/1/012026

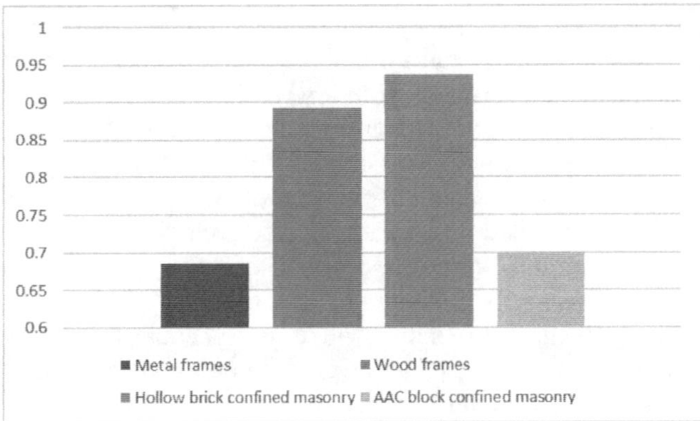


**Figure 8.** The degree of compliance with the reference carbon footprint value [kg $CO_2$ eq./$m^2$].

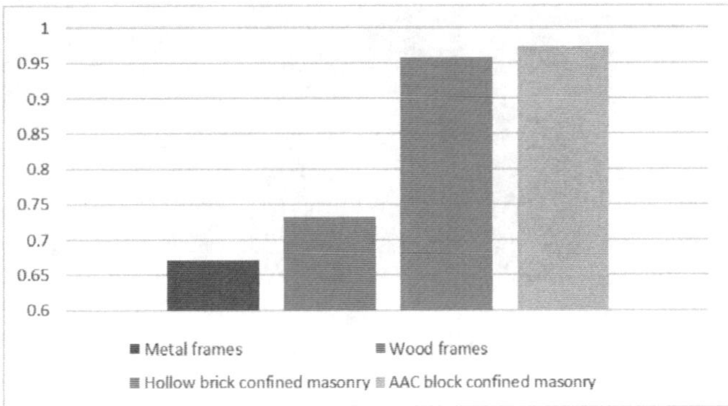


**Figure 9.** The degree of compliance with the reference costs value [RON].

Figure 10 shows a ranking of the solutions for a score in which the coefficient of importance with a higher value (0.4) was given to the "price" criterion, and the indicators related to environmental protection had equal coefficients of importance (0.3), but lower than the price-related one. It can be seen that the solutions that use confined masonry are ranked in the first two positions, the next position is for the solution with wooden frames, and the solution with metal profile frames is in fourth position. Using these importance coefficients, the scores accumulated by the four solutions are not far apart. Practically, the maximum difference between the analysed solutions is 1.64 points for a score from 1 to 10.

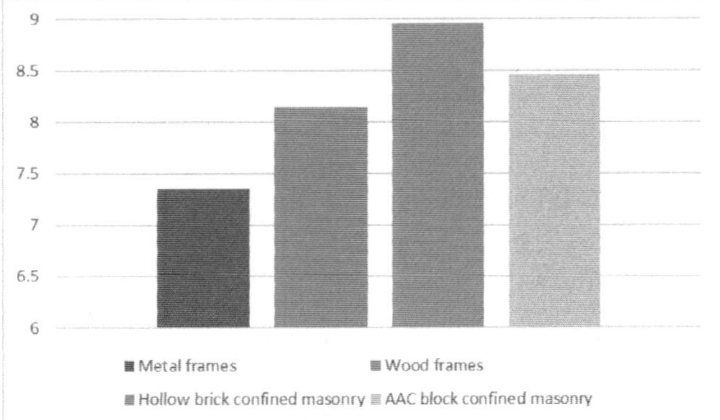


**Figure 10.** Ranking according to the degree of compliance of the reference values with relationship ($0.3*R_{mean}'+0.3*$carbon footprint$+0.4*$cost$)*10$ (points).

This can lead to the conclusion that on a ranking based more on cost, the four debated solutions do not present major differences, but the solutions that assume confined masonry, for the types of walls considered, represent a (slightly) more efficient solution (under the conditions of prices shown above). Also, there is a small difference between the points accumulated by the solution with masonry confined with AAC blocks and the solution with wooden frames (the solutions on position 2 and position 3, respectively). Taking into account the major restrictions for AAC blocks masonry from P100-1/2013 and from CR6/2013, it can be concluded that the most effective solution (for zones characterized by a ground acceleration, ag≥0.25g and in the price conditions presented above) remains the "classic" one of confined masonry, with burnt clay bricks, followed by the solution with a wooden frame structure.

Figure 11 shows a ranking of the solutions that take into account the impact on the environment, more than the financial impact. It can be seen that the environmental indicators ($R'_{mean}$, carbon footprint) were given a higher importance coefficient (0.4) than the cost indicator (0.2).

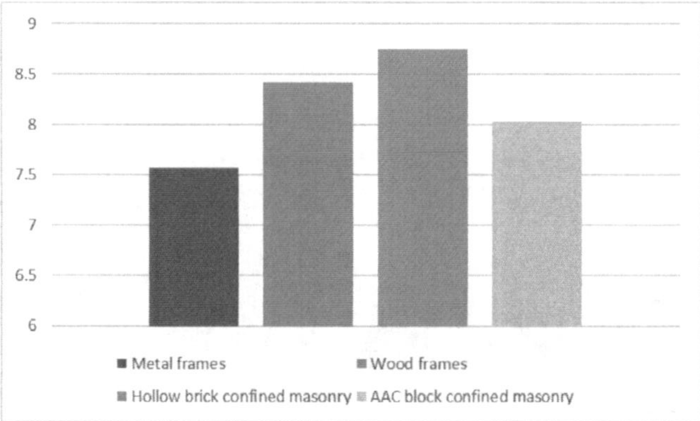


**Figure 11**. Ranking according to the degree of compliance of the reference values with relationship ($0.4*R_{mean}'+0.4*$carbon footprint$+0.2*$cost)*10 (points).

In this case, the ranking has changed as follows: in the first two positions, there are the solutions with hollow brick confined masonry and wood frames and, in the 3rd position, but with a little difference, there is the structure of the AAC block confined masonry. On the 4th position is the solution with metal frames.

It can be seen that on a ranking based predominantly on the indicators related to environmental protection, the difference between the score for the solution with wood frames (from the 2nd position) and that with confined masonry made of AAC blocks (from the 3rd position) has increased, compared to the situation in the case of the ranking based predominantly on price.

In this case, the score difference between the solution in the first position and that in the last position is smaller than in the previous ranking, namely 1.24 points for a score from 1 to 10.

## 4. Conclusions

The current situation on the construction market (in the context of the chaotic increase in prices and the lack of continuity of supply with certain materials but also the need to apply environmentally friendly solutions), requires a work supported by research to compare existing solutions and to innovate new solutions.

This paper presented a comparative analysis of some existing solutions on the market, for the walls of dwellings with one height level. Some of these solutions are used on a larger scale (AAC block masonry, ceramic brickwork or wooden frames), others at a smaller scale (metallic profiles frames). The aim of the paper was to identify the best solutions, by comparing the performance of each, in relation to standard indicators of economic and environmental sustainability.

As a general conclusion of the present study, it could be stated that, taking into account the considered environmental and economic sustainability indicators, the solution "confined masonry with ceramic bricks" is the most effective, being closely followed by the solution "frames made of resinous wood".

## 5. References

[1] *Directive (EU) 2018/844 of the European Parliament and of the Council of 30 May 2018 amending Directive 2010/31/EU on the energy performance of buildings and Directive 2012/27/EU on energy efficiency*
[2] Dumitrescu L, Pescaru R A, Baran I, Bliuc I 2020 *Increasing the energy performance of non-residential buildings by thermal insulation of walls on the interior surface* Proc. of Internat. Conf. Building Services and Energy Efficiency, Modernizing and increasing performance of Building Services 184-196
[3] Damineli B L, Kemeid F M, Aguiar P S, John V M 2010 *Measuring the eco-efficiency of cement use* Cem. Concr. Compos. 32 555-62
[4] Fantilli A P, Chiaia B 2013 *The work of fracture in the eco-mechanical performances of structural concrete* J. Adv. Concr. Tech. 11 282–90
[5] Hanuseac L, Dumitrescu L, Barbuta M, Baran I, Bejan G 2020 *Eco-Mechanical Index of Lightweight Concrete Mixtures with Recycled Materials* Procedia Manufacturing 46 667-674
[6] D'Agostino D, de'Rossi F, Marigliano M, Marino C, Minichiello F 2019 *Evaluation of the optimal thermal insulation thickness for an office building in different climates by means of the basic and modified "cost-optimal" methodology* J. Build. Eng. 24 100743
[7] Tadeu S, Rodrigues C, Marques P, Freire F 2022 *Eco-efficiency to support selection of energy conservation measures for buildings: A life-cycle approach* J. Build. Eng. 61 105142
[8] Inter-CAD Kft., AxisVM, Structural analysis & design software, www.axisvm.eu, 2022
[9] Debard I, RDM 7.04 2020 *Résistance des Matériaux, Calcul des structures par la méthode des éléments finis*, Institut Universitaire de Technologie du Mans
[10] Plag R, Ubakus, U-value calculator, https://www.ubakus.de
[11] Design code for masonry structures, indicative CR 6-2013 (in Romanian)
[12] Seismic Design Code, Part I indicative P 100-1/2013 (in Romanian)
[13] Eurocode 3: Design of steel structures (in Romanian)
[14] Normative on the calculation of global thermal insulation coefficients for residential buildings, indicative C107/1-2005 (in Romanian)
[15] Normative regarding the calculation of the thermoenergetic performances of the building elements, indicative C107/3-2005 (in Romanian)
[16] Normative regarding the design and verification of wooden buildings, indicative NP 005-2022
[17] SR EN 1995-1-1:2004 Eurocode 5: Design of wooden structures. Part 1-1: Generalities. Common rules and building rules (in Romanian)

# Demolition of a reinforced concrete overpass. Case study

**G Boacă[1], M C Scutaru[1], C S Iriciuc[1], G R Gimiga[1], D N Dima[1]**
[1] Gheorghe Asachi Technical University of Iasi-Romania, Faculty of Civil Engineering and Building Services, Department of Transportation Infrastructure and Foundations, Blvd. Mangeron, No. 1, 700050, Iasi, Romania

E-mail: gheorghita.boaca@academic.tuiasi.ro

**Abstract.** Bridges are built to fulfil different roles depending on the needs of the site. One of the most important role is to provide easy and non-life-threating access to different areas of community interest. Over the period of their operation, structures suffer various degradations whose extension may endanger the integrity of the construction, or the changing needs of the users. This raises the problem of finding the most efficient methods of removing the structure from use and returning the land to other uses adapted to the period needs. The structure to which we have turned our attention in this article is located in Bucharest. It is represented by an overpass built in a frame solution with inclined pillars, having the shape of the letter "V". The construction material consists mainly of monolithic cast reinforced concrete. The present article presents a modern methodology for the demolition of the reinforced concrete overpass in discussion, through the progressive dismantling of the component construction elements. This method has many advantages, particularly in terms of the possibility of recycling the construction materials used and in terms of minimising traffic congestion on the railway.

## 1. Introduction

Bridges are built to facilitate user access to different areas of community interest. Their construction leads to an increase in the standard of living through economic development [1]. In addition, different administrative or economic entities may have a need for easy access to two different sites of the same entrepreneur. Those area are, sometimes, separated by the existence of an obstacle such as another road or water. By analysing the possible solutions, in some cases, it will be more economical to build a passage.

A relevant example, in this case, is the construction, in 1963, by the Romanian Railway Company (C.F.R.), of a passage in Sector 6 of Bucharest. This structure provide access from Drumul Săbăreni Street, to the warehouses of C.F.R. and to the whole area located on the right side of the railway. The passage is located at km 6+645 of the railway line 100 Bucharest North - Videle. At that time, railway transport was at its peak and the crossing was a real challenge and a dangerous point for the life of C.F.R. warehouses employees. Since it was the same institution, several journeys had to be made each day, increasing the risk of accidents.

Throughout the operation, managers invest time and financial resources to maintain the structure in optimal conditions for normal operation [2]. Today, one of the biggest challenges for administrators is maintaining existing infrastructure [3, 4]. Thus, administrators and designers seek the best solutions for performing maintenance and repair work on bridges and overpasses on time [5, 6]. Although a lifespan of 100 years is taken into account at the design stage, and despite the best efforts of managers, there are situations where, given the development of an area's economy and the new habits of the population, various structures no longer meet the needs of users after a shorter period [7].

Once this point has been reached, the manager examines from both technical and economic point of view the possibility of maintaining the bridge or to allow other types of investment on the site.

This was also the case for the passage that was under consideration here. In 2020, the manager decided to decommission it, following a technical expert's report which highlighted the serious defects and deterioration of the structure.

At the time of the project to authorise the dismantling work, the problem arose of finding an intervention solution that would not disrupt the traffic on the railway lines crossed. This was because there was no possibility of closing the traffic due to the increased importance of the rail traffic served. Several options for intervention were studied, and the option of dismantling the various components one by one was chosen and it is the focus of this article.

## 2. Presentation of the railway passage structure

Located at km 6+645 of the 100 Bucureştii Noi - Videle line, the overpass was built according to the technological level of 1963. Its main purpose is to ensure the access of the Romanian Railway staff to the units in the area, crossing 3 railway lines. This lines reach the Y end (the end of the line from the station pointing in the opposite direction to kilometre 0 - Bucharest) of the New Bucharest station. The lines crossed are the Bucharest - Rosiori line, the I and II lines (the two most important railway lines in the station, representing the continuation of the current track in front of the station), and the 301Z Bucuresti Noi - Rudeni line. A new connection with Drumul Poiana Pietrei Street in Sector 1 was created with the development of the area after the construction of the overpass [8].

The bridge we are going to focus on is made up of a frame solution with inclined piers in the shape of the letter "V" (see figure 1), made of monolithic cast reinforced concrete. The total length of the crossing is 34.40 m, consisting of three spans of different lengths (9.10 m + 14.80 m + 9.10 m) [8].

**Figure 1.** Side view of the passageway.

The bridge superstructure consists of two wide girders with a constant height of 60 cm and width of 1.30 m (figure 2). The distance between the girders is 3.40 m. To ensure load carrying and load distribution, the tops of the girders have a constant thickness (approximately 30 cm). In addition, to optimize the connection between the two elements, a central truss was placed at the intersection between the beams and the slab, and another truss was placed at the center of each span. The edge truss was replaced by increasing the size and shape of the beams towards the bearing sections, boxing the section, and using a reinforced concrete slab to connect the beams at the bottom [8].

The passage infrastructure consists of 4 inclined piers arranged in a "V" shape (figure 3), made of reinforced concrete. These pillars have a rectangular section, with plan dimensions of 1.30*0.60 m and a length of 7.00 m. The loads transmitted by the superstructure through the infrastructures are transferred to the ground by 4 direct foundations, one for each pillar, with plan dimensions of 2.20*3.85m [8].

**Figure 2.** View of the frame.

**Figure 3.** View of the frame pillar.

**Figure 4.** View of the path on the passage.

Traffic in the bridge area is served by a 7.00 m wide carriageway with 2 lanes. The bridge also has 2 footpaths, each 1.10 m wide, due to its location within the town. Designed and built to meet all the technical requirements for 1963, the carriageway is separated from the footpaths only by concrete curbs [8].

IOP Conf. Series: Materials Science and Engineering 1304 (2024) 012027 doi:10.1088/1757-899X/1304/1/012027

## 3. Demolition of the passage by controlled detonation

Over time, any structure exposed to the effects of weathering, self-weight, and payload will deteriorate. Some of these will affect the strength of the structure, resulting in the need to demolish the bridge or passage in service.

Thus, the manager of the structure, being the same entity that is responsible for the maintenance and management of the obstacle crossed, has imposed several restrictions on the design process. This restrictions was the minimization of the closure time of rail traffic on the crossed lines and the duration of the whole work execution.

Taking these constraints into account, the design team considered several possibilities for achieving the required objectives. In the end, two demolition methodologies were developed, neither of which fully met the beneficiary's requirements.

The first technology analysed involved placing a quantity of explosives at the lower level of 2 inclined pillars on the left side of the structure. These decks were chosen because they represent points whose destruction would immediately lead to the total collapse of the structure.

The detonation could only be carried out safely with a total closure to rail traffic, extended by a minimum of 1 day, which was necessary to completely clear the area of debris from the structure. In this case, the closure to rail traffic was to be for a minimum of 36 hours, with the possibility of extension up to 48 hours.

Also, from the point of view of environmental conditions, the use of controlled detonation is not an optimal method. It leads to large amounts of small waste and a high pollution of the area immediately after detonation. The resulting materials will be difficult to recycle.

Considering the location of the structure, in an urban area, there would have been significant vibrations at the detonation time. This could have exacerbated the deterioration of nearby buildings and endangered people's safety.

After presenting this solution to the beneficiary and analysing the arguments put forward, it was concluded that the solution was not feasible for the Bucharest railway passage. This is because the impact of the disadvantages far outweighed the solution's strengths.

## 4. Demolition of the overpass by phased dismantling of the structure

The second option under consideration involves dismantling the components. This method does not involve the closure of traffic for long periods, but only the replacement of the usual locomotives with diesel locomotives.

The optimal stages of the structure's dismantling were calculated during the design. At that moment it was taking into account the areas of minimum stress and the transport possibilities of the resulting elements. In the areas where the lowest stresses are developed, the elements will be cut, lifted with specially chosen cranes according to the weight of each resulting element and moved to special platforms. They will be transported to specialised warehouses where they will be dismantled and the component building materials will be sorted for recycling.

The projected stages of the structure demolition (figure 5) are:

0. Preliminary stage in which both road and pedestrian traffic on the bridge is completely closed and all contact wires on the railway lines are to be uncovered. Rail traffic will be diverted onto the central line for the duration of the works. Only diesel locomotives will be allowed to run. This decision has been taken mainly for the safety of workers and to avoid accidents on the railway or the site. The diversion will be signposted, thus avoiding the creation of a bottleneck that could jeopardize railway safety.

1. The removal of pedestrian parapets and guardrail panels (figure 5) is intended to remove additional permanent weights that are loading the structure. In this stage, the unravelling of the path elements from the bridge to the monolithic plate level on the whole surface of the passage also takes place.

2. In the next step (figure 5 and figure 6) the cutting and unravelling of the sidewalk brackets and the central plate panels were carried out, thus relieving the main beams of all extra weight. To ease the

IOP Conf. Series: Materials Science and Engineering 1304 (2024) 012027 doi:10.1088/1757-899X/1304/1/012027

demolition process, the cutting of the elements has been dimensioned so that their weight is approximately equal. The resulting dimensions allowed these elements to be handled from the walkway in whole pieces. In order not to break during lifting, the length of the panels will not exceed 5.00 m. In the longitudinal direction, the brackets shall be cut approximately 2.10 m from the edge of the parapet slab (figure 7). To remove the central panels, the elements shall be cut into panels 5.00 m long. In the transverse direction they shall be cut 55 cm from the axis of the structure. The cutting of the elements will be done with a diamond blade with an outside diameter of 900 mm for reinforced concrete. To ensure safety in the work by avoiding the panel falling, before starting the cutting procedures, the panels will be supported by a crane of appropriate capacity at 4 endpoints (figure 8). To minimize the execution time, 2 cranes will be used simultaneously, one for each half of the structure.

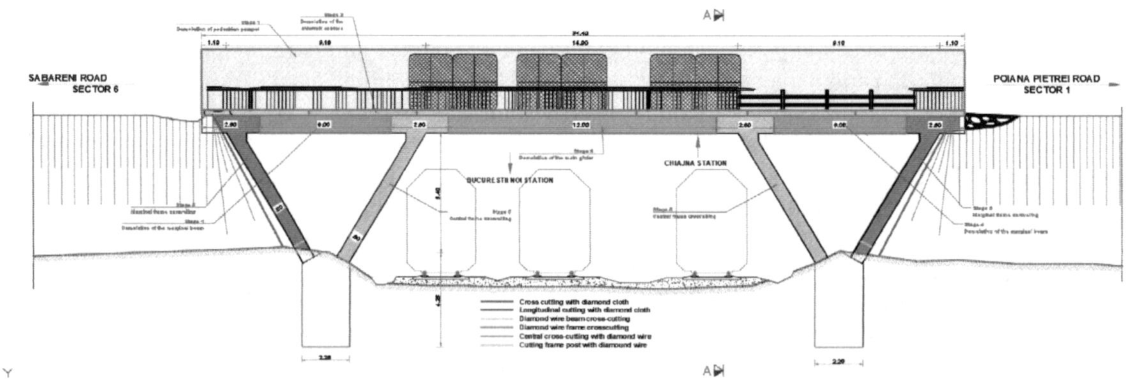


**Figure 5.** Stages of demolition of the structure by dismantling the elements (Side view).

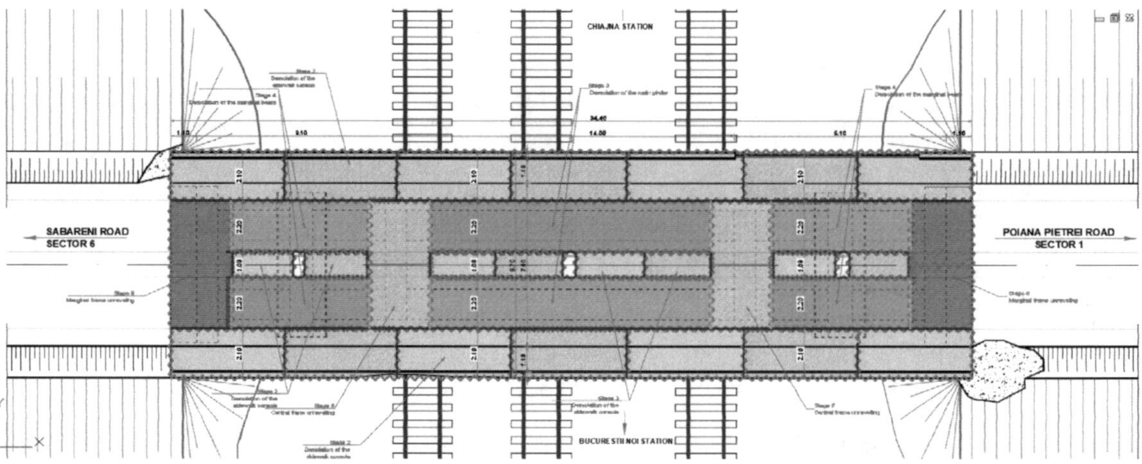


**Figure 6.** Stages of demolition of the structure by dismantling the elements (Plan view).

For the central panel, cross-cuts are made (figure 9) at 5.00 m intervals, paying particular attention to the location of the brackets. Two transverse cuts shall be made on either side of them. Because of their considerable weight, the central panels must also be cut lengthways from one end of the structure to the other. The method of supporting the resulting panels (figure 8) in 4 points will be maintained, allowing also a reduction of the execution time of the work, as immediately after the detachment the panels will be moved on special wagons and transported to the warehouses. The work will be carried out from one end of the bridge, using the retraction method, a crane and 2 cutting machines.

IOP Conf. Series: Materials Science and Engineering 1304 (2024) 012027 doi:10.1088/1757-899X/1304/1/012027

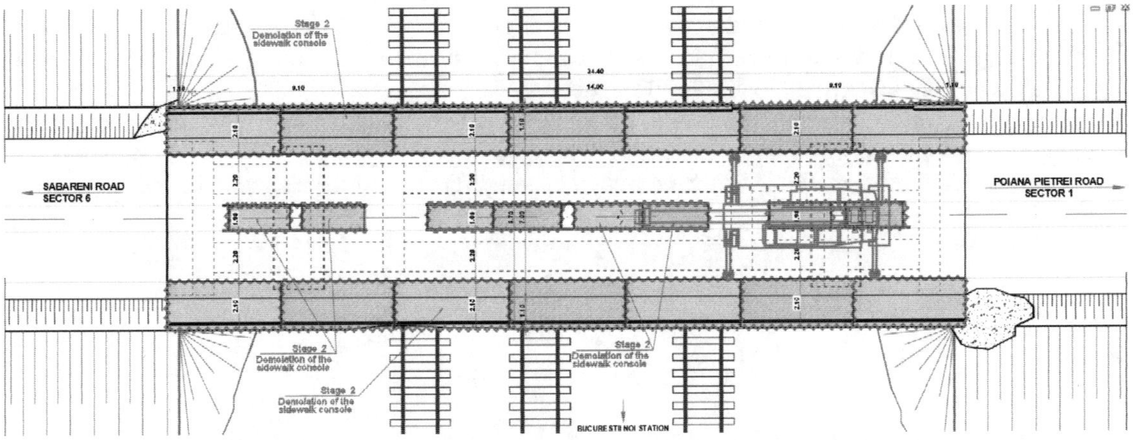


**Figure 7.** Demolition of the pavement brackets and central plate panels (Plan view).

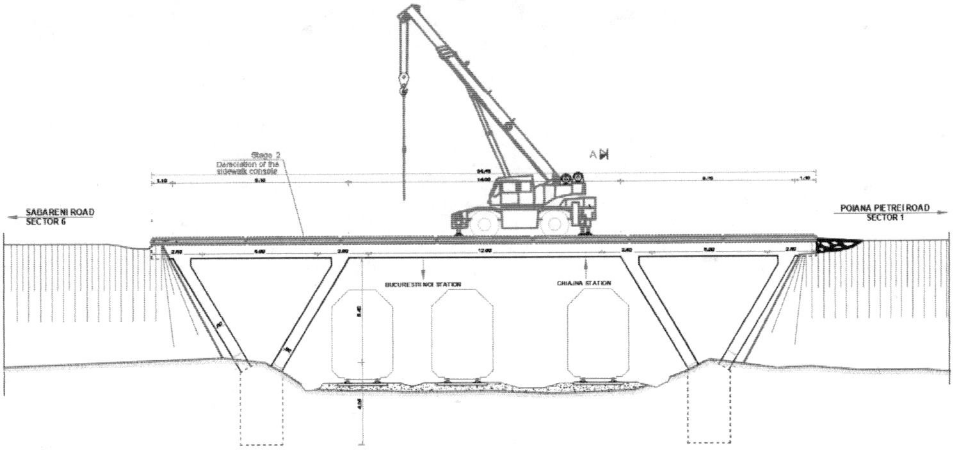


**Figure 8.** Demolition of the pavement brackets and central board panels (Side view).

3. The work will continue with the unravelling and removal of the main beams placed in the central opening (figure 10). In the previous stage, the longitudinal joints were executed, resulting in two independent beam elements. In this stage, the beam ends near the piers, and the middle section of the central brace is cut with a diamond wire. A railway crane will be used to support the resulting elements (figure 11). The maximum dimensions of this elements will be 12.00 m length, 2.20 m width, and 1.00 m height. Before starting the cutting procedure, the elements will be anchored in 4 corners. When the cutting is completed, the crane will lift the reinforced concrete beam element and place it on a railway wagon (figure 11) for transport to the warehouse. Particular attention will be paid to the arrangement and mounting of the fixings, mainly because of the heavy weight of the elements, with a single beam weighing around 45 t. This also leads to the need for a special crane to lift and handle such a weight. In compliance with the conditions imposed by the beneficiary, the entire removal operation of the beam center sections will be carried out with the complete closure of rail traffic. This closure will be outside the heavy traffic intervals, in time windows of approximately 4 hours.

4. The following works were recommended as part of the demolition of the beams for the marginal openings:

- anchoring of the V-frame railway poles to prevent them from falling and destabilizing the frame
- cutting off the central truss in the marginal openings

IOP Conf. Series: Materials Science and Engineering 1304 (2024) 012027 doi:10.1088/1757-899X/1304/1/012027

- cutting the ends of the beam from the marginal openings. To prevent falls, the ends of the beams will be anchored at 4 corners and lifted by a crane at the level of the access ramps. This position of the crane has been chosen to avoid blocking rail traffic
- the elements resulting from the felling will be transported by road to the depots by means of trailers, avoiding blocking the railway wires
- the operation shall be carried out 4 times for each beam, 2 repetitions for each end of the passage

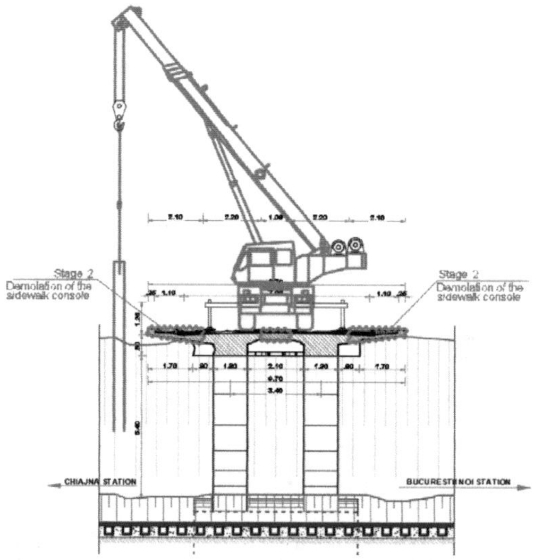


**Figure 9.** Stages of demolition of the structure by dismantling the elements (Cross section).

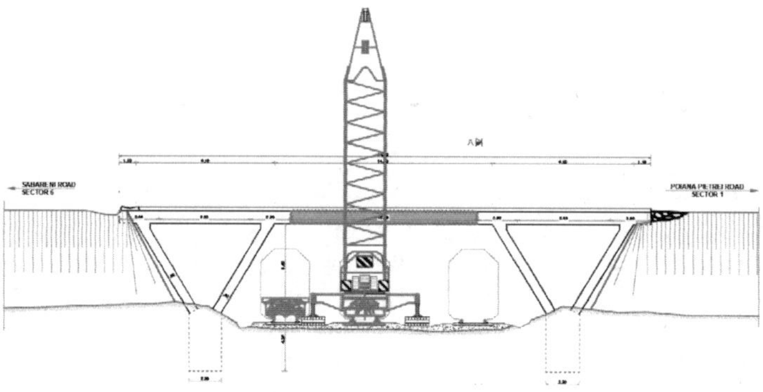


**Figure 10** Demolition of the main beams in the central opening (Cross section).

5. The penultimate stage involves the dismantling of the V-frame columns (figure 13) supporting the superstructure of the passage. To facilitate the dismantling and transport of the elements, the frame will be sectioned into 2 symmetrical segments along the central axis of the marginal slab.
- in order to avoid any destabilization, before cutting the railhead, support the pillar from the railway side with a railway crane positioned on the line next to the pillar
- the first segment of the pole is cut from the railway side, the element is lifted and placed on a railway wagon to be transported to the warehouse
- the demolished segments have a weight of approximately 34 t, leading to increased attention being paid to the strength and location of the fastening and anchoring elements
- operations will be repeated for each of the 4 frame segments
- the beneficiary has approved a closure of rail traffic of maximum 4 hours per frame segment

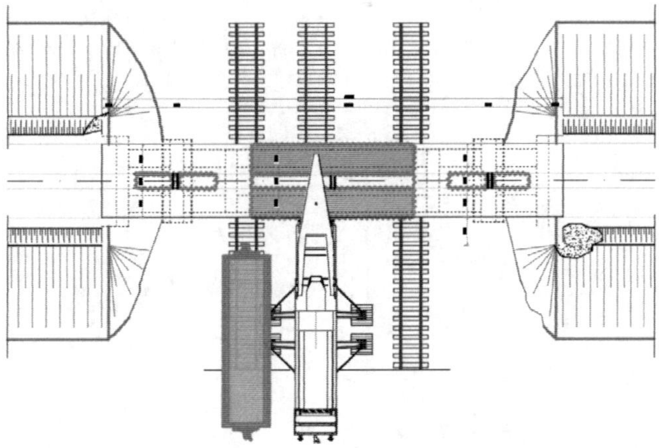

**Figure 11.** Demolition of the main beams in the central opening (Plan view).

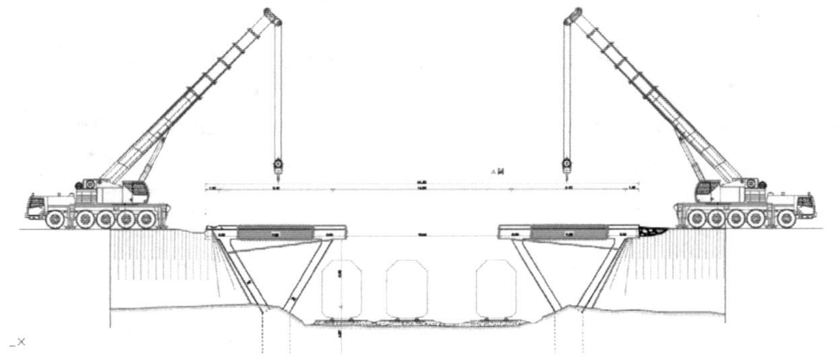

**Figure 12.** Demolition of beams in marginal openings.

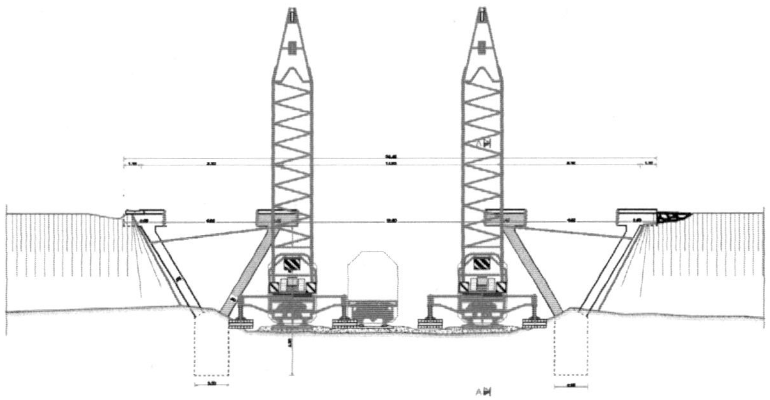

**Figure 13.** Demolition of frame pillars.

6. Finally, the marginal costs of the frames are dismantled (figure 14). The need for this stage arose from the construction of the entire passage, i.e. the burial of the end slabs in the embankment under the access ramps. To reach the end slabs, the road surface on the ramps of the passage will be removed over a length of 1.50 m and the end slabs will be cut in the central section with a diamond-reinforced concrete disc. This will remove the joint between the 2 pillars on each ramp. The resulting

frame segments will be supported by a road crane positioned on the ramps. To remove each segment a cut will be made at the base of the post using a diamond wire, followed by lifting the segment, placing it on the road trailer next to the access ramp, and transporting it to the warehouse.

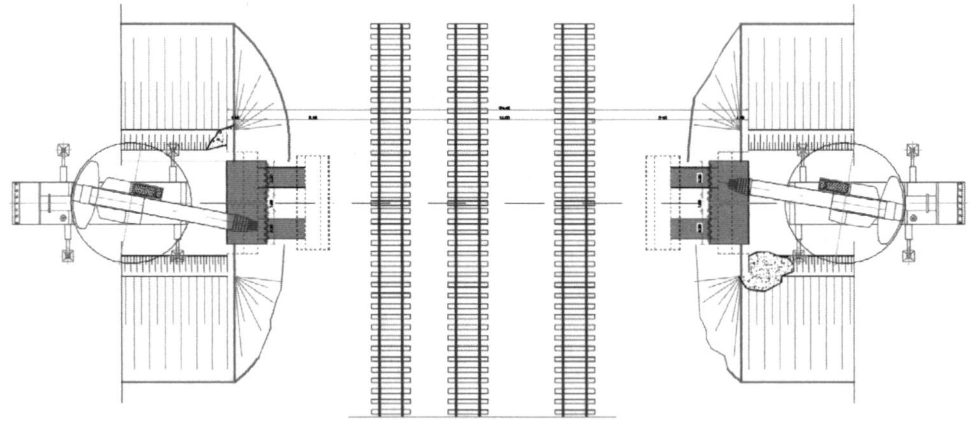

**Figure 14.** Demolition of the marginal pillars of the frames

After completion of the road overpass decommissioning works, the track rectification works will be carried out, together with the laying of contact wires for all 3 tracks. The embankments will be laid out to fit in with the landscape of the area, disregarding the foundation blocks which will remain buried in the site.

For the safety of road traffic on the roads that were served by the passage will be placed barriers and road signs to warn traffic participants.

## 5. Conclusions

When the standard service life of a structure is exceeded, deterioration endangers the safety of users, or the purpose for which the bridge was built is no longer fulfilled, the administrator may decide to decommission the structure. Such works raise problems, especially when it comes to the possibility of recycling materials and carrying out the works in the shortest possible time, without affecting other buildings in the area.

This article focuses on how to carry out the demolition of an overpass built over a railway. The restrictions imposed by the beneficiary led to the development of a special methodology for dismantling the parts of the structure and transporting them to specially equipped warehouses. Within the warehouses, the elements were recycled and the component materials being reused in different ways, depending on their specificities.

This technology has many advantages, the most important of which are the possibility of carrying out the work without closing rail traffic except for extremely limited periods. Another advantage is the fact that it does not cause vibrations that would endanger the safety of adjacent buildings.

This work is part of an extensive research program developed by the Faculty of Construction and Installations of the Technical University "Gheorghe Asachi" of Iasi. The research focuses on the study and development of maintenance works necessary to maintain bridges at optimal parameters for normal operation and on the optimization of interventions on them.

## 6. References

[1] Iliescu M, Viorel G 2012 *Repair and strengthening of concrete bridges* (Cluj-Napoca: U.T.Press) p 1

[2] Nandal M, Sood H, Gupta P K, Haq M Z U 2022 *Morphological and physical characterization of construction and demolition waste,* Materials Today: Proceedings. Preprint https://doi.org/10.1016/j.matpr.2022.11.013

[3] Comisu C C, Țăranu N, Boacă G, Scutaru M C 2017 *Structural health monitoring system of bridges,* Procedia Engineering. 199 2054-2059

[4] Scutaru M C, Țăranu N, Comisu C C, Chingălată C 2017 *Development of bridge structural health monitoring*, Bult. Inst. Polit. Iasi. 63 (67) 105-120

[5] Gimiga G R, Iriciuc S C 2012 *Workability of storable asphalt mixtures, measuring devices*, Bult. Inst. Polit. Iasi. t. LVIII(LXIII) 175-180

[6] Scutaru M C, Țăranu N, Comisu C C, Boacă G, Ungureanu D, Hudișteanu I 2019 *Numerical study of the dynamic characteristics of a hybrid steel-concrete bridge*, IOP Conf. Series: Materials Science and Engineering. 591 012030

[7] Rezaei N, Garber D 2019 *Study of bridge demolition DOT survey and available standard specifications*, Advances in Civil Engineering. 4896717 1-6

[8] Roca C 2022 *Project for authorising the execution of dismantling works - P.A.D. Decommissioning of overpass km 6+645 Bucuresti - Roșiori* (Iasi)

IOP Conf. Series: Materials Science and Engineering 1304 (2024) 012028 doi:10.1088/1757-899X/1304/1/012028

# Ecologization of Waste from the Steel Industry in Civil Engineering Works

**S C Iriciuc[1], G R Gimiga[1], N I Găluşcă[1], D N Dima[1], M C Scutaru[1], Gh Boacă[1]**
[1] "Gheorghe Asachi" Technical University of Iasi-Romania, Faculty of Civil Engineering and Building Services, Department of Transportation Infrastructure and Foundations, Blvd. Mangeron, No. 1, 700050, Iasi, Romania

E-mail: silviu-cristian.iriciuc@academic.tuiasi.ro

**Abstract**. The massive industrialization of the last century has come with the drawback of creating large waste deposits. Over the past century, steel mills have generated significant quantities of steel slag, deposited in heaps that occupy vast land areas. The desire to recycle steel slag arises from both ecological and economic interests. To achieve this goal in the steel industry, it was necessary to recycle steel slag (waste from the technological process of steel production) by using them as aggregates in civil constructions. In this work, the optimal option will be determined based on the mathematical modelling of the main physical-mechanical characteristics of the asphalt mixtures used in wear layers made with steel slag aggregates in comparison with classic asphalt mixtures.

## 1. Introduction

The processing of steel slag is a technique practiced for over 30 years in Europe or the USA, in fact the transformation of these solid wastes into valuable mineral granular materials is a concern, worldwide.
Ferrous metallurgical slags are perhaps the only materials, in the wider category of metallurgical slags, which, through use, do not have negative influences on the environment. The processing technology is part of the "clean technologies" because it does not involve chemical transformations on the slags, but only physical-mechanical ones, and this does not differ much from natural rock processing facilities. The processing of ferrous metallurgical plant slag brings benefits from the point of view of environmental protection, from the community or agricultural point of view, as well as from the technical point of view. Ferrous metallurgical plant slags, being non-toxic materials, but having physical-mechanical properties similar to rocks, are a viable alternative, in certain applications even more valuable than natural variants for the construction field. Steel slags with varied chemical composition are processed as lumps or as fine-grained products. The processing steps during the production of these slag granules correspond to the processing applied to natural stone. Crushed steel slag is used in constructions for traffic (parking platforms, runway strips, bicycle tracks, beltway, etc.). In addition, steel slag is also used for road construction and for mineral seals. In our country, blast furnace slag has been used in road construction for almost 30 years, with a pozzolanic role in the construction of road systems [1].
It has been found, particularly in developed countries concerned with limiting industrial waste, that steel slag left over after the extraction of ferrous waste, sorted and cleaned of impurities, successfully replaces natural aggregates specific to road and/or hydro-engineering works [1]. In road construction, steel slag aggregates are used either as a foundation layer, base layer or, in recent years, mainly for bituminous surfacing.
It has been estimated that worldwide, since 1978, more than half of the steel slag production has been used in road construction, as follows:

- base layers: 57%;
- foundation layers: 6%;
- bituminous layers: 13%
- deposited in landfills 24% [1].

In terms of production, approximately 700 000 tonnes/year of new slag from steel production alone is processed annually. In the dumping ground, excavation is done only for the search of iron to be recovered.

Steel slag, as a waste by-product of steel-making procedure, mainly comprises CaO, SiO2, Al2O3, MgO, and Fe2O3. Although it is prone to swelling due to the hydration of free lime, some of its properties such as hardness, wear resistance, adhesiveness, roughness, and toughness have made it a raw material in civil engineering applications such as embankment fills, road construction, and concrete building. Substitution of natural aggregates with steel slag aggregates can improve the properties of the asphalt mixture and reduce steel production wastes [2].

However, 13 million tonnes of slag are "produced" annually in Romania from steel and cast iron production processes. Their disposal in dumping grounds poses serious environmental problems, with large areas of land taken up and unstable slag layers that can be tens of metres thick. There has always been a concern in steel to use this product as efficiently as possible. These efforts have led to the almost total use of metallurgical slag, especially as a mineral material in construction and as a fertiliser. Through various technologies, waste from slag can become primary approved products used in various branches of the economy such as roads, land improvements, and railway infrastructures [3]. The recycling percentage of this waste makes it one of the most recoverable materials used in modern industry.

## 2. Mechanical and physical characteristics for the steel slag

Slag or metallurgical by-products are chemically and mineralogical different from natural rocks, but have similar physical and mechanical (technical) characteristics.

The widely varying chemical compositions of slags, which result in differences in their physical-chemical and mineralogical properties, make their processing very difficult. The short solidification interval means that steel slag is mainly processed in the solid state.

Following sampling of the Buituri slag dumping ground, 10 samples were considered significant and have been presented in table1 [4].

**Table 1**. The chemical composition of the slag from the Buituri slag dumping ground

| Composition / Specimens | Fe(I) | FeO | $Fe_2O_3$ | Fe(II) | $SiO_2$ | MnO | $Al_2O_3$ | CaO | MgO | S | P | CaO / $SiO_2$ |
|---|---|---|---|---|---|---|---|---|---|---|---|---|
| 1 | 21.24 | 12.22 | 8.40 | 5.95 | 17.20 | 6.42 | 8.92 | 31.26 | 9.63 | 0.22 | 0.50 | 1.82 |
| 2 | 20.21 | 14.80 | 4.20 | 5.87 | 19.46 | 6.80 | 7.62 | 34.14 | 8.72 | 0.13 | 0.40 | 1.75 |
| 3 | 17.84 | 10.21 | 4.13 | 7.87 | 19.18 | 7.15 | 6.17 | 36.25 | 8.05 | 0.17 | 0.55 | 1.85 |
| 4 | 22.17 | 12.51 | 4.38 | 9.74 | 18.81 | 7.14 | 5.68 | 35.70 | 8.21 | 0.12 | 0.48 | 1.89 |
| 5 | 19.56 | 9.88 | 5.18 | 8.32 | 17.56 | 6.08 | 8.01 | 36.52 | 9.11 | 0.18 | 0.53 | 2.08 |
| 6 | 21.35 | 11.42 | 7.47 | 7.32 | 17.01 | 7.02 | 7.64 | 39.3 | 9.10 | 0.13 | 0.46 | 2.31 |
| 7 | 20.08 | 12.35 | 6.12 | 6.28 | 18.56 | 6.86 | 7.08 | 35.36 | 8.37 | 0.14 | 0.51 | 1.90 |
| 8 | 15.37 | 9.25 | 5.86 | 4.15 | 19.35 | 6.36 | 3.81 | 47.16 | 11.05 | 0.35 | 0.75 | 2.43 |
| 9 | 16.45 | 10.12 | 6.35 | 4.22 | 18.06 | 7.37 | 4.06 | 48.56 | 10.73 | 0.38 | 0.81 | 2.68 |
| 10 | 21.03 | 11.52 | 7.35 | 6.39 | 18.62 | 7.11 | 6.35 | 38.77 | 9.01 | 0.15 | 0.58 | 2.08 |
| **Average** | **19.53** | **11.39** | **5.94** | **6.66** | **18.38** | **6.83** | **6.53** | **35.30** | **9.20** | **0.20** | **0.56** | **2.08** |

## 3. Laboratory tests

Laboratory tests were conducted to determine the mechanical and physical properties of the two types of asphalt mixtures. The tests were performed at the Road Laboratory of the Faculty of Civil Engineering and Building Services from Iasi.

*3.1 Grain size*

Both natural and artificial aggregates are utilized in the preparation of the asphalt mixtures as stated by SR EN 13043-2003 [5].

These asphalt mix formulas required the following types of aggregates:

- crushed slag aggregate 0-4 mm
- 0-4 mm chippings
- crushed slag aggregate 4-8 mm
- 4-8 mm chippings
- crushed slag aggregate 8-16 mm
- chippings 8-16 mm
- natural sand 0-4 mm
- filler.

For determining the grain size of natural and artificial aggregates, the following test sieves are used in accordance with SR EN 933-2-2020 [6], [7].

Following the screenings, the obtained sieve passages are shown in Tables 2-9 [5], [7].

**Table 2**. Grading characteristics for natural sand 0-4 mm

| No. | Characteristics | Unit | Results obtained in the laboratory | | Standard | Condition AND. 605/2016;SR EN 13043/2003/AC 2004 | |
|---|---|---|---|---|---|---|---|
| 1 | Grading composition of the mineral aggregate (passing) | aperture sieve | retained % | passing % | SR EN 933/1-2012 | | Category G |
| | | 8 | 0 | 100,0 | | | |
| | | 4 | 2.9 | 97.1 | | | |
| | | 2 | 18.5 | 78.6 | | | |
| | | 0.125 | 69.4 | 9.2 | | | |
| | | 0.063 | 7.6 | 1.6 | | | |
| | | T 0.063 | 1.6 | | | | |
| 2 | Content of particle outside the grading class - retained on the upper sieve (d max), | % | 3.8 | | | max. 10 | |
| 3 | Content of fine particle (passing through the sieve 0,063 mm) max. | % | 1,6 | | | max. 10 | $f_3$ |
| 4 | Passing through the upper sieve (weight) Db | % | 96.2 | | SR EN 933/1-2012 | | 90-99 |
| | Passing through the lower sieve d | % | 0 | | | | 0-10 |
| | Passing through the sieve d /2[a] | % | 0 | | | | 0-2 |
| | Grading | - | | | | continuous | $G_c$ 90/10 |

IOP Conf. Series: Materials Science and Engineering 1304 (2024) 012028 doi:10.1088/1757-899X/1304/1/012028

**Table 3.** Grading characteristics for chippings 0-4 mm

| No. | Characteristics | Unit | Results obtained in the laboratory | | Standard | Condition AND. 605/2016;SR EN 13043/2003/AC 2004 | |
|---|---|---|---|---|---|---|---|
| 1 | Grading composition of the mineral aggregate (passing) | aperture sieve | retained % | passing % | SR EN 933/1-2012 | | Category G |
| | | 8 | 0 | 100 | | | |
| | | 4 | 1.2 | 98.8 | | | |
| | | 2 | 29.1 | 69.7 | | | |
| | | 0.125 | 47.5 | 22.2 | | | |
| | | 0.063 | 21.1 | 1.1 | | | |
| | | T 0.063 | 1.1 | 0 | | | |
| 2 | Content of particle outside the grading class - retained on the upper sieve (d max), | % | 1.2 | | | max. 10 | |
| 3 | Content of fine particle (passing through the sieve 0,063 mm) max. | % | 1.1 | | | max. 10 | $f_{10}$ |
| 4 | Passing through the upper sieve (weight) Db | % | 98.8 | | SR EN 933/1-2012 | | 90-99 |
| | Passing through the lower sieve d | % | 0 | | | | 0-10 |
| | Passing through the sieve d /2[a] | % | 0 | | | | 0-2 |
| | Grading | - | | | | continuous | $G_c$ 90/10 |

**Table 4.** Grading characteristics for steel slag aggregates 0-4 mm

| No. | Characteristics | Unit | Results obtained in the laboratory | | Standard | Condition AND. 605/2016;SR EN 13043/2003/AC 2004 | |
|---|---|---|---|---|---|---|---|
| 1 | Grading composition of the mineral aggregate (passing) | aperture sieve | retained % | passing % | SR EN 933/1-2012 | | Category G |
| | | 8 | 0 | 100 | | | |
| | | 4 | 0.9 | 99.1 | | | |
| | | 2 | 27.5 | 71.6 | | | |
| | | 0.125 | 48.15 | 23.45 | | | |
| | | 0.063 | 21.8 | 1.65 | | | |
| | | T 0.063 | 1.65 | 0 | | | |
| 2 | Content of particle outside the grading class - retained on the upper sieve | % | 0.9 | | | max. 10 | |

| | | | | | | |
|---|---|---|---|---|---|---|
| | (d max), | | | | | |
| 3 | Content of fine particle (passing through the sieve 0,063 mm)max. | % | 1.65 | | max. 10 | $f_{10}$ |
| 4 | Passing through the upper sieve ( weight) Db | % | 99.1 | SR EN 933/1-2012 | | 90-99 |
| | Passing through the lower sieve d | % | 0 | | | 0-10 |
| | Passing through the sieve d /2[a] | % | 0 | | | 0-2 |
| | Grading | - | | | continuous | $G_c$ 90/10 |

**Table 5**. Grading characteristics for chippings 4-8 mm

| No. | Characteristics | Unit | Results obtained in the laboratory | | Standard | Condition AND. 605/2016;SR EN 13043/2003/AC 2004 | |
|---|---|---|---|---|---|---|---|
| 1 | Grading composition of the mineral aggregate (passing) Content of particle outside the grading class - retained on the upper sieve (d max), | aperture sieve | retained % | passing % | SR EN 933/1-2012 | | Category G |
| | | 16 | 0 | 100 | | | |
| | | 8 | 8.1 | 91.9 | | | |
| | | 4 | 70.4 | 21.5 | | | |
| | | 2 | 17.2 | 4.3 | | | |
| | | 0.125 | 3.5 | 0.8 | | | |
| | | 0.063 | 0.4 | 0.4 | | | |
| | | T 0.063 | 0.4 | 0 | | | |
| 2 | Content of fine particle (passing through the sieve 0,063 mm) max. | % | 8.1 | | | max. 10 | |
| 3 | Passing through the upper sieve ( weight) Db | % | 0.4 | | | max. 1 | $f_{0.5}$ |
| 4 | Passing through the lower sieve d | % | 91.9 | | SR EN 933/1-2012 | | 90-99 |
| | Passing through the sieve d /2[a] | % | 21.5 | | | | 0-10 |
| | Grading | % | 4.3 | | | | 0-2 |
| | Grading composition of the mineral aggregate (passing) | - | | | | continuous | $G_c$ 90/10 |

**Table 6.** Grading characteristics for steel slag aggregates 4-8 mm

| No. | Characteristics | Unit | Results obtained in the laboratory | | Standard | Condition AND. 605/2016;SR EN 13043/2003/AC 2004 | |
|---|---|---|---|---|---|---|---|
| 1 | Grading composition of the mineral aggregate (passing) Content of particle outside the grading class - retained on the upper sieve (d max), | aperture sieve<br>16<br>8<br>4<br>2<br>0.125<br>0.063<br>T 0.063 | retained %<br>0<br>8.9<br>71.8<br>15.6<br>2.6<br>0.7<br>0.4 | passing %<br>100<br>91.1<br>19.3<br>3.7<br>1.1<br>0.4<br>0 | SR EN 933/1-2012 | | Category G |
| 2 | Content of fine particle (passing through the sieve 0,063 mm) max. | % | 8.9 | | | max. 10 | |
| 3 | Passing through the upper sieve (weight) Db | % | 0.4 | | | max. 1 | $f_{0.5}$ |
| 4 | Passing through the lower sieve d | % | 91.1 | | SR EN 933/1-2012 | | 90-99 |
| | Passing through the sieve d /2[a] | % | 19.3 | | | | 0-10 |
| | Grading | % | 3.7 | | | | 0-2 |
| | Grading composition of the mineral aggregat (passing) | - | | | | continuous | $G_c$ 90/10 |

**Table 7.** Grading characteristics for chippings 8-16 mm

| No. | Characteristics | Unit | Results obtained in the laboratory | | Standard | Condition AND. 605/2016;SR EN 13043/2003/AC 2004 |
|---|---|---|---|---|---|---|
| 1 | Grading composition of the mineral aggregate (passing) Content of particle outside the grading class | aperture sieve<br>22.4<br>16<br>8<br>4<br>2<br>0.125<br>0.063 | retained %<br>0<br>1.6<br>81<br>17<br>0.1<br>0.2<br>0 | passing %<br>100<br>98.4<br>17.4<br>0.4<br>0.3<br>0.1<br>0.1 | SR EN 933/1-2012 | Category G |

| | | | | | | | |
|---|---|---|---|---|---|---|---|
| | - retained on the upper sieve (d max), | | 0.1 | | | | |
| | Content of fine particle (passing through the sieve 0,063 mm)max. | T 0.063 | | 0 | | | |
| 2 | Passing through the upper sieve ( weight) Db | % | 1.6 | | | max. 10 | |
| 3 | Passing through the lower sieve d | % | 0.1 | | | max. 0.5 | $f_{0.5}$ |
| 4 | Passing through the sieve d /2[a] | % | 98.4 | | | | 90-99 |
| | Grading | % | 17.4 | | | | 0-20 |
| | Grading composition of the mineral aggregat (passing) | % | 0.4 | | SR EN 933/1-2012 | | 0-5 |
| | | - | | | | continuous | $G_c$ 90/20 |

**Table 8**. Grading characteristics for steel slag aggregates 8-16 mm

| No. | Characteristics | Unit | Results obtained in the laboratory | | Standard | Condition AND. 605/2016;SR EN 13043/2003/AC 2004 | |
|---|---|---|---|---|---|---|---|
| 1 | Grading composition of the mineral aggregate (passing) | aperture sieve | retained % | passing % | SR EN 933/1-2012 | | Category G |
| | | 22.4 | 0 | 100 | | | |
| | | 16 | 2.8 | 97.2 | | | |
| | Content of particle outside the grading class | 8 | 79.1 | 18.1 | | | |
| | | 4 | 17.9 | 0.2 | | | |
| | | 2 | 0.1 | 0.1 | | | |
| | - retained on the upper sieve (d max), | 0.125 | 0 | 0.1 | | | |
| | | 0.063 | 0 | 0.1 | | | |
| | | | 0.1 | 0 | | | |
| | Content of fine particle (passing through the sieve 0,063 mm)max. | T 0.063 | | | | | |
| 2 | Passing through the upper sieve ( weight) Db | % | 2.8 | | | max. 10 | |
| 3 | Passing through the lower sieve d | % | 0.1 | | | max. 0.5 | $f_{0.5}$ |
| 4 | Passing through the sieve d /2[a] | % | 97.2 | | SR EN 933/1-2012 | | 90-99 |
| | Grading | % | 18.1 | | | | 0-20 |

| | | | | | |
|---|---|---|---|---|---|
| | Grading composition of the mineral aggregate (passing) | % | 0.2 | | 0-5 |
| | | - | | | continuous $G_c$ 90/20 |

**Table 9**. Grading characteristics for filler

| No. | Characteristics | Unit | Results obtained in the laboratory | | Standard | Condition AND. 605/2016;SR EN 13043/2003/AC 2004 | |
|---|---|---|---|---|---|---|---|
| 1 | Grading composition of the mineral aggregate (passing) | aperture sieve | retained % | passing % | SR EN 933/1-2012 | | Category G |
| | | 2 | 0 | 100 | | | |
| | | 0.125 | 7.83 | 92.17 | | | |
| | | 0.063 | 20.4 | 71.77 | | | |
| | | T 0,063 | 71.77 | 0 | | | |
| 2 | Passing through the upper sieve ( weight) Db | % | 7.83 | | | max. 10 | |
| 3 | Passing through the lower sieve d | % | 71.77 | | | max. 0.5 | $f_{0.5}$ |
| 4 | Content of particle outside the grading class - retained on the upper sieve (d max), | % | 100 | | SR EN 933/1-2012 | | 90-99 |
| | Content of fine particle (passing through the sieve 0,063 mm)max. | % | 100 | | | | 0-10 |
| | Passing through the upper sieve ( weight) Db | % | 100 | | | | 0-2 |
| | Passing through the lower sieve d | - | | | | continuous | $G_c$ 90/20 |

*3.2. Assessment of the asphalt mix recipes*

Thus, according to Table 10 and 11 [1], the following percentages of aggregates were established for the manufacturing of asphalt recipes BA16 with chippings and BA16 with steel slag aggregates, taking into consideration the granularity of each aggregate as a result of the laboratory sieving:

**Table 10**. Analysing the mineralogical mixture of asphalt mix BA16 with chippings

| **Aggregates** | **%** | **retained fraction ( % )** | | | | | |
|---|---|---|---|---|---|---|---|
| | | **16** | **8** | **4** | **2** | **0.125** | **0.063** |
| Chippings 8…16 | 39.50 | 0.63 | 32.00 | 6.72 | 0.04 | 0.08 | 0.00 |
| Chippings 4…8 | 21.60 | 0.00 | 1.75 | 15.21 | 3.72 | 0.76 | 0.09 |
| Chippings 0…4 | 24.10 | 0.00 | 0.00 | 0.29 | 7.01 | 11.45 | 0.27 |
| Natural sand 0…4 | 5.60 | 0.00 | 0.00 | 0.07 | 1.63 | 3.44 | 0.04 |

| | | | | | | | |
|---|---|---|---|---|---|---|---|
| Filler | 9.20 | 0.00 | 0.00 | 0.00 | 0.00 | 0.72 | 6.60 |
| Total retained | 100.00 | 0.63 | 33.74 | 22.28 | 12.40 | 16.44 | 6.99 |
| Total passing | | 99.37 | 65.62 | 43.35 | 30.95 | 14.51 | 7.51 |
| Limits AND 605 | min | 90 | 61 | 39 | 27 | 8 | 7 |
| | max | 100 | 82 | 64 | 48 | 15 | 11 |

**Table 11.** Analysing the mineralogical mixture of asphalt mix BA16 with steel slag aggregates

| **Aggregates** | % | **retained fraction ( % )** | | | | | |
|---|---|---|---|---|---|---|---|
| | | **16** | **8** | **4** | **2** | **0.125** | **0.063** |
| slag aggregates 8…16 | 39.50 | 1.11 | 31.24 | 7.07 | 0.04 | 0.00 | 0.00 |
| slag aggregates 4…8 | 21.60 | 0.00 | 1.92 | 15.51 | 3.37 | 0.56 | 0.09 |
| slag aggregates 0…4 | 24.10 | 0.00 | 0.00 | 0.22 | 6.63 | 11.60 | 0.40 |
| slag aggregates 0…4 | 5.60 | 0.00 | 0.00 | 0.05 | 1.54 | 3.54 | 0.04 |
| filler | 9.20 | 0.00 | 0.00 | 0.00 | 0.00 | 0.72 | 6.60 |
| total retained | 100.00 | 1.11 | 33.17 | 22.85 | 11.58 | 16.43 | 7.13 |
| total passing | | 98.89 | 65.73 | 42.88 | 31.30 | 14.87 | 7.75 |
| limits AND 605 | min | 90 | 61 | 39 | 27 | 8 | 7 |
| | max | 100 | 82 | 64 | 48 | 15 | 11 |

### *3.3. Assessment of the asphalt mix density*

The assessment will be conducted in accordance with SR EN 12697-6 /2020 [9].

This testing method is used to assess the bulk density of bituminous asphalt after compaction. The test is conducted on laboratory-compacted specimens.

To determine the bulk density for the two types of asphalt mixtures, tests were carried out for each mixture with 5 different bitumen dosages. For each percentage of bitumen 6 Marshall specimens were made. The average bulk density values for each bitumen dosage are shown in Tables 12 and 13.

**Table 12**. Bulk density assessment for BA16 with chippings

| bitumen percentage % | bitumen used | bulk density g/cm$^3$ |
|---|---|---|
| 5.75 | | 2,371 |
| 6.00 | | 2,373 |
| 6.25 | D 50/70 | 2,378 |
| 6.50 | | 2,375 |
| 6.75 | | 2,365 |

**Table 13**. Determination of bulk density for BA16 with steel slag

| bitumen percentage % | bitumen used | bulk density g/cm$^3$ |
|---|---|---|
| 5.75 | | 2,629 |
| 6.00 | | 2,654 |
| 6.25 | D 50/70 | 2,696 |
| 6.50 | | 2,691 |
| 6.75 | | 2,680 |

As seen in Figures 1 and 2, the findings were analyzed using the mathematical analysis tool Curve Expert Professional, based on a variety of mathematical functions, to calculate the ideal value of the bitumen % in terms of bulk density [7].

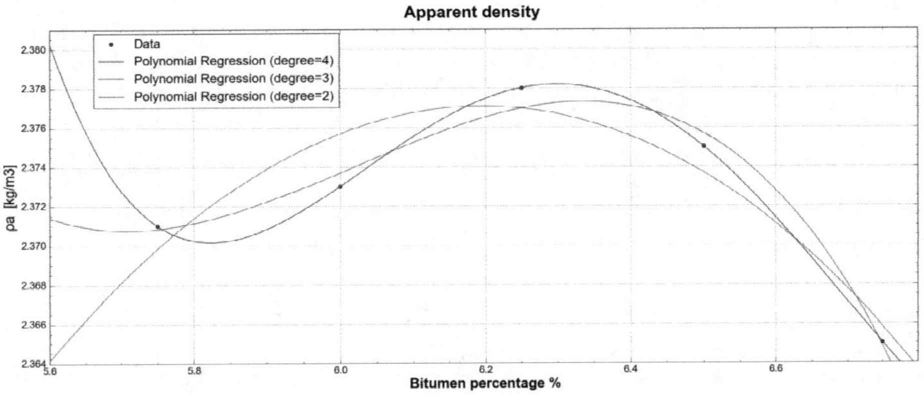


**Figure 1.** Interpretation of bulk density evolution mathematically for BA16 with chippings

Using a variety of mathematical functions and the mathematical analysis software Curve Expert Professional, the findings were evaluated to compute the ideal bitumen % in terms of bulk density.
The polynomial equation of degree 4 is as follows:

$207 - 130.34*x + 30.93*x^2 - 3.25*x^3 + 12.80*x^4 = 0$

The standard error is 0, meanwhile the correlation coefficient is 9.999999999999998E-01.
It can be seen that the maximum bulk density value corresponds to a bitumen dosage between 6.00% and 6.30%.
After interpreting the data, we find that, for asphalt mixtures containing steel slag, the polynomial equation of degree 4 best describes how the values of bulk densities evolve with respect to bitumen %.
The polynomial equation of degree 4 is as follow:

$1723.69 - 1102.47*x + 264.35*x^2 - 28.12*x^3 + 1.12*x^4 = 0$

The standard error is 0, meanwhile the correlation coefficient is 9.999999999999998E-01.

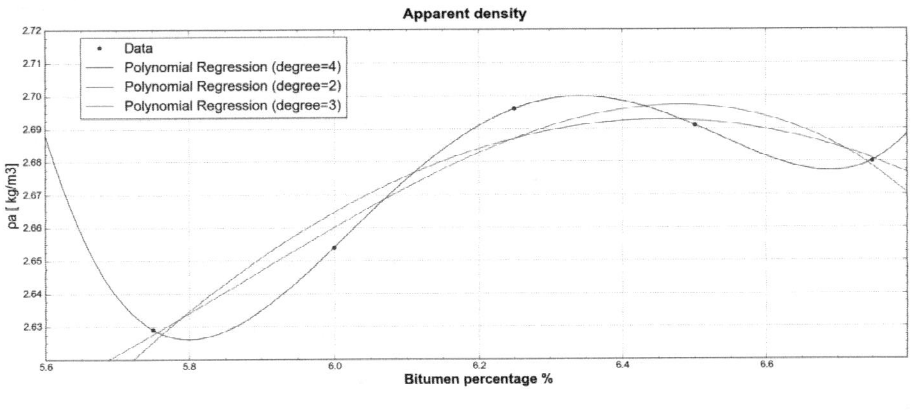


**Figure 2.** Interpretation of bulk density evolution mathematically for BA16 with steel slag

We note that the maximum bulk density value corresponds to a bitumen dosage between 6.00% and 6.30%.

### *3.4. Calculating the water absorption*

When an asphalt mix specimen is held in water under vacuum, the amount of water absorbed by its externally exposed voids is measured and reported as a percentage of the specimen's initial mass or volume. This process is known as water absorption. Tables 14 and 15 show the average water absorption values for each bitumen dosage [7].

**Table 14**. Determination of water absorption for BA16 with with chippings

| bitumen percentage, % | bitumen used | water absorption,% | Condition AND 605 |
|---|---|---|---|
| 5.75 | | 4.823 | 1.5…5.0 |
| 6.00 | | 2.395 | 1.5…5.0 |
| 6.25 | D 50/70 | 1.816 | 1.5…5.0 |
| 6.50 | | 1.650 | 1.5…5.0 |
| 6.75 | | 1.520 | 1.5…5.0 |

In terms of water absorption, the results were interpreted using a mathematical analysis program (Fig. 3, Fig. 4) to determine the optimum percentage of bitumen.

The polynomial equation of degree 4 best reflects, the evolution of the water absorption values in relation to the asphalt percentage for asphalt mixtures with steel slag.

The polynomial equation of degree 4 is as follows:

$$19705.84 - 12197.74*x + 2831.40*x^2 - 292.07*x^3 + 11.30*x^4=0$$

The standard error is 0, meanwhile the correlation coefficient is 1.

We note that with increasing bitumen percentage, the water absorption value decreases, but we observe that for bitumen dosage higher than 6.25% its influence on water absorption is insignificant.

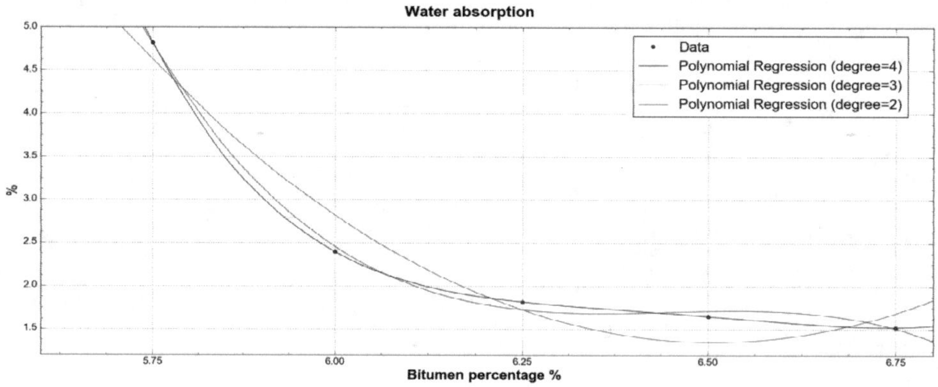


**Figure 3.** Mathematical interpretation of water uptake evolution for BA16 with chippings

**Table 15**. Determination of water absorption for BA16 with steel slag

| bitumen percentage, % | bitumen type | water absorption,% | Condition AND 605 |
|---|---|---|---|
| 5.75 | | 3.642 | 1.5…5.0 |
| 6.00 | | 3.125 | 1.5…5.0 |
| 6.25 | D 50/70 | 2.952 | 1.5…5.0 |
| 6.50 | | 2.480 | 1.5…5.0 |
| 6.75 | | 1.979 | 1.5…5.0 |

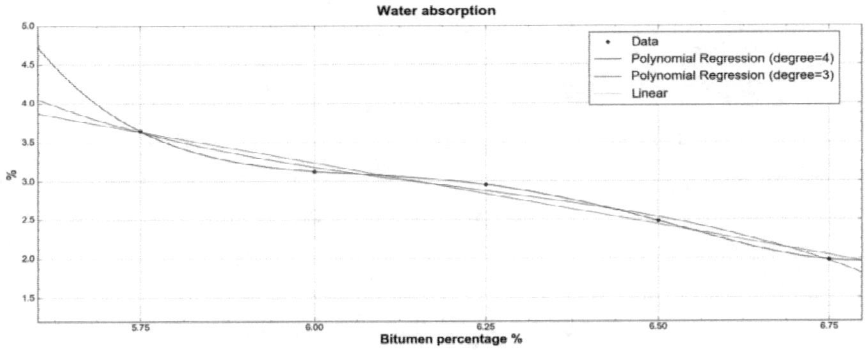


**Figure 4.** Mathematical interpretation of water absorption evolution for BA16 with steel slag

The polynomial equation of degree 4 best reflects, the evolution of the water absorption values in relation to the asphalt percentage for asphalt mixtures with steel slag.
The polynomial equation of degree 4 is as follows:

$$15238.73 - 9707.20*x + 2316.80*x^2 - 245.46*x^3 + 9.74*x^4=0$$

The standard error is 0, meanwhile the correlation coefficient is 1.

It has been noted that the water absorption value reduces as bitumen percentage rises. Steel aggregates have less of an impact on water absorption than chipped aggregates do, and at bitumen concentrations greater than 6.00%, water absorption is negligible.

### 3.5. *Assessment of Marshall stability*

The assessment method is according to SR EN 12697-34:2020 - Asphalt mixtures. Test methods. Part 34: Marshall test [9].

**Table 16**. Marshall test for BA16 with chippings

| bitumen percentage, % | bitumen type | S, kN | I,mm | Ratio S/I. |
|---|---|---|---|---|
| 5.75 | | 10.4 | 2.75 | 3.78 |
| 6.00 | | 10.9 | 3.45 | 3.16 |
| 6.25 | D 50/70 | 10.6 | 3.65 | 2.90 |
| 6.50 | | 11.1 | 3.99 | 2.78 |
| 6.75 | | 10.3 | 3.91 | 2.63 |
| Limit SR EN 12697-34:2020 | | 6.5...13 | 1.5...4.0 | min 1.6 |

A mathematical analysis tool was used to interpret the results in terms of the Marshall test. Multiple mathematical functions were established to best follow the evolution of these values, as indicated in figures 5, 6, and 7 for BA16 with chippings and figures 8, 9, and 10 for BA16 with steel slag.
The polynomial equation of 4 degree that best reflects, the evolution of Marshall Stability values in relation to the asphalt percentage for asphalt mixtures with steel slag is as follows:

$$-59216.90 + 38118.90*x - 9191.13*x^2 + 984.00*x^3 - 39.47*x^4=0$$

The standard error is 0, meanwhile the correlation coefficient is 9.99999999999962E-01.

It is observed that the ideal range for the bitumen percentage needed to obtain Marshall Stability values in accordance to SR EN 12697-34:2020 is between 5.90% and 6.60% [9].

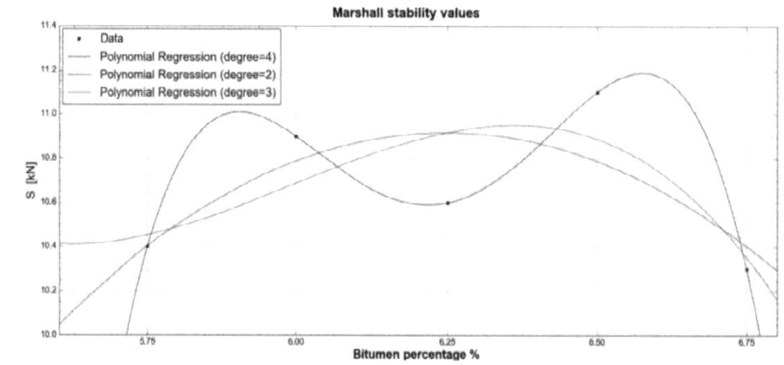


**Figure 5.** Mathematical interpretation of Marshall Stability evolution for BA16 with chippings

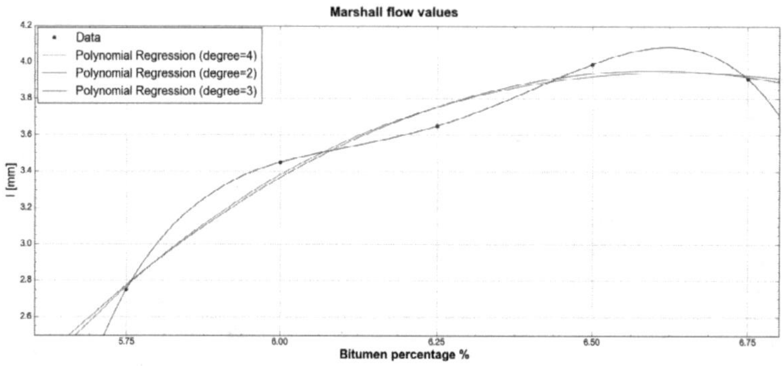


**Figure 6.** Mathematical interpretation of the evolution of the flow for BA16 with chippings

The polynomial equation of 4 degree that best represents the evolution of the flow values in relation to the asphalt percentage for asphalt mixtures with chippings is:

$$-19563.35 + 12527.05*x - 3006.08*x^2 + 320.43*x^3 - 12.80*x^4=0$$

The standard error is 0, meanwhile the correlation coefficient is 9.99999999999962E-01.

It is observed seen that the optimum range of the percentage of bitumen required to obtain Marshall flow values according to SR EN 12697-34:2020 is between 5.75% and 6.40% [9].

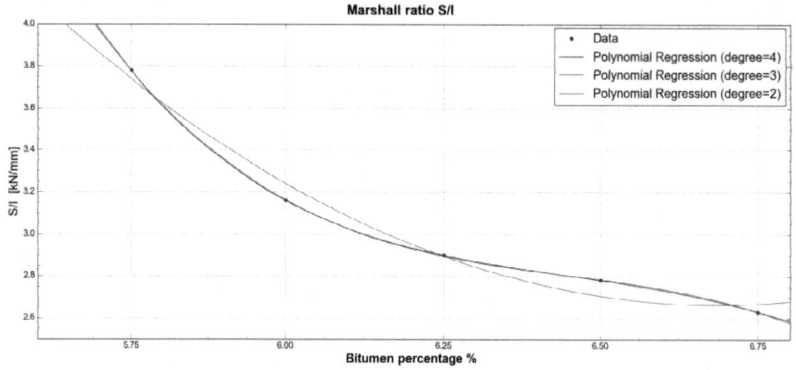


**Figure 7.** Mathematical interpretation of the S/I ratio evolution for BA16 with chippings

The polynomial equation of 4 degree that most accurately depicts the development of the Stability/ Immersion (S/I) ratio values in relation to the asphalt percentage for asphalt mixtures with chippings is:

$$1370.90 - 778.80*x + 165.09x^2 - 15.4*x^3 + 0.53*x^4=0$$

The standard error is 0, meanwhile the correlation coefficient is 1.

To obtain Marshall S/I ratio values according to SR EN 12697-34:2020 the percentage of bitumen must be less than 6.75%[9].

**Table 17**. Marshall test for BA16 with steel slag

| bitumen percentage, % | bitumen used | S, kN | I,mm | Ratio S/I. |
|---|---|---|---|---|
| 5.75 | | 12.1 | 3.45 | 3.51 |
| 6.00 | | 11.8 | 3.78 | 3.12 |
| 6.25 | D 50/70 | 12.9 | 3.81 | 3.39 |
| 6.50 | | 12.3 | 3.62 | 3.40 |
| 6.75 | | 11.9 | 3.55 | 3.35 |
| Limit SR EN 12697-34:2020 | | 6.5...13 | 1.5...4.0 | min 1.6 |

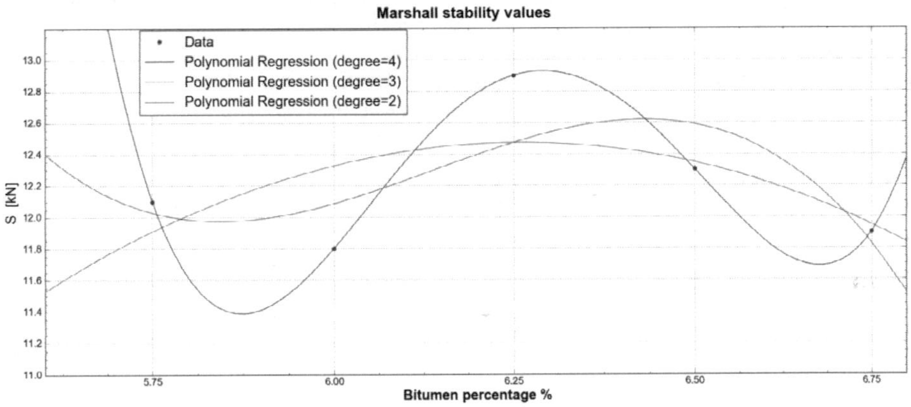


**Figure 8.** Mathematical interpretation of the Marshall Stability evolution for BA16 with steel slag

The polynomial equation of 4 degree that most accurately depicts the development of Marshall Stability values in relation to the asphalt percentage for asphalt mixtures with steel slag is:

$$82285.40 - 52620.27*x + 12603.07*x^2 - 1339.73*x^3 + 53.33*x^4=0$$

The standard error is 0, meanwhile the correlation coefficient is 9.99999999999971E-01.

It is observed that the optimum range of the percentage of bitumen required to obtain Marshall Stability values according to SR EN 12697-34:2020 is between 6.00% and 6.50% [10].

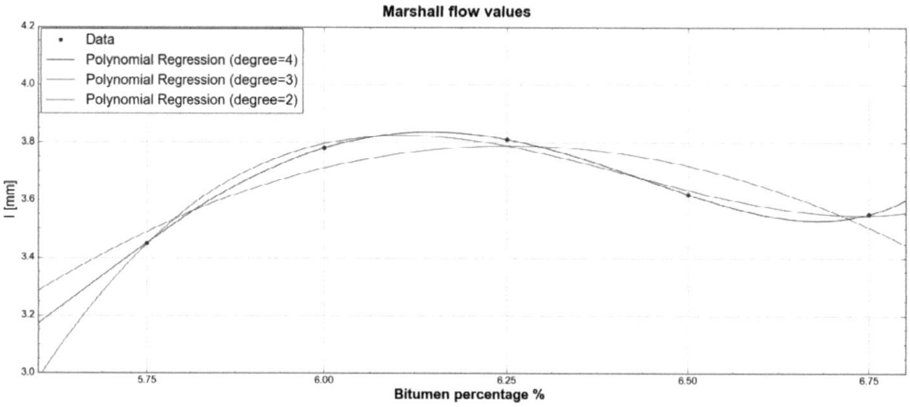


**Figure 9.** Mathematical interpretation of the evolution of the flow for BA16 with steel slag

The polynomial equation of degree 4 best reflects, the evolution of the Marshall flow values in relation to the asphalt percentage for asphalt mixtures with steel slag is:

$3616.06 - 2422.13*x + 606.07*x^2 - 67.09x^3 + 2.77*x^4=0$

The standard error is 0, meanwhile the correlation coefficient is 1.

It is observed that the optimum range of the percentage of bitumen required to obtain Marshall flow values according to SR EN 12697-34:2020 is between 5.75% and 6.00% [7, 10].

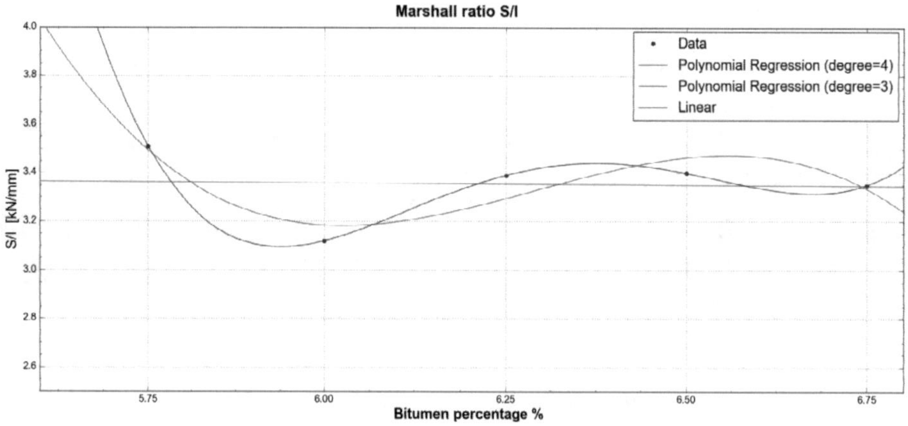


**Figure 10.** Mathematical interpretation of the evolution of the S/I ratio for BA16 with steel slag

The polynomial equation of 4 degree best reflects, the evolution of the values of the S/I ratio in relation to the asphalt percentage for asphalt mixtures with steel slag is:

$19054.64 - 12080.53*x + 2869.17*x^2 - 302.51x^3 + 11.95*x^4=0$

The standard error is 0, meanwhile the correlation coefficient is 9.99999999999980E-01.

It is noted that the optimum range of bitumen percentage required to obtain S/I Marshall Ratio values according to SR EN 12697-34:2020 is between 6.25% and 6.75% [10].

## 4. Conclusions

After mathematically analysing the results obtained for the two types of asphalt mixtures, one with crushed stone (chippings) and the other with slag, we can conclude that:

- From the standpoint of bulk density, we observe that the optimal range of bitumen percentage is. is 6.0% -6.35%, for the two types of asphalt mixtures;
- Comparing the evolution of the parabolas of degree 4 for water absorption in relation to bitumen percentage we observe that, for both asphalt mixtures, the influence of bitumen on absorption is lower for values higher than 6.00%
- Analysing the results obtained for the three parameters of the Marshal test - stability, flow, and S/I ratio - for the asphalt mixture with aggregates, it is observed that the optimal bitumen percentage falls within the range of 5.75%-6.00%.
- Analysing the results obtained for the three parameters of the Marshal test - stability, flow, and S/I ratio - for the asphalt mixture with slag aggregates, it is observed that the optimal bitumen percentage falls within the range of 6.25%-6.50%.
- After laboratory tests and data processing with mathematical analysis program we observe that the bitumen percentage for asphalt mixture with chippings is 6.00% and for asphalt mixture with steel slag aggregates is 6.25%.
- The upper bitumen percentage used for asphalt mix with steel slag is influenced by the degree of coating of the steel slag aggregates. The degree of coating is determined by the affinity of the bitumen to these aggregates [1]. The bitumen affinity to aggregates can be improved by using modified bitumen;
- The advantage of modelling the results using mathematical analysis programs helps us to optimise the bitumen dosage, using the predictability of polynomial functions through calculations as well as graphically;
- By using steel slag aggregates, numerous economic and technical advantages are achieved for both the user and the manufacturer. These advantages include diversification of road infrastructure materials, cost reduction through shorter transport distances and the use of local materials, but most importantly, the elimination of steel slag waste disposal and afforestation of the lands occupied by them.

**Acknowledgments**

This paper was realised with the support of project EFECON – ECO-INNOVATIVE PRODUCTS AND TECHNOLOGIES FOR ENERGY EFFICIENCY IN CONSTRUCTION, POC/71/1/4 - Knowledge Transfer Partnership, Cod MySMIS: 105524, ID: P_40_295, Project co-financed by the European Regional Development Fund.

## 5. References

[1] Iriciuc SC, 2011 Use of crushed steel slag in asphalt mixtures used for roads, *PhD thesis Gheorghe Asachi" Technical University of Iasi-Romania*

[2] Goli A,, 2022 The study of the feasibility of using recycled steel slag aggregate in hot mix asphalte, *https://doi.org/10.1016/j.cscm.2021.e00861*

[3] Iriciuc SC, Galusca NI, 2017 Laboratory Test Based on Specimens of Asphalt Mixtures Made of Crushed Steelworks Slag Aggregate, *Bul. Polyt. Inst. Jassy 2 57-67*

[4] www.anpm.ro/documents/21661/2135

[5] SR EN 13043/2003/AC 2004 Aggregates for bituminous mixtures and surface treatments for roads, airfields and other trafficked areas

[6] SR EN 933-2-2020 - Tests for geometrical properties of aggregates - Part 2: Determination of particle size distribution - Test sieves, nominal size of apertures

**[7]** Iriciuc SC, Gimiga GR, Gălușcă NI, 2023 Asphalt mixtures used in binder layers using steel slag Waste, XXIII International Scientific Conference VSU'2023, pag 23-239

[8] SR EN 933/1-2012 Tests for geometrical properties of aggregates - Part 1: Determination of particle size distribution - Sieving method

[9] SR EN 12697-6 /2020 – Bituminous mixtures - Test methods - Part 6: Determination of bulk density of bituminous specimens
[10] SR EN 12697-34:2020– Bituminous mixtures - Test methods - Part 34: Marshall test

# Impact of the infrared properties of the backsheet on the efficiency of an encapsulated photovoltaic silicon cell

**A. Khouzam[1,2], P.O. Logerais[1*]**
[1] Univ Paris-Est, CERTES, IUT de Sénart-Fontainebleau, 36 rue Georges Charpak, 77567 Lieusaint, France.
[2] ICAM Grand Paris Sud, 34 Points de Vue, 77127 Lieusaint, France.

E-mail: pierre-olivier.logerais@u-pec.fr

**Abstract.** The efficiency of an encapsulated photovoltaic (PV) silicon cell is studied by modeling the radiative exchanges between its layers, focusing on the influence of the radiative properties of the backsheet on the cell temperature. The model uses the Monte-Carlo method to solve the radiative transfer equation. Two-dimensional simulations were performed, showing how the radiative properties of the backsheet affected the silicon cell temperature, which was then used to calculate the current-voltage characteristic and the efficiency of two PV modules. The results revealed a significant sensitivity of the model to the emissivity of the top-side backsheet, with a 19.4°C temperature difference for the silicon cell and, a 9.5% and 13.3% increase in the maximum power between a fully reflective and a fully absorptive backsheet for the two PV modules. Overall, the study highlights the importance of considering radiative properties in the design and optimization of PV panels, as they can have a substantial impact on their performances.

## 1. Introduction

In order to foster the use of renewable energy sources, the implementation of photovoltaic (PV) generation systems has widely been endorsed, notably by integrating these systems into buildings. The PV technology has greatly been diversified, with the emergence of devices exhibiting tailored aspects and functionalities [1]. Colored cell modules, photovoltaic tiles and facade elements represent notable examples of solar energy systems offering conventional building functionalities coupled with electricity generation. Solar modules commonly consist of solar cells, a glass layer on the front, an encapsulant and a protective backsheet on the rear side. The backsheet, usually composed of a polymer material, serves as a barrier against moisture and ultraviolet radiation, while also acting as an electrical insulator. Moreover, a reflective foil backsheet could be utilized to improve the PV module efficiency by diffusely or non-diffusely reflecting the incident light with possible gains in the range of 1 to 3% [2]. White is the most widely used color for backsheets, whereas black backsheets are frequently encountered for aesthetic purposes [3]. According to research [4], optimizing the gap between cells and using white backsheets lead to 1 to 2% of improvement in the maximal power compared with the usage of black ones. Luxembourg et al. demonstrated that the application of black powder coatings with low IR absorption on standard films used in the photovoltaic industry can significantly reduce cell-to-module (CtM) losses for all the black modules [5]. The infrared radiation properties of the underlying substrate are largely preserved when such films are applied. Compared with conventional absorbent black foils, an improvement of the short-circuit current of 1.4 to 1.8% at the PV module level was achieved. McIntosh et al. reported that around 25% of the photons that fall on the backsheet between the PV cells are internally reflected back to the cells, resulting in a 2.4% increase in the module power [6]. They also found that black foil modules have 2.0 to 3.0% less power than white foil modules. Burgers and al. showed that adding mirrors to the front and back surfaces reduces the amount of light emitted from these surfaces [7]. However, some of the emitted light is absorbed by the mirrors due to their imperfect reflectivity.

Additionally, the temperature is one of the fundamental parameters that affects the efficiency of the photovoltaic module and hence, the thermal properties of the backsheets can have a significant impact on the efficiency of the PV module. The temperature of a PV module is influenced by the ambient

temperature (directly proportional), the cell irradiance (directly proportional), and the wind speed (inversely proportional) [8]. Some possible modifications in the structure of a PV module to maximize the efficiency are discussed in [9]. This can be realized by augmenting the reflection of light for PV conversion and dissipating heat for diminishing the operating cell temperatures with conductive and radiative-adapted materials. The use of encapsulating materials with high thermal conductivity enables to lower the operating temperature of a PV module [10-12]. Vaishak et al. conducted a study where they examined the efficiency of a refrigerant-based PV/T system using three different backsheet materials: glass, TPT, and copper [13]. They employed a validated numerical model for their analysis, and the results indicated that the PV/T collector with the TPT backsheet exhibited lower cell temperature, while the collector with the copper backsheet demonstrated the highest coefficient of performance.
In this study, we address the issue of heat flux knowledge within a PV module by examining the impact of the radiative properties of the backsheet on the output parameters of an encapsulated PV cell. For this purpose, a model was developed to understand the radiative transfers inside an encapsulated silicon cell by accounting for the infrared radiations and their influence on the temperature and efficiency of the PV cell. The radiative exchanges between the semi-transparent layers of an encapsulated silicon cell are determined by CFD simulations taking into account the spectral properties of the surfaces for wavelengths ranging from 0.1 to 100 μm. The obtained results are analyzed.

## 2. Modeling

### *2.1. Photovoltaic system*

The subject of this study is an encapsulated silicon photovoltaic (PV) cell composed of the following layers: a backsheet, an ethylene vinyl acetate (EVA) encapsulant, a silicon (Si) cell, and a cover glass, as depicted in Figure 1, which also indicates their sizes. The case of exposure under the lamp of a solar simulator with controlled temperature conditions is considered.

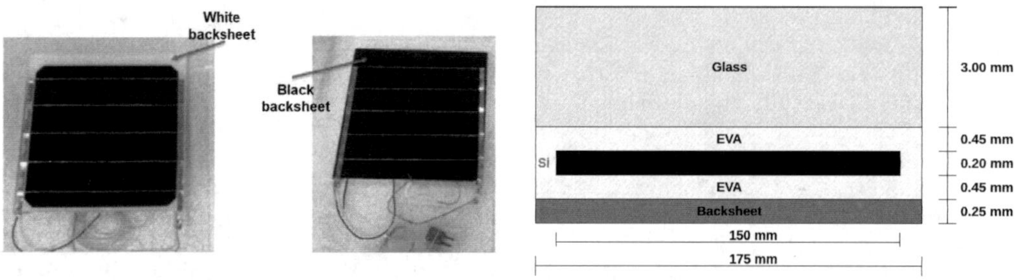


**Figure 1.** Encapsulated photovoltaic PV cell with its dimensions.

### *2.2. Geometry and meshing*

In view of performing the numerical simulations of the photovoltaic system, we used the commercial code CFD-Ace. The geometry was drawn according to the dimensions specified in Figure 1, and the PV cell domain was meshed using "quadrilateral" elements with a size of 0.2×0.2 mm$^2$. The choice of quadrilateral elements was made knowingly, as these elements can accurately capture the shape of the PV cell domain without introducing excessive computational cost. By carefully selecting the mesh resolution, we were able to achieve reliable and accurate results in our simulations. The resulting mesh is shown in Figure 2.
The subdomain of the PV cell is surrounded by two others, one above having a height of 17 cm corresponding to the distance between the emitting lamp of the solar simulator and the PV cell, and the second, down, having the same dimensions. Both allow to model the incident radiations and the heat exchanges with the ambient air and the environment. Special attention was given to the subdomain of the PV cell, where high accuracy was required, the mesh being refined with shrinkage in this area. The presence of the electrodes of the PV cells is not taken into consideration in the present model.

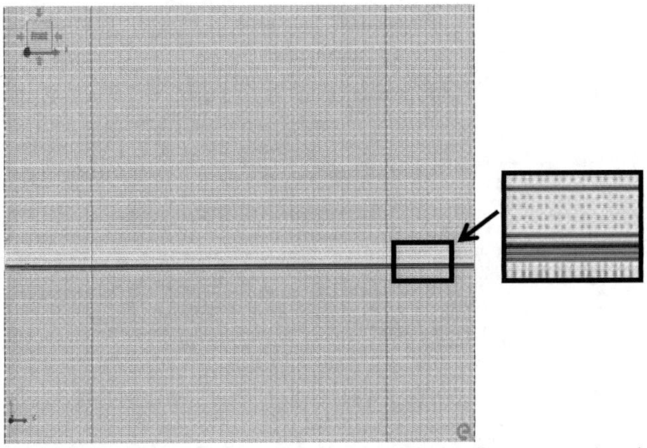

**Figure 2.** Meshing of the system with tight mesh for the encapsulated PV cell.

### *2.3. Governing equations*

To analyze the thermal behavior of the PV cell, two governing equations were employed here [14]. First, the heat conservation equation considering the different heat transfer processes, such as conduction in solids or convection in gases, is used to determine the temperature distribution:

$$div(\rho V h_0) = div(k \cdot grad\, T) + \left[\frac{\partial(u\tau_{xx})}{\partial x} + \frac{\partial(u\tau_{yx})}{\partial y}\right] + \left[\frac{\partial(v\tau_{xx})}{\partial x} + \frac{\partial(v\tau_{yx})}{\partial y}\right] + s_h \quad (1)$$

where the total mass enthalpy is:

$$h_0 = i + \frac{P}{\rho} + \frac{1}{2}\,(u^2 + v^2) \quad (2)$$

$\boldsymbol{i}$ is the internal mass energy. It is a variable that depends on the density $\boldsymbol{\rho}$ and the temperature $\boldsymbol{T}$ ;
$\boldsymbol{u}$ and $\boldsymbol{v}$ are the components of the fluid velocity $\boldsymbol{V}$ according to the directions $\boldsymbol{x}$ and $\boldsymbol{y}$ ;
$\boldsymbol{k}$ is the thermal conductivity;
$\boldsymbol{P}$ is the static pressure;
$\boldsymbol{\tau_{ij}}$ is the viscosity tensor;
$\boldsymbol{s_h}$ represents a variable containing terms from additional heat sources such as a radiative source or chemical reaction.

The second governing equation used was the radiative heat transfer equation so as to study the behavior of a medium that emits, absorbs, and scatters energy through radiation.

$$\Omega.grad(I(r,\Omega)) = -(\kappa + \sigma)I(r,\Omega) + \kappa I_b(r) + \frac{\sigma}{4\pi}\int_{\Omega'=4\pi} I(r,\Omega) \cdot \phi(\Omega' \rightarrow \Omega)\, d\Omega' \quad (3)$$

where:

$\Omega$ is the propagation direction of the beam;
$I(r,\Omega)$ is the intensity of the radiation. It is a function of the position $r$ and the direction $\Omega$ ;
$\kappa$ et $\sigma$ are respectively the absorption and diffusion coefficient;
$I_b(r)$ is the intensity of the blackbody at medium temperature;
$\phi(\Omega)$ is a phase function of the energy transfer from the direction of arrival $\Omega$ to the direction of exit $\Omega'$;
$\Omega.grad(I(r,\Omega))$ is the intensity gradient in the direction of propagation $\Omega$ ;
$-(\kappa + \sigma)I(r,\Omega)$ represents respectively the losses due to the absorption $\kappa$ and the diffusion $\sigma$ ;
$\kappa I_b(r)$ corresponds to the radiative emission of the body;
$\frac{\sigma}{4\pi}\int_{\Omega'=4\pi} I(r,\Omega) \cdot \phi(\Omega' \rightarrow \Omega)\, d\Omega'$ is the gain due to diffusion.

IOP Conf. Series: Materials Science and Engineering 1304 (2024) 012029 doi:10.1088/1757-899X/1304/1/012029

On the surface, the intensity is:

$$I(r,\Omega) = \varepsilon I_b(r) + \frac{\varrho}{\pi}\int_{n.\Omega'} |n \cdot \Omega'| \, I(r,\Omega')\, d\Omega' \tag{4}$$

with

$I(r,\Omega')$ : intensity of the energy of the radiation leaving the surface;
$\varepsilon$ : surface emissivity;
$\varrho$ : surface reflectivity;
$n$ : unit vector normal to the surface.

The Monte-Carlo method was used to solve the radiative transfer equation by simulating the heat exchanges by radiation. This method traces the rays emitted by different surfaces until they are absorbed by another surface, allowing for the calculation of the radiative heat flux. The method is described in [15] and the details of the photon packet tracing are explained in [16]. The surfaces were discretized into patches, and the radiative heat flux for each patch $(i)$ was determined by the incident radiation from all the other patches $(j)$ as well as its own emission:

$$Q_i = q_i A_i = \sum_{j=1}^{N_S} (M_{ij} - \delta_{ij}\varepsilon_j)\sigma T_j^4 A_j \tag{5}$$

where

$Q_i$ is the heat transfer rate;
$q_i$ is the heat flux density;
$A_i$ is the area of the surface under consideration;
$M_{ij}$ is the exchange matrix;
$\delta_{ij}$ is the symbol of Kronecker;
$\varepsilon_j$ is the emissivity of the patch $j$ ;
$\sigma$ is Stephan Boltzmann's constant of $5.669.10^{-8}\ W.m^{-2}.K^{-4}$ ;
$T_j$ is the average temperature of patch $j$.

### 2.4. Photovoltaic conversion

Upon computing the values of the cell temperature from the model of the photovoltaic module, the diode equivalent circuit of a photovoltaic panel was applied to simulate the changes in the current-voltage characteristics and to assess the variations in the output power as a function of the cell temperature [17]. The basic equation that mathematically describes the current-voltage $I(V)$ characteristic of an ideal photovoltaic cell in the semiconductor theory is expressed by the following equation:

$$I = I_{PV} - I_{0,cel}\left[exp\left(\frac{qV}{akT}\right) - 1\right] \tag{6}$$

where $I_{PV}$ is the current generated by the incident light (it is directly proportional to the irradiance of the Sun) $[A]$, $I_{0,cel}$ is the saturation or inverse leakage current of the diode $[A]$, $q$ is the electronic charge $[1.60217646 \times 10^{-19}\ C]$, $k$ is the Boltzmann's constant $[1.3806503.10^{-23} J/K]$, $T$ is the temperature of the pn junction $[K]$, and a is the ideality constant of the diode.

The current generated by the incident light on the photovoltaic cell is indeed directly proportional to the solar irradiance and is also influenced by the temperature according to equation (7):

$$I_{PV} = (I_{PV,n} + k_I \Delta T)\frac{G}{G_n} \tag{7}$$

where $I_{PV,n}$ is the current generated by the light in its nominal state (usually for 25°C and 1000 $W/m^2$) with $\Delta T = T - T_n$, $T$ and $T_n$ being the actual and nominal temperatures $[K]$, $G$ is the irradiance at the surface of the device, and $G_n$ is the nominal irradiance $[W/m^2]$. The diode saturation current $I_0$ and its temperature dependence can be expressed as:

$$I_0 = I_{0,n}\left(\frac{T_n}{T}\right)^3 exp\left[\frac{qE_g}{ak}\left(\frac{1}{T_n}-\frac{1}{T}\right)\right] \tag{8}$$

where $E_g$ is the gap energy of the semiconductor ($E_g \approx$ 1,12 eV for crystalline Si at $25°C$), and $I_{0,n}$ is the nominal saturation current:

$$I_{0,n} = \frac{I_{sc,n}}{exp\left(\frac{V_{oc,n}}{aV_{t,n}}\right)-1} \tag{9}$$

To account for the electrical losses, equation (6) can be written with the series and parallel resistances $R_s$ and $R_p$:

$$I = I_{PV} - I_0\left[\exp\left(\frac{V+R_sI}{V_t a}\right)-1\right] - \frac{V+R_sI}{R_p} \tag{10}$$

where

$I_{PV}$ and $I_0$ are the photovoltaic and saturation current of the PV cell, respectively;
$V_t = \frac{N_skT}{q}$ is the thermal voltage $[K]$ for $N_s$ cells connected in series.

## 3. Simulation

### *3.1. Volume properties*

Table 1 presents the thermal properties applied to the different subdomains. A Tedlar-based backsheet was considered, although other stack variants are possible with varying thermal properties [20,21]. The initial temperature is 20°C. The air above and below the encapsulated solar cell is initially considered with a velocity of $u = v = 0$.

**Table 1.** Materials and air properties.

| Designation | Density $\rho$ $(kg.m^{-3})$ | Specific heat $c_p$ $(J.kg.K^{-1})$ | Thermal conductivity $k$ $(W.m^{-1}.K^{-1})$ | Reference |
|---|---|---|---|---|
| Air | 1.204 | 1004 | 0.0262 | [18] |
| Glass | 2500 | 800 | 1.4 | [18] |
| EVA | 960 | 2090 | 0.35 | [19] |
| Silicon | 2329 | 700 | 148 | [18] |
| Backsheet | 1370 | 1760 | 0.35 | [19] |

### *3.2. Boundary conditions*

To model the photovoltaic system when being exposed to irradiance, the upper and lower parts were subjected to Neuman-type boundary conditions (Figure 3.a). Applied heat flux values were found to achieve an incident irradiance of 1000 W/m² on the top surface of the encapsulated PV cell while obtaining the silicon cell at a temperature of 25°C as per the STC conditions. An emissivity of 1 was considered for the upper limit representing the emission area of the lamp to obtain a blackbody emission so as to approximate an AM1.5 spectral distribution. A blackbody property was also taken for the bottom surface of the backsheet and the bottom boundary of the subdomain. The sides of the system were assumed to be fully reflective and were subjected to a convective flux condition to account for convective heat transfer. Spectral properties for a wavelength range from 0.1 µm to 100 µm were introduced for the interfaces involving the glass, EVA, and silicon layers (Figure 3.b and c). The backsheet was considered as an opaque material with a graybody property for its upper surface. A fully absorptive condition (emissivity $\varepsilon_\lambda = 1$, reflectivity $R_\lambda = 0$) was initially taken for the latter.

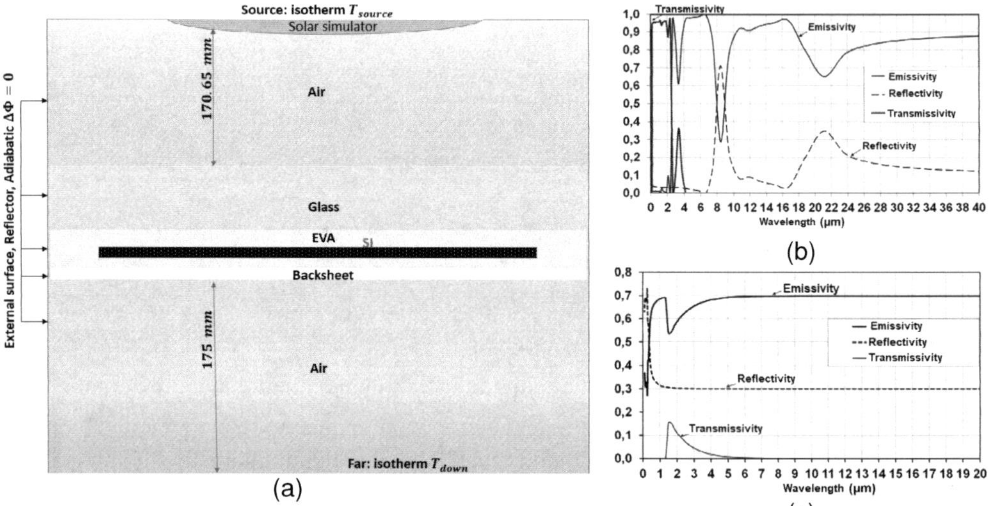


**Figure 3.** (a) Boundary conditions. (b) Radiative properties of the glass and of the encapsulant. (c) Radiative properties of the silicon [18,22,23].

### 3.3. Calculations

Equations (1) to (5) were solved using the finite volume method with the CFD-Ace code considering the Algebraic Multi Grid solver for faster convergence, and the solutions of equations (6) to (10) were determined utilizing a Matlab program. Parametric calculations of the temperature field were carried out varying the emissivity $\varepsilon_\lambda$ of the top side of the backsheet from 1 to 0 with a step of 0.1. At the same time, the reflectivity $R_\lambda = 1 - \varepsilon_\lambda$ was scaled from 0 to 1. The average temperature of the silicon cell was calculated for each step and the value found was introduced as the cell temperature in the single diode model for finding the output parameters of a photovoltaic module based on the radiative properties of the backsheet. The input parameters of two commercial PV modules having silicon PV cells of size 156 mm × 156 mm close to the one in the present model were entered (Table 2).

**Table 2.** Input parameters of the two commercial PV modules considered.

| Parameters | KC200GT | Q6LM |
|---|---|---|
| $I_{sc,n}$ [A] | 8.21 | 7.61 |
| $V_{oc,n}$ [V] | 32.90 | 0.61 |
| $I_{mp}$ [A] | 7.61 | 7.11 |
| $V_{mp}$ [V] | 26.30 | 0.51 |
| $P_{max}$ [W] | 200.14 | 3.64 |
| $K_v$ [V/K] | -0.123 | -0.002 |
| $K_i$ [A/K] | 0.0032 | 0.0381 |
| $N_S$ | 54 | 1 |

## 4. Results and discussion

### *4.1. Photovoltaic cell temperature*

The temperature field obtained for an emissivity of the top side of the backsheet of 1 is shown in Figure 4. The temperature within the silicon solar cell was found to be homogeneous and its mean value was calculated and introduced into the single diode model to calculate the output parameters for the two PV modules as previously mentioned.

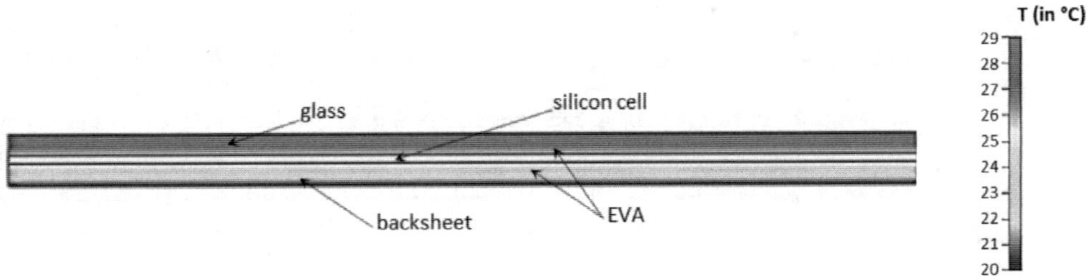


**Figure 4.** Temperature contours of the PV encapsulated cell (top side backsheet with $\varepsilon_\lambda = 1$).

The variation of the mean PV cell temperature according to the emissivity of the upper surface of the backsheet is plotted in Figure 5. The temperature of the silicon cell is increased from 26.2°C to 45.6°C when the emissivity is reduced from 1 to 0. This temperature difference of 19.4°C between a fully reflective and a fully absorptive backsheet shows that the PV cell temperature is sensitive to the change in the radiative properties in the infrared domain. When the emissivity is reduced, the absorptivity is also decreased and the reflectivity is augmented, so the radiative heat flux from the upper surface of the backsheet towards the silicon cell becomes more significant. A second-degree polynomial is found as a fitting curve with a determination coefficient $R^2$ close to 99%. This means that the derivative of the temperature with respect to the emissivity is an affine function of equation $21{,}192.\,\varepsilon_\lambda - 28{,}262$. The variation appears to be more significant for lower emissivities. When backsheets are exposed to UV radiation [24] or to various weather conditions [25], they can be degraded by chemical mechanisms or cracked, changing their reflectance and possibly inducing variations in the cell temperature. The ideal surface property for the backsheet would be to be fully absorptive in the IR domain and fully reflective in the visible range. The aim is to maintain the operating temperature of the cell as low as possible in order to improve its performance and to reflect the maximum of light towards the PV cell.

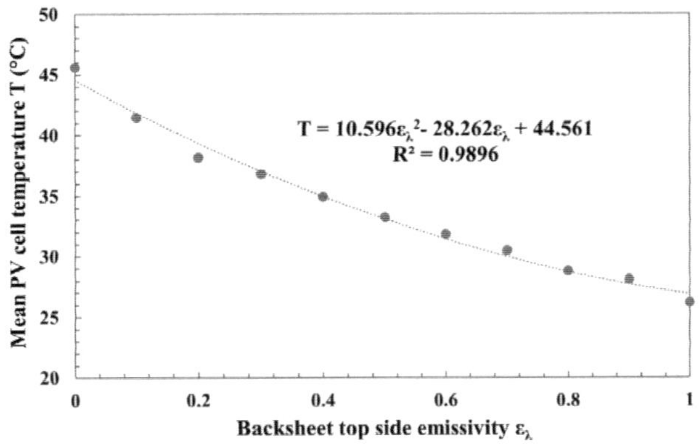


**Figure 5.** Mean silicon PV cell temperature versus top-side backsheet emissivity.

### 4.2. *Variations of the electrical performance*

The I-V curves of both the PV modules are modified when the emissivity of the top-side backsheet is varied. The example of the Q6LM PV module is displayed in Figure 6. As the emissivity increases, the cell temperature is lowered and hence both the open-circuit voltage and the maximum power are raised. We observe a noticeable improvement in the open-circuit voltage and no significant effect on the short-circuit current. The fill factor is slightly increased when the temperature becomes more important.

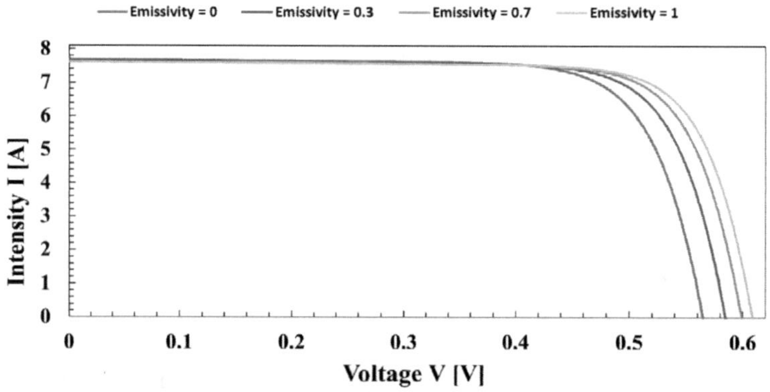


**Figure 6.** I-V curves obtained for several emissivities of the backsheet for the Q6LM model.

The efficiency of the PV module can be calculated as follows:

$$\eta = \frac{P_{max}}{S \times G} \tag{11}$$

where $S$ is the illuminated surface. When the emissivity increases, the efficiency is improved. Its relative variation for the two cases is 9.5% for the KC200GT module and 13.3% for the Q6LM one (Figure 7). This discrepancy is due to the temperature coefficients which are different for both the PV modules. This result highlights the importance of absorbing infrared radiations at the bottom of the PV module to reduce the cell temperature at its best.

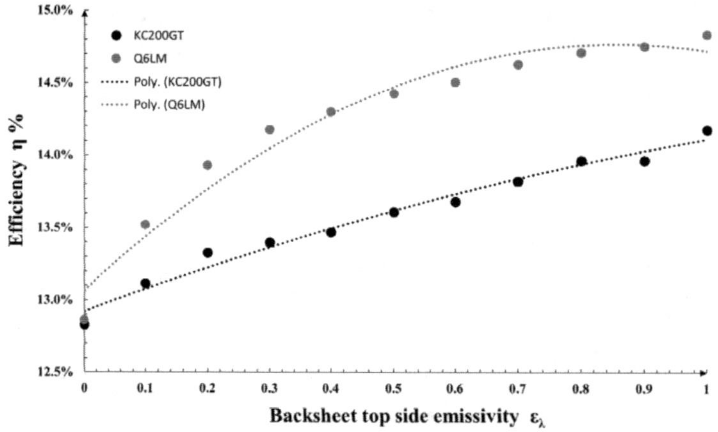


**Figure 7.** Efficiency of the PV versus top-side backsheet emissivity for the two PV modules.

IOP Conf. Series: Materials Science and Engineering 1304 (2024) 012029 doi:10.1088/1757-899X/1304/1/012029

## 5. Conclusion and perspectives

The purpose of this study was to investigate the radiative transfers within the layers of an encapsulated photovoltaic cell to gain a better understanding of its properties and to improve its performance. To achieve this, a two-dimensional model of the encapsulated silicon cell was constructed, and the conservation equations accounting for the heat exchanges were specified. The radiative transfer equation was then solved using the Monte-Carlo method. The simulation results showed that the thermal emissivity of the backsheet had a significant impact on the temperature of the solar cell, with a 13% variation in the efficiency observed when the emissivity was changed from 0 to 1. The sensitivity of the cell temperature to the IR properties of the backsheet was also ascertained. A backsheet that is fully reflective in the wavelength range corresponding to the spectral response of the PV cell and strongly absorbent in the IR range was identified as the ideal setting.

A performance calculation was carried out on two PV modules, showing that a temperature difference of 19.4°C between a totally reflective and a totally absorptive backsheet could lead to a significant gain of 13% on the maximum power. This highlights the importance of considering the radiative properties of the backsheet when designing and manufacturing encapsulated PV cells.

Further research could include a comparison between simulation results and experimental tests by manufacturing an encapsulated cell prototype. Additionally, a coupling of the PV conversion could be considered by modeling the PN junction while considering the spectral and scattering properties of the carriers with the finite volume code. A parametric study varying the optical indices of the backsheet could also be handled to minimize the cell temperature and maximize the power gain. Future calculations will consider the spectral properties of the backsheet.

Overall, this study provides valuable insights into the radiative properties of encapsulated photovoltaic cells and highlights the importance of considering the radiative properties of the backsheet in improving their performance.

## 6. References

[1] Shukla AK. Sudhakar K. and Baredar P. 2017 Energy and Buildings 140 188.
[2] Mittag M. Grünzweig A. Wiese M. Mahmoud N. Schmid A. and Heinrich M. 2017 Proc. 33rd European PV Solar Energy Conference and Exhibition (Amsterdam).
[3] López-Escalante MC. Navarrete-Astorga E. Gabás Perez M. Ramos-Barrado JR. and Martín F. 2020 Applied Energy 279 115741.
[4] Lim H. Cho SH. Moon J. Jun DY. and Kim SH. 2022 Applied Sciences 12(1) 443.
[5] Luxembourg SL. Kloos M. Danzl FJK. Theewis J. and Roosmalen JAM. 2016 Proc. 32nd European PV Solar Energy Conference and Exhibition (Munich).
[6] McIntosh KR. Swanson RM. and Cotter JE. 2005 Progress in Photovoltaics: Research and Applications 14 167.
[7] Burgers AR. Slooff LH. Kinderman R. and van Roosmalen JAM. 2005 Proc. 20th European PV Solar Energy Conference and Exhibition (Barcelona).
[8] Kurnik J. Jankovec M. Brecl K. and Topic M. 2011 Solar Energy Materials and Solar Cells 95(1) 373.
[9] Silverman TJ. Deceglie MG. Subedi I. Podraza NJ. Slauch IM. Ferry VE. and Repins I. 2018 IEEE Journal of Photovoltaics 8 532.
[10] Kim N. Kim D. Kang H and Park YG. 2016 Energy 113 515.
[11] Lee B. Liu JZ. Sun B. Shen CY. and Dai GC. 2008 Express Polymer Letters 2(5) 357.
[12] Pavgi A. Oh J. and TamizhMani G. 2021 Energies 14(5) 1252.
[13] Vaishak S. and Bhale PV. 2021 Renewable Energy 168 160.
[14] CFD Research Corporation 2014 CFD'ACE (U) Module Manual (Version 2014 Huntsville).
[15] Modest MF. 1993 Radiative Heat Transfer (McGraw-Hill New York).
[16] Mazumder S. and Kersch A. 2000 Numerical Heat Transfer, Part B: Fundamentals 37 185.
[17] Villalva MG. Gazoli JR. and Ruppert EF. 2009 Brazilian Journal of Power Electronics 14(1) 35.

[18] Logerais PO. Riou O. Delaleux F. Durastanti JF. and Bouteville A. 2015 Applied Thermal Engineering 77 76.
[19] Hoang P. Bourdin V. Liu Q. Caruso G. and Archambault V. 2014 Solar energy materials and solar cells 125 325.
[20] Mahmood FI. and TamizhMani G. 2023 Solar Energy 252 20.
[21] Geretschläger KJ. Wallner GM. and Fischer J. 2016 Solar Energy Materials and Solar Cells 144 451.
[22] Palik ED. 1998 Handbook of Optical Constants of Solids (Academic Press, Orlando).
[23] Sharma BK. Desai U. Singh A. and Singh A. 2020 Journal of Applied Polymer Science 137 48268.
[24] Liu F. Jiang L. and Yang S. 2014 Solar Energy 108 88.
[25] Lyu Y. Fairbrother A. Gong M. Kim J.H. Gu X. Kempe M. Julien S., Wan K.T. Napoli S. Hauser A. O'Brien G. Wang Y. French R. Bruckman L. Ji L and Boyce K. 2020 Solar Energy 199 425.

**Acknowledgments**

The authors would like to extend their sincere thanks to Mrs. Wilhelmina Logerais, a native English speaker, for her assistance in proofreading this paper.

IOP Conf. Series: Materials Science and Engineering 1304 (2024) 012030 doi:10.1088/1757-899X/1304/1/012030

# The impact of the embankment and the recreational lake on capturing front with drilling wells

**N Marcoie, D Toacă, D Toma, V Boboc, T A Hrăniciuc, R Mitroi**
"Gheorghe Asachi" Technical University of Iaşi, 65 Prof.dr.docent D. Mangeron Street, 700050, Iaşi, Romania

E-mail: raluca.mitroi@academic.tuiasi.ro

**Abstract.** This paper presents the influence on the levels from the well caption front Gherăeşti-Bacău, located on the right side of the Bistriţa River, through the construction, on the right shore, of a 2.25 km long perimetral embankment, and a recreational lake with a surface of 3.18 hectares and average depth of 5.5 m (Figure 1). Through numerical modelling, using the modelling software of Visual Modflow aquifer, the influence on the hydrostatic level from drilled wells was determined, as well as the influence on the phreatic flow through the making of the recreational lake, which captures a part of the underground water. The use of the mathematical model allows the determination of the optimal parameters for an extraction from the aquifer known through classical methods. To achieve the mathematical model the method of filtration resistances was used, which is applied in the case of a sudden curvature of the streamlines (in the proximity of the imperfect permeable/impervious borders: channels, drains, well systems, screens), hence they are applied during pressure/level drops. This method allows the following equalizations: the imperfect wells can be considered as perfect; the lines of perfect/imperfect wells can be considered as perfect trenches (channels) with a flow equal to the sum of all wells; a stratified aquifer (up to 3 layers) can be equivaled with a homogenous one. Aspects regarding the flow were taken into consideration: with a free level or under pressure, the dimensioning of the model, the initial conditions of the aquifer layer, the lithostratigraphic characteristics, homogeneity, anisotropy, the transfer mechanisms in the domain's interior, the studied hydrogeological structure was schematized, having done a monolayer model. The calibration of the model consists of adjusting, in reasonable conditions, of the domain data (the permeabilities), boundary conditions, so that through the running of the model and obtaining the corresponding piezometric lines and the obtained data will overlap the levels of the existing drillings.

## 1. Case study. The Gherăeşti-Bacău catchment front

The Gherăeşti-Bacău catchment front has 57 boreholes, 8 executed in 1966 and 49 executed in 1982 (Figure 1). Currently, 44 boreholes are in operation. The catchment front is located in the municipality of Bacău, on a winding alignment that follows the old bed of the Bistriţa River. The boreholes are located at an average distance of approximately 75 m from the main riverbed, being capture boreholes with infiltration from the bank.

The neighbourhoods of the catchment front are represented by private homes, the Bacău II reservoir, the Bistriţa River bed and the offtake of UHE Bacău I.

The boreholes have depths of 8.00 – 12.00 m, are completed with $\phi$= 324 mm (12 3/4") steel column with flared slot filters and are spaced at an average distance of 90 m from each other.

The average thickness of the collector is $H$ = 6.58 m. The total flow rate exploited from the 44 wells is 190.00 l/s, and the maximum flow rate exploited by one borehole is: $Q$ = 6.0 l/s (0.006 $m^3$/s), with an average flow $Q_{med}$ = 4.31/s (0.0043 $m^3$/s).

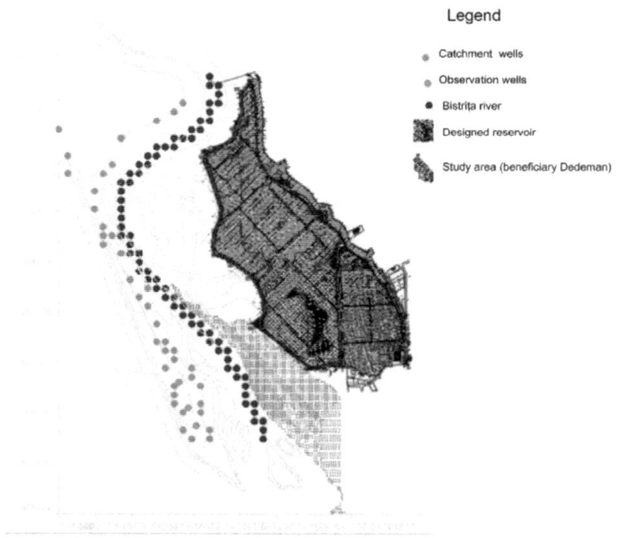


**Figure 1.** Gheräești-Bacău study area

## 2. Numerical modelling using Visual Modflow aquifer modelling software

The equation underlying the modelling of aquifers used by the Modflow program is [1, 2, 3]:

$$\frac{\partial}{\partial x}\left(K_{xx}\frac{\partial h}{\partial x}\right)+\frac{\partial}{\partial y}\left(K_{yy}\frac{\partial h}{\partial y}\right)+\frac{\partial}{\partial z}\left(K_{zz}\frac{\partial h}{\partial z}\right)+W=S_s\frac{\partial h}{\partial t} \quad (1)$$

where,

$K_{xx}$, $K_{yy}$, and $K_{zz}$ are the hydraulic conductivity values on the $x$, $y$, and $z$ axis (m/s); $h$ is the potentiometric head (m); $W$ - volumetric flux ($s^{-1}$); $t$ - time (s); $S_s$ – specific storage of the porous material ($m^{-1}$).

Equation (1) combines the initial conditions imposed on the model, describes the distribution of groundwater in a homogeneous and anisotropic medium and determines the main axes of the hydraulic conductivity.

### *2.1. The method of filtering resistances for the calculation of catchments with wells*

The method of filtering resistances is applied in the case of sudden curvature of stream lines (near imperfect permeable/impermeable boundaries: channels, drains, well systems, screens etc.), so when level/pressure drops occur [4].

This method allows the following equations: imperfect wells can be considered as perfect; lines of perfect/imperfect wells can be considered as perfect trenches (channels) with flow equal to the sum of all wells; a stratified aquifer (up to three layers) can be considered a homogeneous aquifer. Thus, at distances of the same order of magnitude as the thickness of the layer, the movement of underground water is predominantly horizontal - with the distribution of pressures similar to that in the case of perfect drains.

$$\Phi=\Phi_c+\Phi_x \quad (2)$$

where:

$\Phi$ - internal/additional resistance;

$\Phi_c$ - the resistance of the imperfect well according to order, in the case of non-stationary movement depends also on time (it is considered constant in terms of time when $\alpha=\frac{r^2}{a\cdot t}<5\cdot10^{-5}$);

$\Phi_x$ - resistance according to the character of the opening of the layer; depends on the following factors: filter construction; modification of the structure of the stream in the area before the well, with processes of erosion (suffusion), corrosion/clogging of the filter. Even at the full opening of the aquifer layer filter, when we can consider $\Phi_c = 0$, the action of the resistance $\Phi_x$ does not decrease (it even increases) and must be considered in the calculations.

$\Phi$ can be considered implicitly in the resistance $F$ of a perfect well by considering:

$$r_c^0 = r_c \cdot e^{-2\pi\phi} \quad (3)$$

where:

$r_c^0$ - equivalent radius of a perfect well;

$r_c$ - radius of the imperfect well.

### *2.2. Identifying the calculation relations for $\Phi$ for a simple situation, a terrain with a homogeneous layer (Figure 2)*

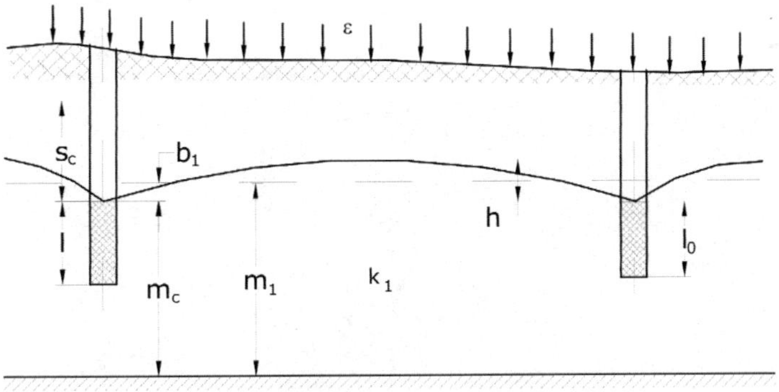


**Figure 2.** Filtering scheme for one-layer terrain

$$\phi = f \quad (4)$$

The dimensionless resistance $f_i$ from a homogeneous ground corresponds to the thicknesses $m_i = m$, $m_c$, $m_1$ and is equal to:

$$f_i = f_{ci} + f_{xi} \quad (5)$$

where:

$f_{ci}$ - the resistance due to the imperfection according to the degree of opening in a homogeneous layer, with waterproof sole and ceiling, of thickness $m_i$, for which these resistances are determined;

$f_{xi}$ - resistance depending on the nature of the opening in a homogeneous layer of thickness $m_i$.

The relations established for aquifers under pressure can be used to determine the resistance $\Phi$ in the upper layer of aquifers with a free level. In this case, the thickness of the stream in the upper layer $m_1$ and the length of the filter $l$ are averaged, with a small approximation as follows. When placing the well filter in the upper layer of the well with a non-submersible filter, at which the hydrodynamic level reaches below the length of the filter in the upper layer $l_0$ ($l < l_0$), $m_1 = m_c + 0.5 \cdot S_c, b_1 = 0.5 \cdot S_c, l = m_1 - a_1$.

For wells with a submersible filter, where the hydrodynamic level is located above the filter: $m_1 = m_c + 0.5 \cdot S_c, b_1 = m_1 - l - a_1, l = l_0$.

When placing the filter in the bottom layer, $m_1 = m_c + 0.5 \cdot S_c$.

In the case of systematic vertical drainage, when averaging the sizes $m_1$, $b_1$ and $l$, instead of the dislevelment $S_c$ it is recommended to use the ascent $h$ in the upper layer.

### *2.3. Identification of the relations for the calculation of the resistance $\overline{\phi}$ for terrain with a homogeneous layer*

$$\overline{\phi} = \frac{1}{m} \cdot \overline{f} \quad (6)$$

In the previous formula $\overline{f}$ is determined with equation (4), in which $\overline{f_{ci}}$ is evaluated according to the relationship (5).

The solid lines for the filter placement near the top and bottom boundaries of the layer, the dotted line (Figure 2): for filters of length l< 0,5·$m_i$, located approximately in the middle of the layer $b_i + \frac{l}{2} = (0{,}35 \ldots 0{,}65) \cdot m_i$.

### 3. Development of the numerical model for the phreatic aquifer in the Gherăești - Bacău area

Mathematical modelling of groundwater flow involves knowledge of the entire system of groundwater flow, geological and hydrogeological knowledge (hydraulic conductivity, transmissivity, porosity etc.) [5, 6].
Based on the level measurements carried out in the field, the lithological descriptions made during the construction of the boreholes of the catchment front, the hydrogeological sections through these boreholes made available by the beneficiary, it is concluded that the aquifer system presents itself as a continuous environment within its natural limits, on the entire study area.
Figure 3 shows the boreholes and observation points where level measurements were made. Using them, the map of the initial piezometric surface was developed.

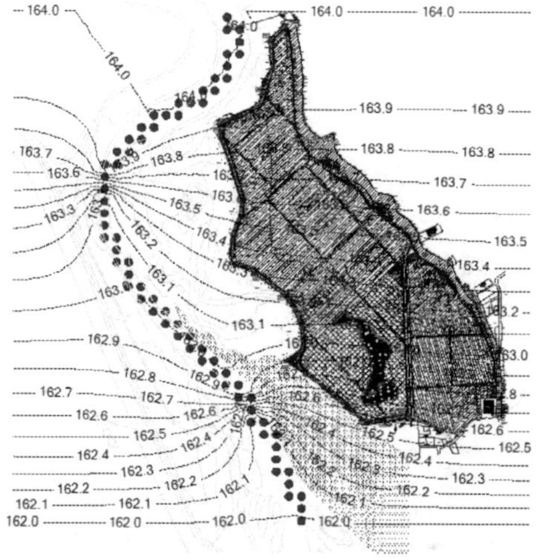


**Figure 3.** Map of the initial piezometric surface

Aspects related to the flow regime were taken into account: with free level or under pressure, model sizing, boundary geometry, boundary conditions for the aquifer water flow system, initial conditions of the aquifer layer, litho-stratigraphic characteristics, homogeneity, anisotropy, the transfer mechanisms within the domain. The studied hydro structure was schematized, developing a unilayer model.
Based on the presented data, the modelling area was established and the boundary conditions were defined [7]:

- Dirichlet type condition – imposed hydraulic elevation, upstream elevation being given by the 164.00 m hydroisohypses, respectively the downstream elevation corresponding to the 162.00 m hydroisohypses;
- Cauchy-type conditions (potential-dependent flow) on the Bistrița river, the imposed elevations are those representing the free surface of the water in the river;
- Limit type conditions in depth, given by the impermeable layer resulting from the drillings within the catchment front;
- Lines of constant or variable potential, respectively streamlines, correspond from a mathematical point of view to mixed-type boundary conditions (Dirichlet, von Neuman) (Figure 4).

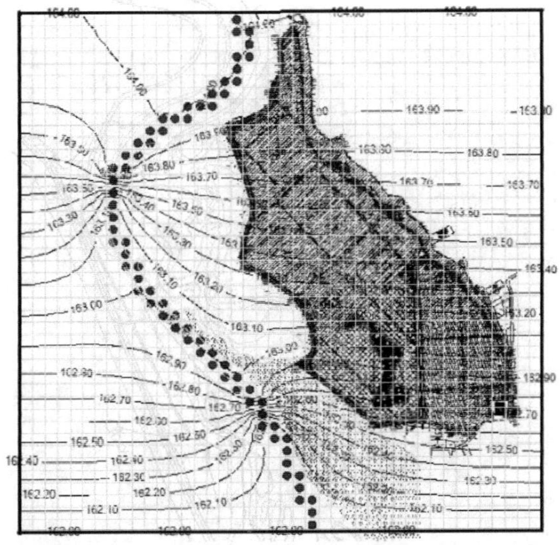

**Figure 4.** Defining the boundary of the modelled domain in the study area

The data can be entered interactively, with the possibility of modifying certain parameters such as, for example, permeability discretization, base elevations, layer granularity, hydraulic conductivity.
The perimeter of the hydraulically significant area was included in the active area of the numerical model, the adjacent areas being eliminated. In the horizontal plane, a discretization *dx* = 2100 m, *dy* = 2100 m was used (Figure 5).

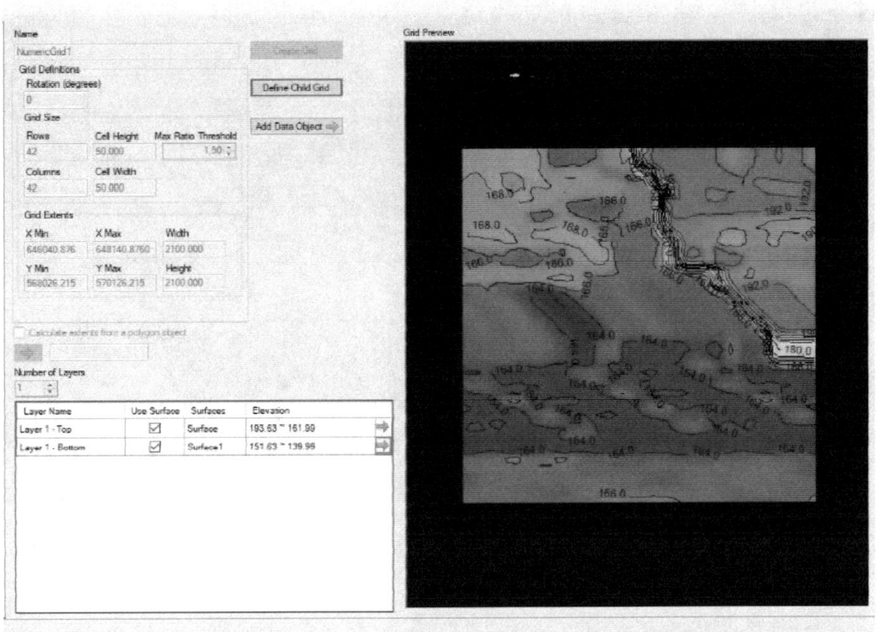

**Figure 5.** Discretization of the studied domain

The elements used to determine the land surface in the studied area are presented in figure 6, Figure 7 and Figure 8.

IOP Conf. Series: Materials Science and Engineering 1304 (2024) 012030 doi:10.1088/1757-899X/1304/1/012030

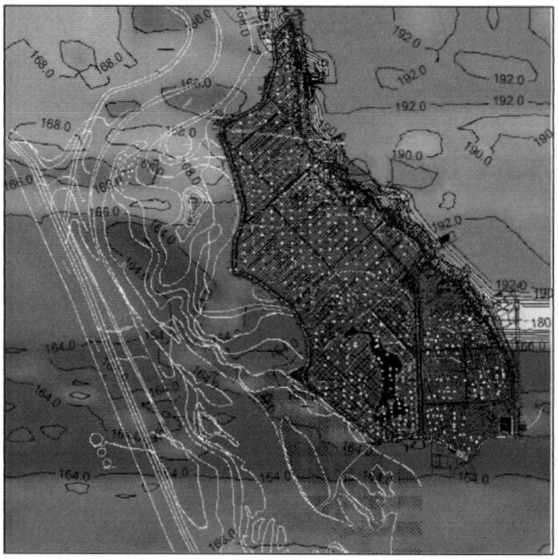


**Figure 6.** The modelled surface with the highlighted land elevations

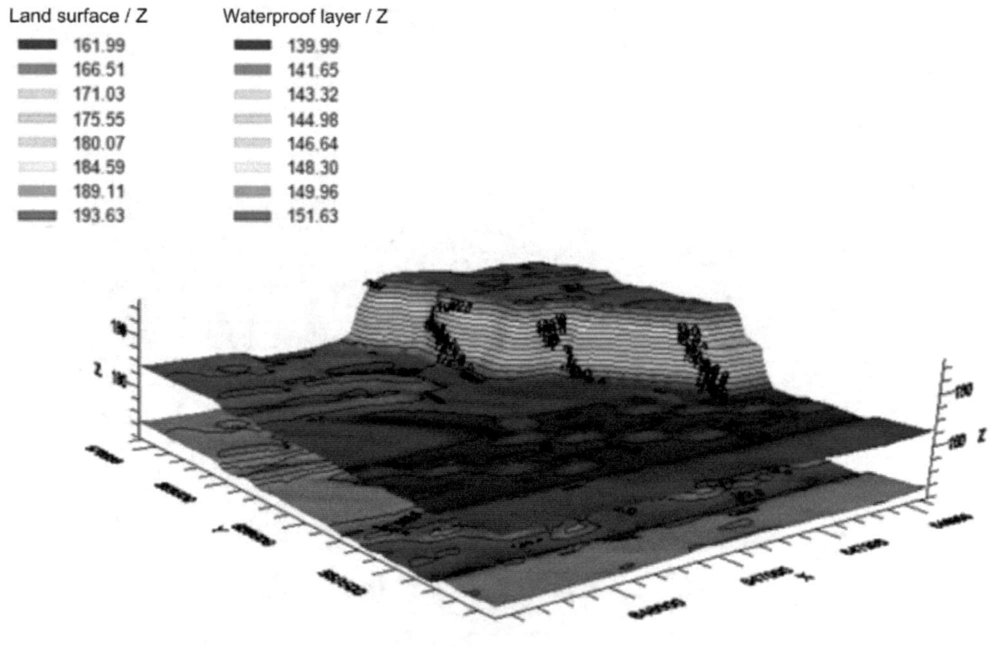


**Figure 7.** 3D representation of the land surface and waterproof layer

In the vertical plane, a vertical discretization was used. This discretization allows the appropriate vertical placement of the modelled objects at different elevations, the different zoning of permeability as well as different thicknesses of the aquifer (Figure 8).

IOP Conf. Series: Materials Science and Engineering 1304 (2024) 012030 doi:10.1088/1757-899X/1304/1/012030

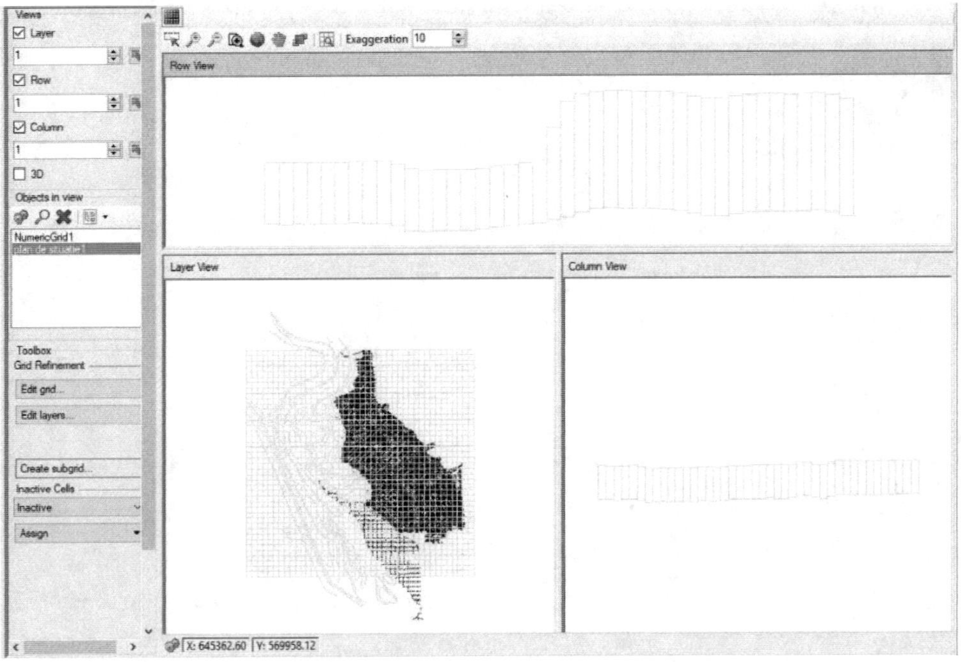


**Figure 8.** Discretization of the model in plan, cross-section and longitudinal section

Based on these modelling parameters, the numerical flow model and its calibration will be designed. The model is materialized through hydroisohypses, flow directions and hydraulic gradients. These parameters are overlayed on the real situation through a basic hydrogeological map.

The boundary conditions were expressed from considerations regarding the schematization of the area of interest.

In the mathematical modelling of groundwater flow processes the calibration process includes the adaptation, possibly the adjustment of the input data, of the model so that the results obtained through the numerical simulation correspond to the situation in the field [8].

The calibration of the model consists in adjusting the field data (permeabilities) and the boundary conditions in reasonable ways so that by running the model and obtaining the corresponding hydroisohypses they overlap the levels from the existing boreholes (Table 1, Figure 9).

**Table 1.** Coordinates and measured levels of observation wells

| Bore no. | x | y | Elevation | Well bottom | Logger id | Screen_top | Screen_bottom | Observed head |
|---|---|---|---|---|---|---|---|---|
| F 11 | 647837.51 | 568530.86 | 163.6 | 157.9 | 1 | 163 | 159 | 162.68 |
| F1 | 646742.3558 | 568373.2083 | 162.81 | 157.11 | 1 | 162 | 158 | 161.4 |
| F2 | 646795.3066 | 568228.6014 | 164.76 | 158.06 | 1 | 163 | 159 | 161.34 |
| F3 | 646840.8212 | 568125.062 | 165.8 | 158.1 | 1 | 163 | 159 | 161.68 |

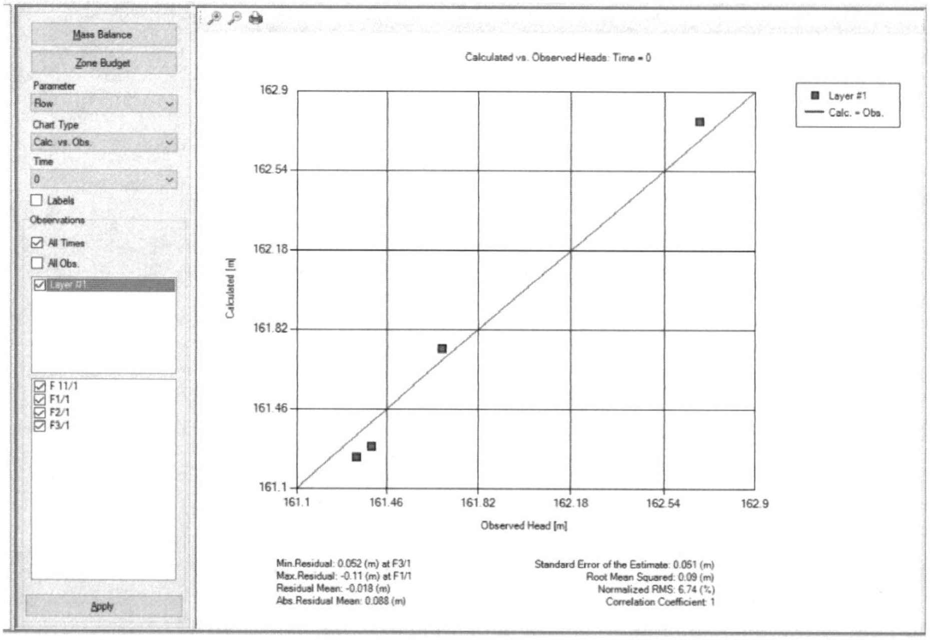


**Figure 9.** Calculated piezometric levels vs. measured piezometric levels

The total error obtained was 0.051 m. Since the size of the domain is not very large, the result can be considered as a very good one.
The map of piezometric surfaces resulting from the calibration process with the distribution of capture wells on the discretized domain is presented in Figure 10.

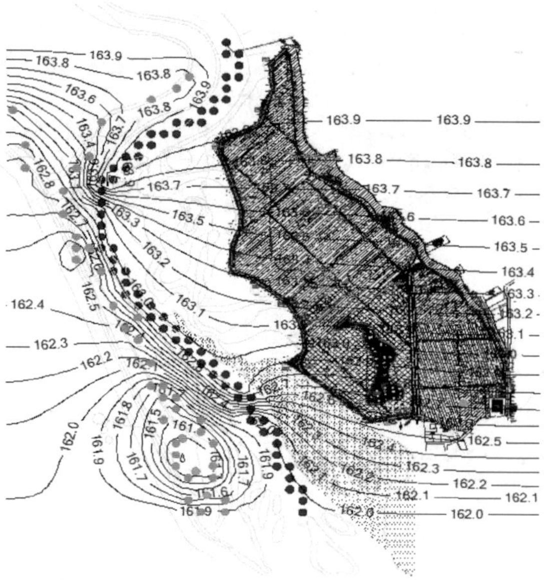


**Figure 10.** Map of piezometric surfaces with capture wells

The 2D and 3D representation of the evolution of the piezometric elevations in the case of drilling wells in the Gherăești area is shown in Figure 11.

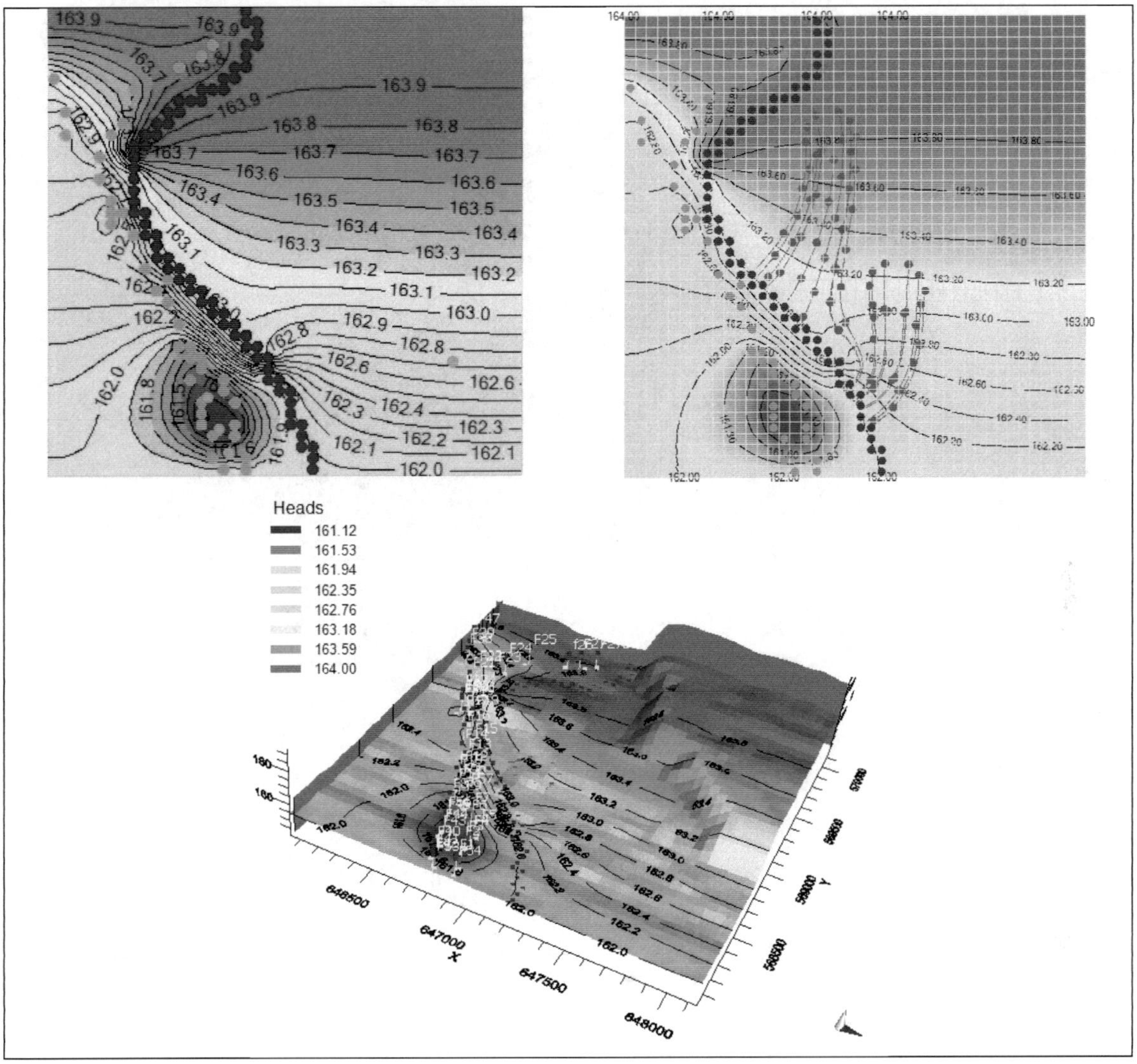

**Figure 11.** 2D and 3D representation of the evolution of the piezometric elevations in the case of drilling wells in the Gherăești area

According to the mathematical model, the aquifer is mainly supplied from the high area (terrace), the groundwater inflow being approximately 500 l/s, and secondary from precipitation over the entire surface with a flow rate of 125 l/s.

There is also an important connection between the aquifer and the Bistrița River, which drains the aquifer with a flow rate of 370 l/s.

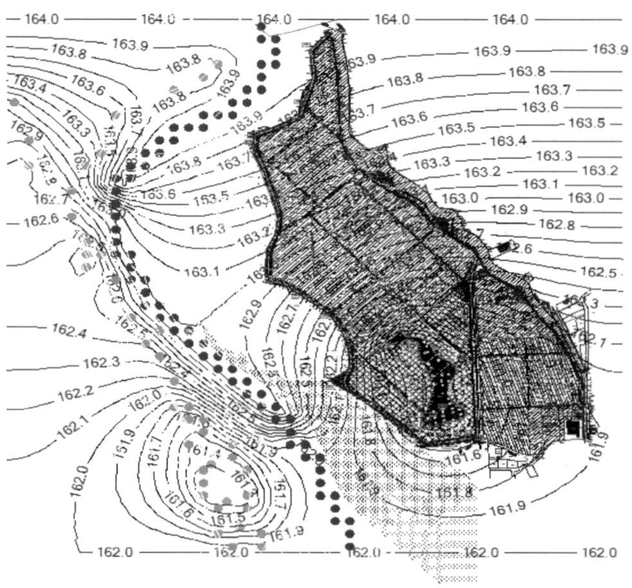

**Figure 12.** Map of piezometric surfaces in the case of drilling wells in the Gherăești area and the designed recreational lake

## 4. Conclusion

There are a multitude of models for the simulation of groundwater dynamics in the case of catchments with wells, varying from simple to complex. A realistic approach requires a good knowledge of the classic modelling structure, mainly because every groundwater modelling system must be flexible enough to be integrated into a model of the entire system.

The hydrodynamic spectra determined on the model give a complete picture of the movement of water towards the catchment fronts. Based on them, the specific flow along the catchment fronts can be determined, the main element in establishing the optimal distances between the water catchment wells and the designed recreational lake.

By running the flow model, the map of hydroisohypses and the velocity field in the study area were obtained. Furthermore, the simulation of the flow and transport model in the groundwater was carried out.

According to the mathematical model, the aquifer is mainly supplied from the high area (terrace), the inflow of underground water being considerable. There is also an important connection between the aquifer and the Bistrița River.

The flow captured by the wells in the Gherăești area represents approximately 20% of the total transited flow (at a maximum flow exploited by a borehole of $Q$=6.0 l/s), and the location of the designed recreational lake, according to the layout plan in the study area, does not affect the water levels in the catchment area.

## 5. References

[1] McDonald, M. G. and A. W. Harbugh. – *A modular three-dimensional finite-difference ground-water flow model.* U.S. Geological Survey Techniques of Water-Resources Investigations Book 6, 1988.

[2] Wen-Hsing Ciang, Kinzelbach, *Processing MODFLOW, A Simulation system for modelling grounwater flow and pollution*, 2005.

[3] x x x - *Visual Modflow Pro,* U.S. Geological Survey Modular Ground-water model (2007).

[4] Toacă, D., Bartha I., Neagu, M. R., Marcoie, N., - *Aquifers Hydrodynamic Modeling using Analogical and Numerical Model, for Underground Water's Capture.* Bul. Inst. Politehnic, Iaşi, Tomul LII (LVI), Fascicula 3-4, s. Hidrotehnică, 2007.

[5] Danchiv, A., Stematiu, D., *Metode numerice în hidrogeologie*. Ed. Didactică şi Pedagogică, R.A. Bucureşti, 1997.

[6] Zidaru, Gh., *Mişcări potenţiale şi hidrodinamica reţelelor de profile*. Ed. Didactică şi Pedagogică, Bucureşti, 1981.

[7] Reichard, E. G., *Groundwater – surface water management with stochastic surface water supplies: A simulation optimization approach*, Water Resources Research, 1995

[8] Riefler R.G., D. P. Ahifeld, *The impact of numerical precision on the solution of confined and unconfined optimal hydraulic control problems*, Hazardous Waste & Hazardous Materials, 1996.

IOP Conf. Series: Materials Science and Engineering 1304 (2024) 012031 doi:10.1088/1757-899X/1304/1/012031

# Improving pressure monitoring and control in order to reduce water loss in water urban public systems

**F Stătescu[1], V Boboc[1] , G Tatu[1] , G C Sârbu[1] , N Marcoie [1] , D Toma[1]**
[1] "Gheorghe Asachi" Technical University of Iasi, *Romania*

E-mail: valentin.boboc@academic.tuiasi.ro

**Abstract.** Water losses from distribution systems are present in all situations. Reducing them to zero is impossible. Modern technologies for operating water distribution networks are an effective means of reducing water losses. In this sense, the conducted research aimed at testing some solutions to improve the monitoring and control of pressure in a distribution network. The mathematical model of the water distribution network was developed and calibrated (Mike Urban application) and the hydraulic modeling of the network was carried out. The results were obtained through: periodic simulations with the help of the monitoring application; comparisons for flow and pressure values, data collected from field measurements, data obtained from SCADA application. These complex analyzes allow the identification, in the shortest possible time, of existing problems and areas with losses.

## 1. Introduction

Drinking water is often considered an element of nature that we have access to in a natural way, just as we benefit from the air we breathe. We lose sight of the fact that drinking water is a product, and its production in a sufficient quantity is achieved through a complex process [1]. If we also consider the inadequate management of water distribution, which is present in many situations, a less visible but crucial problem is highlighted, that of water losses, part of non-revenue water (NRW).

Reducing all components of water loss to zero is neither technically possible nor economically viable. Components of water loss must be accurately assessed and prioritized for the reduction process.

Operating a drinking water distribution network is no easy task. Such a network serves a large area, consisting of a complex system of pipes, equipment and accessories. For these reasons, it is very difficult to detect damages, modifications and illegal connections.

Water losses are caused by multiple causes: broken pipes and water leaks, poor water management, illegal connections and unauthorized consumption. Fortunately, just as varied are the solutions that can be applied with considerable benefit.

Pipe breaks and water leaks affect both the water company and the environment. This is due to the fact that lost drinking water no longer reaches consumers and also negatively affects the environment. These pipe breaks and water leaks can occur suddenly or gradually as a result of corrosion, wear and tear or lack of maintenance.

Mismanagement of drinking water leads to the impossibility of billing by water companies for the actual consumption. Competent and appropriate management considers the permanent maintenance of water distribution networks in order to ensure normal operating parameters and permanent checks on the customers served.

Illegal connections and unauthorized consumption can in some cases represent a substationary percentage of water losses in distribution networks.

## 2. Reducing water losses through advanced strategies and technologies

Losses in drinking water distribution networks can be reduced by several methods [2,3,4]:

IOP Conf. Series: Materials Science and Engineering 1304 (2024) 012031 doi:10.1088/1757-899X/1304/1/012031

- Dividing the distribution network into districts. A technique for an efficient operation of drinking water distribution networks is to divide it into sectors, also called district metering areas (DMA). Thus, water losses can be assessed for each sector, and operators can plan and prioritize their actions.
- Rapid damage assessment and repair. Quick actions lead to saving resources, protecting the environment and causing less inconvenience to customers who need a continuous supply of water.
- Monitoring network activities. Different techniques can be applied to measure and manage water pressure at different points in the distribution network.
- Pressure management. This is the most effective method of managing potable water losses. Most pipe breaks occur not so much due to high pressures, but mainly due to pressure fluctuations, which cause continuous expansion and contraction of pipes, resulting in their rupture. The higher the network pressures, the higher the amount of water losses.
- Collection of data on network operation and taking the correct decisions. The data collected in real time is an essential support in making sound decisions regarding the reduction of water losses.
- Fighting illegal consumption. There are effective practical ways to detect and monitor illegal drainage and water theft.
- Use of quality products and solutions. Any effective drinking water distribution system is based on the use of high-quality resources and solutions. The research carried out on the „automated installation for the verification of flow transducers and compact and combined thermal energy meters, Code MY SMIS: 122085", from the Faculty of Hydrotechnics, Geodesy and Environmental Engineering, Technical University "Gh. Asachi" Technical of Iasi, can offer modern and efficient solutions for the optimal exploitation of water distribution networks [5].
- High level vocational training and education. Scientific research, knowledge sharing and access to good techniques are essential to ensure that the water supply sector is able to meet future challenges.

## 3. Research performed on the zonal distribution system: Iași Municipality and the Metropolitan Area

### *3.1. System characteristics and the need for pressure control*

The distribution system of Iași City (Fig. 1) has the peculiarity of the need to pump water in stages, from ground level +40 to ground level +400. Hence, this system includes a series of intermediate pumping stations, storage tanks, compensation devices for the maximum hourly consumption and a series of local booster pumping stations.

IOP Conf. Series: Materials Science and Engineering 1304 (2024) 012031 doi:10.1088/1757-899X/1304/1/012031

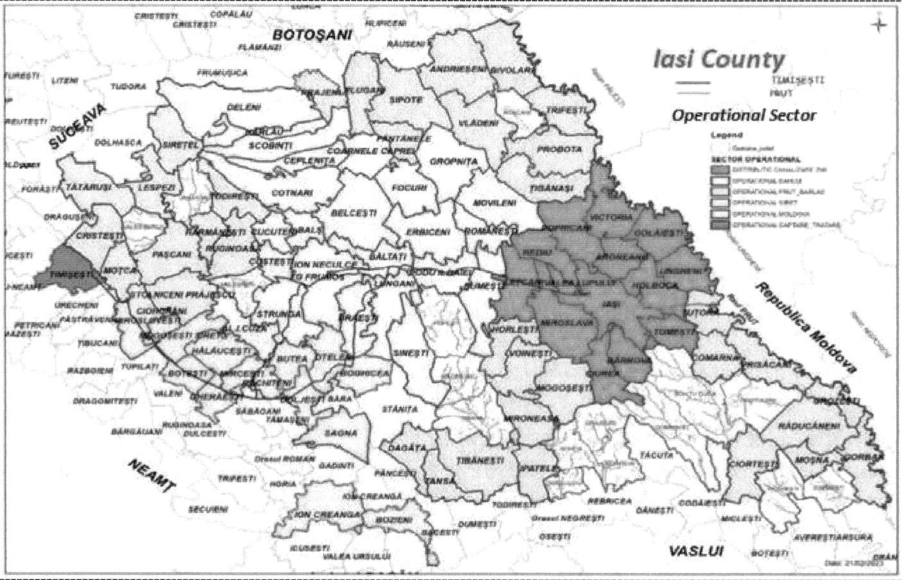


**Figure 1.** Location of water distribution system: Iași City and the Iasi Metropolitan Zone.

The main constructive elements of the system are shown in Table 1 and Fig.2.

**Table 1.** Constructive features of Iasi City and Iasi Metropolitan Zone water systems.

| Year | Length | Pumping stations | Re-chlorination station | Tank | | Connections | |
|---|---|---|---|---|---|---|---|
| | | | | No | Storage volume | Population | Economic agents |
| | (km) | (pcs) | (pcs) | (pcs) | (m$^3$) | (pcs) | (pcs) |
| 2022 | 1426 | 38 | 17 | 61 | 118.825 | 49.280 | 6215 |

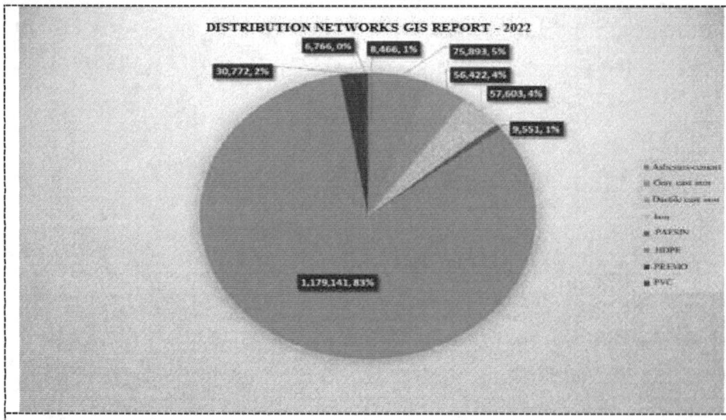


**Figure 2.** Pipes Materials: asbestos-cement / gray cast iron / ductile cast iron / steel.

The Iasi system, like most water supply systems, is designed to ensure a minimum working pressure at all points of the system throughout the day. It therefore represents the minimum pressure at a critical point in the system, which is often either the highest point in the system or the furthest from the source.

The system features significant fluctuations in water demand throughout the day, with peaks in the morning and evening, along with periods of low demand at night and sometimes in the early afternoon. In many residential areas, especially the newly established ones, there are also seasonal fluctuations, caused by climatic factors with an impact on irrigation needs or migration during vacations, for periods of days or weeks.

In accordance with applicable design standards, the water supply system is designed to meet the pressure requirement during peak demand periods when head losses are at their highest and internal pressures are at their lowest. As a result of this design methodology, the system experiences higher pressures than necessary during the off-peak period. This is clear from the fact that in most areas, major breakdowns occur in the late evening and early morning when system pressures are the highest in order to meet water demand.

The pressures at which the distribution network operates must be rigorously monitored and controlled, in order to reach the optimal level, which ensures the delivery of water to consumers in a sufficient and efficient manner, according to the supply contract. The benefits of this process are reducing the number of breakdowns due to high static/transient pressure and thus, reducing the magnitude of actual losses. However, a high-performance design and operation of the network is necessary to ensure consumers the requested and contracted level of service. Active pressure control has as its main objective the reduction of excessive pressure in a water distribution system, which results in the reduction of water losses and implicitly the frequency of failures in the network.

*3.2. Work methodology*

*3.2.1. Research directions* In order to improve pressure monitoring and control, the mathematical model of the water distribution network (Mike Urban application) was developed and calibrated, and a network hydraulic modeling was carried out.

The established research plan includes: periodic simulations using the application for monitoring; comparisons for flow and pressure values, resulting from field measurements, with data from the SCADA application. These comparisons allow the identification, in the shortest possible time, of existing problems and loss areas.

The recalibration of the mathematical model of the network is carried out at a time interval of 5 years, or when important structural changes of the network are implemented, which can be:

- expanding the network in neighboring areas;
- alternation of measurement areas on existing neighborhoods;
- new measurement areas by DMAs;
- the construction of new supply/transport pipelines.

*3.2.2. Implementation of new methods for pressure control* The improvement of the monitoring and control of pressure in the system was carried out by implementing a plan of measures, which includes:

- Installation of valves and creating a system zoning, an activity that involves the partial or total closing of the valves in the system, having the effect of reducing the transport capacity in the network, or moving a supply area to a lower pressure. Particular attention was paid to the activity of defining the size of the zones, so that the water demand for fire hydrants and for refilling the system after a major breakdown on a main water pipeline are ensured.
- The control and reduction of the pumping height was applied in the case of areas where the water demand decreased.
- The installation of pressure reducing valves is the most frequently used method of reducing pressure in distribution networks (Fig. 3). Various types of valves have been used, such as: valves that produce a constant pressure between inlet and outlet, valves that provide a constant

downstream pressure, valves that provide a variable downstream pressure allowing pressure drops during the night. Due to their sensitivity in operation, the pressure reducing valves (PRV) are periodically checked and provided with a by pass.

**Figure 3.** Pressure decreasing device, AVK, Dn 1000 mm, fitted with Pilot timer for a Dn 1000 pressure adjusting valve used for setting two pressure thresholds, including a retrofit (CHA KIT) and an electronic e-Timer-33 module.

- Programming the frequency converters for preset pressure maintenance hourly intervals through which the reduction of the pressure supplied by pumping equipment was achieved in the hourly interval 23:00-03:00, when water consumption was at minimum values.

In order to carry out the research, the following elements were taken into account:

- the number of connections;
- the length of distribution pipes;
- the number of properties;
- the population served and type of consumption: domestic, and non-domestic;
- the probable rates of physical losses;
- the pressure exponent for the system;
- details of any commercial/industrial consumers.

The conveyed data are the same as those used in the minimum night flow (MNF) analysis. In addition, three types of pressure profile and the water demand of the area in 24 hours have been identified. The average hourly values must be known at the following characteristic points:

- the pressure at the entry point ($p_{inlet}$);
- the pressure at the middle point of the zone ($p_{center}$);
- the pressure at the critical point, at the most disadvantaged consumer ($p_{critical}$);
- water demand in the area or DMA ($Q_{hr.med}$).

These four sets of hourly values are continuously measured, using water meters and electromagnetic flowmeters to measure flow and pressure loggers to measure pressure. All measuring equipment data is entered into the SCADA BlueMonitor application, they provide data at 15-minute intervals and are permanently monitored. For critical values, alarms have been set and are automatically transmitted to the operating personnel on field.

- The complete solution implemented in the distribution system of Aroneanu commune, Iași county (Fig. 4), includes the Pegasus 2 system, with a control unit with bidirectional communication system, MultiLog data-loggers and the PressView software platform. Thanks to the two-way communication system, the equipment can remotely control the pressure regulator valve using the latest GPRS communication technologies: LTE-M, NBIoT, 4G, and 3G with downgrade to 2G.

**Figure 4.** The Pegasus system, connected to an automatic pressure control valve, a MUT 1222 insertion flow meter and a MultiLog2 data logger (featuring an integated telemetry system).

### *3.3. Obtained results*

*3.3.1. Simulation of network operation* Work was performed under the MikeUrban DHI application, a specialized program, compatible with the program related to the GIS application, so that common data is exported to the modeling program. Mathematical models of the networks were performed for the localities within operating area, which include: the network graph with all the basic data (Fig. 5): flow rate and supply mode (tank, pumping); the method of extracting water from network – directly to consumers or through pumping stations - repumping (characteristics); network graph (bar connection, bar lengths, diameters, material, roughness, topographic elevations at nodes).

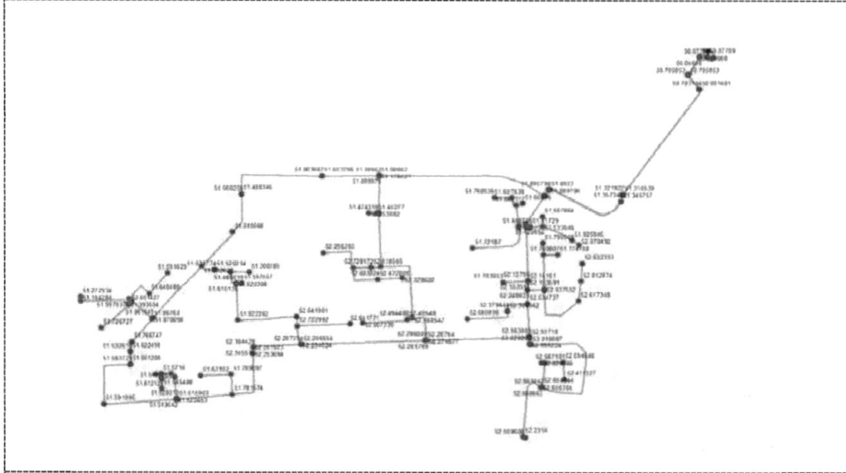

**Figure 5.** Pressure simulation within hourly interval 05:00 – 23:00 : maximum and minimum values inside nodes according to the MikeUrban modelling application.

In order to calibrate the hydraulic model, flow/pressure measurements were performed at predetermined points, such as: supply nodes, storage points and critical points. The model thus calibrated can simulate any section of a day or period of time. It offers the possibility of studying the behavior of the network in various operating situations, the scenarios being adapted in direct relation to the operational needs.

*3.3.2. Comparative analysis: pressure vs. number of failures* The research on this issue was carried aut in several localities, and the results are presented in Fig. 6. It was studied whether there is a connection between the minimum pressure in the network and the number of failures, the study being made for 10 Km of network. It can be seen that a general relationship cannot be defined only between these two parameters. The causes that determine the occurrence of failures are much more numerous.

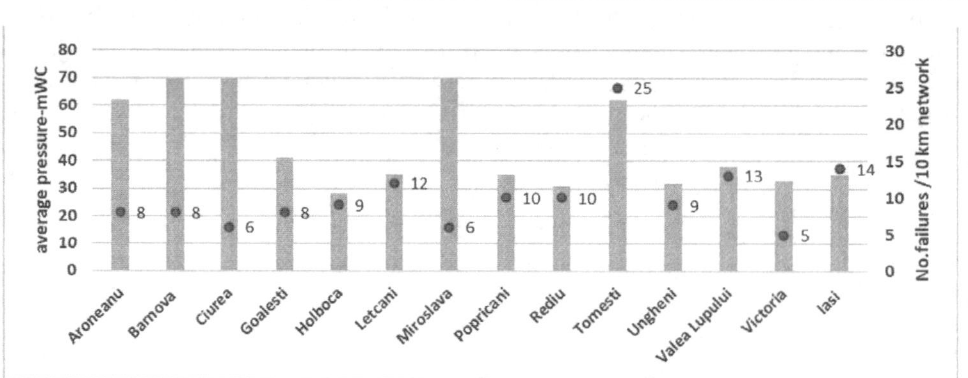


**Figure 6.** Pressure analysis vs. No. of failures.

Collected data were processed in GIS ArcMap, and hence thematic maps were generated (Fig. 7), maps very useful for the network operation process.

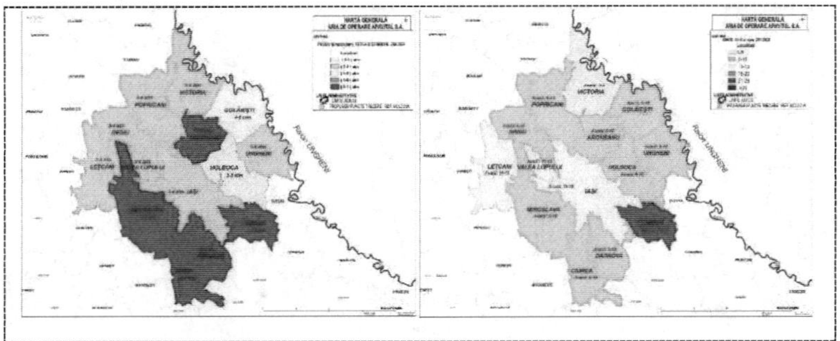

**Figure 7.** Thematic maps (average pressure, number of failures).

*3.3.3. Behavior of the automatic pressure controlling devices* The tested solution included: the Pegasus 2 system, with a control unit, cu featuring a bidirectional communication system, the MultiLog data loggers and the PressView platform. The results of a test for this system are shown in Fig. 8.1 and Fig. 8.2.

a) Graph: the pressure before activating the control as a flow function

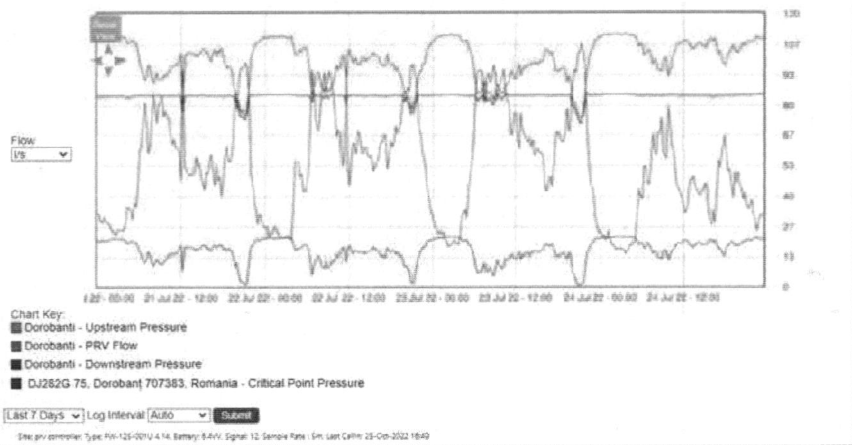


**Figure 8.1.** Pressure before activating the control as a flow function.
Red: pressure upstream of the pressure adjsting valve
Blue: pressure downstream of the pressure adjsting valve
Purple: pressure in critical point
Green: flow

b) Graph: the pressure after activating the control as a flow function

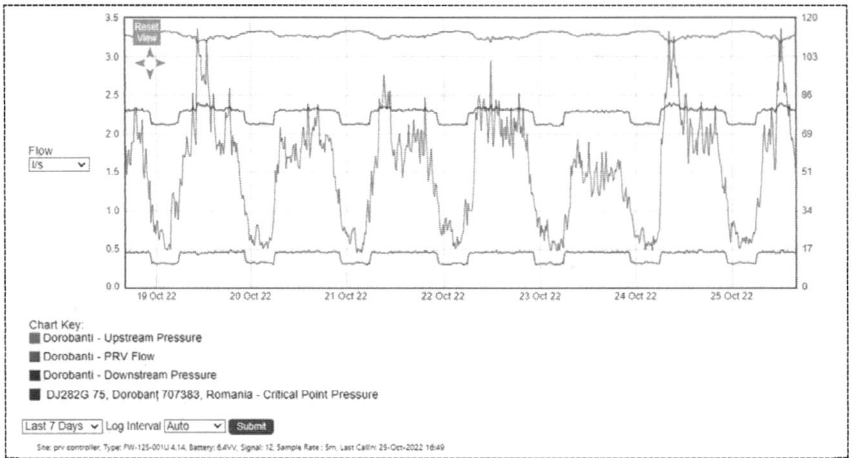


**Figure 8.2.** Pressure after activating the control as a flow function.
Red: pressure upstream of the pressure adjsting valve
Blue: pressure downstream of the pressure adjsting valve
Purple: pressure in critical point
Green: flow

A good pressure control as a function of flow has been provided, after the activation of this system.

## 4. Conclusions

The pressure management is a complex process by means of which the pressures at which a distribution network operates are monitored and controlled in order to ensure sufficient and efficient water delivery to consumers. The benefits of this process are: reducing the number of network failures due to high static/transient pressure and reducing the amount of real losses. However, the importance of design and other operating conditions in ensuring the required service level should not be underestimated.

The measurement of pressure and volume must be carried out with metrologically verified measuring instruments, which ensure confidence in the measured values. Determining measurement errors is a mandatory step in testing the functionality of water distribution systems.

Accurate metering of transited water volumes is an essential measure of water loss reduction plans.

Advanced scientific research is a guarantee for the correct and efficient solution of current and future challenges in the field of water distribution.

## 5. References

[1] Berardi, L.; Giustolisi, O. Calibration of Design Models for Leakage Management of Water Distribution Networks. Water Resour. Manag. 2021, 35, p2537–2551

[2] Chauhan, R.K.; Chauhan, K.; Dewal, M. Implementation of an Intelligent System in Process Control. In Proceedings of the National Conference on Trends in Instrumentation and Control Engineering (TICE 2009), Patiala, India, 28–29 October 2009; p 156–160

[3] Morosini, A.F.; Caruso, O.; Costanzo, F.; Savic, D. Emergency Management of Water Distribution Systems: The Nodal Demand Control. J. Procedia Eng. 2017, 186, p 428–435

[4] Ramos, H.; Mcnabola, A.; López-Jiménez, P.; Pérez-Sánchez, M. Smart Water Management towards Future Water Sustainable Networks. Water 2019, 12, 58

[5] Roy, U. Leak Detection in Pipe Networks Using Hybrid ANN Method. Water Conserv. Sci. Eng. 2017, 2, p 145–152

[6] x x x Financing contract: 351/390027/08.09.2021, Competitiveness Operational Program 2014 – 2020, Axis 1 – Research, technological development and innovation (CDI) in support of economic

IOP Conf. Series: Materials Science and Engineering 1304 (2024) 012031 doi:10.1088/1757-899X/1304/1/012031

competitiveness and business development, Action: POC/163/1/3/Stimulation the demand of companies for innovation through CDI projects carried out by companies individually or in partnership with CD institutes and universities, in order to innovate processes and products in economic sectors with growth potential

IOP Conf. Series: Materials Science and Engineering 1304 (2024) 012032 doi:10.1088/1757-899X/1304/1/012032

# Hydrodynamic modelling and simulation of the variation of the water quality parameters of the Bahlui River

**L Tătaru[1], T A Hrăniciuc[2], F Stătescu[3], N Marcoie[4], C M Vârlan (Toma)[5], R. Mitroi[6],**
"Gheorghe Asachi" Technical University of Iași, 65 Prof. dr.docent D. Mangeron Street, 700050, Iași, Romania

E-mail: nicolae.marcoie@academic.tuiasi.ro

**Abstract.** For the simulation of hydrodynamic changes and pollutant dispersion in the study area, an appropriate model was applied using the DHI series software. Flows were modelled using the hydrodynamic model and the hydrological model, and water quality was modelled using MIKE 11 ECO LAB. The paper presents a simulation for the water quality of the Bahlui River. Mike 11 contains a GIS map and simulation editors. Water quality parameters such as temperature, dissolved oxygen and biochemical oxygen demand were simulated. Modelling results can be used as a planning tool to help decision makers during different phases.

## 1. Introduction

In the context of climate change, in order to bring the quality indicators to an ecologically acceptable level, it is essential to constantly monitor the quality and quantity of surface waters. For the management, conservation and rational use of water resources, decisions for water quality management must be made in accordance with the WFD objectives. Some studies highlight the need for new advances and appropriate practices in strategies to improve and monitor the quality of the water in rivers and lakes [21; 29; 34; 3].

Janssen conducted a study in 2015 that presented the diversity and evolution of models on a global scale. Modelling programs have been and are being developed for various approaches, formulations, complexity and applications, being divided into two categories: statistical models and process-based models. Their analysis resulted in 133 models, packages or programming languages, divided as follows: ecological, socio-economic, physical, hydrological, hydrodynamic, hydraulic and geochemical models. In general, modelling programs need large amounts of data to simulate parameters [34; 10] and the choice of the appropriate model for each situation is limited by the available input data [5] the accuracy of the model, the type of calculation or the complexity of the mathematical description [30].

## 2. Development of modelling programs

Deterioration of water quality in surface waters is caused primarily by the development of industry and agriculture, and secondly by population growth [26], especially in lower-middle countries, which is why modelling programs are constantly being improved and widely used [33; 23]. They are effective scientific tools for environmental impact assessment [13; 24] and facilitates decision-making regarding the improvement of water resource management [14]. Mike 11 is one such example used in the present research.

The Danish Hydraulic Institute created the professional engineering software suite known as Mike 11 for analysis, dynamic modelling and simulation of pollutant dispersion in surface waters [8; 32].

The main computational component of the program is the hydrodynamic module that uses the Saint-Venant equations [27] utilized to determine the flow and water level's space-time variation [16; 32].

Other available modules, frequently applied for rivers, are: the hydrological module (NAM – Nedbør – Afströmmings Model), the water quality module (AD-WQ – Advection Dispersion-Water Quality

Model), the sediment transport module, the area analysis module flood forecast (FF – Flood Forecast) [2].

## 3. Results and discussion

The data necessary to carry out the simulation were obtained from the Prut-Bârlad Water Basin Administration, SGA Iași.

Three of the six modules of the Mike 11 programming package were used: the hydrodynamics module, the hydrological module and the water quality module.

Data processing for simulation involves the creation of the river network and cross sections, the definition of boundary conditions and hydrodynamic parameters, the introduction of experimental concentrations. Every 60 minutes rainfall, water level and flow data are generated as time series in a separate file as input to the simulation editor. Graphical representation and interpretation of results is done through Mike View [2; 19].

### *3.1. Realization of the Bahlui River Network*

The network editor is used for digitizing the river and allows accessing and editing data from other editors. Data visualization is done graphically or tabularly [2; 15; 12]. The transversal sections of the Bahlui river, have been inserted in this editor corresponding to the river section between the source of the river and downstream of the Pârcovaci reservoir (Figure 1).

**Figure 1.** The network and cross-sections of the Bahlui River, the Izvoare section - the Pârcovaci reservoir.

### *3.2. File of cross sections*

The file shown in Figure 2 includes information about lateral roughness, hydraulic radius, values that define the level of the river bed (X,Z) [2; 12].

54 cross-sections of river (54 sections) and lake (11 sections) cross-sections collected from Bahlui riverbed topography data were entered.

Each cross-section is unique and the following information can additionally be included: topographical identification, mileage and name of the cross-section [15].

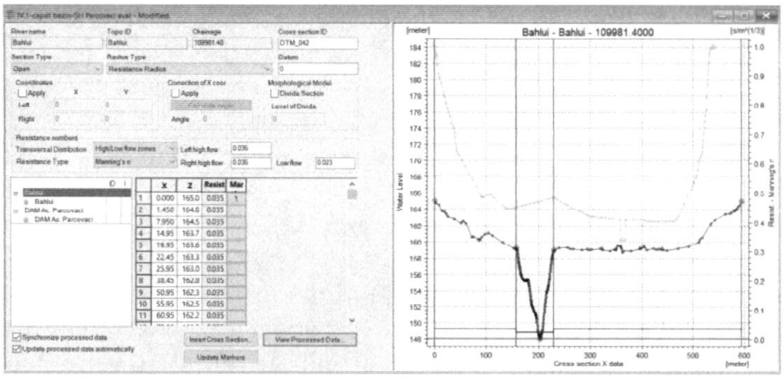


**Figure 2.** Defining and viewing data in a cross section.

*3.3. Boundary conditions file*

This file is used to specify the water level, inlets and outlets along the river, solution concentration in a hydrograph, various meteorological data and boundary conditions related to the structures imposed on the model [15; 12].

According to the information gleaned from the limnimetric keys, the limiting condition at the upstream end was the level recorded in the first segment — the Bahlui river's source — and the one at the downstream end was the flow observed in the last section — the Pârcovaci reservoir. (Figure 3).

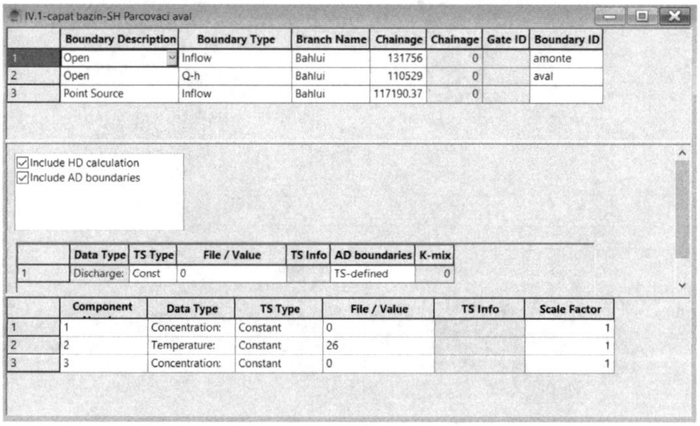


**Figure 3.** Definition of boundary conditions.

Q-h values are calculated at alternate points along the river (Kamel, 2008), then will be used in the water quality module (AD-WQ) to assess pollutant dispersion [2].

*3.4. Water-quality module*

The WQ module deals with the basic aspects of water quality [12], which described the processes of transformation and transport of pollutants along rivers [32]. The AD module is based on the one-dimensional mass conservation equation (for dissolved or suspended substances) [31; 2] and has the following form:

$$\frac{\partial(AC)}{\partial t} + \frac{\partial(QC)}{\partial x} - \frac{\partial}{\partial x}\left(AD\frac{\partial C}{\partial x}\right) = AKC + C_2 q \tag{1}$$

IOP Conf. Series: Materials Science and Engineering 1304 (2024) 012032 doi:10.1088/1757-899X/1304/1/012032

in which: C – the concentration of the analysed substance in the river, D – the dispersion coefficient, A – cross-sectional area, K – the linear reaction coefficient, C2 – the concentration of the pollution source, Q – lateral flow, x – spatial coordinate and t – time coordinate.

For the coupling of the two modules, data from the hydrodynamic module is required [31; 2].

### *3.5. The hydrodynamic module*

The hydrodynamic (HD) model solves the mass and momentum conservation equations (Saint Venant equations) using the implicit six-point finite difference Abbot-Ionescu scheme [31; 2; 28]. The equations have the following form:

$$\begin{cases} \dfrac{\partial A}{\partial t} + \dfrac{\partial Q}{\partial X} = q \\ \dfrac{\partial Q}{\partial t} + \dfrac{\partial}{\partial X}\left(\dfrac{\alpha Q^2}{A}\right) + qA\dfrac{\partial h}{\partial X} + \dfrac{gn^2 Q|Q|}{AR^{4/3}} = 0 \end{cases} \quad (2)$$

where: $Q$ – flow, $A$ – flow area, $q$ – the lateral inlet flow, $h$ – water depth, $R$ – hydraulic range, $\alpha$ – the moment distribution coefficient and $n$ – Manning's coefficient.

The hydrodynamic module and the water quality module together, offers the following simulation possibilities: current water quality assessment, information to improve water quality, warning of pollution accidents [11] or the prediction of the evolution over time of the water quality parameters (information on the changes that take place in the watershed in the long term) [18; 33].

The module can work directly on standard templates (Figure 4), covering different combinations of water quality processes (eutrophication, heavy metals, sediments and vegetation, biochemical oxygen demand or dissolved oxygen) or can be adjusted according to the specific needs of each aquatic environment [34; 11] and can be applied to simple cases, where the major problem is oxygen depletion, as well as to more complex cases, where the effects of biochemical oxygen consumption must be considered [20; 6] or that can describe nutrient cycling through quantitative scenario assessments [17].

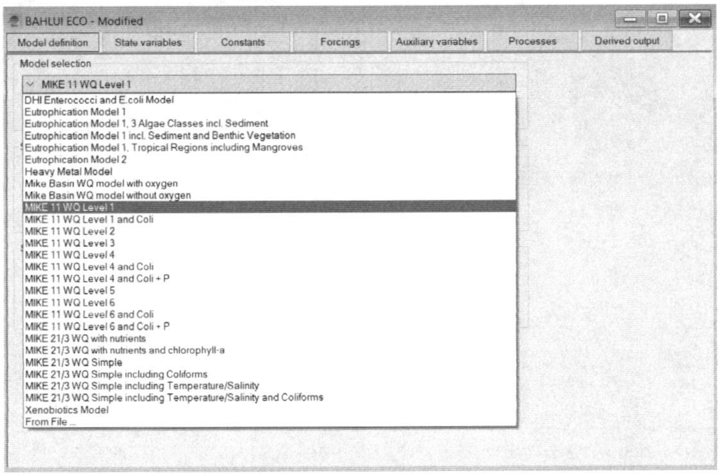


**Figure 4.** Variables and types of processes included at the different levels of the model.

Level 1 of the model was used in this study and includes for dissolved oxygen reaeration processes and for biochemical oxygen consumption it includes processes of oxidation or decomposition of organic matter [25].

### *3.6. The hydrological module*

Model inputs included meteorological data of the study area, water level and flow [33]. The module allows the user to specify the type of simulated hydrological process [9].
Model calibration was performed using data from the Bahlui river basin. In Figures 5 and 6, the primary hydrological characteristics of this time period are depicted.

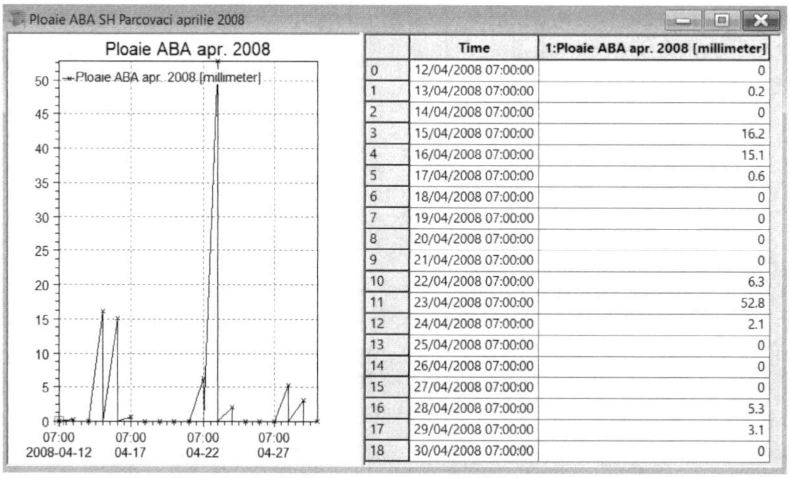


| | Time | 1:Ploaie ABA apr. 2008 [millimeter] |
|---|---|---|
| 0 | 12/04/2008 07:00:00 | 0 |
| 1 | 13/04/2008 07:00:00 | 0.2 |
| 2 | 14/04/2008 07:00:00 | 0 |
| 3 | 15/04/2008 07:00:00 | 16.2 |
| 4 | 16/04/2008 07:00:00 | 15.1 |
| 5 | 17/04/2008 07:00:00 | 0.6 |
| 6 | 18/04/2008 07:00:00 | 0 |
| 7 | 19/04/2008 07:00:00 | 0 |
| 8 | 20/04/2008 07:00:00 | 0 |
| 9 | 21/04/2008 07:00:00 | 0 |
| 10 | 22/04/2008 07:00:00 | 6.3 |
| 11 | 23/04/2008 07:00:00 | 52.8 |
| 12 | 24/04/2008 07:00:00 | 2.1 |
| 13 | 25/04/2008 07:00:00 | 0 |
| 14 | 26/04/2008 07:00:00 | 0 |
| 15 | 27/04/2008 07:00:00 | 0 |
| 16 | 28/04/2008 07:00:00 | 5.3 |
| 17 | 29/04/2008 07:00:00 | 3.1 |
| 18 | 30/04/2008 07:00:00 | 0 |

**Figure 5.** Upstream Pârcovaci reservoir boundary conditions time series.

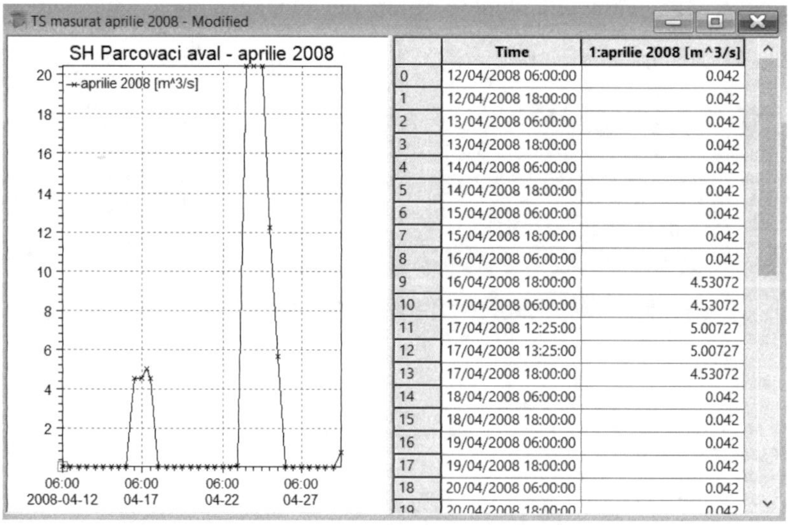


| | Time | 1:aprilie 2008 [m^3/s] |
|---|---|---|
| 0 | 12/04/2008 06:00:00 | 0.042 |
| 1 | 12/04/2008 18:00:00 | 0.042 |
| 2 | 13/04/2008 06:00:00 | 0.042 |
| 3 | 13/04/2008 18:00:00 | 0.042 |
| 4 | 14/04/2008 06:00:00 | 0.042 |
| 5 | 14/04/2008 18:00:00 | 0.042 |
| 6 | 15/04/2008 06:00:00 | 0.042 |
| 7 | 15/04/2008 18:00:00 | 0.042 |
| 8 | 16/04/2008 06:00:00 | 0.042 |
| 9 | 16/04/2008 18:00:00 | 4.53072 |
| 10 | 17/04/2008 06:00:00 | 4.53072 |
| 11 | 17/04/2008 12:25:00 | 5.00727 |
| 12 | 17/04/2008 13:25:00 | 5.00727 |
| 13 | 17/04/2008 18:00:00 | 4.53072 |
| 14 | 18/04/2008 06:00:00 | 0.042 |
| 15 | 18/04/2008 18:00:00 | 0.042 |
| 16 | 19/04/2008 06:00:00 | 0.042 |
| 17 | 19/04/2008 18:00:00 | 0.042 |
| 18 | 20/04/2008 06:00:00 | 0.042 |
| 19 | 20/04/2008 18:00:00 | 0.042 |

**Figure 6.** Downstream of the Pârcovaci reservoir, time series of the boundary conditions.

### *3.7. The simulation editor*

The simulation editor is the main editor in Mike 11, where the corresponding files created previously are entered. It consists of five tabs: Models, Inputs, Simulation, Results and Start (Figure 7) [2].

IOP Conf. Series: Materials Science and Engineering 1304 (2024) 012032 doi:10.1088/1757-899X/1304/1/012032

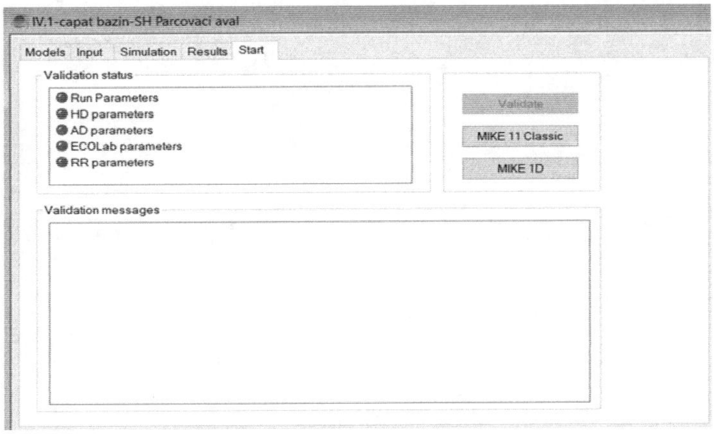


**Figure 7.** The simulation editor.

Mike 11 was able to determine the discharge at each location using hydrological data as well as cross-section information from the Bahlui River [9]. The hydrodynamic model's longitudinal profile, as produced by the Mike 11 software, is shown in Figure 8.

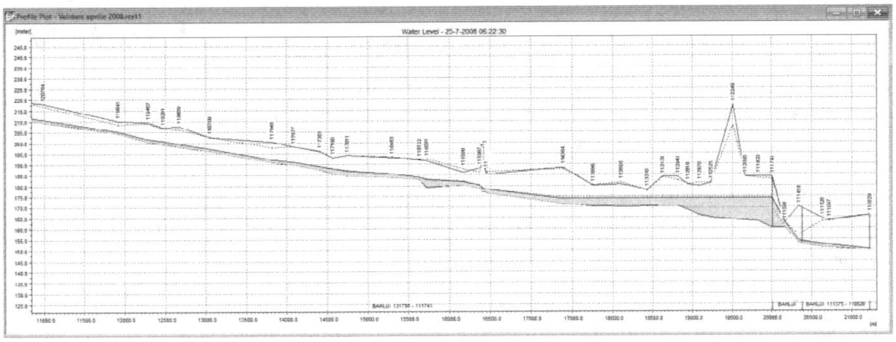

**Figure 8.** Longitudinal profile of the Bahlui River.

The Prut-Bârlad Water Basin Administration, SGA Iasi, provided information on the water quality, including temperature, dissolved oxygen content, and biochemical oxygen consumption.
With the help of the mathematical model used, the variation of temperature, dissolved oxygen and biochemical oxygen consumption along the Bahlui River was simulated, in accordance with the particularities of the river, and the results of the simulated series are presented below (Figures 9, 10, 11).

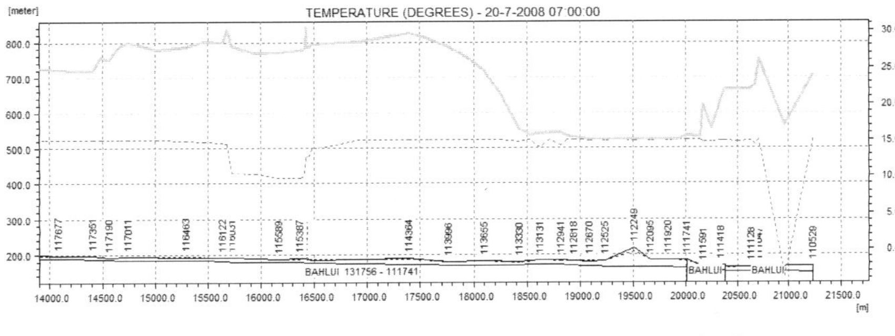


**Figure 9.** Temperature variation along the Bahlui River.

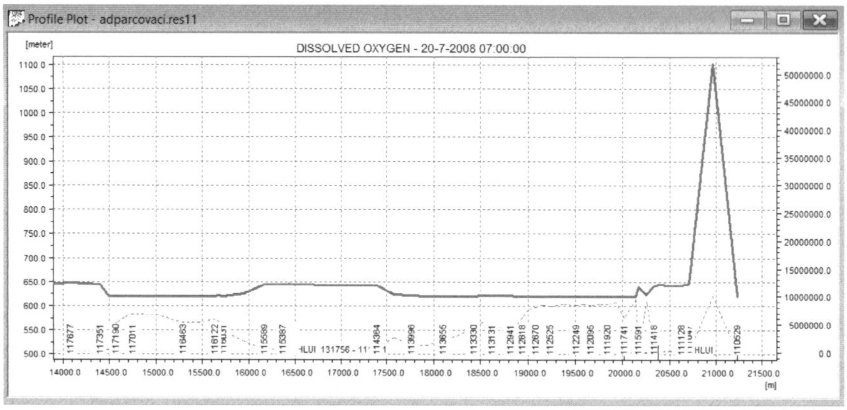


**Figure 10.** Dissolved oxygen variation along the Bahlui River.

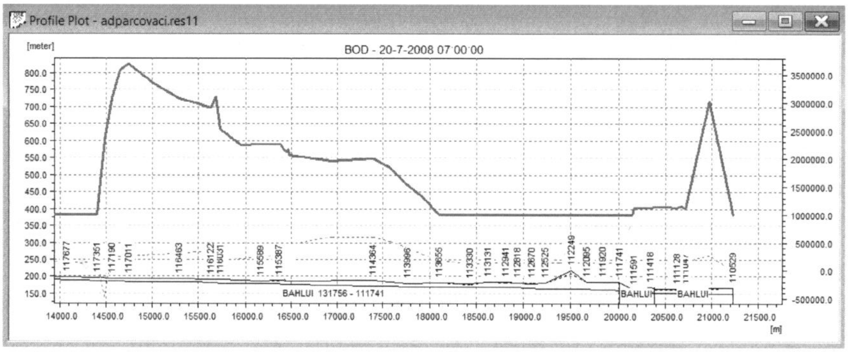


**Figure 11.** Variation in biochemical oxygen consumption along the Bahlui River.

## 4. Conclusion

The Bahlui River's hydrodynamic model, which is linked to the water quality simulation model, is a useful tool for managing surface water resources and has produced excellent results.

The water quality study module has six different levels of complexity. In this study, level 1 was used, which limits the number of analysed variables to 3 parameters, namely: temperature, dissolved oxygen and biochemical oxygen consumption.

The input data to the hydrodynamic model, obtained by direct measurements in the field, consisted of flow and level hydrographs. The measurements of flows and levels from April 2008 were used to calibrate the model.

The findings of this study can be applied to a variety of tasks, such as developing a plan for Bahlui River water quality protection measures, better predicting hydrodynamic changes in the riverbed and water quality parameters, supporting decisions, and managing surface water resources more effectively.

## 5. References

[1] www.ysi.com

[2] Benchea RE 2012 Sistem expert pentru monitorizarea, evaluarea calității și gestionarea resurselor de apă. Studiu de caz: râul Bahlui, Teză de doctorat, Iași.

[3] Butts M et al 2012 Mike She-Ecolab—an integrated catchment-scale eco-hydrological modelling tool, XIX international conference on water recourses, CMWR 2012, University of Illinois at Urbana-Champaign, 17–22.

[4] Cheng K et al 2022 An Urban Water Pollution Model for Wuhu City, *Water* (Switzerland), 14, 3, 386.

[5] Chu ML et al 2013 Impacts of urbanization on river flow frequency: A controlled experimental modeling-based evaluation approach, *Journal of Hydrology*, 495, 1-12.
[6] Crețescu I et al 2013 Development of an expert system for surface water quality monitoring in the context of sustainable management of water resources, *Environmental Engineering and Management Journal*, 12, **8**, 1721-1734.
[7] DHI, Mike 11, Reference Manuals: Water Quality (ECO LAB) Module, Danish Hydrologic Institute, Denmark, 2008c.
[8] DHI, Mike 11, User Manual, Danish Hydraulic Institute, Denmark, 1992.
[9] Eisakhani M et al 2012 Validation of MIKE 11 Model Simulated Data for Biochemical and Chemical Oxygen Demands Transport, *American Journal of Applied Sciences*, 9, 3, 382-387.
[10] El Saadi AM 2015 Simulation adequacy assessment of water quality of Rosetta Branch, *Water Quality Research Journal*, **50** (4), 359–368.
[11] Gao F et al 2021 Eutrophication regulation of Wuxing Lake based on eco-dynamics model, Arabian Journal of Geosciences, 14, 1321.
[12] Girbaciu A et al 2015 Water Quality Modelling Using Mike 11, REV. CHIM. (Bucharest), 8, 66.
[13] Gunawardena A et al 2017 Industrial pollution and the management of river water quality: a model of Kelani River, Sri Lanka, *Environ Monit Assess*, 189: 457.
[14] Hanington P et al 2017 A hydrological model for interprovincial water resource planning and management: A case study in the Long Xuyen Quadrangle, Mekong Delta, Vietnam, *Journal of Hydrology*, **547**, 1-9.
[15] Hrăniciuc TA et al 2012 Flood mapping with Mike Flood Model for a flood event reconstitution, *Journal of Environmental Protection and Ecology*, **13**, 2, 756–763.
[16] Jahandideh-Tehrani M et al 2020 Hydrodynamic modelling of a flood-prone tidal river using the 1D model MIKE HYDRO River: calibration and sensitivity analysis, *Environ Monit Assess*, **192**, 97.
[17] Janssen ABG et al 2015 Exploring, exploiting and evolving diversity of aquatic ecosystem models: a community perspective, *Aquatic Ecology*, **49**, 513–548.
[18] Jayawardena A and Lui P.H., Numerical solution of the dispersion equation using a variable dispersion coefficient: method and applications, Hydrological Sciences – Journal – des Sciences Hydrologiques, 29, 293-309, 1984.
[19] Kamel AH 2008 Application of a hydrodynamic Mike 11 model for the Euphrates River in Iraq, *Slovak Journal of Civil Engineering*.
[20] Keller V 2006 Risk assessment of bdown-the-drainQ chemicals: Search for a suitable model, *Science of the Total Environment*, **360**, 1-3, 305 – 318.
[21] Khwairakpam E et al 2019 Monitoring and modelling water quality of Loktak Lake catchment, SN *Applied Sciences*, **1**, 491.
[22] Li Q, Wang T, Zhu Z, Meng J, Wang P, Suriyanarayanan S, Zhang Y, Zhou Y, Song S, Lu Y, Yvette B 2017 Using hydrodynamic model to predict PFOS and PFOA transport in the Daling River and its tributary, a heavily polluted river into the Bohai Sea, China. *Chemosphere*, **167**, 344-352.
[23] Liang J et al 2015 MIKE 11 model-based water quality model as a tool for the evaluation of water quality management plans, *Journal of Water Supply*: Research and Technology – AQUA, 6, 64.
[24] Liu L and You Xy 2021 Study of water quality response to water transfer patterns in the receiving basin and surrogate model, *Environmental Science and Pollution Research*.
[25] Milišić H et al 2019 Mathematical Modelling of Surface Water Quality, *Advanced Technologies, Systems, and Applications* III, **60**, 138-156.
[26] Nada A et al 2021 Water quality modelling and management for Rosetta Branch, the Nile River, Egypt, *Environ Monit Assess*, **193**, 603.
[27] Ngo LL et al 2008 Implementation and Comparison of Reservoir Operation Strategies for the Hoa Binh Reservoir, using the Mike 11 Model, *Water Resources Management*, **22**, 457–472.
[28] Ngo LL et al 2005 Application of MIKE 11 in managing reservoir operation, International

Conference on Reservoir Operation and River Management, China, 17-23.
[29] O'Grady J et al 2021 A comprehensive review of catchment water quality monitoring using a tiered framework of integrated sensing technologies, *Science of the Total Environment*, 765, 142766.
[30] Ostojski M 2013 Application of hydrological and hydraulic models for hydrological data transfer, *Acta Geophys*, **61**, 690–705.
[31] Shakibaeinia A et al 2017 A numerical framework for modelling sediment and chemical constituents transport in the Lower Athabasca River, *Journal of Soils and Sediments*, 17, 1140–1159.
[32] Thu Minh HV et al 2022 A Model-Based Approach for Improving Surface Water Quality Management in Aquaculture using MIKE 11: A Case of the Long Xuyen Quadangle, Mekong Delta, Vietnam, *Water*, **14**, 412.
[33] Wang Q et al 2011 Water quality model establishment for middle and lower reaches of Hanshui River, China, *Chinese Geographical Science*, **21**, 646–655.
[34] Waseem M et al 2018 Suitability of common models to estimate hydrology and diffuse water pollution in North-eastern German lowland catchments with intensive agricultural land use, *Frontiers of Agricultural Science and Engineering* **5**, 4, 420-431.

IOP Conf. Series: Materials Science and Engineering 1304 (2024) 012033 doi:10.1088/1757-899X/1304/1/012033

# Digital solutions in the determination of bearing capacity for straight and stepped mat foundations on sloping ground

**F Bejan, M Aniculăesi, C V Gafincu and I Lungu**
"Gheorghe Asachi" Technical University of Iasi, Faculty of Civil Engineering and Building Services, 1 Prof. D. Mangeron Blvd., 700050, Iasi, Romania

E-mail: florin.bejan@academic.tuiasi.ro

**Abstract**. On sloped sites where a mat foundation is to be used, because of the limited bearing capacity of the ground, it is necessary to cut into the sloped surface to provide a level base for the mat foundation. To reduce the amount of excavation and backfill a stepped mat may be used down the slope. Traditionally, determining the bearing capacity values for spread foundations on sloping ground has been a complex and time-consuming process. Several factors must be considered, including various variables (soil type, soil profile, slope angle and geometrical configuration of the infrastructure). The digital solution used in this study involves the use of advanced software that takes into account various input parameters and performs complex calculations to determine the bearing capacity of the foundation on horizontal and sloped ground. For this research, the analyses were done for three different slope angles ($\beta = 5°, 10°, 15°$) and a comparison is made between the bearing capacity of the ground for straight and stepped mat foundations using both analytical methods and finite element method.

## 1. Introduction

Building on sloping ground can present a number of challenges, including issues with foundation stability or ground general instability, but with proper design, planning and execution, it is possible to create safe and stable buildings that are visually appealing and blend in with the surrounding landscape. One way to address these challenges is through the use of set-back foundation configurations. Set-back foundations are designed that each level of the building is set back from the one below it, following the slope of the ground [1]. These types of foundations allow the building to follow the natural slope and the ground will provide adequate support and stability. However, the bearing capacity of these foundations must be carefully considered.

The computation of the soil bearing capacity is one of the most important problems in geotechnical engineering and has a decisive role in the design of safe and economic spread foundations. Terzaghi was the first to evaluate the bearing capacity of a strip footing on a horizontal and homogeneous ground [2]. The computation relationship includes the influences of the soil properties (internal friction angle, cohesion and unit weight) and the features of the foundation (foundation depth and the width of the footing). This well-known relationship was updated by different authors [3][4][5][6] taking into account new contribution factors: base inclination ($b_c$, $b_q$, $b_\gamma$), embedment depth ($d_c$, $d_q$, $d_\gamma$), ground surface inclination ($g_c$, $g_q$, $g_\gamma$), load inclination ($i_c$, $i_q$, $i_\gamma$), foundation base shape ($s_c$, $s_q$, $s_\gamma$) and load eccentricity. This paper focuses on the ground surface inclination factors named also slope factors. The first estimation of the bearing capacity factors for a strip foundation on sloping ground was proposed by Meyerhof in 1957 [4]. After that, Brinch Hansen [5] and Vesić [6] proposed new relationships for the slope factors. These slope factors are independent from the internal friction angle. Later, some researchers have published research about the bearing capacity of foundations on sloping ground, but was mostly limited to purely cohesive massifs [7], [8], [9] or to purely cohesionless massifs [10]. Some investigations showed that, in the case of cohesionless soils, the bearing capacity is always governed by the foundation soil failure, while in cohesive soils the bearing capacity is dictated by the slope stability

[11]. Van Baars proposed new slope factors based on the numerical calculations, taking into account the influence of the internal friction angle [12].

Considering the increased number of buildings positioned on natural slopes but also the multitude of situations caused by architecture design and asymmetric shapes, the evaluation of the soil bearing capacity becomes more complex. In this paper the bearing capacity for buildings' foundations located on sloping grounds is investigated using the finite element method and compared with the existing analytical equations.

## 2. Context and description of the foundation design conditions

One footing width (B = 30 m) in two configurations was considered. The geometries of the problem analysed is shown in Figure 1. Cases ($a_1$ to $a_4$) represent a straight mat foundation that transmits a uniform vertical pressure to the foundation ground and cases ($b_1$ to $b_4$) represent a stepped mat foundation. A step with the height of 3.0 m was considered, which divides the foundation into a 10 m and 20 m openings.

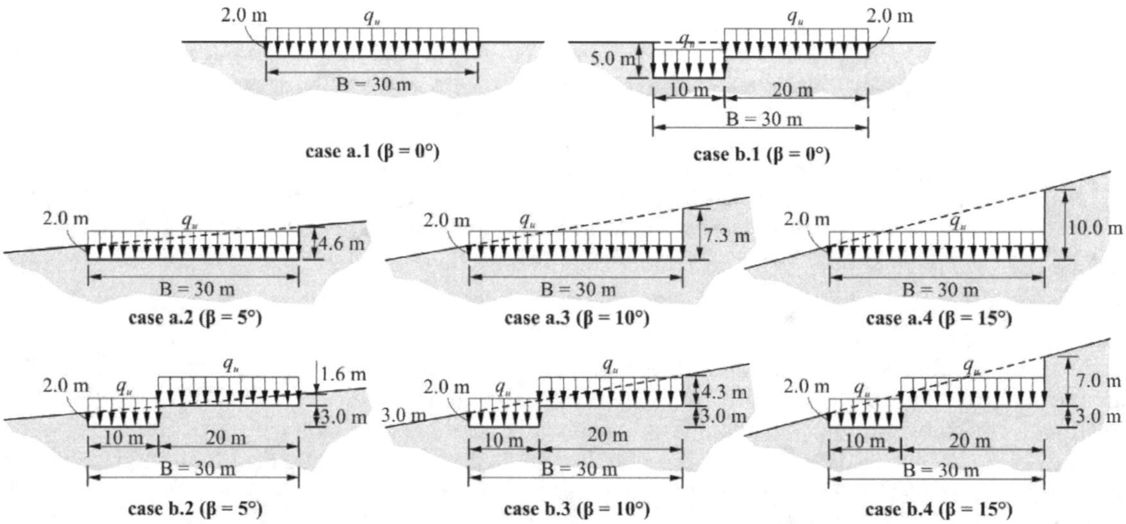


**Figure 1.** The geometries of the analysed footing's cases: a – straight mat, b – stepped mat

Four slope angles were considered (β = 0°, 5°, 10°, 15°). For each geometry, six soil types with different geotechnical characteristics are considered, traced from NP 122:2010 [13]. The detailed geotechnical parameters for the six types of soils (S1 to S6) are presented in Table 1.

**Table 1.** Soil parameters

| Geotechnical characteristics | Units | S1 | S2 | S3 | S4 | S5 | S6 |
|---|---|---|---|---|---|---|---|
| Internal friction angle of the soil, $\phi$ | ° | 16 | 14 | 11 | 13 | 11 | 8 |
| Soil cohesion, c | kPa | 12 | 10 | 9 | 27 | 24 | 21 |
| Soil unit weight, $\gamma$ | kN/m³ | 18 | 18 | 18 | 18.5 | 18.5 | 18.5 |

The ultimate bearing capacity of the ground is determined using both analytical methods and finite element method.

## 3. Methods

### 3.1. Analytical methods

A total of 72 analytical calculation has been performed for comparison purpose. This calculation has been done using Brinch Hansen, Vesic and Van Baars methods [5; 6; 12]. The general equation used for evaluating the ultimate bearing capacity $q_u$ is:

$$q_u = c'N_c b_c d_c g_c i_c s_c + q'N_q b_q d_q g_q i_q s_q + 0.5\gamma' B' N_\gamma b_\gamma d_\gamma g_\gamma i_\gamma s_\gamma \tag{1}$$

where $B$ is the width of the foundation; $B'$ is the reduced width of the foundations; $\gamma$ is the soil unit weight; $c$ is the cohesion of the soil; $q'$ is the effective overburden lateral to the footing level; $N_c$, $N_q$, $N_\gamma$ are the bearing capacity factors which are functions of the friction angle $\phi$; $b_c$, $b_q$, $b_\gamma$ are the base inclination factors; $d_c$, $d_q$, $d_\gamma$ are the depth factors; $g_c$, $g_q$, $g_\gamma$ are the ground inclination factors; $i_c$, $i_q$, $i_\gamma$ are the load inclination factors; $s_c$, $s_q$, $s_\gamma$ are the shape factors for the footing geometry.

In this study the values for base inclination factors (*b*), depth factors (*d*), load inclination factors (*i*) and shape factor (*s*) are set to one. The values for the bearing capacity factors and slope factors are determined the equations from Table 2.

**Table 2.** Factors for determining $q_u$ of foundations on sloping ground

| Brinch Hansen | Vesić | Van Baars |
|---|---|---|
| $N_q = \tan^2(45+\phi/2)\cdot e^{\pi\tan\phi}$ | $N_q = \tan^2(45+\phi/2)\cdot e^{\pi\tan\phi}$ | $N_q = \tan^2(45+\phi/2)\cdot\cos^2\phi\cdot e^{\pi\tan\phi}$ |
| $N_c = (N_q-1)\cdot\cot\phi$ | $N_c = (N_q-1)\cdot\cot\phi$ | $N_c = (N_q-1)\cdot\cot\phi$ |
| $N_\gamma = 1{,}5(N_q-1)\cdot\tan\phi$ | $N_\gamma = 2(N_q+1)\cdot\tan\phi$ | $N_\gamma = 4\tan\phi\cdot(e^{\pi\tan\phi}-1)$ |
| $g_c = 1-\beta°/147°$ | $g_q = g_\gamma = (1-\tan\beta)^2$ | $g_q = 1-(\beta/\phi)^{3/2}$ for $\beta\le\phi$ |
| $g_q = g_\gamma = (1-0.5\tan\beta)^5$ | $g_c = \dfrac{N_q g_q - 1}{N_q - 1}$ for $\phi > 0$ | $g_\gamma = 1-(\beta/\phi)^{2/3}$ for $\beta\le\phi$ |
| | $g_c = 1-\dfrac{2\beta}{\pi+2}$ for $\phi = 0$ | $g_c = \cos\beta\cdot\left(e^{-2\beta\tan\phi} - \dfrac{2\beta}{2+\pi}\cdot e^{-\pi\tan\phi}\right)$ for $\beta\le\phi$ |

$N_c$, $N_q$ and $N_\gamma$ - bearing capacity factors; $g_c$, $g_q$ and $g_\gamma$ - slope factors; $\beta$ - slope angle in degrees; $\phi$ - internal friction angle of the soil.

### 3.2. Finite element method (FEM)

The FEM involves the application of the Mohr-Coulomb model, which considers a linearly elastic-perfect plastic behaviour for the soil. This model requires five input parameters, which include two parameters related to elasticity (i.e. soil modulus $E_s$ and Poisson's ratio $\nu$) and three parameters related to strength (internal friction angle $\phi$, cohesion $c$ and the dilatancy angle $\psi$). The values for the soil modulus and Poisson's ratio were kept constant ($E_s$ = 20 MPa and $\nu$ = 0.4990) for all types of analyzed soils.

This study includes finite element analyses that are exclusively plane strain analyses and were carried out using the Plaxis2D software. The dimensions of the model have been selected in such a way that the model boundaries do not intersect with the significant isobars, as shown in Figure 2. The finite element model has implemented the *"standard fixity"* condition, whereby the lateral vertical boundaries are prevented from moving in the horizontal direction, and the bottom boundary of the model is restrained from both vertical and horizontal movements.

The discretization of the model was accomplished using triangular elements with 15 nodes. The meshing was performed through a fully automated finite element generation technique that was integrated into the Plaxis2D software. The mesh coarseness was set to *"medium"*. In order to obtain the most accurate results, a mesh with medium element distribution was generated, with a coarseness factor of 0.10 for the area next to the foundations (60.00 m to the right, 30.00 m to the left and 45.00 m on the bottom). Figure 2 displays the standard meshing that was generated for the numerical model.

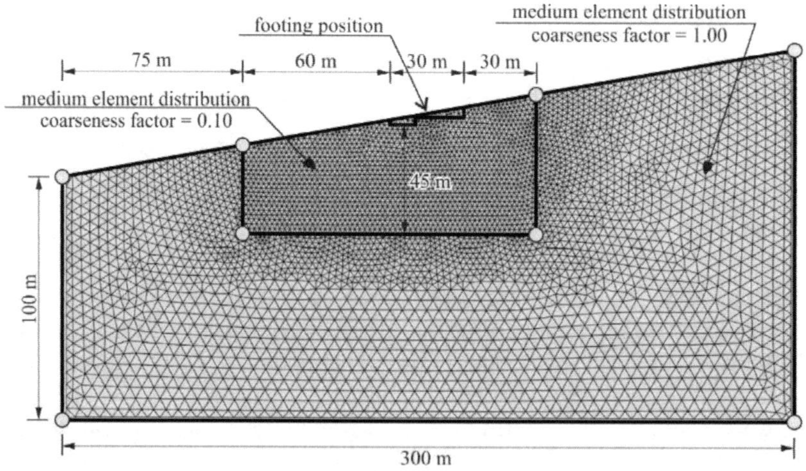


**Figure 2.** A model generated in Plaxis2D

In the *Initial Phase* zero initial stresses are generated by using the gravity loading procedure with $\Sigma M_{weight}$ equal to zero. The prescribed load is activated in a separate phase. The calculation type is *Plastic Analysis* and a tolerated error of 0.01 is defined. The applied pressure was selected so as to obtain the collapse of the ground. The bearing capacity $q_u$ was obtained by multiplying the considered pressure with the reached phase proportion $\Sigma M_{stage}$.

## 4. Results and discussions

Four different methods to evaluate the bearing capacity of spread foundations have been used, and their effectiveness has been studied. The results obtained using different methods may differ significantly, leading to different values on the bearing capacity.

### 4.1. Analytical methods

The bearing capacity values determined using analytical methods are presented in Table 2. For horizontal ground ($\beta = 0°$) the smallest values for $q_u$ are obtained using Brinch Hansen method for all type of analysed soil. The largest values for $q_u$ are obtained in the case of Vesić method. The percentage difference of $q_u$ between Vesić and Brinch Hansen methods is between 35.30% and 45.05%.

**Table 3.** Soil bearing capacity values (in kPa) determined using analytical methods

| β | Brinch Hansen method | | | | | | Vesić | | | | | | Van Baars | | | | | |
|---|---|---|---|---|---|---|---|---|---|---|---|---|---|---|---|---|---|---|
| | S1 | S2 | S3 | S4 | S5 | S6 | S1 | S2 | S3 | S4 | S5 | S6 | S1 | S2 | S3 | S4 | S5 | S6 |
| 0° | 683 | 494 | 311 | 603 | 450 | 296 | 1122 | 850 | 566 | 932 | 712 | 473 | 723 | 537 | 345 | 633 | 477 | 313 |
| 5° | 569 | 412 | 262 | 526 | 395 | 263 | 927 | 701 | 464 | 756 | 572 | 369 | 485 | 347 | 209 | 453 | 333 | 207 |
| 10° | 473 | 343 | 220 | 460 | 347 | 234 | 748 | 564 | 369 | 595 | 443 | 273 | 308 | 198 | 90 | 300 | 201 | - |
| 15° | 390 | 283 | 184 | 403 | 306 | 209 | 582 | 437 | 282 | 445 | 324 | 184 | 133 | - | - | - | - | - |

For sloped ground the smallest values for $q_u$ are obtained using Van Baars method. Here is the largest rate of decrease of the bearing capacity with the increase of the slope angle.

### 4.2. *Finite element method*

The detailed results obtained in the finite element analyses are presented in Appendix A. For soil type S2, when the slope increases from 0° to 15° the bearing capacity decreases from 702 to 265 kPa (Figure A2). For soil type S3, when the slope increases from 0° to 10° the bearing capacity decreases from 475 to 270 kPa (Figure A3). For soil type S4, when the slope increases from 0° to 15° the bearing capacity decreases from 883 to 485 kPa (Figure A4). For soil type S5, when the slope increases from 0° to 15° the bearing capacity decreases from 450 to 306 kPa (Figure A5). For soil type S6, when the slope increases from 0° to 10° the bearing capacity decreases from 480 to 256 kPa (Figure A6). This shows that the decrease of the bearing capacity of the soil varies considerably for different soil types with the increase of the slope degree.

### 4.3. *The impact of the analysis method on the bearing capacity*

The values of the bearing capacity on horizontal ground for the six types of soils determined using one numerical method (FEM) and three analytical methods (Brinch Hansen, Vesić and Van Baars) were analysed (Figure 3). It was observed that the bearing capacity values varied significantly between different types of soils and methods. The main reason for these differences is that the quantitative evaluation for $N_\gamma$ are different in all the analytical methods used. The FEM and Vesic method produced similar values for the bearing capacity in all 6 soil types with higher values, in general, obtained using Vesic method. Additionally, Brinch Hansen and Van Baars methods also produce similar values for the bearing capacity in all 6 soil types, but their values were generally lower than those produced by FEM and Vesic method.

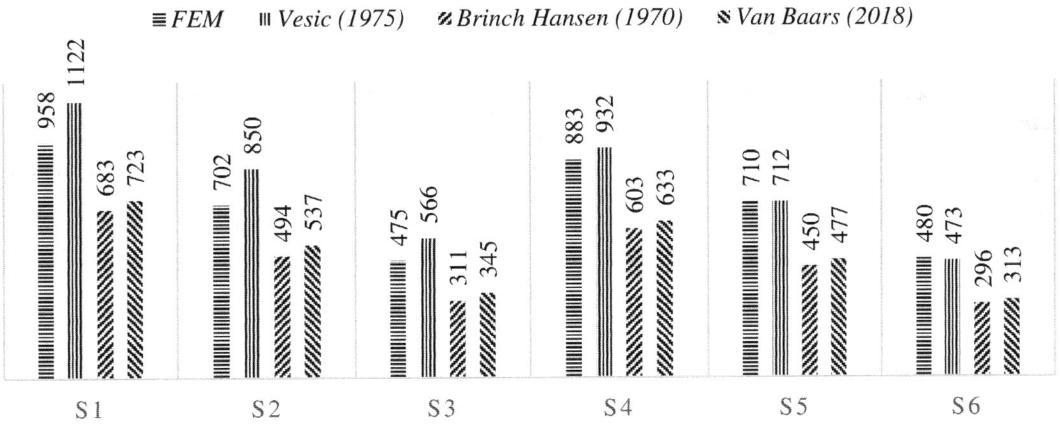


**Figure 3.** The values for the bearing capacity of foundations on horizontal ground for different soil types (in kPa) using FEM and analytical methods

### 4.4. *The influence of the slope inclination angle on the bearing capacity*

Variations in the slope inclination angle ($\beta$) significantly affect both the stability conditions and the bearing capacity values of a footing founded on sloping terrain. As the inclination of the slope increases, the bearing capacity of the footing decreases. This study considers three distinct slope angles: $\beta = 5°$, 10°, and 15°, and Appendix A illustrates that $q_u$ decreases as the slope angle increases. This phenomenon occurs because steeped slopes result in a smaller passive zone, which in turn provides less resistance from the soil adjacent to the slope face (as shown in Figure 4). In addition, when a footing is situated on a slope, the passive zone formation is one-dimensional and confined by the slope face, resulting in a significant reduction of the confinement pressure and a corresponding decrease in bearing capacity.

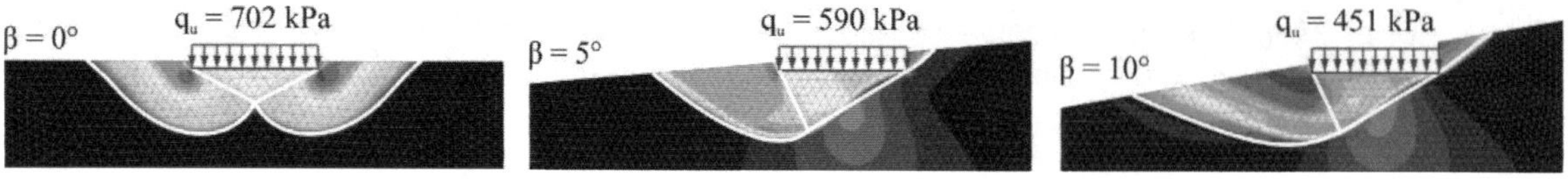


**Figure 4.** Comparison between passive zones for foundations placed on horizontal ground (left) and foundations placed on slopes of 5° (middle) and 10° (right)

The rate of decrease in bearing capacity for foundations on a 5° slope, when compared to foundations on horizontal ground, is presented in Figure 5. The results showed that the rates of decrease in bearing capacity using Van Baars method are approximatively double than those from the other three methods. A possible reason for the Van Baars method giving values for the bearing capacity lower than other methods is that the factors of the bearing capacity were calibrated using only two values for the internal friction angle (0° and 30°) while the beta angle has three values (10°, 20° and 30°).

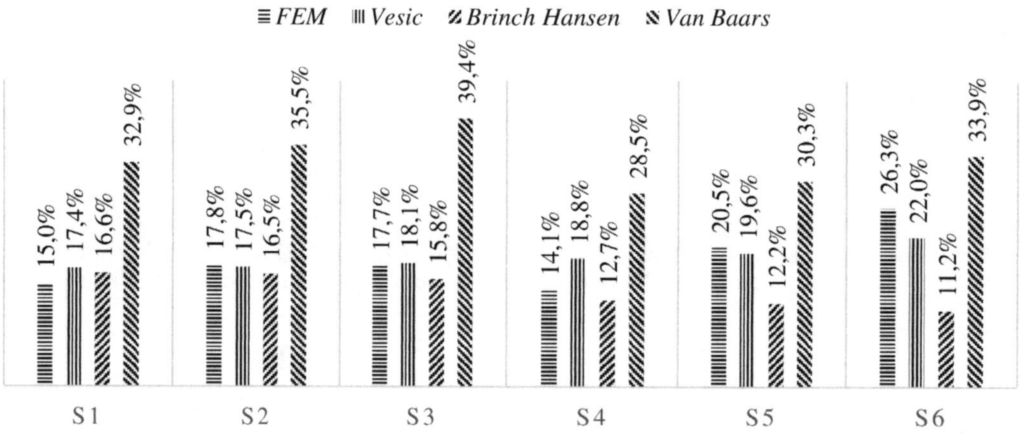


**Figure 5.** The rate of decrease of bearing capacity for the foundation on sloping ground with a slope angle of 5° compared with the foundation on horizontal ground

The rates of decrease of the bearing capacity for foundations on a 10° slope, when compared to foundations on horizontal ground, are presented in Figure 6.

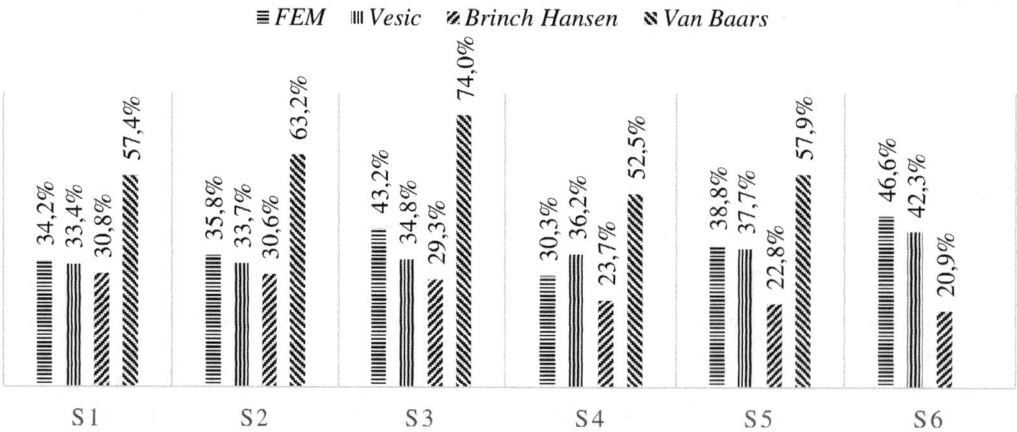


**Figure 6.** The rate of decrease of bearing capacity for the foundation on sloping ground with a slope angle of 10° compared with the foundation on horizontal ground

The results showed that the rates of decrease of the bearing capacity using Van Baars method are higher than those from the other three methods.

The rate of decrease of the bearing capacity for foundations on a 15° slope, when compared to foundations on horizontal ground, is presented in Figure 7. In the case of Van Baars method, the rate of decrease was computed only for soil type S1. In the case of soil types S2, S2, S3, S4, S5 and S6, it was not possible to calculate the bearing capacity because the Van Baars method can only be applied when the internal friction angle is higher than the slope angle. It was also observed that, in the cases where FEM was used for soil type S1 and S6, the bearing capacity could not be calculated due to the slope collapsing before the application of pressure from foundation.

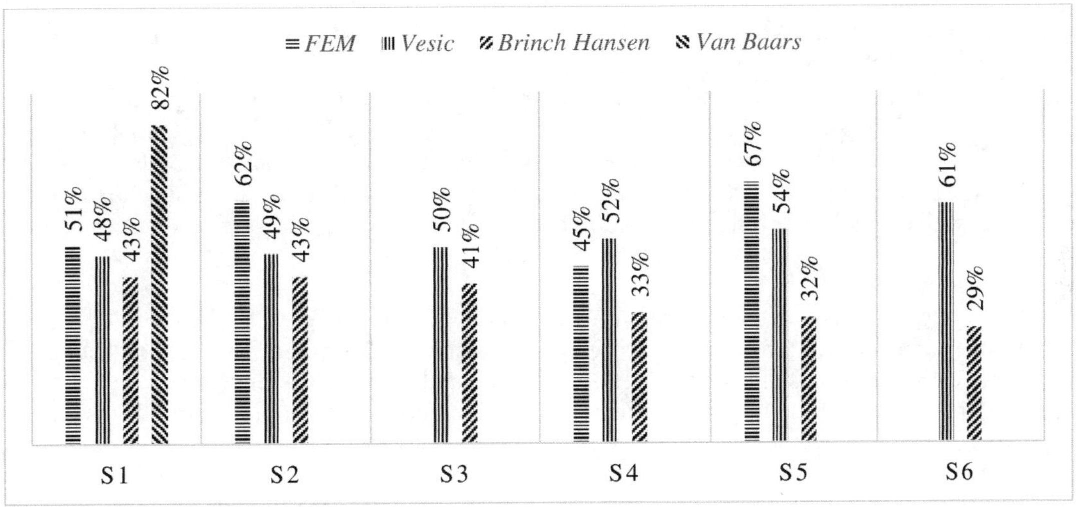


**Figure 7.** The rate of decrease of bearing capacity for the foundation on sloping ground with a slope angle of 15° compared with the foundation on horizontal ground

It was observed from the results that, in general, the rates of decrease in bearing capacity are higher when Van Baars method is used. This indicates that the utilization of Van Baars method leads to a higher reduction in the overall bearing capacity of the foundation on sloping ground compared to horizontal ground.

### *4.5. Bearing capacity for stepped footings*

It was observed that, for horizontal ground ($\beta = 0°$), the bearing capacity of the stepped mat is approximately equal to that of the straight mat, as shown in Table 4. This suggests that the dependence of steps of the foundation mat does not significantly affect the overall bearing capacity when the ground is horizontal. This can be explained by the fact that, on horizontal ground, the load is evenly distributed across the foundation, regardless of whether it is a stepped or straight mat. Therefore, it can be concluded that, for horizontal ground conditions, the use of stepped mat does not provide any significant advantage in terms of bearing capacity over a straight mat.

In the case of slopes of 5° and 10°, it was observed that the bearing capacity of the stepped mat is lower than that of the straight mat, and the rate of decrease is between 16% and 27%. This indicates that, on slopes, the use of a stepped mat can lead to a significant reduction in bearing capacity. The reason for this decrease can be explained by the fact that, on slopes, the load distribution is uneven, and the stepped mat may not be able to effectively redistribute the load. Therefore, it can be concluded that the effectiveness of the stepped mat in increasing bearing capacity is dependent on the slope angle and soil type.

**Table 4.** Soil bearing capacity values (in kPa) determined using FEM for straight mat and stepped mat

| β | Levelled mat | | | | | | Stepped mat | | | | | |
|---|---|---|---|---|---|---|---|---|---|---|---|---|
| | S1 | S2 | S3 | S4 | S5 | S6 | S1 | S2 | S3 | S4 | S5 | S6 |
| 0° | 958 | 702 | 475 | 883 | 710 | 480 | 978 | 716 | 483 | 888 | 677 | 450 |
| 5° | 815 | 590 | 391 | 759 | 564 | 354 | 633 | 452 | 286 | 615 | 455 | 292 |
| 10° | 631 | 451 | 270 | 616 | 434 | 256 | 500 | 370 | 198 | 500 | 363 | 210 |
| 15° | 471 | 265 | - | 485 | 237 | - | 461 | 343 | - | 440 | 328 | - |

It was observed that there are no analytical methods available for evaluating the bearing capacity of stepped foundations. To address this issue, one possible approach is to use the base inclination factors ($b$), where the selected angle is the one between the line joining the extremities of the footing and the horizontal direction. The use of the base inclination factors can provide a useful approximation of the bearing capacity of a stepped foundation, without the need for complex analytical methods. The reason for this is that the base inclination factors take into account the effect of the steps on the load distribution, and provide a means of estimating the reduction in bearing capacity due to the presence of the steps.

## 5. Conclusions

In this study, numerous finite element simulations of straight and stepped mat foundations on sloping ground have been made in order to compare the bearing capacities with the common analytical equations for the bearing capacity proposed by Brinch Hansen, Vesić and Van Baars. The embedding of digital solutions in the determination of bearing capacity for foundations on sloping ground has greatly improved the accuracy and efficiency of the process. By utilizing advanced software, engineers are able to quickly and accurately determine the bearing capacity of a foundation, resulting in safer and more reliable structures. This is demonstrated by the fact that the values for the bearing capacity obtained from the model footing tests from the literature are all on the upper bound of various proposed theoretical solutions [14].

Overall, these results indicate that it is important to consider multiple methods when determining the bearing capacity of soil for foundation design, as different methods may give different results for different soil types.

Further research will be conducted to deal with different types of foundation configurations and different types of soils. The results presented in this paper provides a global understanding of the bearing capacity of foundations on sloping ground and offer useful data and references for the bearing capacity computation.

IOP Conf. Series: Materials Science and Engineering 1304 (2024) 012033 doi:10.1088/1757-899X/1304/1/012033

**Appendix A** - FEM incremental displacement plots at failure

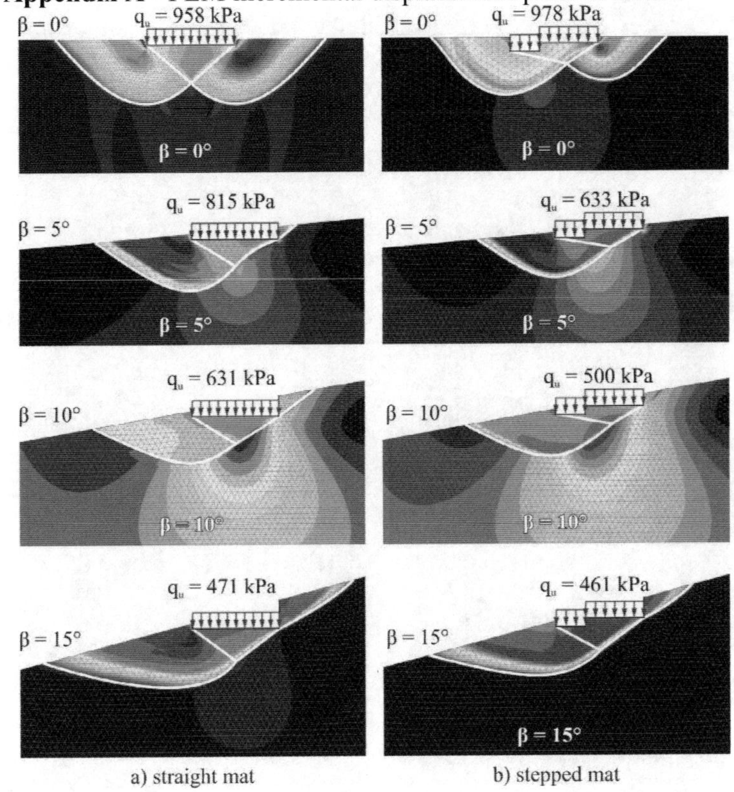


**Figure A1.** FEM incremental displacement plots at failure for soil type S1

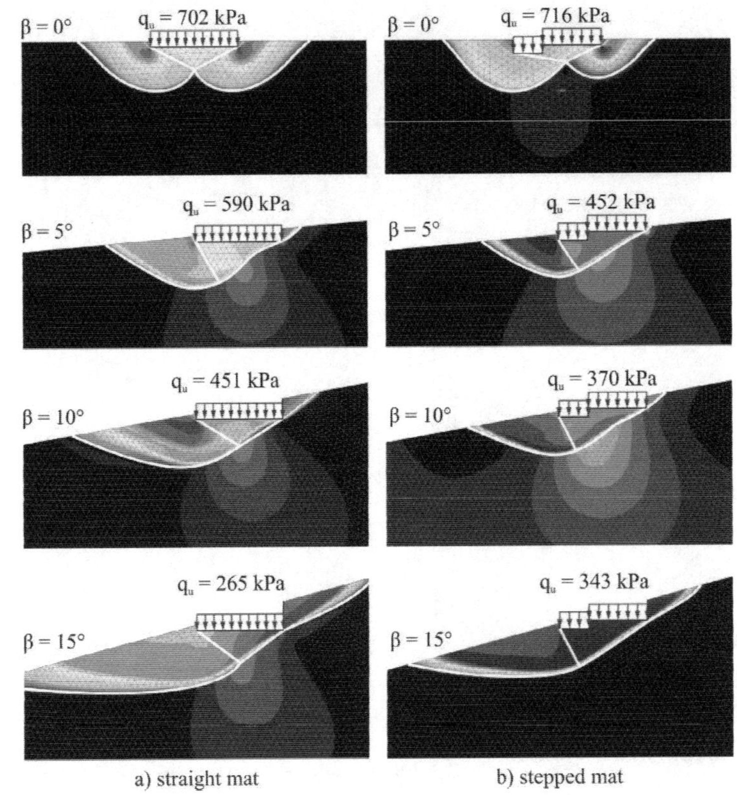


**Figure A2.** FEM incremental displacement plots at failure for soil type S2

IOP Conf. Series: Materials Science and Engineering 1304 (2024) 012033 doi:10.1088/1757-899X/1304/1/012033

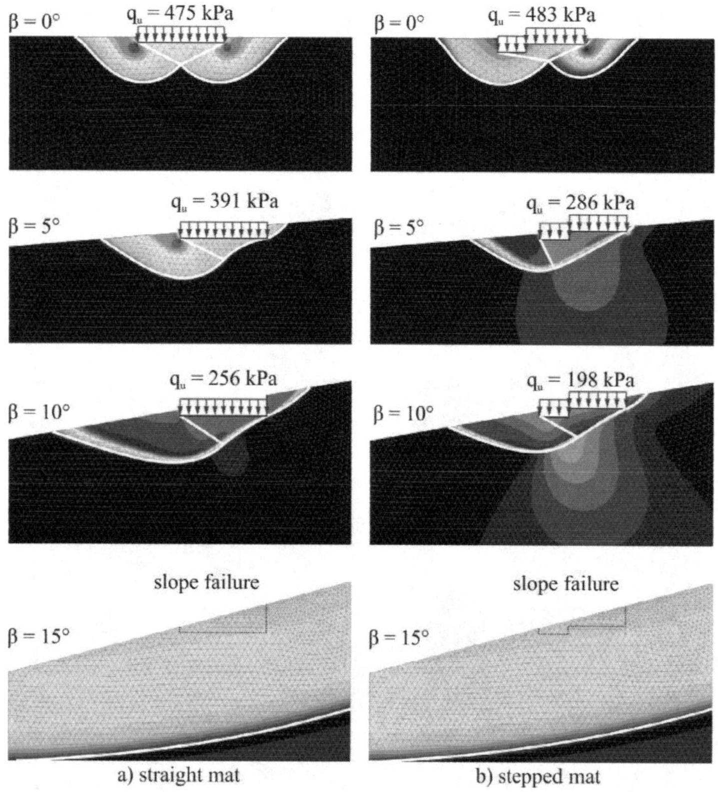


**Figure A3.** FEM incremental displacement plots at failure for soil type S3

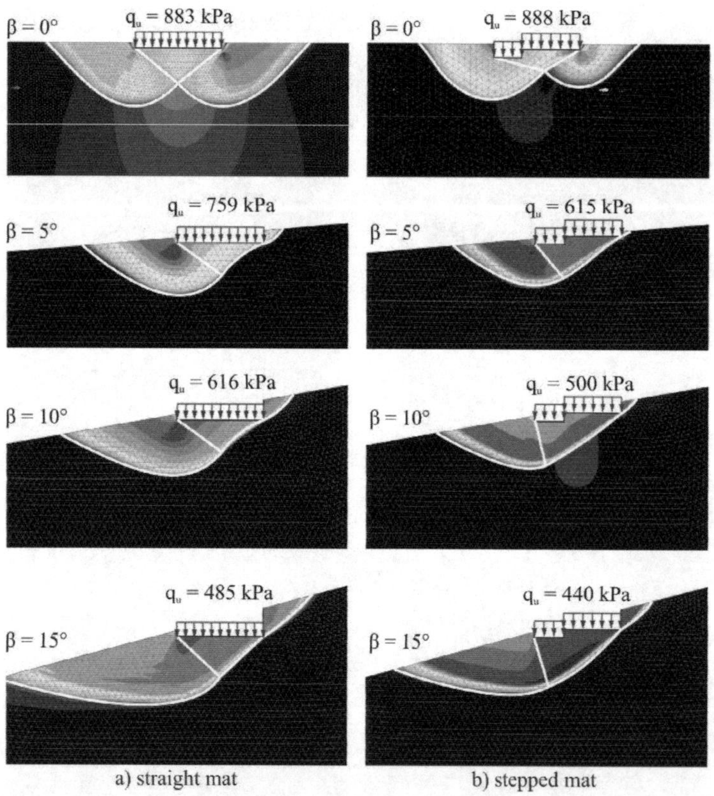


**Figure A4.** FEM incremental displacement plots at failure for soil type S4

IOP Conf. Series: Materials Science and Engineering 1304 (2024) 012033 doi:10.1088/1757-899X/1304/1/012033

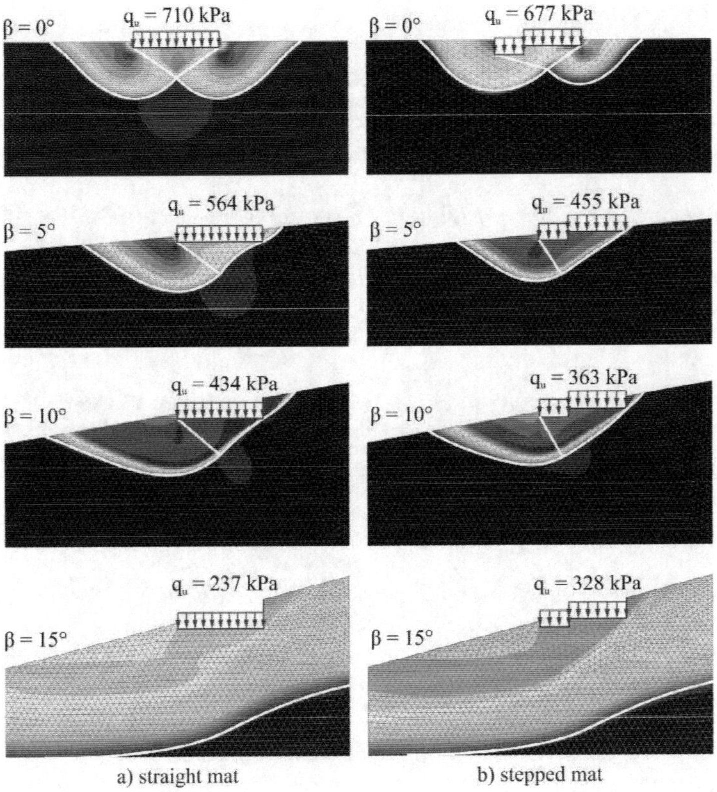


**Figure A5.** FEM incremental displacement plots at failure for soil type S5

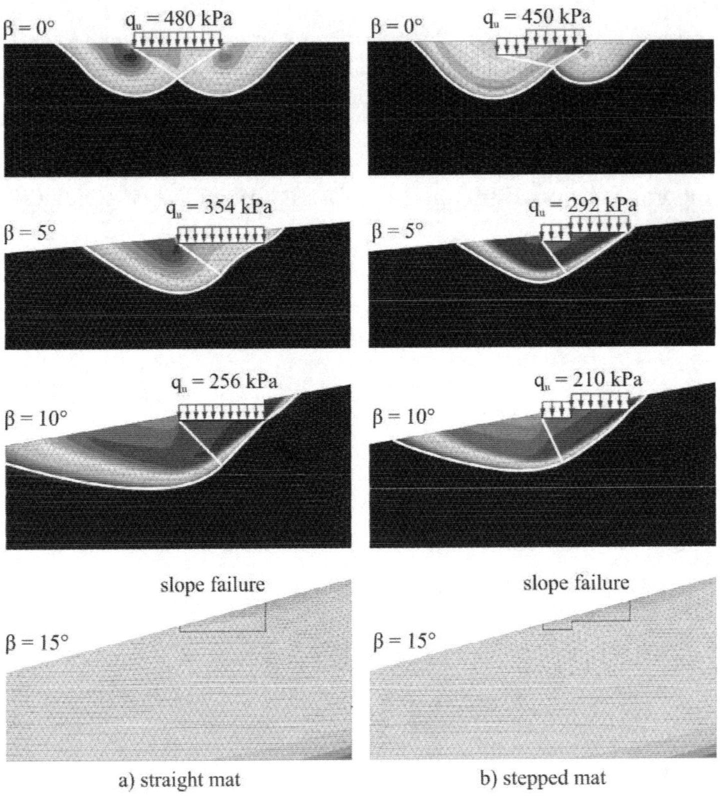


**Figure A6.** FEM incremental displacement plots at failure for soil type S6

## References

[1] Gafincu C V, Mihai P, Lungu I 2021 Seismic response of setback buildings with multiple underground levels, *IOP Conf. Series: Materials Science and Engineering* **1141** 012015
[2] Terzaghi K 1943 *Theoretical soil mechanics* (New York: Wiley)
[3] Meyerhof G G 1953 *The bearing capacity of foundations under eccentric and inclined loads* Proceeding of the 3rd International Conference on Soil Mechanics and Foundations Engineering, vol. I, pp. 440
[4] Meyerhof GG 1957 *The ultimate bearing capacity of foundation on slopes* Proceeding of the 4th International Conference on Soil Mechanics and Foundations Engineering, vol. I, pp 384-386
[5] Brinch Hansen J 1970 *A revised and extended formula for bearing capacity* Danish Geotechnical Institute, Bulletin no. 28, Copenhagen
[6] Vesić A S 1973 *Analysis of ultimate loads of shallow foundation* J Soil Mech Found Div ASCE 99(SM1):45-73
[7] Azzouz A S and Baligh M M 1983 *Loaded areas on cohesive slopes*, Journal of Geotechnical Engineering, vol. 109, pp. 724-729
[8] Georgiadis K 2010 *Undrained Bearing Capacity of Strip Footings on Slopes*, J. Geotechnical and Geoenvironmental Engineering, 136(5): 677-685
[9] Shiau J, Merifield R, Lyamin A. and Sloan S 2011 *Undrained Stability of Footings on Slopes*, Int. J. Geomech., ASCE, 11(5), 381-390
[10] Graham J, Andrews M and Shields D H 1988 *Stress Characteristics for Shallow Footings in Cohesionless Slope* Canadian Geotechnical Journal, 25(2): 238-249
[11] Dhiraj R, Bharathi M 2013 Bearing capacity of shallow foundation on slope: a review Proceedings of GGWUIP-2013 / IGS Ludhiana Chapter (Paper ID: NC-45)
[12] Van Baars S 2018 *The bearing capacity of shallow foundations on slopes* Numerical Methods in Geotechnical Engineering IX – Cardoso et al. (Eds), Taylor & Francis Group, London.
[13] NP 122:2010 – Regulations on the determination of characteristic and design values of geotechnical parameters (in Romanian)
[14] Cerato A B 2005 *Scale Effect of Shallow Foundation Bearing Capacity on Granular Material*, PhD Thesis, University of Massachusetts, MA, USA.

**Acknowledgements**

This paper was realized with the support of project EFECON – ECO-INNOVATIVE PRODUCTS AND TECHNOLOGIES FOR ENERGY EFFICIENCY IN CONSTRUCTION, POC/71/1/4 - Knowledge Transfer Partnership, Cod MySMIS: 105524, ID: P_40_295, Project co-financed by the European Regional Development Fund.

# AUTHOR INDEX